Electronics for Radiation Measurements Volume 2

Gad Shani, Ph.D.

Department of Nuclear Engineering
Ben-Gurion University
Beer-Sheva, Israel

CRC Press
Boca Raton New York London Tokyo

Library of Congress Cataloging-in-Publication Data

Gad Shani
Electronics for radiation measurements / Gad Shani.
p. cm.
Includes bibliographical references and index.
ISBN 0-8493-4995-8 (Vol. 1) ISBN 0-8493-9495-3 (Vol. 2)
1. Electronic instruments—Design and construction. 2. Nuclear counters—Design and construction.
I. Title
TK7878.4.S43 1996
621.381—dc20 95-44689
CIP

International Standard Book Number 0-8493-9495-3
Library of Congress Card Number 95-44689
Printed in the United States of America 1 2 3 4 5 6 7 8 9 0
Printed on acid-free paper

Preface

Electronics for nuclear measurements is the branch of the wide range of electronics, which deals with the electric signals obtained from the nuclear radiation detector. The purpose of the present book (Volumes 1 and 2) is to be a source of information for electronic systems and methods used in nuclear measurements. These books cover the most common subjects in this field. Volume 1 starts with an introductory chapter on the basic concepts of electronics then goes on to describe and discuss the most advanced developments of the electronic systems used. Nuclear electronics (like other fields in electronics) has developed very rapidly in the last two decades. The basic ideas developed 20 years ago are, in many cases, still valid and useful today. These concepts were the springboard for the development of modern electronic systems. These books attempt to give the reader as much of the basic ideas as possible as well as the latest developments. In the last decade high energy physics made a tremendous development and very large scale radiation detectors are used, requiring large complicated electronic data acquisition systems. These systems are the direct offspring of the nuclear electronic systems of 20 years ago.

Although these volumes are not textbooks, the first chapter gives a short outline of the very basic concepts of electronics for the benefit of readers who lack this knowledge. This is the basis on which of electronic systems are built. When a nuclear particle or quantum enters a detector, an electric signal is created. The signal is transfered to the electronic system for analysis. The simplest analysis is just counting the number of events in the detector (or number of events per unit time). More complicated analyses are: energy distribution, time distribution, particle identification, etc., in particular in high energy physics where large quantity of data has to be handled simultaneously. Generally the signal from the detector is transfered into the preamplifier, then to an amplifier, then to any data-analysis system, according to the measurement purpose. The preamplifier is the most important link in the nuclear electronics chain of instruments. The preamplifier's purpose is to obtain the small signal from the detector, separate it from the noise and shape it in a form that will keep the timing, energy, pulse shape (particle identification) and any other information obtained from the detector. Because of its importance, the first chapter of these two volumes (after the introduction) is devoted to a detailed discussion of preamplifiers. The main feature of the preamplifier is its low noise. Many detection events create very small signals so that the noise added by the preamplifier must be as low as possible, to maintain undistorted information carried by the detector signal. The other important feature of the preamplifier is matching the detector output impedance and the kind of signal delivered, namely: charge sensitivity, current sensitivity, or voltage sensitivity. The preamplifier input configuration determines its sensitivity to the kind of signal obtained from the detector. Following the discussion of these important features of the preamplifiers, some examples of various preamplifiers are given in Chapter 2.

The next link in the nuclear electronics chain is the amplifier. The amplifier receives the pulse from the preamplifier, amplifies it, and shapes it for further analysis. These two functions of the amplifier must be done without losing the information carried by the pulse. Several techniques are used for amplification and shaping of the preamplifier pulse. The amplifier must be stable (gain independent of pulse height, pulse rate, temperature etc.) and of low noise. Chapter 3 deals with the techniques and methods used to fulfill those goals of the amplifier.

The following chapters of the book deal with different ways of handling data obtained from amplifiers, for various kinds of analysis. If counting only is needed, then the amplifier pulses are fed into a scaler, generally via a discriminator of some kind. Time and energy analyses are done with a multichannel analyzer (MCA). The basic subsystem in the MCA is the analog-to-digital converter (ADC). These circuits are discussed in the chapters following amplifiers. Several methods for analog-to-digital conversion are described. Along with the ADC, other kinds of conversions are discussed for cases when a different kind of analysis is sought or when the signal is obtained in another form than a simple amplifier output pulse. For precise-time analysis, constant fraction timing discrimination is used (Chapter 8). Pulse timing can sometimes be used for particle (or radiation) identification. Different kinds of radiation produce different pulse timing (pulse shape) in some detectors. This difference in pulse timing can be used for identification in a pulse shape discriminator (PSD). The various methods of particle identification using PSD are discussed in Chapter 9.

Because of its importance in nuclear measurements, a short chapter is devoted to application of photomultiplier tubes. Scintillation detectors can provide important information regarding the radiation detected, therefore the application of a photomultiplier tube must be done in a proper way. Examples of voltage application to the PM tube are given in Chapter 10.

An important field in modern nuclear electronic system is data acquisition. In high energy physics very large detectors are used to detect very high energy particles. Stopping of these high energy particles creates very large amount of data in a very short period of time. This vast amount of data must be handled by the electronic system. Relatively new methods were developed for this data acquisition, the CAMAC and FASBUS. The basic principles of these methods are explained in Chapter 12. It should be noted that in many electronic systems, computers are used for data storage. Computers are not discussed in this book because they cannot be defined as electronic systems for nuclear measurements.

High-voltage supplies and transmission lines are integral part of every nuclear measurement systems. A brief discussion of those two functions are given in the final chapters of Volume 2.

The Author

Gad Shani, Ph.D., is a Professor in the Department of Nuclear Engineering at Ben-Gurion University, Beer-Sheva, Israel.. Professor Shani received his B.Sc. degree in electrical engineering and M.Sc. degree in nuclear science at the Technion Israel Institute of Technology in 1964 and 1966, respectively, and his Ph.D. in nuclear engineering from Cornell University, Ithaca, New York. He has been on the faculty at Ben-Gurion University since 1970, where he was one of the founders of the Department of Nuclear Engineering and Head of the department from 1980 to 1984. In 1993, he became the Head of the Biomedical Engineering Department. Professor Shani is the incumbent Chair in the Nuclear Engineering Department.

In the course of his academic carreer, Professor Shani has been awarded numerous grants and awards. Among them are The European Community Award as a Visiting Scientist at the Nuclear Research Center in Ulrich, Germany and the British Royal Society Award at the Atomic Research Establishment at Harwell, England. He has been a Visiting Professor at McMaster University, Canada, The University of California Los Angeles, and at the University of California Santa Barbara, where he spent two years, 1977 to 1979. He was a Visiting Professor at The Ohio State University in 1990 and a Visiting Scientist at Brookhaven National Laboratories in 1991 and continues his collaboration there.

Professor Shani is a member of the Israel Nuclear Society, where he was a member of the Council from 1980 to 1984, Israel Society of Bio-Engineering, Israel Society of Radiation Protection, Israel Society of Aerosols Research, Israel Ecology Society, Cornell Society of Engineers, the American Nuclear Society, and the International Society of Neutron Capture Therapy.

His current interests include neutron physics, nuclear reactor physics, fusion reactor neutronics, nuclear medicine, health physics, neutron capture therapy, application of nuclear radiation in science, industry, and medicine, nuclear detection and instrumentation, and magnetic properties of materials. He has published more than 120 papers and authored or co-authored five books.

Dedication

To my Parents

Table of Contents

VOLUME 2

Chapter 5
Discriminators 1
5.1 General Considerations 1
5.2 Integral Discriminator 2
5.3 Timing Discriminator 10
5.4 Differential Discriminator (Single-Channel Analyzer) 18
References 26

Chapter 6
Scalers and Rate Meters ....
6.1 Two-State Circuits 27
6.2 Counting Systems 29
6.3 Synchronous Counters 31
6.4 Data Readout 32
6.5 Rate Meters 32
References 35

Chapter 7
Multichannel Analyzers (MCA) 37
7.1 Introduction 37
7.2 Memory 37
7.2.1 Bistable Multivibrator 37
7.2.2 Memory Circuits 40
7.3 Single-Parameter Multichannel Analyzers 44
7.3.1 Early Models 44
7.3.2 CAMAC-Aided MCA 49
7.4 Multiparameters Multichannel Analyzers 65
7.5 Multiscaler 69
7.6 Multiplexer 72
References 77

Chapter 8
Constant-Fraction Timing Discriminators 79
References 106

Chapter 9
Pulse Shape Discriminators 107
9.1 Introduction 107
9.2 Zero Crossing 107
9.3 Other Methods 113
References 130

Chapter 10
Photomultiplier Tube Voltage Divider 131
References 147

Chapter 11
Auxiliary Circuits 149
11.1 Timing Circuits 149
11.1.1 Pulse Resolving 149
11.1.2 Time Spectroscopy 149
11.1.3 Pulse Shape Timing 150
11.1.4 Coincidence Circuits 151
11.2 Linear Gate 159
11.3 Sample-and-Hold Circuits 164
11.4 Pulse Stretcher 173
References 174

Chapter 12
Data Acquisition 175
12.1 CAMAC 175
12.1.1 Basic Concepts 175
12.1.2 CAMAC Systems 177
12.2 FASTBUS 201
12.2.1 Introduction to FASTBUS 201
12.2.2 FASTBUS Applications 203
12.3 Other Data Acquisition Systems 214
References 232

Chapter 13
High-Voltage Supply 233
13.1 Introduction 233
13.2 Basic Methods for Direct Current Supply 234
13.2.1 Full-Wave Power Supplies 234
13.2.2 Full-Wave Bridge Rectifier Power Supplies 235
13.2.3 Voltage-Multiple Power Supplies 235
13.2.4 Regulated Power Supplies 236
13.2.5 Shunt-Regulated Power Supplies 237
13.2.6 Series-Regulated Power Supplies with Zener Reference 237
13.2.7 Series-Regulated Power Supplies with Feedback 238
13.3 High-Voltage Supply Circuits 238
References 252

Chapter 14
Transmission and Delay Lines 253
1. Transmission Lines 253
2. Delay Lines 270
References 275

Index 277

VOLUME 1

Chapter 1
Introduction 1
1.1 Basic Electronic Concepts 1
1.1.1 Elements 1
1.2 Semiconducting Devices 4
1.2.1 Diodes 4
1.2.2 Transistors 6
1.2.3 Field-Effect Transistors 10
1.2.4 Tunnel Diodes 12
1.3 Basic Transistor Circuits 13
1.3.1 Bipolar Junction Transistors 13
1.3.2 Field-Effect Transistor Amplifiers 13
1.4 Logical Circuits 14
1.5 Logic Signals 17
1.6 Definitions 18

Chapter 2
Preamplifiers 21
2.1 Introduction 21
2.2 Preamplifier Noise Analysis 25
2.2.1 Basic Considerations 25
2.2.2 Low Noise Preamplifiers 35
2.3 Charge-Sensitive Preamplifiers 64
2.3.1 Characteristics 64
2.3.2 Frequency Dependence 87
2.4 Current Preamplifier 92
2.5 Voltage-Sensitive Preamplifier 104
2.6 Monolithic Preamplifier 110
2.7 Miscellaneous Preamplifiers 131
References 176

Chapter 3
Amplifiers 179
3.1 Introduction 179
3.2 Pulse Shaping 182

3.2.1 Basic Pulse Shaping 182
3.2.1.1 Unipolar Pulse Shaping 182
3.2.1.2 Bipolar Pulse Shaping 183
3.2.1.3 Differentiation-Integration Pulse Shaping 185
3.2.2 Filter Pulse Shaping 186
3.2.3 Semi-Gaussian Pulse Shaping 191
3.2.4 Transients and Short Pulses 201
3.2.4.1 Differentiators' Cascade Response 201
3.2.4.2 Integrators' Cascade Response 202
3.3 Feedback 202
3.3.1 Negative Feedback 204
3.3.2 Parallel Feedback 204
3.3.3 Single Transistor Amplifier Feedback 204
3.3.4 Series Feedback 204
3.3.5 Shunt Feedback 205
3.4 Gain Stability 206
3.5 Pole-Zero Cancellation Network 211
3.6 Sum Effects 213
3.7 Baseline Restorer 214
3.8 Pileup Rejection 227
3.9 Operational Amplifiers 233
3.9.1 Differentiator 239
3.9.2 Integrator 240
3.10 Other Amplifiers 240
3.11 Amplifier Coupling 247
References 252

Chapter 4
Converters 255
4.1 Analog-to-Digital Converters 255
4.1.1 Types 255
4.1.2 Development 264
4.2 Digital-to-Analog Converters 307
4.3 Time-to-Amplitude Converters 308
4.4 Time-to-Digital Converters 325
4.5 Other Converters 347
References 350

Index 353

Chapter 5

Discriminators

CONTENTS

5.1 General Considerations 1
5.2 Integral Discriminator 2
5.3 Timing Discriminator 10
5.4 Differential Discriminator (Single-Channel Analyzer) 18
References 26

5.1 GENERAL CONSIDERATIONS

A pulse-height discriminator is a circuit which has an output pulse when the input pulse exceeds a certain level. The output pulse is a square standard pulse with short rise and fall times and with defined height and width. The output pulse timing depends on the input pulse height since lower pulses reach the discriminator level later. The time difference between the input and output pulses is called walk (Figure 5.1). The discriminator level should be independent of pulse rate. This is the reason for the DC coupling between the input stage and the discriminator. The buffer (Figure 5.2) is an amplifier of unit gain. The discriminator feeds the pulse shaper every time there is an input pulse of a proper height. The bias circuit provides the threshold voltage for the discriminator. There are several ways to achieve pulse-height discrimination. One of them is the use of a multivibrator. Any positive feedback amplifier can be a multivibrator. A positive feedback increases the gain by increasing the slope:

$$\frac{dv_{out}}{dv_{in}} = \frac{A}{1-BA} \tag{5.1}$$

As long as BA < 1 the gain slope is positive (greater than A). If BA > 1, the slope becomes negative (see Figure 5.3). When v_{in} is between v_1 and v_2, there are three possibilities. v_{out} jumps from v_{min} to v_{max} when v_{in} passes v_2, and in the opposite direction from v_{max} to v_{min} when v_{in} goes down through v_1. An example of a positive feedback is the Schmitt trigger shown in Figure 5.4. When v_{b1} is low Q_1 is cut off. If $R_{e1} \gg R$, R_{e1} is a current source, the current flows through Q_2. Ignoring the base current, the base voltage of Q_2 is

$$v_{b2} = v_1 R_b / (R_{c1} + R_a + R_{b2}) \tag{5.2}$$

If v_{b1} rises, Q_1 starts conducting when v_{b1} reaches $\sim v_{b2}$. By the feedback $R_a R_{b2}$ the circuit switches to the second stable state, where Q_1 is conducting and Q_2 is cut off. The collector voltage of Q_1 drops by $R_{c1}I_e$ and the voltage of the base of Q_2 is

$$v_{b2} = (v - R_{c1}I_e) R_b / (R_c + R_a + R_{b2}) \tag{5.3}$$

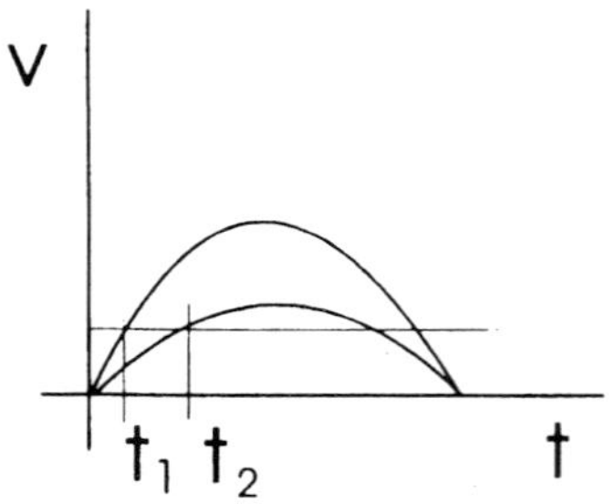

Figure 5.1 Pulse walk.

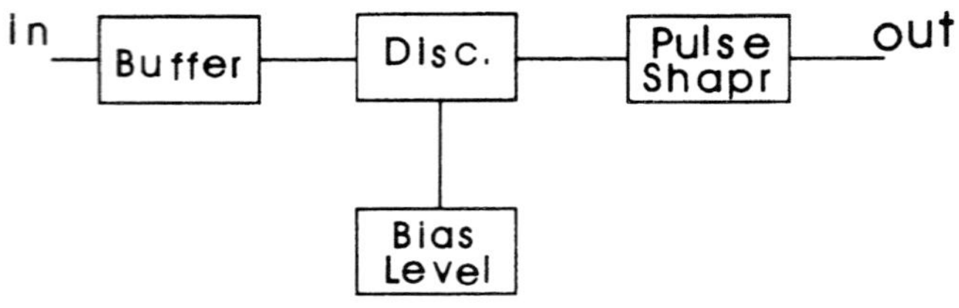

Figure 5.2 Discriminator components.

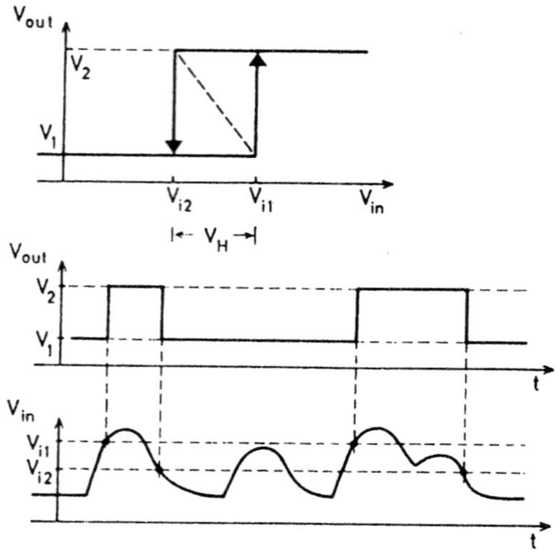

Figure 5.3 Ideal amplifier with positive feedback.

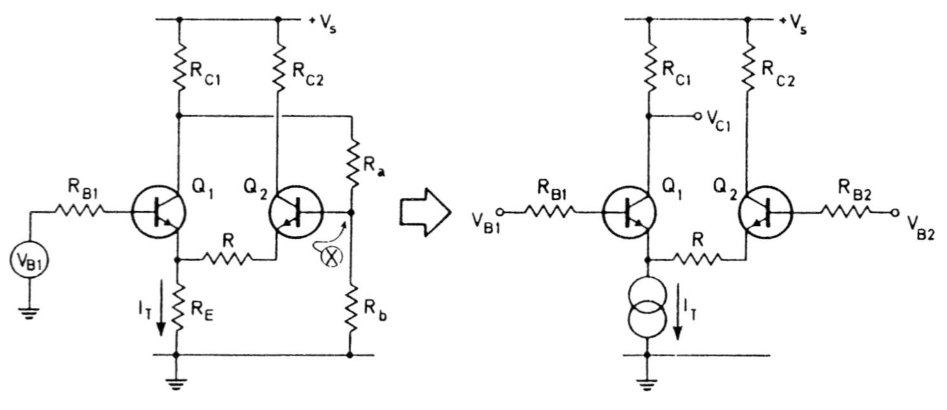

Figure 5.4 The Schmitt trigger (left side) and an equivalent circuit for evaluation of the loop gain (right side).

5.2 INTEGRAL DISCRIMINATOR

An integral discriminator produces a standard pulse for every input pulse of amplitude above threshold. A Schmitt trigger can be used as an integral discriminator. Such a circuit is shown in Figure 5.5. V_{B2} is adjusted by the potentiometer P to a reference level. v_{b2} when Q_1 is cut off[2] determines whether a switch

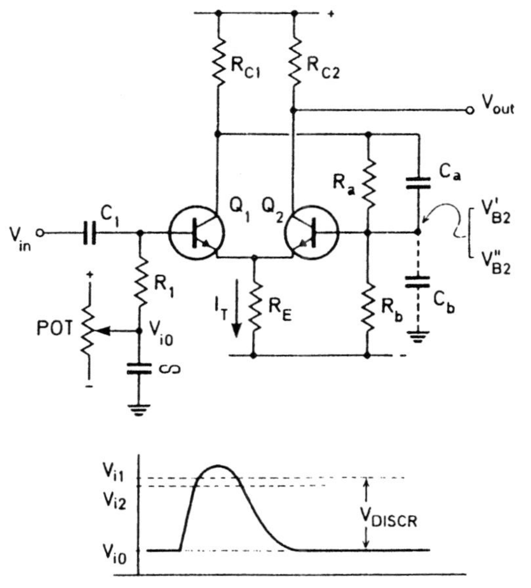

Figure 5.5 Schmitt trigger-type integral discriminator.

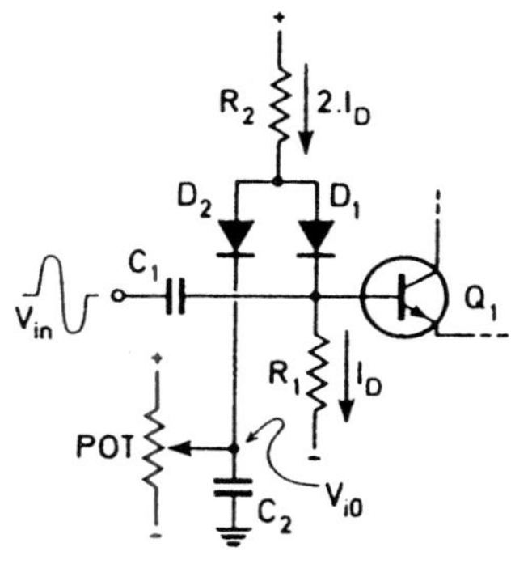

Figure 5.6 Baseline restorer for bipolar pulses. (From Robinson, L.B., *Rev. Sci. Instrum.*, 32, 1057, 1961. With permission.)

over will take place. Every pulse of height greater than the difference ($v_{b2} - v_{b1}$) (set by the potentiometer) will cause the switch over and an output pulse.

The stability of the discriminator level depends on the stability of v_{b1} and v_{b2}. The stability depends on the lower side on the potentiometer stability and on the upper side on the stability of the circuit voltage supply and the resistors R_{b2}, R_a, R_{c1}. To cancel mutual changes both voltage sources are connected. The potentiometer is connected to +v and voltage is supplied to the base of Q_2, by connecting R_{b2} to the point A where the potentiometer is connected to +v.

The circuit is stable against temperature change since the temperature dependence is according to the emitter. Both emitter currents are equal so that temperature related changes are canceled.

To obtain sensitivity to low pulses the circuit should be built so that the difference between the voltage of the base of Q_2 and that at the pot, will be small but not too small so that the stability is not lost. A circuit for low-level discrimination with high feedback and stability is achieved by using diodes. Another problem in the circuit observed above is that the input RC coupling causes baseline shift, thus the discriminator level. This can be solved by connecting a diode parallel to R_{B1}. The base voltage is thus kept. It is not useful for bipolar pulses since the capacitor discharge through R_{B1} will create a positive pulse. For a bipolar pulse the circuit shown in Figure 5.6.[1] is used. The transistor base is kept at v_{bi} by the two diodes.

Another circuit that can be used as a discriminator is a monostable multivibrator as shown in Figure 5.7. The R_{B2}-C_b system is designed to determine the output pulse time. The time is $(v_{b2}/I)C_b$ where v_{b2} is the voltage of the base of Q_2 when Q_2 is cut off. The diode should stabilize v_{b2} during the time Q_2 is conducting, to the voltage determined by the divider R_2, R_3.

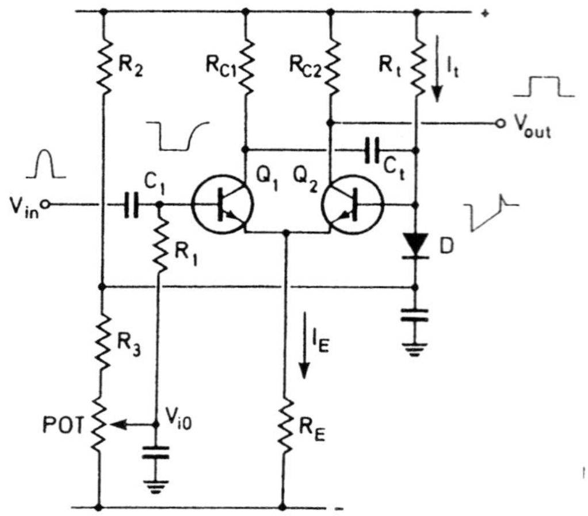

Figure 5.7 Monostable multivibrator used as integral discriminator.

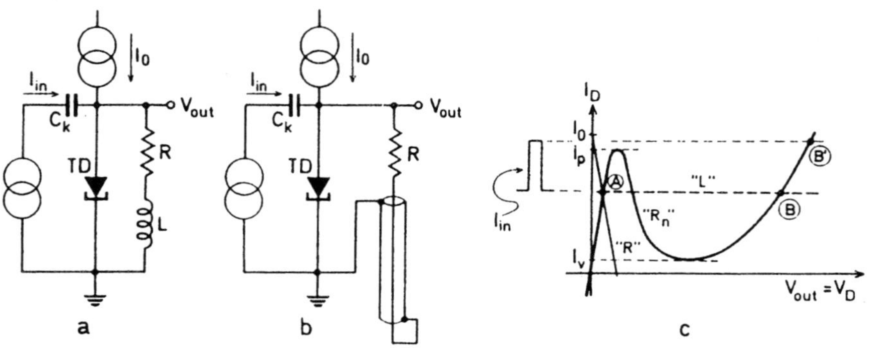

Figure 5.8 Tunnel diode monostable multivibrator with coil L (a) or cable (b), and the related characteristics.

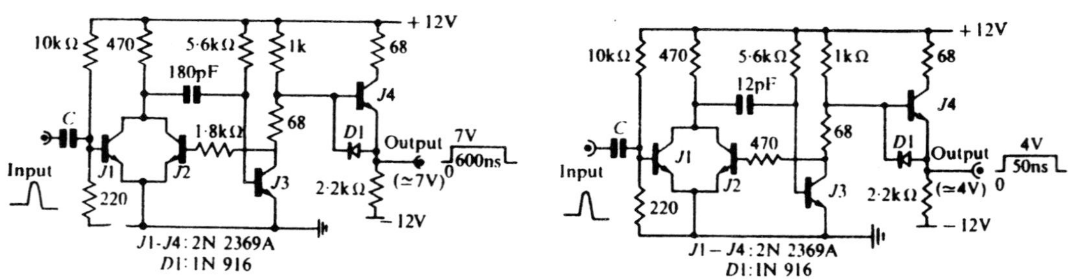

Figure 5.9 (A) NIMS pulse former: slow. (B) NIMS pulse former: fast.

A monostable multivibrator can be achieved with a tunnel diode as shown in Figure 5.8. A coil can be used instead of the cable. The resistor is small so that the load line, Figure 5.8(b), crosses the characteristics at one point only. The cable impedance is R. The input pulse raises the current above I_p and the diode switches to high-voltage state. The pulse reflected from the cable (or current drop in the coil) switches the diode to low-voltage stage back.

Two standard discriminators are shown in Figures 5.9A and B. One is fast and the other is slow. These are saturated monostable with npn transistors. It is preferred to pnp transistors for fast saturated switching since their storage time is shorter. The trigger is positive and the capacitor C can be used for differentiation of the input pulse.

An integrated circuit (IC) discriminator is shown in Figure 5.10. The transistors Q_1 and Q_2 constitute a differential amplifier with a constant current Q_3. A 2-mV change in the input voltage Q_3 is enough to

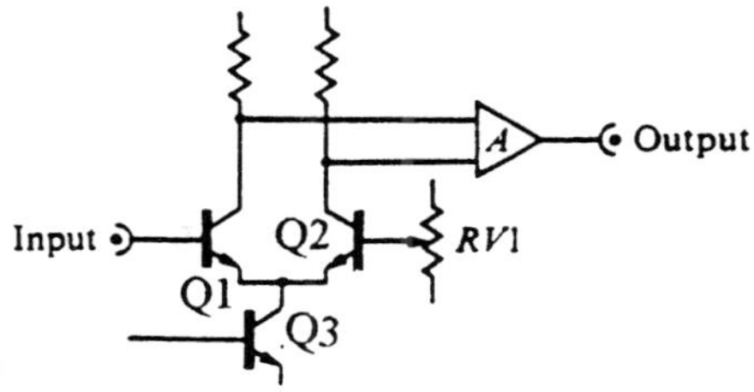

Figure 5.10 Schematic of IC discriminator.

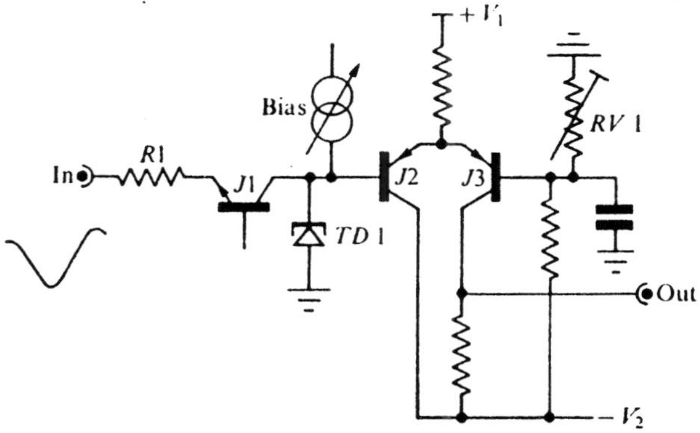

Figure 5.11 Discriminator for bias below I_v.

cause a change in the output by saturating amplifier A. The output voltage is –0.5 and +3 V with time walk of 30 nsec. It is relatively not a fast discriminator.

A tunnel diode (TD) discriminator made of a monostable multivibrator is shown in Figure 5.11. The input resistor R_1 should match the input cable (50 or 100 Ω). The signal enters the diode TD through the common base amplifier used as a high source impedance to the tunnel diode. Q_2 is cut off. Q_3 is conducting; the voltage V_{eb} of Q_2 is set by the variable resistor. This voltage is 150 mV below the conductivity threshold (0.3 V in Ge, 0.7 V in Si). When a pulse comes in the diode switches to high voltage and will conduct current to Q_2 and Q_3. When the diode returns to low voltage, the transistors return to their previous stage. The diode requires inversed voltage so that the current will be below the valley current. The discriminator maximum sensitivity is the diode inverse current.

An improved circuit is shown in Figure 5.12.[2] The transistors Q_4, Q_5 present common base amplifiers, conducting throughout the whole cycle. The npn-pnp combination, Q_5-Q_3 and Q_4-Q_2 are the means to obtain the DC reference level for reset and output pulse. The coaxial cable between Q_2 and Q_4 is of length adjusted to the output pulse length; and used to reset the tunnel diode, it is matched by R_1. The current in R_2 provides zero baseline for the output pulse; without it the baseline will be negative by $50 \times I$ V where I is the quiescent collector current in Q_5. The bias of Q_2 is determined by the potential in the base of Q_3 and the tunnel diode will cause Q_2 and Q_3 to change over when it switches to high-voltage state. The advantages of this circuit are that the common emitter amplifier is the fastest (0.3 nsec delay at 10 mA) and the current from Q_3 into Q_2 is used to reset the tunnel diode; thus the current steered from Q_3 into Q_2 is low. The circuit is capable of handling 10^8 pulses per second.

Valentine et al.[3] used feedback control of the discrimination threshold and added four to six decades to the useful range of otherwise conventional pulse-counting systems. The additional range is obtained at the high end by retarding saturation of the output count rate with higher threshold levels.

The end of the useful range of a conventional level-sensitive counting system is often considered to be the point at which pileup-induced counting losses exceed 10%. In a well-designed system, the output count rate saturates beyond this point and becomes insensitive to further increases in the input rate. However, with properly designed linear circuitry, measurable variation continues to occur in the analog signal for several more decades of input range. A typical system for making average rate measurements of a Poisson-distributed time series of discrete events (e.g., photons interacting in a PIN silicon diode, Figure 5.13) consists of a detector, a pulse-shaping amplifier (PSA), a level-sensitive discriminator (differential comparator), and a count-rate meter. The useful range of this system is limited at the low count-rate end by either signal-to-noise considerations or by constraints on the sampling time; at the high

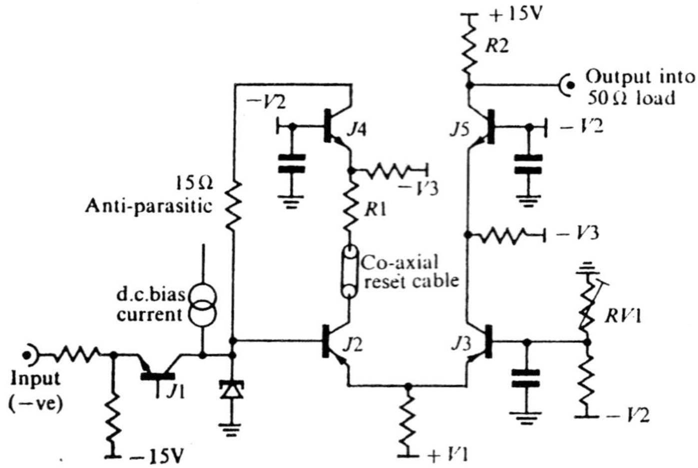

Figure 5.12 Improved pulse former.

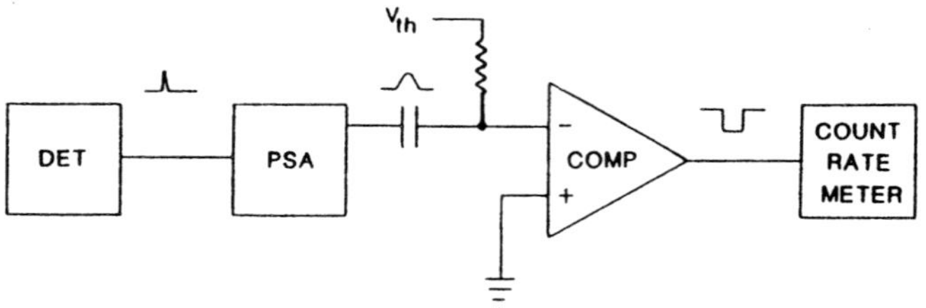

Figure 5.13 In a conventional pulse-counting system, V_{th} must be negative enough to limit the spurious count rate to an acceptable level. (From Valentine, K. H. et al., *IEEE Trans. Nucl. Sci.*, 34, 78, 1987. With permission.)

end it is limited by the bandwidth of the detection system. For very high event rates, the count rate will saturate at:

$$f_o = \frac{1}{2\pi}\left(\frac{\int_0^\infty \omega^2 |T(\omega)|^2 \, d\omega}{\int_0^\infty |T(\omega)|^2 \, d\omega}\right)^{1/2} \tag{5.4}$$

where ω is the radial frequency and T(ω) is the total transfer function of the system from the detector to the comparator input. The count rate vs. event rate [$C(\nu)/f_o$ vs. ν/f_o] characteristic for this system shows rapid saturation of the count rate and attendant loss of sensitivity (i.e., $S = dC/d\nu$) for event rates $\nu/f_o > 0.1$.

For event rates $\nu/f_o > 10$:

$$C\left(V_{th,\nu}\right) = f_o \exp\left(-V_{th}^2/2k\nu\overline{q^2}\right) \tag{5.5}$$

and consequently:

$$\lim_{\nu\to\infty} C\left(V_{th,\nu}\right) = f_o \text{ and } \lim_{\nu\to\infty} S = 0 \tag{5.6}$$

Threshold adjustment function is automatically accomplished by adding negative feedback from the count-rate output of a conventional counting system to the input of its comparator (Figure 5.14). About the simplest realization of the extended range counting (ERC) block of Figure 5.14 is a single-pole

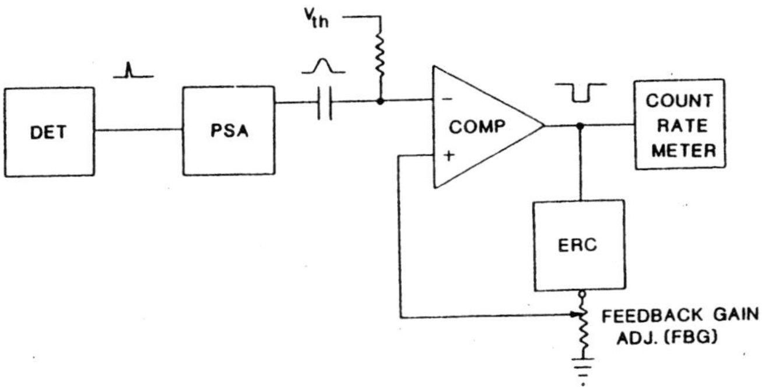

Figure 5.14 Extended range counting is achieved with the addition of an ERC negative feedback loop to a conventional counting system. (From Valentine, K. H. et al., *IEEE Trans. Nucl. Sci.*, 34, 78, 1987. With permission.)

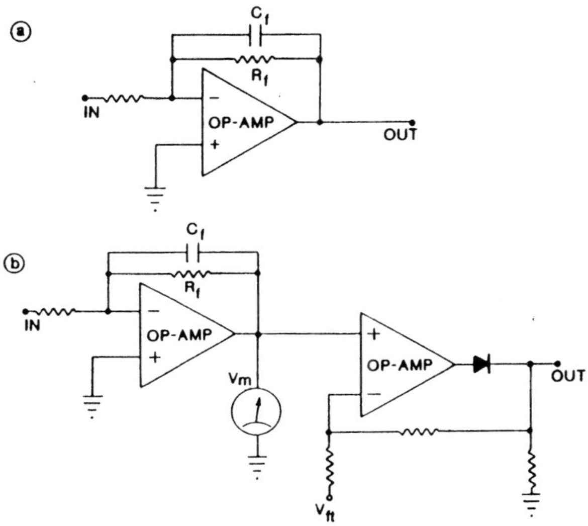

Figure 5.15 Simple realizations of the ERC negative feedback loop include (a) a single-pole integrator and (b) an integrator with an ideal diode to inhibit feedback until a preset value of v is exceeded (determined by V_{ft}). Also shown in the latter realization is a convenient point for driving analog meters. (From Valentine, K. H. et al., *IEEE Trans. Nucl. Sci.*, 34, 78, 1987. With permission.)

inverting integrator (Figure 5.15), which has a long time constant (i.e., $\tau_i = R_f C_f \gg 1/f_o$) so that its response to an isolated pulse from the comparator produces an insignificant change in the discriminator threshold, $V_d = V_{fb} - V_{th}$. However, as the threshold level-crossing rate increases toward f_o, V_{fb} tends toward a level determined by the feedback gain (FBG) adjustment. The amount of range extension provided by this simple negative feedback mechanism depends on the feedback gain setting. A somewhat more versatile realization (Figure 5.15b) is achieved by including a feedback threshold (V_{ft}) and an ideal diode to suppress feedback until a predetermined event rate is exceeded. This version is particularly useful for meter display schemes (e.g., connection of a voltmeter to V_m) because it permits the maximum possible sensitivity to be achieved over the lower decades of the usable range.[3]

The complementary level discriminator (CLD) design goals include the supply current programmability, protection from photocurrent burnout, and the best possible response time from the available transistors.[4] It was also desired that the output voltage swing matches that of the input requirements of various complementary metal oxide semiconductor (CMOS) logic devices operating at $V_{cc} = 5$ V when +/–6 V supply is used for the discriminator. Figure 5.16 shows the schematic of the level discriminator as fabricated. The bias current of the entire circuit is controlled by a current-setting branch consisting of Q_7, Q_8, and Q_9. A current programming resistor between Q_9 emitter and Q_7 collector determines the branch current, which in turn is reflected throughout the rest of the circuits. The circuit has the usual differential input stage of transistors Q_1 and Q_2 with resistive collector loads driving a second differential stage of Q_3 and Q_4. The input stage is conventional without bias current cancellation circuit, as the source impedance driving the input terminal is usually quite low, i.e., tens of ohms presented by the series resistance in the case of the current feedback amplifier.

The collectors of the second differential amplifier stage drive complementary current mirror-multipliers to the output terminal using the multibase and collector geometry transistors Q_6 and Q_{19} as mirrors.

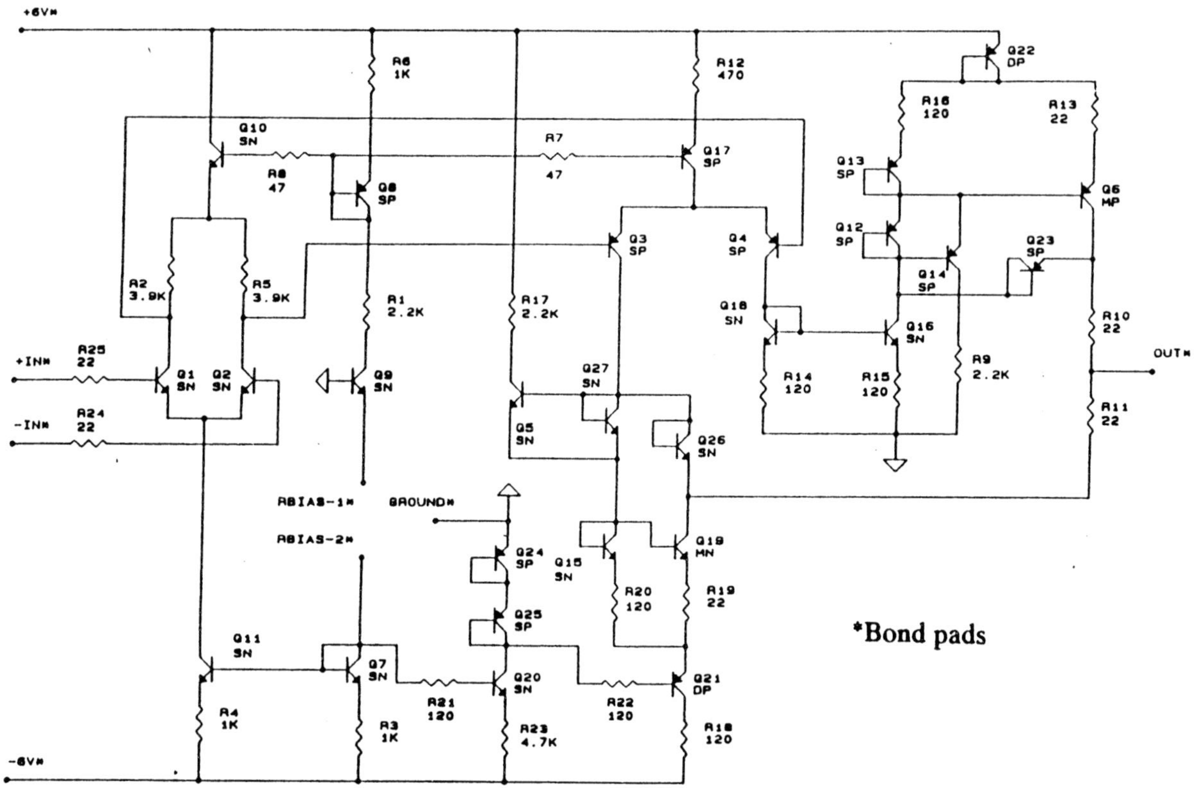

Figure 5.16 Complementary level discriminator schematic. (From Hahn, S. F. and Cafferty, M. M., *IEEE Trans. Nucl. Sci.*, 38, 44, 1991. With permission.)

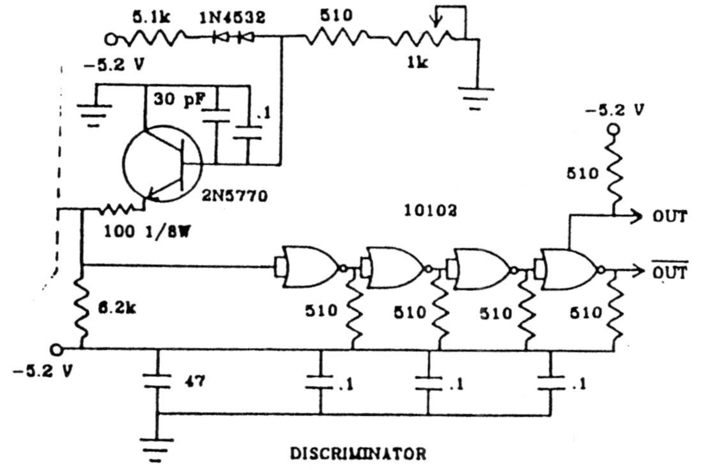

Figure 5.17 The discriminator circuit. (From Yoshida, R. et al., *IEEE Trans. Nucl. Sci.,* 33, 226, 1986. With permission.)

The transistors Q_{23} and Q_{26} are to shunt the excess collector currents from Q_6 and Q_{21}, respectively, back to the current sources for self-balance. Thus in the case of no DC load at the output terminal, the diode currents through Q_{23} and Q_{26} will supply four fifths of the collector currents of Q_{16} and Q_3, respectively. The use of emitter degeneration resistors not only compensates the mismatch of the mirror transistors, but also limits the transient photocurrents.

Transistors Q_{20}, Q_{21}, Q_{24}, and Q_{25} forces the output low state to very near 0 V by maintaining the Q_{21} base voltage at two diode drops below ground level. This circuit is activated only when the output is at low state. Transistor Q_{22} is to reduce the output swing below 5 V (to be compatible with the more common logic input levels) when +/–6 V supplies are used for the device. The gain of the CLD is about 5000 with a load of 10 kΩ, so the device responds to a differential signal as small as a millivolt for a full output voltage.[4]

A discriminator described by Yoshida[5] consists of a threshold-setting network providing temperature compensation and four emitter-coupled logic (ECL) NOR gates (10102). The first three of these gates are operated in the analog region and the last gives complementary ECL signals (Figure 5.17).

The threshold is set by a 1-kΩ potentiometer which is in a remote location from the base. This network mimics the structure of the internal ECL threshold so that the drifts in the DC levels in the ECL NOR gates should be matched by the drifts in the threshold network.

The signal, whose DC level has been set by the threshold network, is amplified further by four ECL NOR gates operating in the analog region (each stage giving a gain of four). The last ECL gate will give complementary ECL signals provided the input to the first gate has an input amplitude sufficiently above the reference voltage of the ECL gate (nominally –1.26 V). This would imply that the signal must be at least 3 mV above the noise at the input of the discriminator to produce clean complementary ECL output.

In asynchronous readout systems, discriminators (or comparators) are best realized as a cascade of low-gain limiting amplifiers. The small-signal gain and speed may be optimized as well as the number of stages of minimum power dissipation. A typical low-gain stage is given in Figure 5.18.[6] The input pair M3a-M3b is preferably operated in weak inversion where the transductance is maximum at given drain current. By preventing the drain current of diode-connected loads M4a-M4b from going down to zero when the input pair is overdriven, current sources αI_q ($\alpha = 0.05$ to 0.1 typically) reduce the time necessary to recover from overload conditions, thus increasing the speed. M4a-M4b are to be biased at moderate gate overdrive voltage ($V_{GS} - V_T = 0.4$ to 0.6 V) so as to realize an optimum small-signal gain of 3 to 5.

Assuming N identical gain stages, the time T_c necessary for comparison is approximately:

$$T_c = N\sqrt[N]{A}\,\frac{C_L}{g_m} \tag{5.7}$$

where A is the overall small-signal gain, C_L is the equivalent load capacitance of on stage, and g_m is the transconductance of the differential pair M3a-M3b.

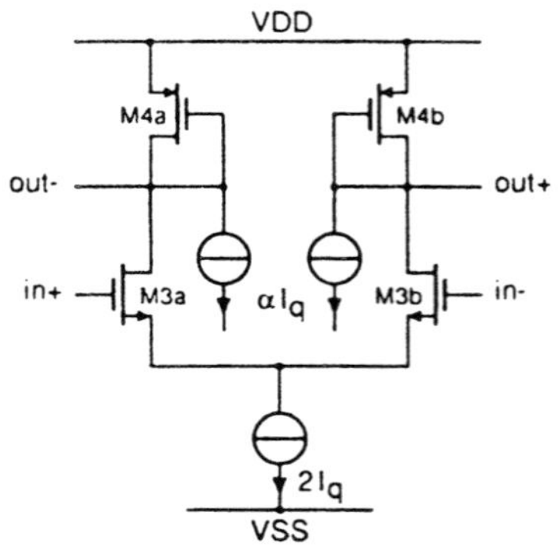

Figure 5.18 Typical MOS implementation of a low-gain, low-power limiting amplifier. (From Krummenacher, F., *Nucl. Instrum. & Methods,* A305, 527, 1991. With permission.)

High dynamic range (2000:1) discriminator (HDRD) measurements with fragmented heavy ions were described by Llacer et al.[7] A block diagram of the HDRD is shown in Figure 5.19. The HDRD has a topology similar to a network developed earlier for Si(Li) X-ray detectors in a Tokamak (magnetically confined fusion reactor) environment. A triangular-shaped waveform with a well-defined return to the baseline is generated by combining an asymmetrical bipolar delay-line pulse shaper with a known RC integrator. The triangular shape waveform will return precisely to the baseline if the amount of asymmetry of the bipolar waveform and the value of the RC integrator are correctly chosen. A convenient solution is to make the RC integrator time constant equal to the delay-line length, and make the amplitude the second lobe of the bipolar waveform to be 1/e of the amplitude of the first lobe. The second (short-term) pole zero control makes this adjustment. This shaper produces a pulse that has a nearly ideal shape, very close to a symmetrical triangle, which is the shape for best signal/noise when delta noise dominates. The resolving time (pulse width) reaches a maximum of only two times the RC integrator time constant for the entire energy range. In order to achieve the large dynamic range, the discrete transistor amplifiers have been made capable of a linear 18-V unipolar excursion. A gated baseline restorer has been used in the circuit, although a low-duty cycle is anticipated. Diode limiters precede the selected high-quality voltage comparator (CMP-05). With these circuits we have measured a dynamic range of 3500:1 on the workbench.

5.3 TIMING DISCRIMINATOR

With application of the arrival time of an event in addition to its pulse amplitude, pulse analysis circuits can be constructed which have improved characteristics. A relatively simple circuit addition results in a variable pulse amplitude discriminator whose output pulse always occurs at the crossover of a double, delay-line-shaped input pulse independent of the discriminator level. This output pulse can then be used to indicate the arrival time of an event with one to two orders of magnitude better time resolution than a leading-edge discriminator over the same dynamic range. This is the antiwalk discriminator.

The basic antiwalk discriminator circuit, shown in Figure 5.20[8] consists of Schmitt trigger (T1 and T2) and a feedback circuit (T3–T7) which changes the discriminator hysteresis so that the discriminator returns to its normal state as the input pulse changes polarity. When the positive portion of the input pulse exceeds the discriminator level, T1 conducts and T2 is cut off. The positive pulse, generated at the collector of T2, causes T7 to stop conducting and the collector voltage of T7 drops until T6 conducts. This negative pulse is applied to the base of T2 by emitter-follower T5. The base of T2 is then set at the same voltage as the base of T6 which is also the baseline voltage of the input pulse. Therefore, the discriminator will return to its normal state when the input pulse changes polarity regardless of where the discriminator level is set. The return of the discriminator to its normal state causes T7 to conduct; and the differentiated negative pulse at the base of T9 causes T9 to saturate, generating a fast positive output pulse.

The discriminator outputs cause positive current pulses to be switched into the shorted delay line which acts as a differentiating and shaping network. When both inputs are switched during the coincidence time (twice the delay time of the delay line), the current pulses sum and a coincidence output pulse is generated.

A high-performance, low-cost discriminator employing both the leading-edge and constant-fraction trigger techniques has been developed by Sanders et al.[9] This hybrid design provided an experiment with

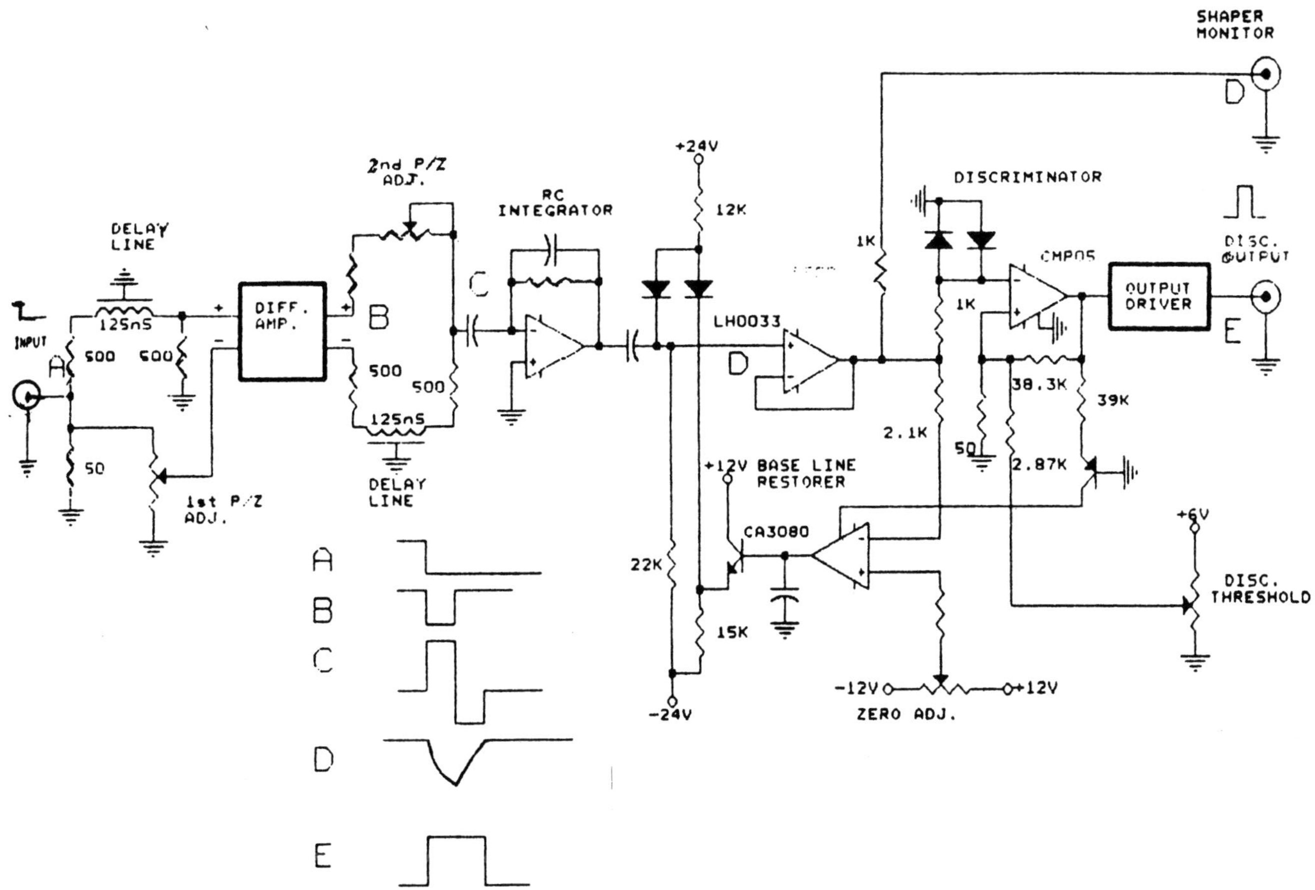

Figure 5.19 Block diagram of the high dynamic range discriminator. (From Llacer, J. et al., *IEEE Trans. Nucl. Sci.*, 37, 457, 1990. With permission.)

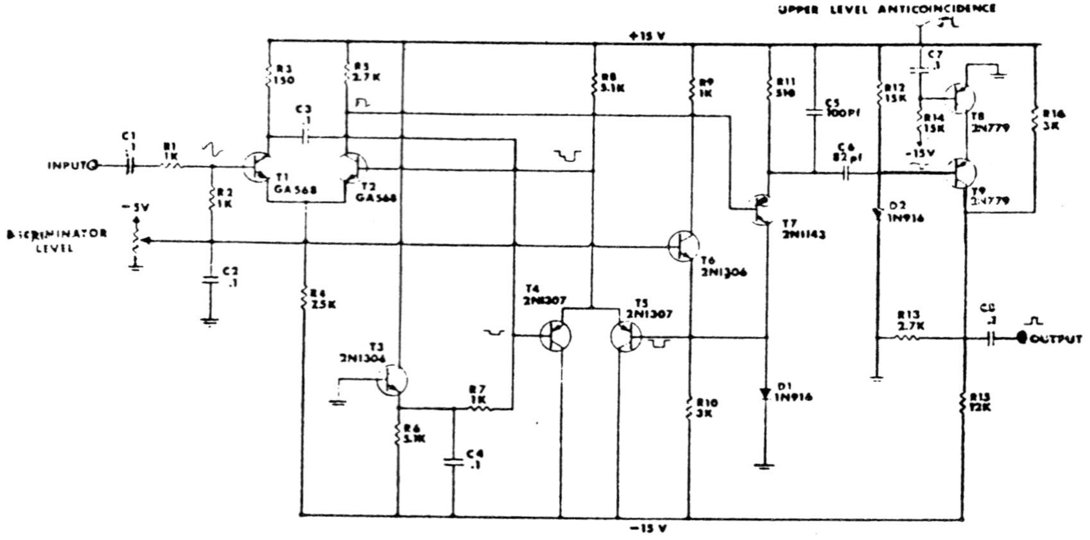

Figure 5.20 Antiwalk discriminator. (From Emmer, T. L., in NAS-NRC Pub. 1184, Monterey, 1963, p. 112. With permission.)

the ultimate in on-line and off-line time resolution. Accurate time resolution, both on-line and after off-line corrections, is crucial in reducing accidental triggers and rejecting backgrounds in the final analysis.

The basic problem is to relate the time of passage of a particle through a detector (scintillator) to the time an output pulse is issued by the discriminator. The uncertainty Δt is the sum of several contributions:

$$\Delta t = \Delta t_{\text{scint}} + \Delta t_{\text{coll}} + \Delta t_{\text{PMT}} + \Delta t_{\text{jitter}} + \Delta t_{\text{walk}} \tag{5.8}$$

where Δt_{scint} is the time variation intrinsic to the scintillation process, Δt_{coll} is the time variation introduced by the light collection process, Δt_{PMT} is the time variation introduced by the photomultiplier tube, and Δt_{jitter} is the variation in the transit time through the discriminator for pulses of a fixed pulse height and rise time. It consists of the statistical variation in the arrival times of the photoelectrons at the selected pulse height fraction, and the switching time variation induced by the fact that a given amount of charge must be collected after the discriminator threshold level is exceeded to cause the circuit state to switch.

Δt_{walk} is the variation in Δt introduced by the fact that the fixed discriminator threshold is exceeded at different times by pulses of identical shape, but different amplitude. This effect is intrinsic to the design of a leading-edge triggered discriminator. The simplest method of reducing timing walk is by using the lowest possible trigger threshold. This risks low-level noise-induced triggers. A second high-level discriminator is then used reducing the rate of spurious noise-induced triggers. It is inferior to the constant-fraction method. A zero-crossing discriminator provides an output related to the same phase point on all pulses of identical shape, regardless of pulse height, canceling the walk effect.[9]

Combining the virtues of the crossover method with the use of the optimal threshold fraction was used by Sanders et al. for the constant-fraction-of-pulse-height trigger. It is illustrated in Figure 5.21.

The photomultiplier (PM) anode current pulse is passed through an attenuation-subtraction network that produces a bipolar pulse with a zero-crossing phase point corresponding to the desired fractional triggering level. If a 10% fractional triggering level is required, part of the anode pulse is attenuated to 10% of its initial amplitude. The delay time t_d is chosen such that the 10% phase point on the delayed pulse arrives at a summing point at the same time as the maximum amplitude of the attenuated pulse. The two pulses are added algebraically to produce the bipolar waveform. The attenuated pulse exactly cancels the delayed and inverted pulse at the 10% phase point on the delayed pulse. This bipolar resultant is presented to a zero-crossing discriminator which triggers, effectively, at the 10% fractional level regardless of initial amplitude. The discriminator output is timed to the zero crossing by suitable arrangement of the internal delays and pulse widths. The bipolar pulse may be formed by a transistor summing circuit,

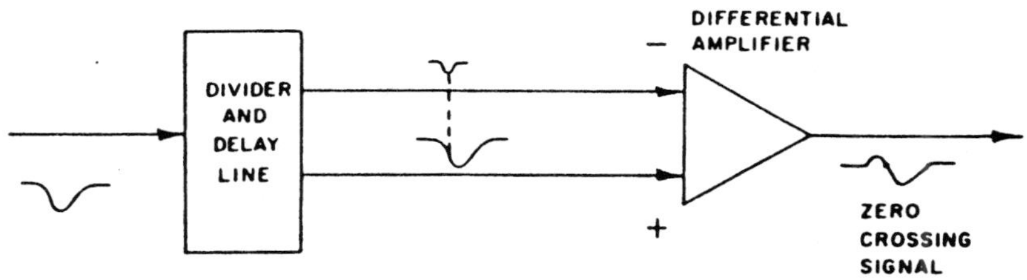

Figure 5.21 Schematic diagram of the constant fraction front-end signal processing. The input signal is converted to a bipolar signal with its zero crossing timed to a fixed point of the original leading edge. (From Sanders, G. H. et al., *Nucl. Instrum. & Methods,* 180, 603, 1981. With permission.)

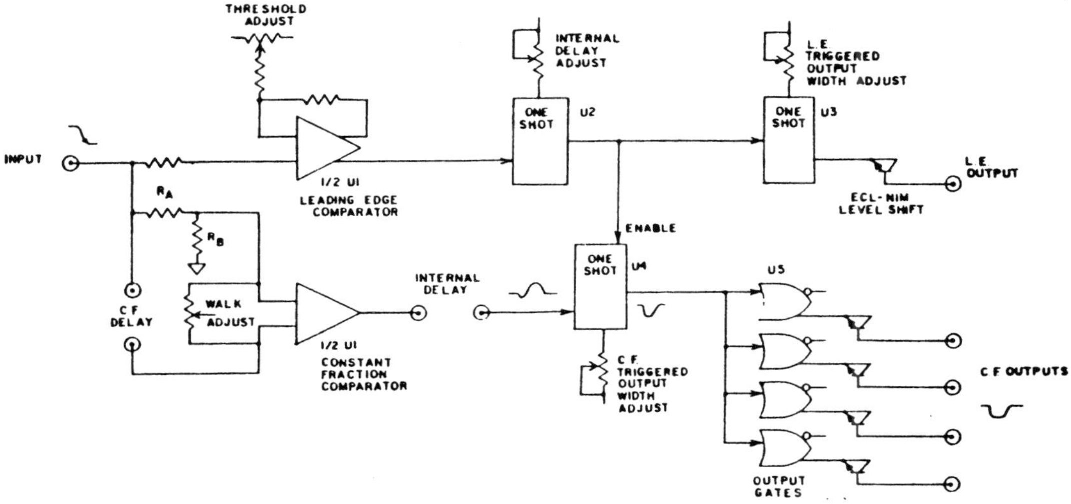

Figure 5.22 Simplified circuit diagram of one discriminator channel. (From Sanders, G. H. et al., *Nucl. Instrum. & Methods,* 180, 603, 1981. With permission.)

an integrated circuit comparator, or a high-quality passive ferrite transformer. A simplified diagram of one discriminator channel is shown in Figure 5.22.

A timing discriminator using leading-edge extrapolation was described by Gottschalk.[10] The circuit recovered the timing information inherent in a slow-rising input pulse, producing an output "timing edge" a fixed time after the starting corner of the input pulse. Output time was nearly independent of input rise time, input pulse height, and discriminator threshold setting. The circuit used two switched current sources which charged a single capacitor.

Figure 5.23(a) is the circuit block diagram. A comparator C fired when the input pulse exceeded the threshold V_T, setting a flip-flop; previously, comparators A and B have fired at one third V_T and two thirds V_T, respectively. Current sources i_1 and i_2 were controlled by A and B so that both were switched on by A and subsequently one was shut off by B. Accordingly the capacitor charged at a certain rate between the crossing of one third V_T and two thirds V_T and at a lower rate thereafter. Figure 5.23(b), drawn with the assumption of a linear rise and $i_1 = i_2$, shows that the capacitor voltage will eventually reach the same trajectory V(t) for a linear rise of whatever slope; the same holds true for a given slope and any threshold. It was necessary to add a fourth comparator D which reset the flip-flop when this ramp reached some fixed voltage. The output trailing edge occurred at a fixed time after the "true" start of the pulse; the leading edge, at the instant the pulse crosses V_T. The capacitor was discharged by a clamp controlled by D.

Figure 5.24 shows a practical circuit for microsecond pulses. It uses standard, inexpensive components. The current generators are transistors F and H switched on by the sharp positive-going 710 output via an inverter. At the two thirds V_T level, transistor E is turned on harder than F, diverting one of the currents. The nonlinearity correlation adjusts the ratio i_1:i_2.

A leading-edge discriminator produces an output logic pulse when the input signal crosses a fixed threshold level. While the most simple to implement, leading-edge timing has the disadvantage in that the time of occurrence of the output pulse is a function of the amplitude and rise time of the input signal.

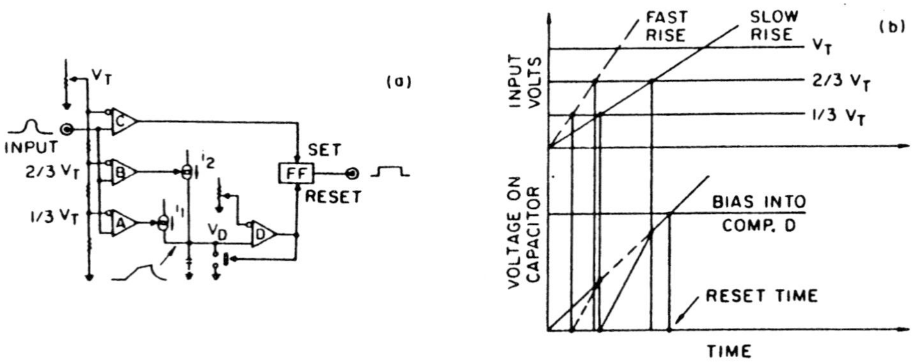

Figure 5.23 (a) Block diagram of the discriminator; (b) graphic construction showing how the circuit compensates from rise-time difference. (From Gottschalk, B., *Nucl. Instrum. & Methods,* 190, 67, 1981. With permission.)

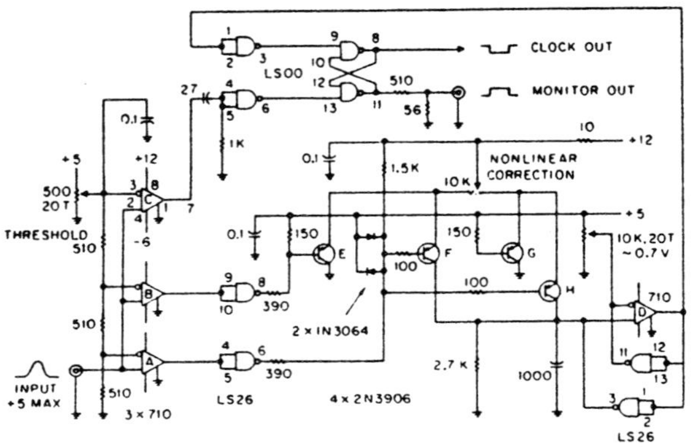

Figure 5.24 Schematic diagram of the discriminator. (From Gottschalk, B., *Nucl. Instrum. & Methods,* 190, 67, 1981. With permission.)

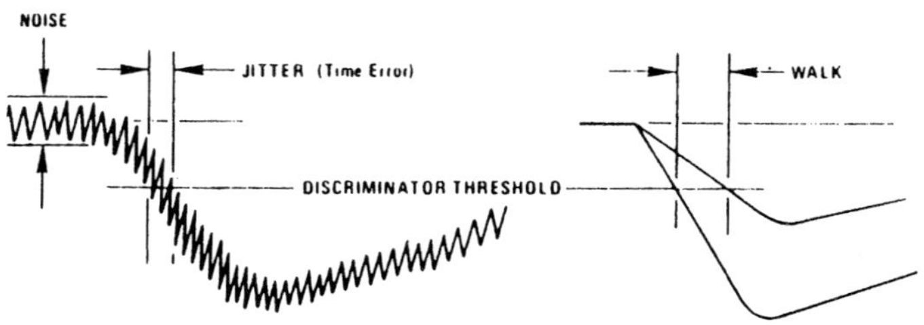

Figure 5.25 Jitter and walk in leading-edge time derivation. (From EG&G ORTEC Catalog, 91/92 Detectors and Instruments for Nuclear Spectroscopy. With permission.)

Figure 5.25 illustrates leading-edge timing. Note the time walk for the two different input pulse amplitudes. Figure 5.25 also illustrates noise-induced jitter. Statistical amplitude fluctuations of the detector signals and noise on the input signal cause jitter or uncertainty in the time at which the input signal crosses the discriminator threshold. The amount of jitter in the timing signal depends on the noise amplitude and the slope of the signal at the threshold level.

Turko et al.[11] described a zero-crossing discriminator that accurately determines the peak of a quasi-Gaussian waveform by differentiating it and detecting the resulting zero crossing. It consists of a fast voltage comparator and two integrating networks: an RC section and an LR section used in a way that keeps the input impedance purely resistive. A time walk of 100 psec in an amplitude range exceeding 100:1 has been achieved for waveforms from 1.5 to 15 nsec FWHM. An arming level discriminator was added to eliminate triggering by noise. Constant fraction discriminators in effect normalize the amplitude variation of input pulses and, providing they have the same shape, may reduce the output time walk to a fraction of a nanosecond over a large amplitude range. Such discriminators, however, require optimization for a given waveform by externally applied delay lines and can be expensive in large experiments using many timing channels. While constant fraction discriminators use a selected point on the rising edge of the pulse as a reference, zero-crossing discriminators differentiate the input waveform and use the

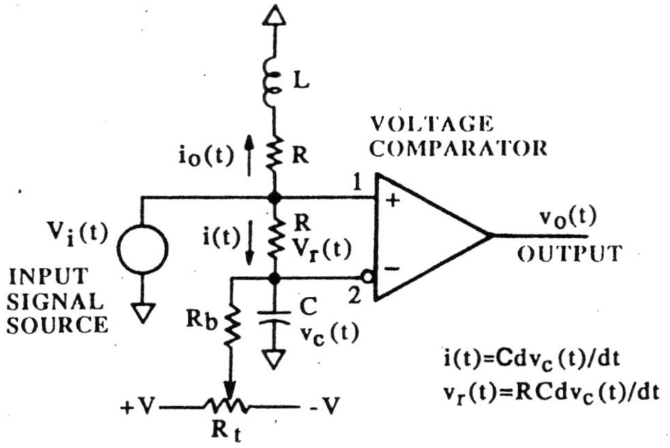

Figure 5.26 Voltage comparator in a zero-crossing detector configuration. (From Turko, B. T. et al., *IEEE Trans. Nucl. Sci.*, 39, 1311, 1992. With permission.)

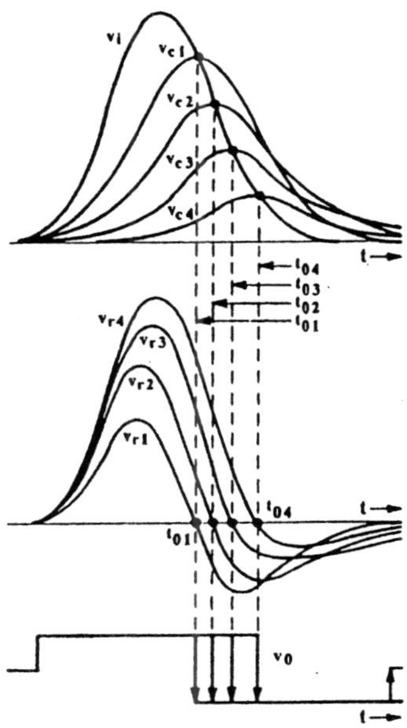

Figure 5.27 Top: comparator noninverting input signal, trailing edge crosses peaks of the signal after integration (v_{C1} to v_{C4}), at the inverting input. Middle: volt difference v_{r1} to v_{r4} across the comparator inputs. Bottom: Comparator output zero crossing. (From Turko, B. T. et al., *IEEE Trans. Nucl. Sci.*, 39, 1311, 1992. With permission.)

polarity change of the derivative as the time reference. Because that occurs at the top of the pulse, in comparing the two types, it can be said that the selection point of a zero-crossing discriminator is fixed at a fraction of 100%.

A simplified circuit (Figure 5.26) shows the voltage comparator noninverting input driven from a pulse source $V_i(t)$. The inverting input signal is integrated by an RC network. Also, this input is biased by a DC current provided by a resistor, R_b, much larger than R. The signal source is also terminated by an LR network. If RC = LR, the net load to the signal source is R all the time. Due to the integration, the voltage at the inverting input, $V_c(t)$, lags the input signal, $V_i(t)$. The voltage difference between the two inputs is

$$V_r(t) = V_i(t) - V_c(t) = RC\,dV_c(t)/dt \tag{5.9}$$

The current charging capacitor C drops to zero when the peak of $V_c(t)$ is reached, i.e., $dV_c(t)/dt = 0$. At this point the voltage differences between the comparator inputs must also be zero. These conditions are illustrated in Figure 5.27. A bellshaped input voltage, $V_i(t)$, is integrated by four different time constants, resulting in curves $V_{c1}(t)$ to $V_{i4}(t)$. According to Equation 5.9, in Figure 5.27 the $V_i(t)$ trailing edge must cross all $V_c(t)$ curves at their maximum. The curves $V_{r1}(t)$ to $V_{r4}(t)$ is the voltage difference across the comparator inputs, and $V_o(t)$ is the comparator output. The oscillogram of a 6 nsec FWHM input signal integrated by time constants from 0.8 to 10.2 nsec is shown in Figure 5.27.

The comparator output response lags due to propagation delay and is also a function of input overdrive. On the other hand, the amount of overdrive is a function of input signal amplitude and the integration time constant, RC. As seen in Figure 5.27 the magnitude of overdrive, after the zero level is crossed at the points t_{01} to t_{04}, decreases with increasing RC. Consequently, the output delay increases.

An arming (voltage over threshold) discriminator has been combined with the zero-crossing discriminator for preventing triggering by noise. A dual (although somewhat slower) comparator (HCMP 96850)

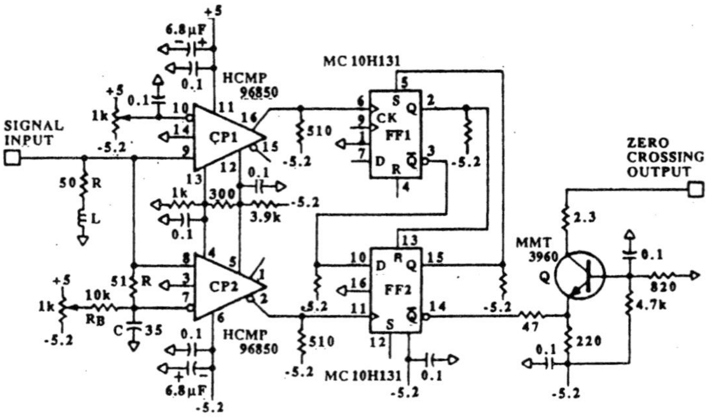

Figure 5.28 Zero-crossing (CP2) and arming discriminator (CP1) circuit diagram with clocked output shaper and dead-time circuit. (From Turko, B. T. et al., *IEEE Trans. Nucl. Sci.*, 39, 1311, 1992. With permission.)

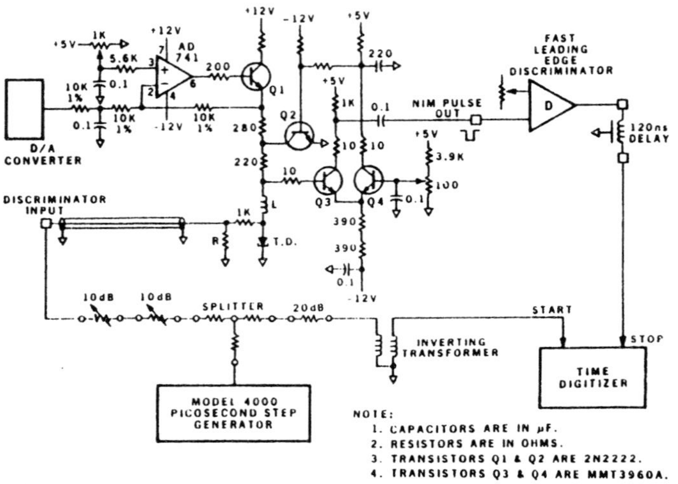

Figure 5.29 Schematic diagram of tunnel diode test circuit. (From Kolbe, W. F. et al., *IEEE Trans. Nucl. Sci.*, 36, 412, 1989. With permission.)

was used along with a dual D-type flip-flop (MC10 H131) for shaping the discriminator output (Figure 5.28), and a fast bipolar transistor (MMT 3960) as 50-Ω line driver. For a 3-V positive input pulse (6 nsec FWHM), and integration time constant RC = 1.8 nsec, the zero-crossing point occurs 2 nsec after the peak. The positive zero-crossing edge appears 2 nsec later and the shaped output 1 nsec thereafter. The discriminator is paralyzed for the duration of the time-over-threshold set by the arming comparator threshold level. The total propagation delay from the input pulse peak and the output leading edge is about 5 nsec.[11]

An automated system for the evaluation of fast rise time voltage discriminators is described by Kolbe et al.[12] A high-performance time digitizer was used to determine the elapsed time between the generation of a test pulse and the response of the discriminator under test. In the design of a properly functioning voltage discriminator, three key components must be considered. These are the input pulse driver amplifier, the nonlinear threshold detection device, and the threshold sensing amplifier and output circuitry. For the adequate discrimination of nanosecond and subnanosecond rise time pulses, tunnel diodes have proved themselves to be reliable threshold detection devices. The tunnel diode is essentially a current-sensing device. When the input current to the diode exceeds a certain threshold, it exhibits a negative resistance and an abrupt transition to a higher voltage stage. This voltage transition can then be detected by the output amplifier and converted to a suitable output pulse. To set the operating threshold of the discriminator, a current source is used to bias the tunnel diode at an appropriate level.

The bandwidth and driving characteristics of the discriminator input amplifier can significantly affect the response of the discriminator system to an input pulse. To better evaluate the performance of the tunnel diode alone, Kolbe et al. have constructed the special test circuit shown in Figure 5.29 in which

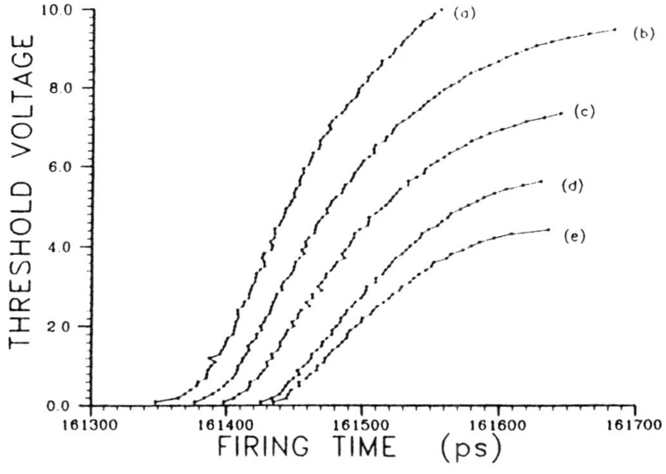

Figure 5.30 Firing time measurements of custom components model 151050G, 10 mA diode for (a) 1 dB, (b) 3 dB, (c) 5 dB, (d) 7 dB, and (e) 9 dB input pulse attenuation. (From Kolbe, W. F. et al., *IEEE Trans. Nucl. Sci.*, 36, 412, 1989. With permission.)

the input amplifier is eliminated. In this arrangement, the coaxial line from the input pulse is carefully terminated with 50 Ω and the pulse current is applied to the tunnel diode through a moderately large resistance of 1000 Ω. The diode output is then coupled to the output sensing amplifier through a small (4-turns, 2-mm diameter) choke to provide additional isolation. The sensing amplifier (Q_2, Q_3, and Q_4) is adjusted to trigger at a point in the voltage switching range of the tunnel diode. A fast leading-edge discriminator is used to condition the pulse output of the sensing amplifier to present a nuclear instrument module (NIM) pulse of constant width to the stop input of the time digitizer.

The results of a typical series of measurements is shown in Figure 5.30 for the Custom Components model 151050G, 10-mA diode with the input pulse attenuated by 1, 3, 5, 7, and 9 dB. When the data are normalized by scaling all the amplitudes to 0 dB, good agreement in the amplitudes of the top of the pulses is obtained; however, at low threshold levels the data exhibit a negative time shift of about 20 psec for each 2-dB increase in the pulse power. This shift is apparently due to pulse overdrive, i.e., amplitude-dependent differences in the time required for the pulse to charge the device up to the threshold level and for the sensing amplifier to fire.[12]

A subnanosecond time discriminator system was described by Benulis and McFarlane[13] which gives an output if two pulses on separate input lines (which overlap) have their leading edges separated by a time greater than a preset value. The time discriminator provides 200-psec resolution, 60-nsec (15-MHz) double-pulse resolution, better than 1% linearity over a 20-nsec range, and ECL and NIM outputs with adjustable width control.

Figure 5.31 shows the circuit and Figure 5.32 the pulse waveforms at various points. The start and stop NIM inputs are converted to ECL levels by Q_{10} and Q_1. The start differential switch consists of Q_6 and Q_7. In the reset state Q_6 is on and Q_7 is off, thereby holding the emitter of Q_8 at a potential of about +2.2 V. When the start pulse arrives, Q_6 is turned off and Q_7 is turned on; and the Q_8 emitter is driven negative, allowing the current source (which is also on if no stop pulse has arrived) to linearly discharge C1, the charge storage capacitor. The constant current source consists of Q_2 and Q_3; Q_2 provides temperature compensation. Use is made of 1% metal film resistors here and in the reference circuits. The stop differential switch is formed by Q_4 and Q_5; Q_4 is off and Q_5 is on for the rest state. When the stop input pulse arrives, the current source is turned off since current now flows in Q_4.

The AM 685 comparator switches when the voltage on C1 falls below the reference voltage (due to a start pulse with no stop pulse). The reference for the comparator is formed by Q_9 and associated resistors. Q_9 gives temperature compensation for Q_8. A front panel potentiometer P1 allows adjustment of this voltage.

Gates A3 and A4 form a triple input AND with the start, stop, and the comparator output; and enforce the requirement of overlap of a start and stop pulse. The output of A4 goes to B2, a one short, whose width is controlled via a front panel potentiometer P3. The output of B2 is buffered by 5 and A6 to provide two front panel ECL outputs, and is also converted by Q_{11}–Q_{14} to provide two NIM outputs and one inverted NIM output.

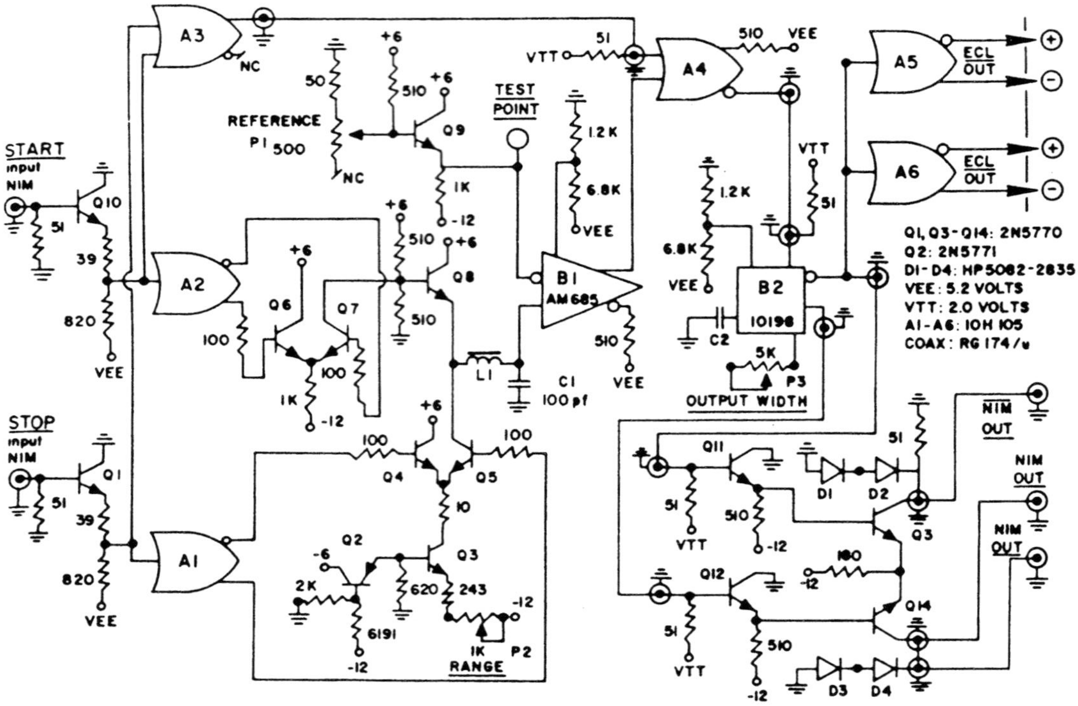

Figure 5.31 Circuit diagram of time discriminator. (From Benulis, C.A. and McFarlane, W. K., *Nucl. Instrum. & Methods,* A240, 130, 1985. With permission.)

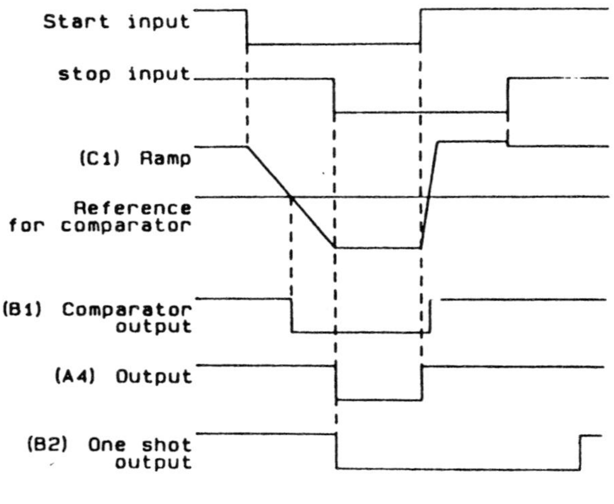

Figure 5.32 Waveforms at various points. (From Benulis, C. A., and McFarlane, W. K., *Nucl. Instrum. & Methods,* A240, 130, 1985. With permission.)

5.4 DIFFERENTIAL DISCRIMINATOR (SINGLE-CHANNEL ANALYZER)

A differential discriminator is a circuit which provides an output pulse for each input pulse of height in the range between V_{th} and $V_{th} + \Delta v$ where V_{th} is a threshold voltage below which the system will not respond; the system will not respond to a pulse above $V_{th} + \Delta V$ either. The system principle is shown in Figure 5.33. It is made of two integral discriminators adjusted to V_{th} and $V_{th} + \Delta v$, respectively, connected to an anticoincidence circuit. There will be an output pulse when only one discriminator, the v_{th}, will respond. Generally the lower discriminator pulse is wider since the input pulse is wider at lower voltage. In such a case the anticoincidence circuit might respond to that part of the pulse which does not overlap the output pulse of the upper discriminator where they both responded. Several methods were used to overcome this problem. The wide pulse can be differentiated and then clamped with a diode, or a delay can be used to obtain a better overlap.

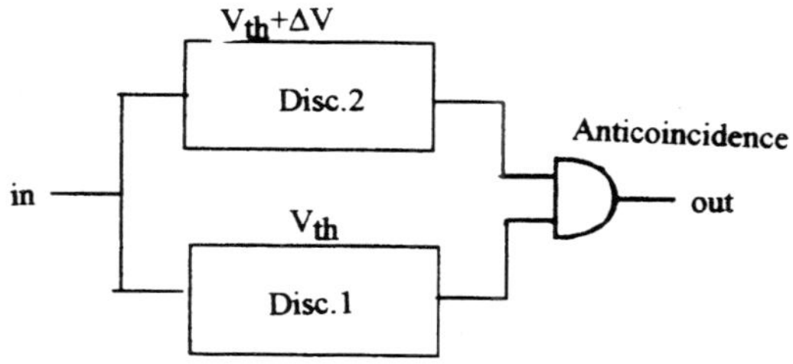

Figure 5.33 Differential discriminator.

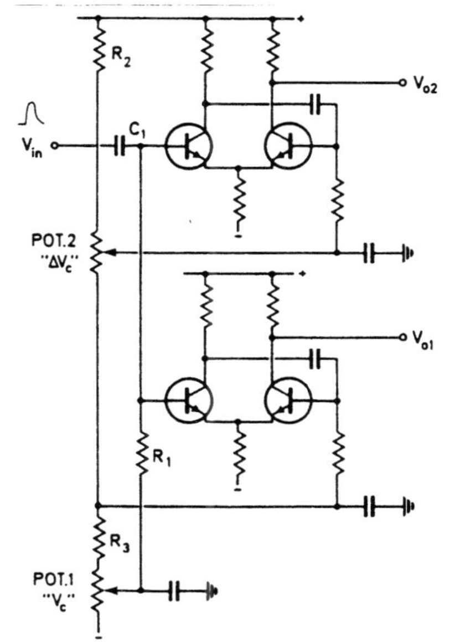

Figure 5.34 Adjusting the channel position and the channel width by two independent potentiometers.

To obtain independence of ΔV two dependent potentiometers are used so that Δv is always a certain voltage above v_{th} whatever this voltage is. This is achieved in the circuit shown in Figure 5.34. P_1 supplies voltage to both bases of Q_1 in circuit 1 and Q_3 in circuit 2, and P_2 adjusts the voltage of circuit 2 to fit ΔV above V_{th}.

Because of the difference in pulse width of the pulse at each circuit, a timing problem arises. Lower pulses will arrive at the threshold later. A bipolar pulse and a zero-crossing discriminator circuit are used to overcome the difficulty. The high and low pulses arrive at the same time. A constant fraction can be used in parallel to the discriminator for time information.

A differential pulse height discriminator was designed by Cole.[14] It was complete with full channel logic and could be used with either symmetrical or asymmetrical expansion of the channel width or operate as two independent discriminators sharing a common input terminal. The advantage of the design was the low-power requirement. A 5-V, 5-mA supply was required. A block diagram is shown in Figure 5.35. A positive input signal was applied to the common input terminals of two Schmitt-type discriminators (D_H, D_L) which are individually back biased by the positive voltages, V_H and V_L. These voltages were derived from two DC amplifiers, A_H and A_L; and defined the upper and lower limits of the channel. The difference between these voltages defined the channel width V_w and their average value at any given magnitude defined the channel center V_c, i.e., $V_c = 1/2(V_H + V_L)$. The output waveform produced by the two discriminators were connected to two monostable circuits, M_H and M_L. These performed the required logic and ensured that the channel output pulse was produced only when the amplitude of the input signal lied within the confines of the channel. The two discriminators and the two DC amplifiers were made of national semiconductor quad comparator type LM339. The two monostable circuits were made from a CMOS quad 2 input NAND gate-type CD4011. The complete circuit is shown in Figure 5.36.

A simple single-channel analyzer using standard components was described by Correia et al.[15] High-temperature stability (0.002%/°C) and resolution (= 100 μV) was achieved. The two-level generator circuits allowed full DC coupling of the input pulses.

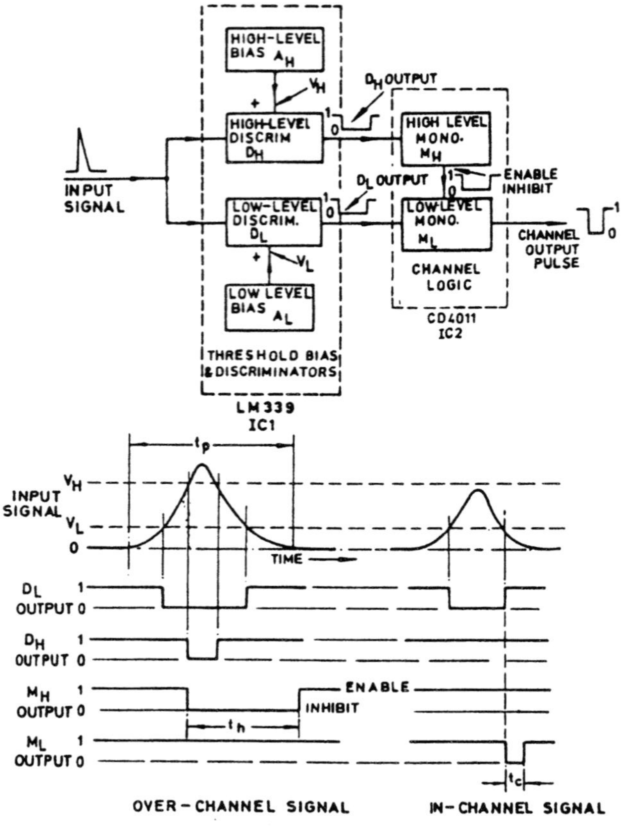

Figure 5.35 Block schematic of the differential discriminator with typical operating waveforms. (From Cole, H. A., *Nucl. Instrum. & Methods,* 138, 551, 1976. With permission.)

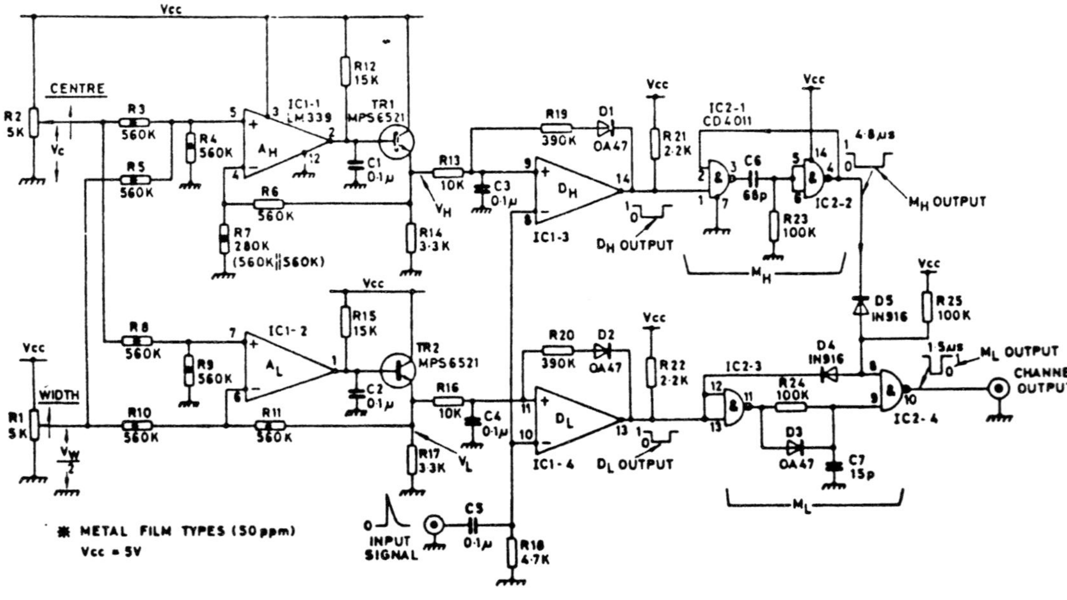

Figure 5.36 Circuit diagram of the differential discriminator (symmetric mode). (From Cole, H. A., *Nucl. Instrum. & Methods,* 138, 551, 1976. With permission.)

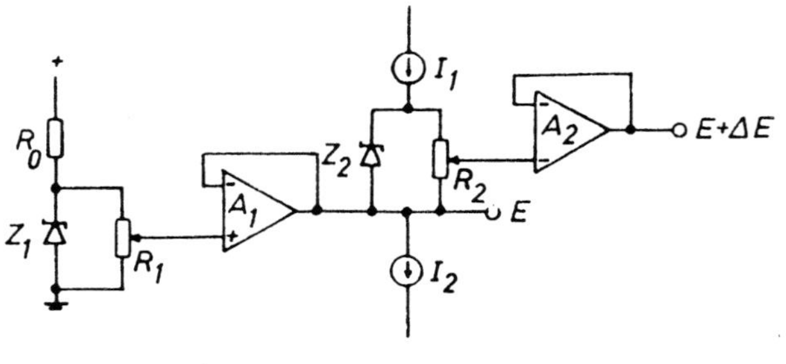

Figure 5.37 DC level generator circuit: first configuration. (From Correia, C. A. B. et al., *IEEE Trans. Nucl. Sci.*, NS25, 59, 1978. With permission.)

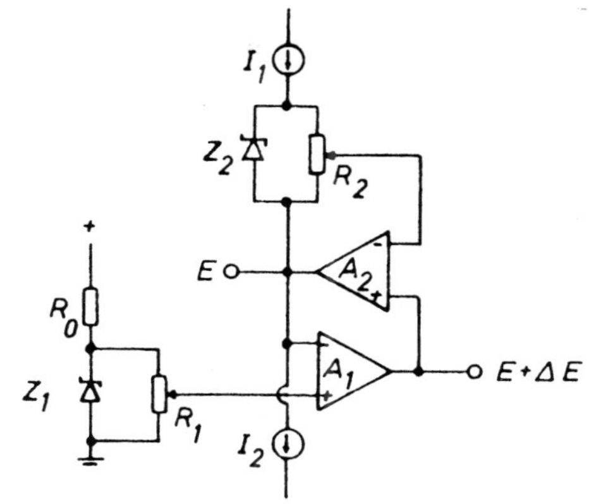

Figure 5.38 DC level generator circuit: second configuration. (From Correia, C. A. B. et al., *IEEE Trans. Nucl. Sci.*, NS25, 59, 1978. With permission.)

The upper and lower DC voltage levels required to define the window of a single-channel analyzer (SCA) were generated by circuits that, unlike the ones of most commercial units which only allow AC coupling, allow full DC coupling of the pulses to be analyzed.

Two alternative DC levels generator circuits are shown in a schematic way in Figures 5.37 and 5.38. In both circuits the window width ΔE was constant and independent on the baseline voltage, E.

In the first one E was derived from a potentiometer, R_1, in parallel with a zener diode, Z_1, which was fed through a resistor R_o. This DC level was available at the output of operational amplifier A_1, connected as a voltage follower. ΔE was generated across potentiometer R_2 in parallel with zener Z_2 which was fed by a current source, I_1. This voltage was added to E in the way shown, and E + ΔE was available at the output of operational amplifier A_2. The use of high-impedance sources to feed the zeners allowed high-temperature stability if the bias current intensity was set for the value (I_{ZT}) given by the manufacturer. Current source $I_2(= I_1)$ did not need to be a high-performance one, since it was used only to sink I_1, thus protecting the output of operational amplifier A_1.[15]

The second circuit (Figure 5.38) used the same basic components as the first one but in a different arrangement: operational amplifier A_2 was in the feedback loop of A_1, and the ΔE generator circuit in the loop of A_2. E was obtained at the output of A_2 and E + ΔE at the output of A_1.

The importance of the input offset voltage and bias currents of the operational amplifiers in the performance of the circuits of Figures 5.37 and 38 can be discussed with the help of Figures 5.39 and 5.40, respectively. In these Figures R_1' and R_2' are the internal resistances for the Thevenin equivalent circuit for the generators of the voltages E and ΔE; e_{0s1}, and e_{0s2} are the offset voltages of A_1 and A_2, respectively; and I_{1+}, I_{1-}, I_{2+}, and I_{2-} are the input bias currents for the operational amplifiers.

To reduce the input current errors the compensation resistances R_{C1} and R_{C2}, in Figures 5.39 and 5.40 were made equal to R_1' and R_2', respectively. In practice since the values of $R_1'$and R_2' depend on the setting of the potentiometers R_1 and R_2 of Figures 5.37 and 5.38, the actual values to be taken are those for an average setting, i.e., $R_{C1} = R_1/4$ and $R_{C2} = R_2/4$. If the input offset currents for the operational amplifiers A_1 and A_2 were $I_{01} \equiv I_{2+} - I_{2-}$, respectively, one can write for the output voltages of the circuit of Figure 5.39:[15]

$$V_2 = E + e_{os1} + R_1' I_{01} \tag{5.10}$$

$$V_1 = E + \Delta E + e_{os1} + e_{os2} + R_1' I_{01} + R_2' I_{02} \tag{5.11}$$

and for the circuit of Figure 40:

$$V_2 = E + e_{os1} + R_1' I_{01} \tag{5.12}$$

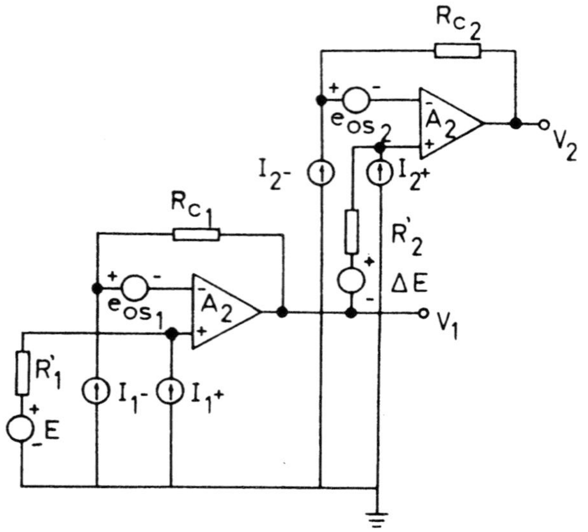

Figure 5.39 First configuration equivalent circuit. (From Correia, C. A. B. et al., *IEEE Trans. Nucl. Sci.,* NS25, 59, 1978. With permission.)

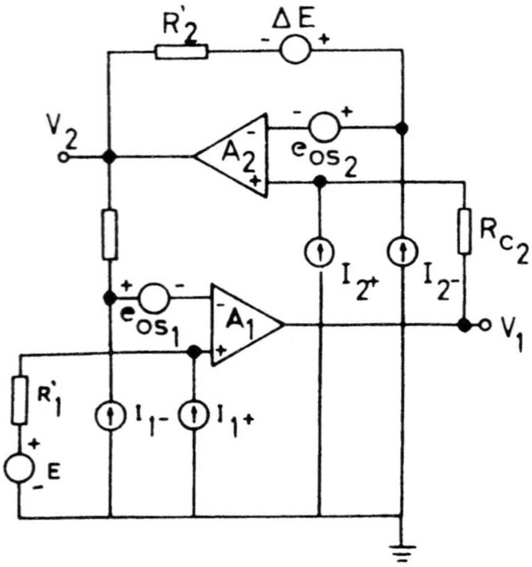

Figure 5.40 Second configuration equivalent circuit. (From Correia, C. A. B. et al., *IEEE Trans. Nucl. Sci.,* NS25, 59, 1978. With permission.)

$$V_1 = E + \Delta E + e_{os1} - e_{os2} + R_1' I_{01} - R_2' I_{02} \tag{5.13}$$

These results show that, since the offset voltages and currents are only given in absolute values, the actual deviations from the exact values E and ΔE are the same for both circuits.

A fast single-channel analyzer with built-in gate and adjustable delay for external timing pulses was described by Bialkowski et al.[16] The goal was to reduce the dead time of data-taking systems and to facilitate the off-line analysis. The SCA was equipped with lower and upper level controls, external and internal modes of operation, and output delay control. It had 1-kΩ input resistance which allowed the use of 50-Ω termination included in constant fraction discriminator (CFD). The output delay control was functional only for external mode, when CFD output is connected to the strobe input. For internal mode of operation the output pulse time position followed the input pulse rise time.

The block scheme of the SCA is shown in Figure 5.41. The lower level discriminator (LLD) sets the flip-flop (FF1) via a fast differentiating network. The upper level discriminator (ULD) is enabled by the LLD and resets the FF1 via the OR gate and the dead-time circuit. The dead time was adjustable from 1 to 5 μsec and protects the SCA from a new setting before the input pulse returned to the initial level. The FF1 was left in set position when the input amplitude is between the lower and upper thresholds. Delay 1 defined the time when the input amplitude was examined and should be adjusted to a value higher than the input rise time. Delay 2 gave the maximum waiting time for an external strobe pulse. The third delay, with the flip-flop 2 (FF2) formed the output delay circuit.

A NIM module with eight low-level, ultra-fast differential discriminator channels was described by Brahy and Rossa.[17] The module was designed to increase the operating intensity of photomultipliers to 5×10^7 pulses per second. It allowed the use of a very long transmission cable (from 200 to 300 m) between the photomultiplier and the module input. The differential discriminator provided high-noise immunity, a triggering threshold from 1 mV, and a double-pulse resolution of 4 nsec (250 MHz).

The circuit is shown in Figure 5.42. The input signals pass through an amplifier (VV 101) with a gain of 10. They are then discriminated by the differential comparator AM 685, the additional output signals of which are shaped and amplified by T_1T_2. The leading edge is then differentiated by the circuit C7 R16, and the signal obtained is discriminated by T_4T_5.

The width of the output pulse is adjusted by the current generator T_7 by discharging the capacitor C9, previously charged by a pulse of constant width and amplitude. The delta signal produced by the current generator is shaped by T_8T_9 and $T_{10}T_{11}$ before being sent to the outputs.

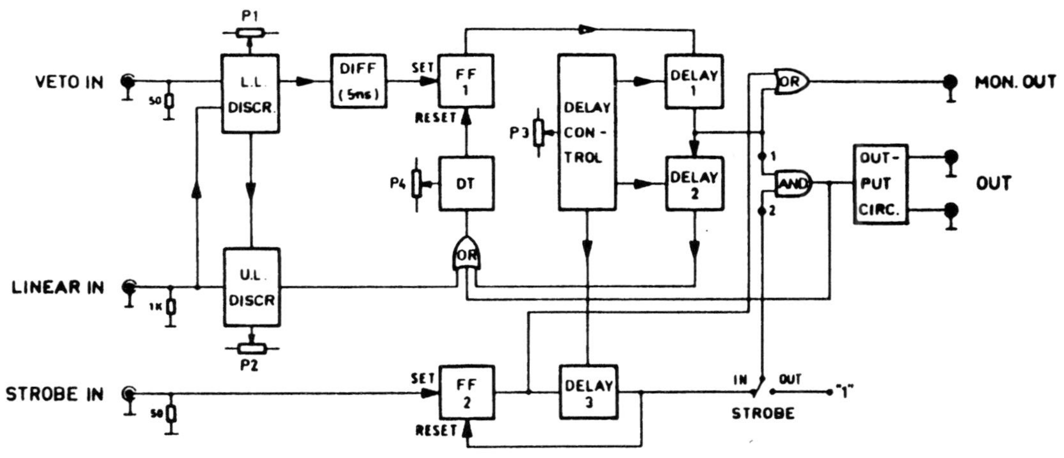

Figure 5.41 Block scheme of the SCA. (From Bialkowski, J. et al., *Nucl. Instrum. & Methods,* 190, 531, 1981. With permission.)

A peak stabilized single-channel analyzer has been designed by Grinberg et al.[18] The system block diagram is shown in Figure 5.43. The input signal is fed through a baseline restorer (BLR) to three discriminators. The levels of the discriminators are LLD, LLD + 1/2W, and LLD + W where LLD and W are the baseline and window settings, respectively. The logic outputs from these three discriminators are connected to operate as a single channel with a split window, or as two single channels with adjacent equal windows (1/2W). For an input pulse in the range of the lower window, the set-reset-flip-flop F1 is not energized and the trailing edge of the LLD output (which is used as a clock) sets the J-K flip-flop F3. For a pulse in the range of the higher window, the output of LLD + 1/2W sets F1, which leaves F3 in the reset position at the trailing edge of the clock.

The two single channels work in the trailing-edge timing mode. F3 triggers the monostable UV1 which resets F3 and gives the logic output of the lower window. The higher window circuit is identical to the lower one. The output of UV1 and UV2 are connected to the OR gate G1 to give a SCA output whenever there is an input pulse within one of the two windows. UV1 and UV2 outputs are also connected through binary units and pumping circuits to a Miller integrator. With the mode switch in the RATE position the signals from the pumping circuits are of the same polarity. The Miller integrator output is equal to the sum of the two channels, and the system works as a single channel and rate meter. In the STAB position, the signals from the pumping circuits are of opposite polarities and the Miller integrator output is equal to the difference in rate between the two channels.

In the internal mode, the Miller integrator output is connected to the baseline and window bias sources. The system has an internal stabilizing loop and works as a stabilized single channel.[18]

Nontiming units produce an SCA output pulse if the input signal is within the window settings. The output occurs when the trailing edge of the input signal recrosses the lower level threshold. Figure 5.44 shows two superimposed output pulses from a main amplifier that meet the window requirements of the single-channel analyzer. The output from the nontiming SCA for each pulse is shown below the pulses. Since the linear input pulses are referenced to the same starting time, it is clear that the output logic signals exhibit "time walk" relative to the input pulses. Figure 5.45 shows two pulses from a main amplifier and the response for a peak-detection single-channel analyzer and timing single-channel analyzer. The conventional zero-crossing technique has been widely used for timing single-channel analyzers. This technique utilizes the zero crossing of the bipolar output signal from a pulse-shaping amplifier to derive timing information, and uses the peak amplitude of the pulse for the energy range information. Figure 5.46 shows two bipolar pulses provided from a main shaping amplifier. Both pulses meet the SCA window requirements. Each output signal is generated when the corresponding input signal crosses the baseline. The bipolar output signal from a double delay-line-shaping amplifier crosses the baseline at a fixed fraction that is effectively 50% of the charge collected from the detector.

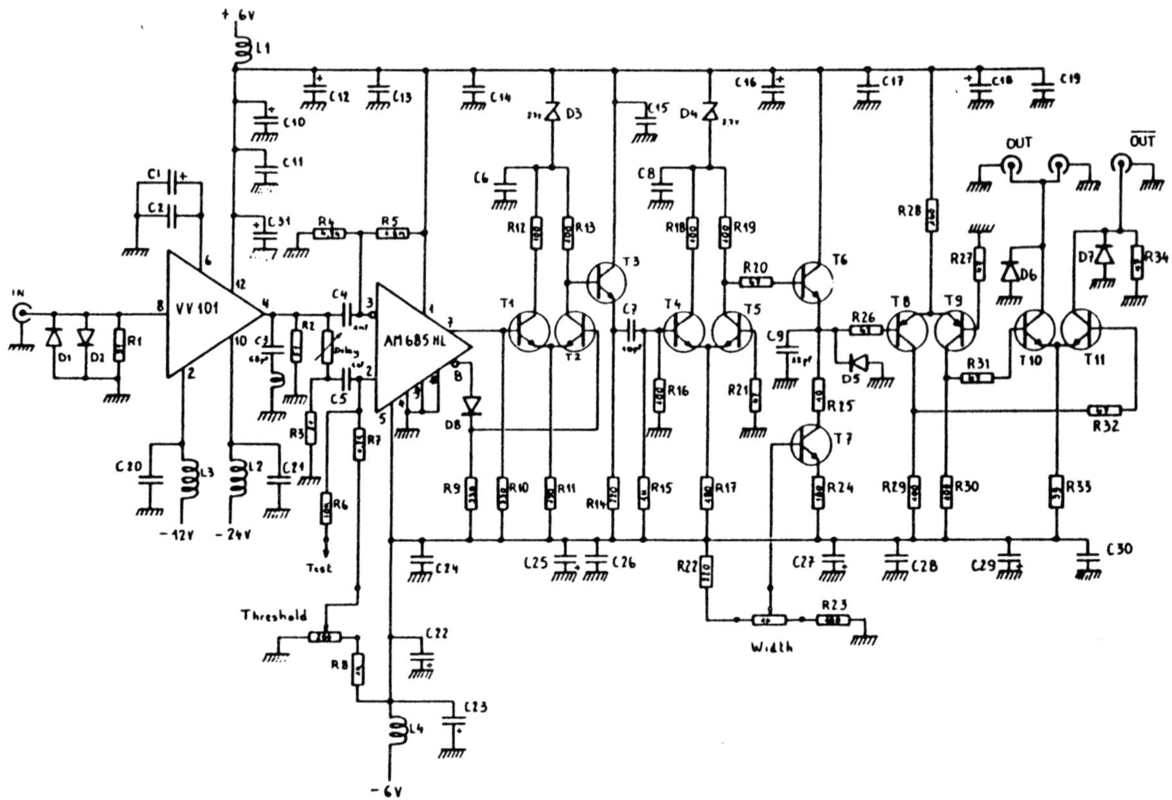

Figure 5.42 Circuit diagram of one of the eight channels of the differential discriminator. (From Brahy, D. and Rossa, E., *Nucl. Instrum. & Methods,* 192, 359, 1982. With permission.)

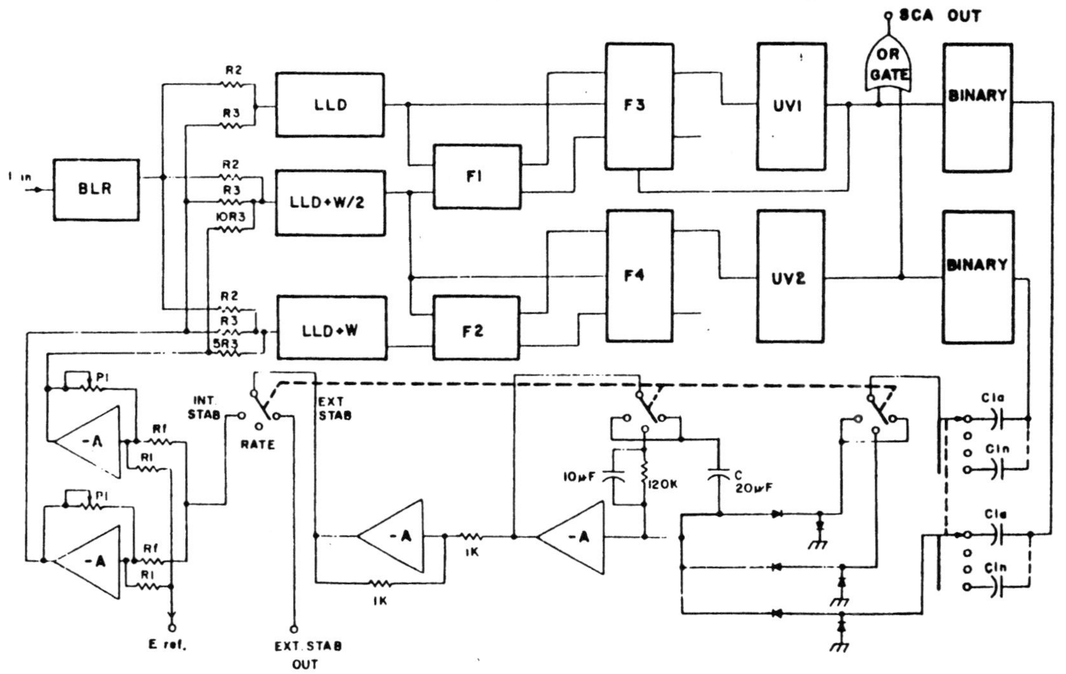

Figure 5.43 System block diagram. (From Grinberg, J. et al., *Nucl. Instrum. & Methods,* 82, 278, 1970. With permission.)

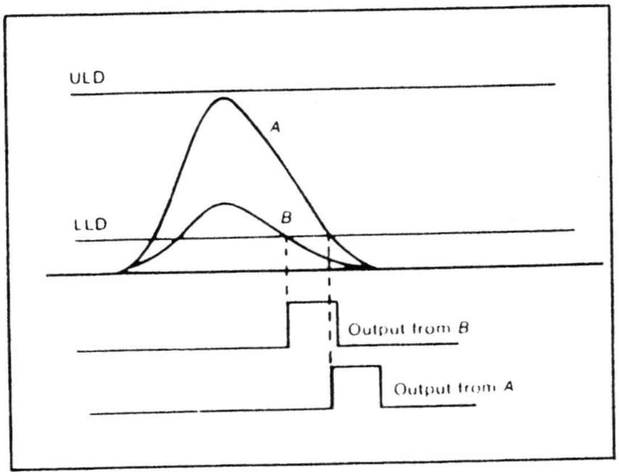

Figure 5.44 Nontiming output. (From EG&G ORTEC Catalog: 91/92 Detectors and Instruments for Spectroscopy. With permission.)

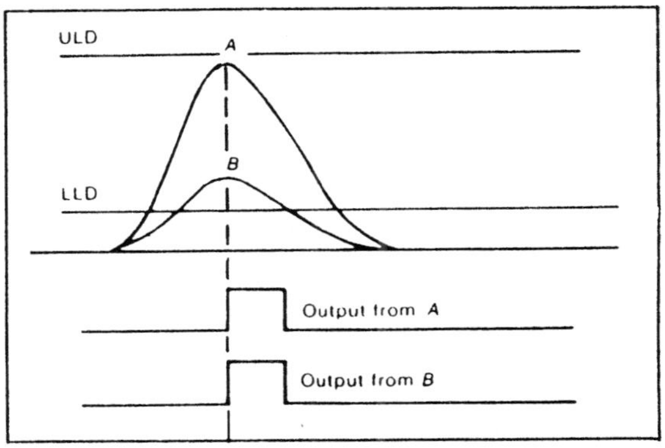

Figure 5.45 Peak-sensing output. (From EG&G ORTEC Catalog: 91/92 Detectors and Instruments for Spectroscopy. With permission.)

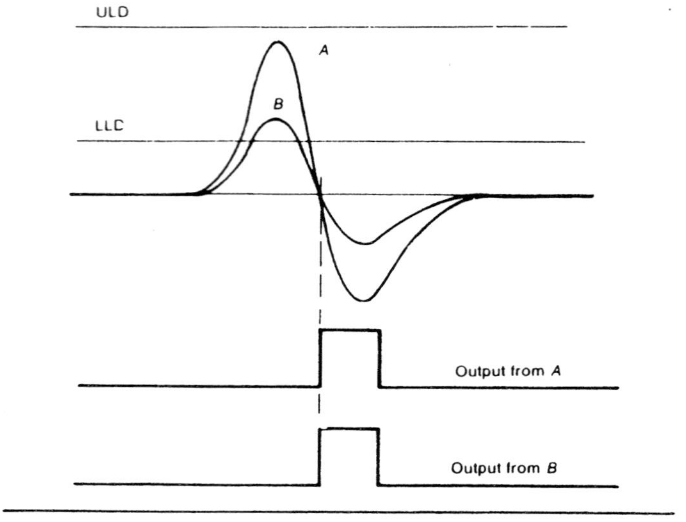

Figure 5.46 Zero-crossing output. (From EG&G ORTEC Catalog: 91/92 Detectors and Instruments for Spectroscopy. With permission.)

REFERENCES

1. **Robinson, L. B.,** *Rev. Sci. Instrum.*, 32, 1057, 1961.
2. **Herbst, L. J.,** *Nucl. Instrum. & Methods,* 70, 189, 1969.
3. **Valentine, K. H. et al.,** *IEEE Trans. Nucl. Sci.,* 34, 78, 1987.
4. **Hahn, S. F. and Cafferty, M. M.,** *IEEE Trans. Nucl. Sci.,* 38, 44, 1991.
5. **Yoshida, R. et al.,** *IEEE Trans. Nucl. Sci.,* 33, 226, 1986.
6. **Krummenacher, F.,** *Nucl. Instrum. & Methods,* A305, 527, 1991.
7. **Llacer, J. et al.,** *IEEE Trans. Nucl. Sci.,* 37, 457, 1990.
8. **Emmer, T. L.,** in NAS-NRC Pub. 1184, Monterey, 1963, p. 112.
9. **Sanders, G. H. et al.,** *Nucl. Instrum. & Methods,* 180, 603, 1981.
10. **Gottschalk, B.,** *Nucl. Instrum. & Methods,* 190, 67, 1981.
11. **Turko, B. T. et al.,** *IEEE Trans. Nucl. Sci.,* 39, 1311, 1992.
12. **Kolbe, W. F. et al.,** *IEEE Trans. Nucl. Sci.,* 36, 412, 1989.
13. **Benulis, C. A. and McFarlane, W. K.,** *Nucl. Instrum. & Methods,* A240, 130, 1985.
14. **Cole, H. A.,** *Nucl. Instrum. & Methods,* 138, 551, 1976.
15. **Correia, C. A. B. et al.,** *IEEE Trans. Nucl. Sci.,* NS25, 59, 1978.
16. **Bialkowski, J. et al.,** *Nucl. Instrum. & Methods,* 190, 531, 1981.
17. **Brahy, D. and Rossa, E.,** *Nucl. Instrum. & Methods,* 192, 359, 1982.
18. **Grinberg, J. et al.,** *Nucl. Instrum. & Methods,* 82, 278, 1970.
19. EG&G ORTEC Catalog, 91/92 Detectors and Instruments for Nuclear Spectroscopy.

Chapter 6

Scalers and Rate Meters

CONTENTS

6.1 Two-State Circuits 27
6.2 Counting Systems 29
6.3 Synchronous Counters 31
6.4 Data Readout 32
6.5 Rate Meters 32
References 35

6.1 TWO-STATE CIRCUITS

The simple electronic counting is done on a binary base. The circuits used are those that can be in one of the two states on-off or 0-1. The most common circuit is the bistable flip-flop circuit. Several variations of the basic circuit exist. One example is shown in Figure 6.1. Resistors R_1, R_2, R_3 and R_4, R_5, R_6 are such that the transistor bases are at positive voltage, i.e., conducting. The base currents are amplified in positive feedback amplifiers and a situation is formed where one transistor is in saturation and the other is cut off. A positive input pulse will cause a switch over. The system can be made to operate with negative pulses by replacing R_8, R_7 with diodes which will let only negative pulses through and a diode that will start the input to ground when it is positive. Pnp transistors can be used instead of npn.

The switch over time depends on the time constants R_8C_1 and R_7C_2. Small time constants enable fast flip-flops (10 to 100 MC). The circuit state is determined by the collector voltage of each transistor. If Si transistors are used, a lower voltage is needed for switch over and the negative voltage supplied to the bases through R_3, R_6 is not needed.

Another way to obtain a bistable flip-flop is using a tunnel diode. A tunnel diode has two stable states. The advantage of using a tunnel diode is its speed — 300 MHz. A simple system with a one tunnel diode and its characteristics are shown in Figure 6.2. The first pulse, set, raises the diode voltage to point B and the reset pulse returns it to the low voltage state A. The two inputs can be connected together and direct the pulse to the point that will be at voltage lower than the input. This is a one input type flip-flop (type T).

A circuit with two tunnel diodes for 100-MHz frequency is shown in Figure 6.3. R_3 and R_4 are the diodes load. R_2 is such that only one diode is at high-voltage low-current state and the second is at low voltage high current. The current difference flows through L. The time constant of L and the equivalent resistor of R_1, R_3, R_4 is longer than the input signal. At zero input TD_1 is at low voltage, when the signal arrives the current in the coil flows in the opposite direction to that before the signal and the current in R_1 after the signal keeps TD_1 at high-voltage low-current state or state 1. The next signal has the opposite effect and returns the circuit to zero where TD_1 is at low voltage state. It is always possible to return the system to zero by applying a positive voltage with a push button at the junction L, R_4, R_3.

The simple flip-flop (FF) circuit is called R-S, i.e., set-reset and can be at state Q(1) or $\overline{Q}(0)$ (complementary). It is described schematically in Figure 6.4. The states of the inputs and outputs are shown in the table of Figure 6.4. Q and $\overline{Q}$ are the output after each input pulse R and S. The simple system

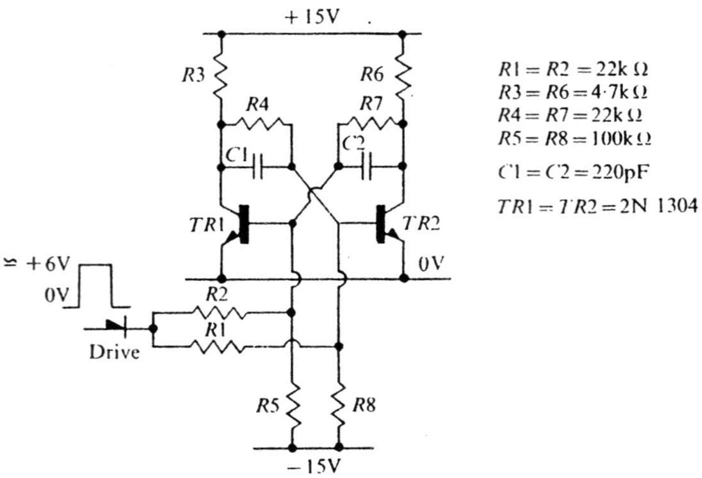

Figure 6.1 Eccles-Jordan bistable: normal drive.

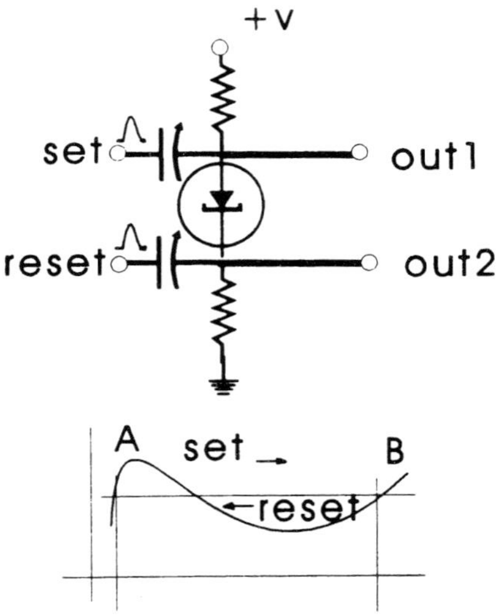

Figure 6.2 Two states tunnel diode circuit.

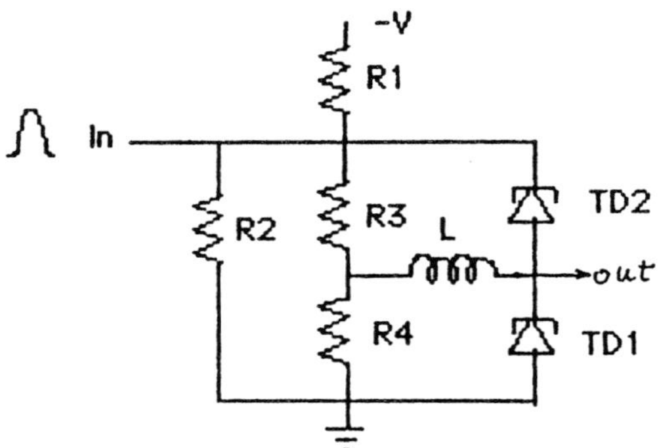

Figure 6.3 Two tunnel diodes, two states circuit.

can be improved by adding auxiliary systems. Generally, R and S inputs are fed through AND gate together with a constant clock pulse (CP). C and P can be connected directly without the AND gate so that clear (C) (Q = 0) or preset (P) (Q = 1) can be obtained. The system is shown in Figure 6.5.

If the outputs Q and $\overline{Q}$ are connected to the AND gates with the input signals a K-J system is obtained as shown in Figure 6.6. A delay can be added to the input for time matching. The output which is at state 1 must appear after that as state 0. The return of the output to the input enables the state T = K = 1 which is used to inverse the system state as shown in Table 6.1. The columns Q and $\overline{Q}$ are given after the J and K signals entered. Q′ and $\overline{Q}'$ are Q and $\overline{Q}$ before K and J entrance, i.e., J = K = 0 did not change the systems state. K = J = 1 did change the system state. This system is capable of binary counting. Another system that can be used for counting on a scale of two is T (trigger or toggle) flip-flop. The system is obtained when the two inputs R-S or J-K are connected together. A simple example is shown in Figure 6.7. In this system every input pulse causes a switch over Q′ = $\overline{Q}$.

A D flip-flop is also used (D = data) with one input to both sides but an inverter is used on one input so that this input obtains an input inverted to that of the other. The output is $Q_{n+1} = D_n$.

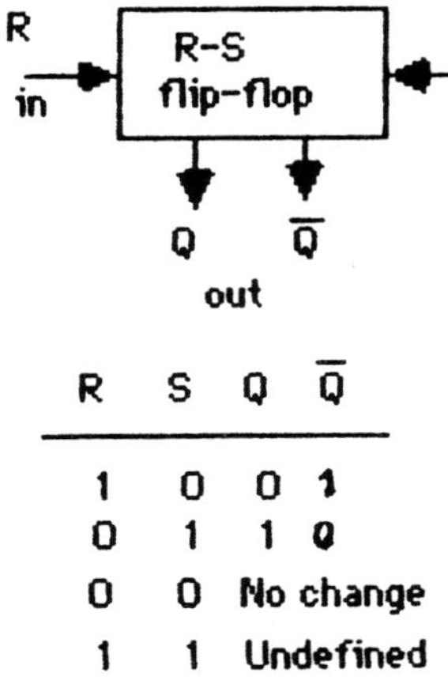

Figure 6.4 Flip-flop states.

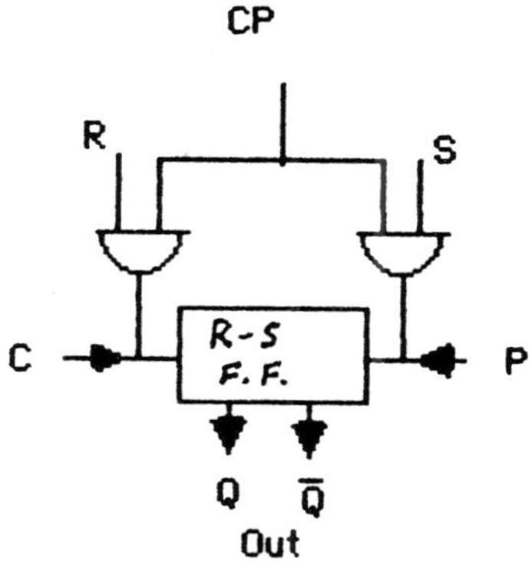

Figure 6.5 Improved flip-flop circuit.

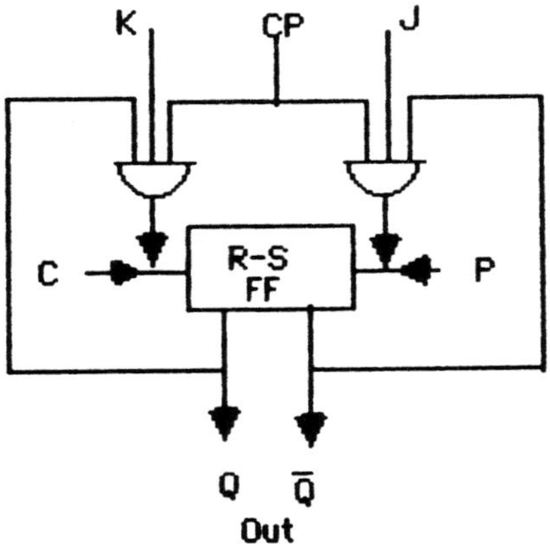

Figure 6.6 K-J circuit.

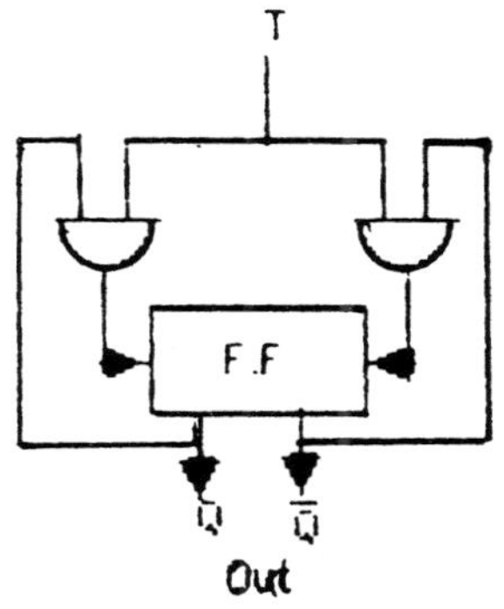

Figure 6.7 T flip-flop.

Table 6.1 J-K States

J	K	Q	$\overline{Q}$
0	1	0	1
1	0	1	0
0	0	Q′	$\overline{Q}$′
1	1	$\overline{Q}$′	Q′

6.2 COUNTING SYSTEMS

The simplest counting system is the binary counting made by T sections in cascade as shown in Figure 6.8. The outputs at Q are the binary counts. Before counting the system can be cleared so that all flip-flops are at zero (Q = 0).

A binary counting system with a J-K flip-flop system is shown in Figure 6.9. For K = J = 1 + V provides input voltage and the counting is done by input to CP. Here too a clear can be added.

Decimal counting can be obtained from the binary system by binary coded decimal (BCD). The simplest way of using this method is with four flip-flops to the count nine as before; then set all FF to zero at ten and start again from one. A simple circuit is shown in Figure 6.10. At the beginning of counting

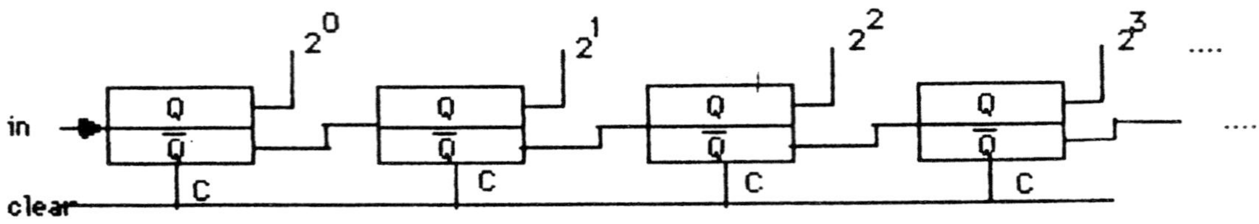

Figure 6.8 Binary counting system.

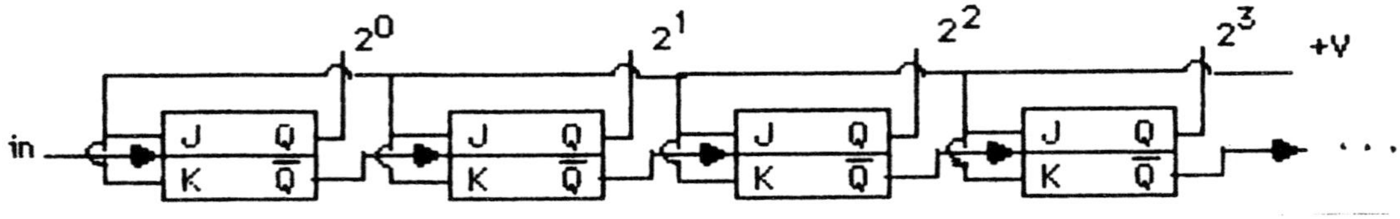

Figure 6.9 J-K counting system.

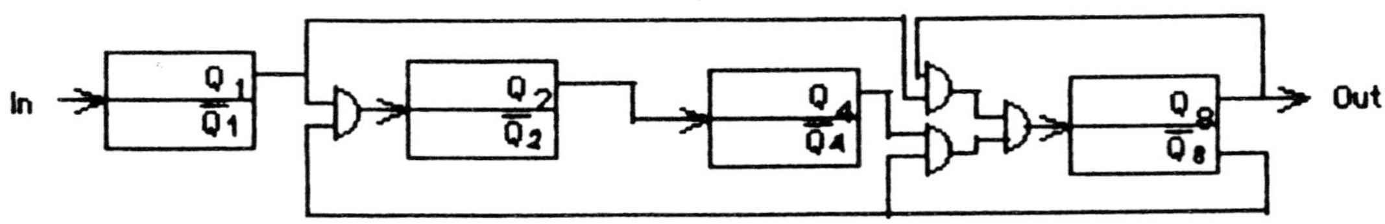

Figure 6.10 BCD counter.

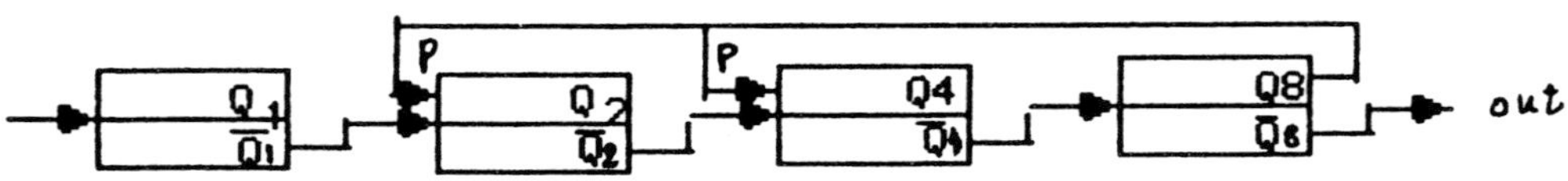

Figure 6.11 Feedback decimal scaler.

all FF are at zero and the AND gate between the first and second FF and between the third and fourth get one input from $\overline{Q}_8 = 1$. The system thus counts the first eight pulses as usual. At the eighth count $\overline{Q}_8$ switches to zero and the gates close. Q_8 switches to one and the second gate opens. The ninth count will switch Q_1 to 1 and the tenth will return everything back to zero.

Another way of decimal counting is by adding artificial six pulses so that the tenth pulse will read as $2^4 = 16$ as shown in Figure 6.11. The first eight pulses are counted as usual. The 8th count switches over Q_2 and Q_4 so that the 10th count is like the 16th and all returns to zero.

Another way for decimal counting is using R-S flip-flop for Q_8 and all the rest T flip-flop. Q_4 is connected to Q_8 and Q_1 to $\overline{Q}_8$. The input to the second FF will be through a gate getting signals from Q_1 and $\overline{Q}_8$; this way the eighth count will disconnect the system after the first FF and the input will go to the fourth FF for the ninth and tenth count which will set the system to zero.

A BCD counting can be done with the system shown in Figure 6.10. Up to the seventh pulse the output of the first FF is connected to the input of the fourth FF (in addition to the cascade) and the output of the second and third FF to the J input of the fourth FF (after disconnecting +V). J_8 will be operated by Q_2 and Q_4. To avoid counting in the second and fourth FF, $\overline{Q}_8$ is connected to J_2 (instead of +V) so that the last counts are counted by the fourth FF. Many other possibilities are available.

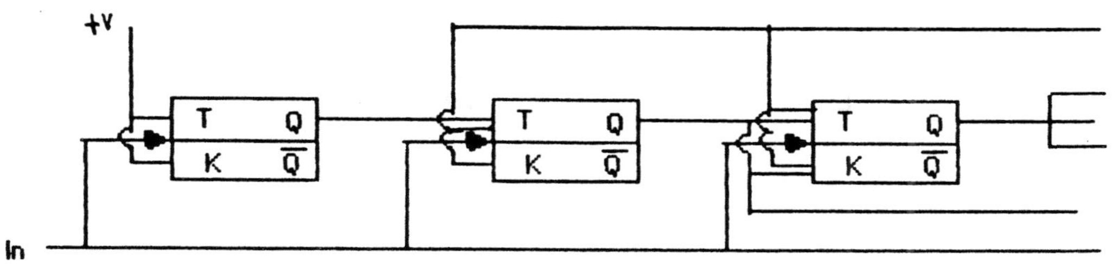

Figure 6.12 Parallel synchronous binary counter.

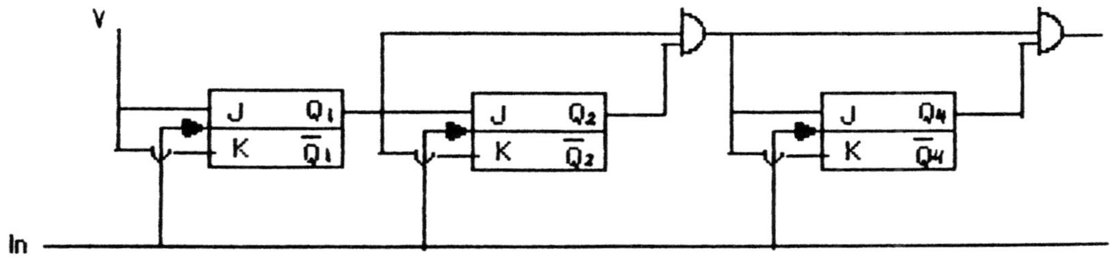

Figure 6.13 Series synchronous binary counter.

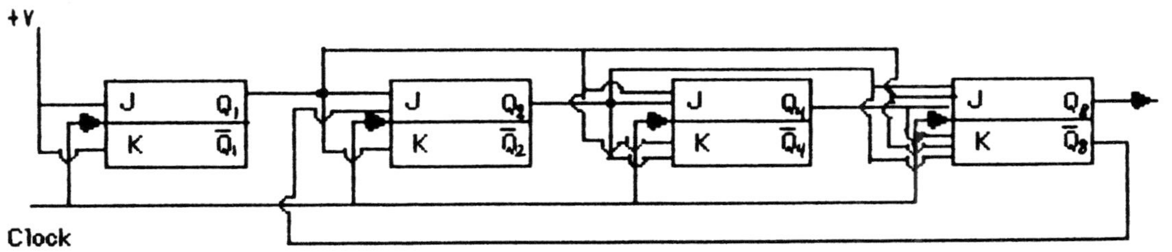

Figure 6.14 Decimal synchronous counter.

6.3 SYNCHRONOUS COUNTERS

A synchronous counter is a counter in which all flip-flops are triggered together, synchronously with the input pulse in parallel as shown in Figure 6.12. The K-J inputs of the first stage are at high voltage all the time; thus toggling on successive input pulses. All K-J inputs are connected in parallel to all previous outputs. All FF are initially at zero and the second pulse will switch both the first and second FF since $Q_1 = 1$ after the first pulse. The third FF is still inhibited by Q_2. The condition for the third FF toggle is fulfilled in the fourth pulse and so on. The system operates as a binary counter.

The system shown in Figure 6.12 is connected in parallel; it can also be connected in series by connecting each output to the previous one through an AND gate as shown in Figure 6.13. The system operates as described above for the parallel system. Decimal synchronous counting is done with a system as shown in Figure 6.14. This is a parallel system similar to the one in Figure 6.12, adjusted for decimal counts. Up to eight, the system counts as the previous one. J_2 is enabled by $\overline{Q}_8$. After the eighth count $\overline{Q}_8 = 0$ and inhibits J_2. The ninth count is counted by the first FF; then $Q_1 = 1$ enables K_8 only since the zero in Q_2 and Q_4 inhibit J_8. The ninth count will cause $Q_1 = 0$, $\overline{Q}_8 = 1$, $Q_8 = 0$; and the whole system is at 0.

The main advantage of the synchronous counter is that it enables simultaneous readout of all states without interruption of the counting.

Another counting method (decimal or other) is the shift register. This is a FF system in which "1" is obtained at the output of each FF in a row according to the number of input pulses. The first input pulse will give "1" at the output of FF1. The second pulse will give an output in FF2 and so on up to the tenth FF which will reset all to 0 and the cycle will repeat itself. The system is shown in Figure 6.15. Before counting starts, the system should be cleared so that $Q_0 = 1$, $Q_1 = Q_2 = ...Q_9 = 0$. In addition to $Q_0 = 1$ there is an input at K_0 since $\overline{Q}_9 = 1$. An input pulse will switch over FF_0 so that $\overline{Q}_0 = 1$ and there will be an input at K. J will have an input because of the delay built in between J and K. After the first pulse

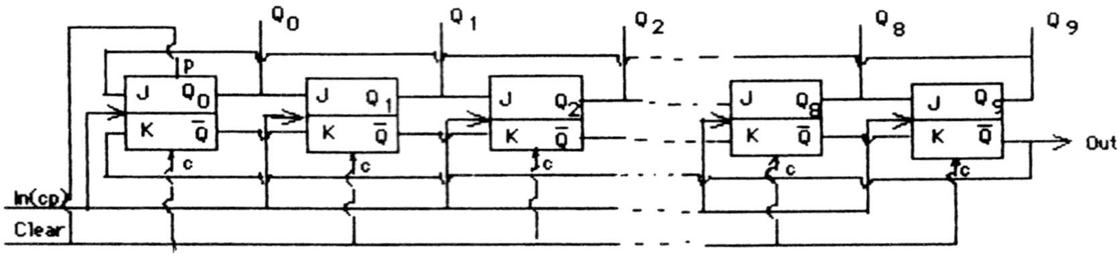

Figure 6.15 Shift register scaler.

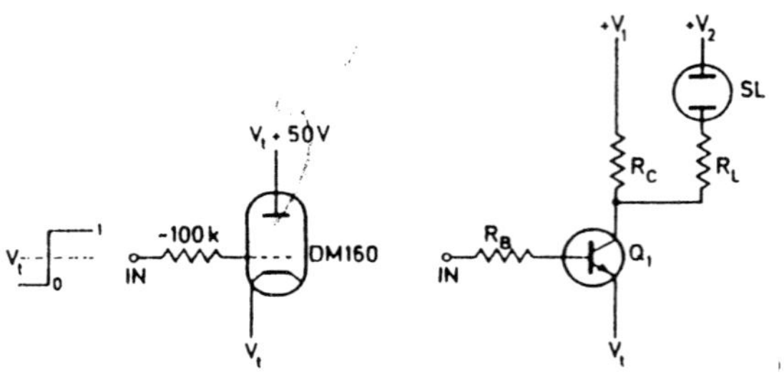

Figure 6.16 Mixie indicator.

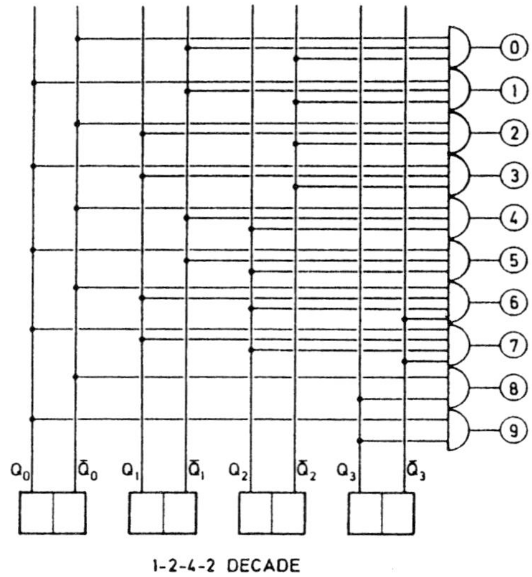

Figure 6.17 Decimal readout system.

$Q_0 = 0$, $Q_1 = 1$, and all the rest 0. When the second pulse comes, there is an input in J_2; Q_1 will switch to 0; and $\overline{Q}_1$ will switch to 1 and provide an input at K_2 so that $Q_2 \rightarrow 1$, $\overline{Q}_2 \rightarrow 0$, $Q_1 \rightarrow 0$, and $\overline{Q}_1 \rightarrow 1$. All the rest do not change. In this way the process will continue; each pulse will cause a count in the next FF. After the ninth FF (actually the tenth) all will be reset to zero and the system will return to initial state with a pulse delivered to the scale of the tens.

6.4 DATA READOUT

Scaler readout is done by turning on pilot lamps (Nixie indicators). The pilot lamps are connected on one side to voltage supply $+V_2$ and on the other side to a transistor collector which is the output side of the Q or $\overline{Q}$ sides of a flip-flop accordingly. The connection is done through a high-voltage transistor which can withstand the voltage needed for the pilot lamp. The voltages are adjusted by the resistors so that the pilot lamp turns on and off as required (Figure 6.16).

The decimal readout is done by ten pilot lamps reading the scaler outputs. The decimal readout connection is shown in Figure 6.17. The combinations of the binary numbers give the decimal digits. A mechanical relay can be used instead of the pilot lamps so that the data can be printed out.

6.5 RATE METERS

A four-decade digital log rate meter using integrated circuits for all active components has been designed and tested by Thorington and Andrews.[1] The rate meter was capable of measuring pulse rates ranging from 1.11×10^{-1} to 2.0×10^{7} pulses per second. With a very simple modification the rate meter can be made to have any desired number of decades in its output.

The input pulse rate was determined by digital counting circuits, and the logarithm of the pulse rate was approximated by seven digital functions which were generated by a series of combinational logic

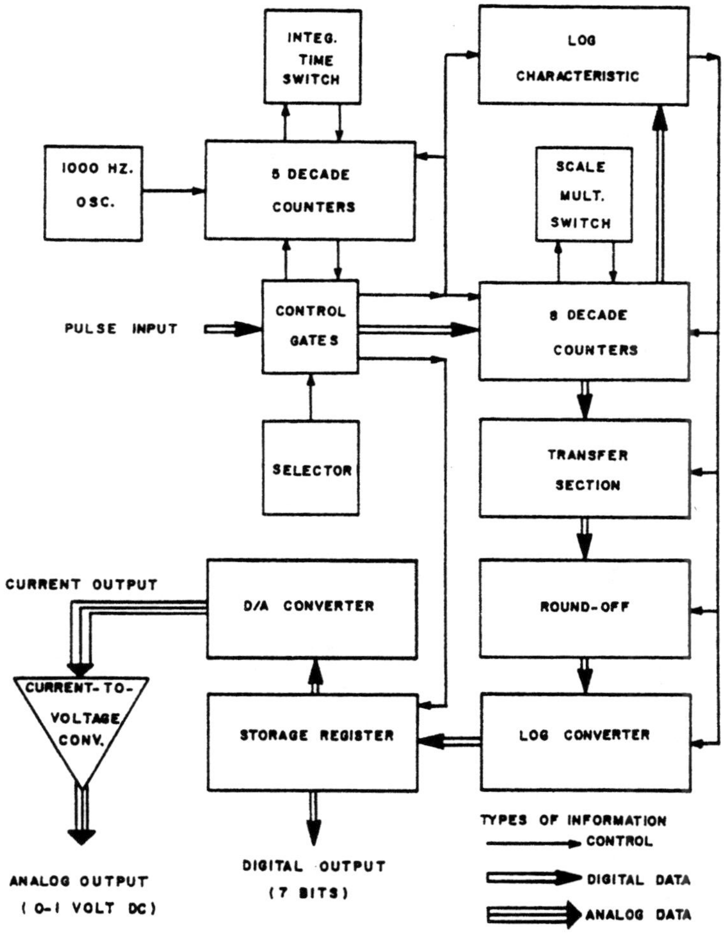

Figure 6.18 Block diagram of digital log rate meter. (From Thorington, J.M., Jr. and Andrews, V. E., *IEEE Trans. Nucl. Sci.,* NS18(1), 148, 1971. With permission.)

networks. The output was available in either digital or analog form. The system was composed of the following sections: timing and control, counting, transfer, round-off, log converter, storage, and digital-to-analog converter.

A digital combinational network was designed that will accept the 24 outputs from six binary coded decimal (BCD) decade counters and outputs from four flip-flops. These flip-flops were sequentially set to logic "1" as each of the four most significant decades was saturated. The combinational network produced seven binary output functions which represent an approximation of the log of the number stored in the counters.

The block diagram of the resulting system is shown in Figure 6.18. The purpose of the timing and control section was to generate three accurately timed control functions and a 500-Hz test frequency.

The transfer section took the outputs from the decade counters and flip-flops in the counting section. It produced eight output functions which were equal to the BCD outputs from the two most significant decimal equivalent numbers in the decade counters.

The round-off section rounded off the BCD outputs from the transfer section to the decimal equivalent of two significant figures. This operation was accomplished by the binary addition of the pseudo round-off function to the binary number. A 10-b precision current source constructed on a single silicon chip was used to convert the seven binary functions to an analog current. An operational amplifier was used to produce a voltage analog of the current output from the D/A converter.

A digital rate meter using a new principle of averaging of input pulse rate was proposed by Vankov and Dimitrov.[2] The general block diagram is shown in Figure 6.19. The averaging time interval T was

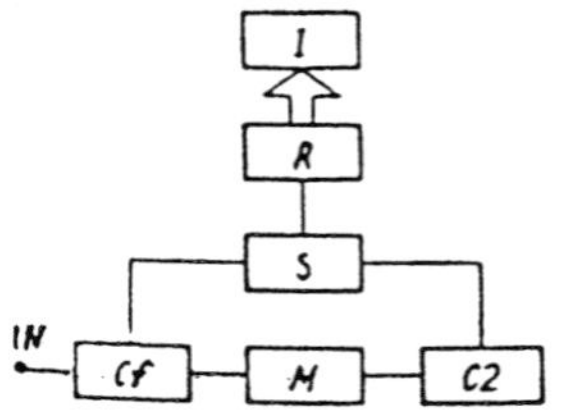

Figure 6.19 General block diagram of the rate meter. (From Vankov, I. D. and Dimitrov, L. P., *Nucl. Instrum. & Methods,* 188, 319, 1981. With permission.)

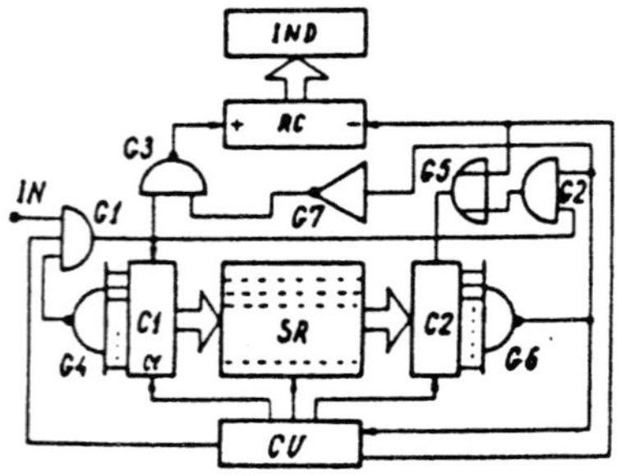

Figure 6.20 Detailed block diagram of the rate meter. (From Vankov, I. D. and Dimitrov, L. P., *Nucl. Instrum. & Methods,* 188, 319, 1981. With permission.)

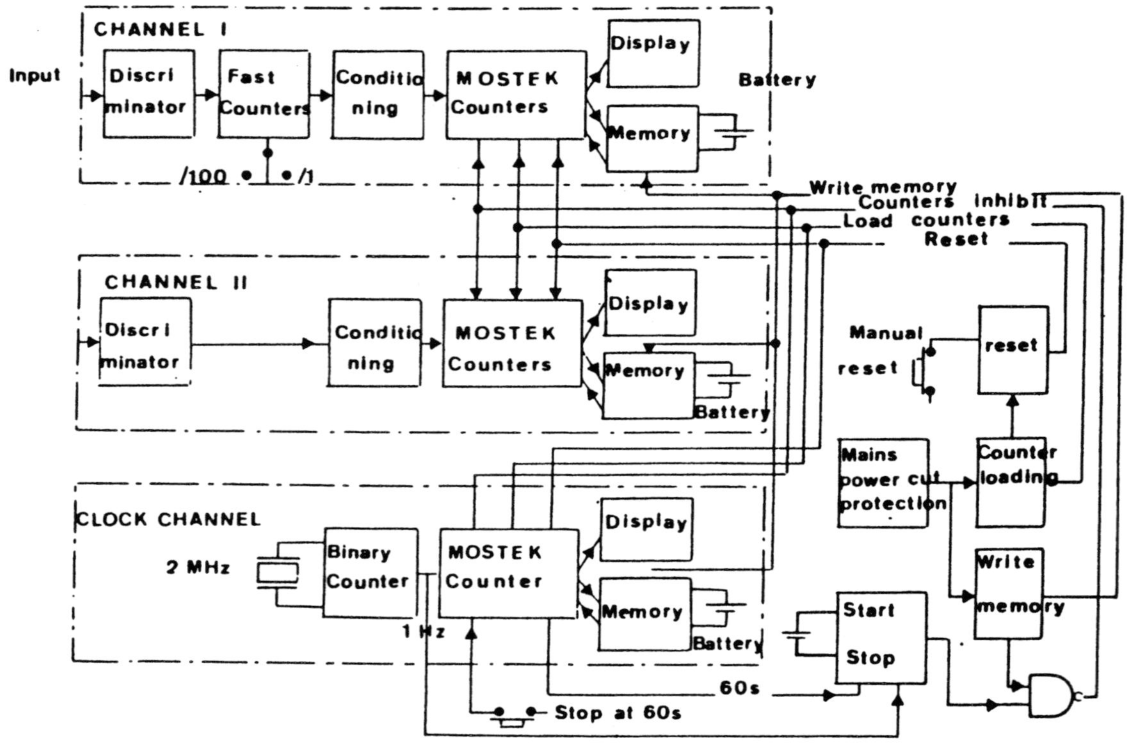

Figure 6.21 Block diagram of high capacity counting system. (From Millet, P. et al., *Nucl. Instrum. & Methods,* 188, 93, 1981. With permission.)

divided into r subintervals each with a duration of $\tau = T/r$. The input pulses of each subinterval were registered in the counter C1. At the end of the subinterval the number registered in C1 was stored in a first-in-first-out (FIFO) memory M and through the summator S was also added to the number in the register R. The number of addresses in the FIFO memory was equal to r. Thus after the end of the first group of r subintervals (i.e., of the averaging period T), all the addresses of the memory were used. The number from the first subinterval which stayed at the outputs of the FIFO memory was now written in the counter C2. The number of all input pulses received during the interval T was stored at the same time in the register R. The input pulses from the next (r + first) subinterval were counted again in C1. At the end of this subinterval the information in R was corrected only by the difference between the numbers registered in C1 and C2. After that the number from C1 was written in the FIFO memory and the number of pulses from the second subinterval came into C2.

The counter C1 should be binary. If the counter C2 is also a binary but subtracting counter, the functions of the summator S and the register R can be realized by a reversible (up/down) counter. The most suitable FIFO memory construction is the use of r-bit shift registers with a number equal to the number of flip-flops in C1 and C2.

The block diagram corresponding to such a design of the rate meter is shown in Figure 6.20. Initially the counters C1 and C2, the reversible counter RC, and the shift registers SR are cleared. The gates G1 and G3 are open and G2 is closed. The input pulses pass through G1 and are registered in C1 and in RC (through G3). After a time τ the first subinterval finishes. The gate G1 is then blocked by a signal from the control unit (CU). A shift pulse is fed from CU to the shift registers (SR), and the information from C1 is stored in their first bits. Then C1 is cleared and G1 is opened again, thus starting the measurement in the second subinterval. This procedure is repeated after each subinterval until the rth subinterval, when all bits of the shift registers SR will be filled and the information of the first subinterval will stay in the last bits of SR, i.e., at their outputs.

An inexpensive, high-capacity counting device has been made by Millet et al.[3] for time analysis of light pulses by time-amplitude conversion. In order to allow for particularly long storage times, it had two channels with a capacity of 10^{12} pulses. The device consisted mainly of two counting channels and a clock. As shown in Figure 6.21 the two counting channels were similar: the pulses were fed into a variable threshold discriminator and they then went through a frequency divider comprising two transistor-to-transistor logic (TTL) decade counters before reaching the MOS-type counting circuit, which could record 10^{12} pulses. This circuit was connected to the display units and also to a memory whose main purpose was to safeguard the results in case of power failure.

The duration of recording was indicated by a clock composed of a crystal quartz oscillator and a counting circuit.

REFERENCES

1. **Thorington, J. M., Jr. and Andrews, V. E.,** *IEEE Trans. Nucl. Sci.,* NS18(1), 148, 1971.
2. **Vankov, I. D. and Dimitrov, L. P.,** *Nucl. Instrum. & Methods,* 188, 319, 1981.
3. **Millet, P. et al.,** *Nucl. Instrum. & Methods,* 188, 93, 1981.

Chapter 7

Multichannel Analyzers (MCA)

CONTENTS

7.1 Introduction 37
7.2 Memory 37
7.2.1 Bistable Multivibrator 37
7.2.2 Memory Circuits 40
7.3 Single-Parameter Multichannel Analyzers 44
7.3.1 Early Models 44
7.3.2 CAMAC-Aided MCA 49
7.4 Multiparameters Multichannel Analyzers 65
7.5 Multiscaler 69
7.6 Multiplexer 72
References 77

7.1 INTRODUCTION

A multichannel analyzer is a system which accepts pulses from the amplifier and sorts them out according to pulse height distribution or time distribution in two or more parameters. The data is stored in a memory according to its distribution and can be retrieved by various means.

7.2 MEMORY

7.2.1. BISTABLE MULTIVIBRATOR

A bistable multivibrator (flip-flop [FF]) is an electronic circuit with two stable states, set = 1 and reset = 0. These two states are defined by two voltage levels determined in the circuit design according to the logic system used. The two voltage levels are opposite of each other designated as Q and $\overline{Q}$. At set state Q = 1, $\overline{Q}$ = 0. At reset state Q = 0, $\overline{Q}$ = 1. The bistable multivibrator can be designated as set-reset flip-flop (SRFF) (Figure 7.1).

A flip-flop circuit is made basically of two active components (transistors) (Figure 7.2). When the circuit is symmetrical, T_{R_1} is identical to T_{R_2}, $R_{C_1} = R_{C_2}$, $R_1 = R_2$ and $R_{B_1} = R_{B_2}$. Writing $R_c = R_{C_1} = R_{C_2}$, and $R_B = R_{B_1} = R_{B_2}$, if T_{R_1} is at cutoff.

$$U_{B_1} = -U_{BB} - \frac{R_1}{R_1 + R_B}; \ I_{B_1} = 0 \quad (7.1)$$

When T_{R_2} is at saturation:

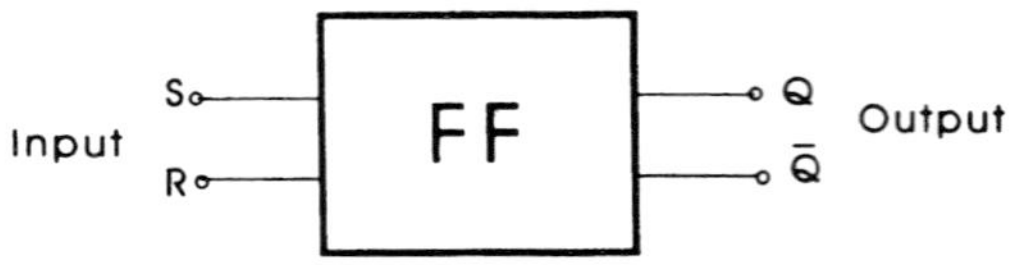

Figure 7.1 Set-reset flip-flop (SRFF).

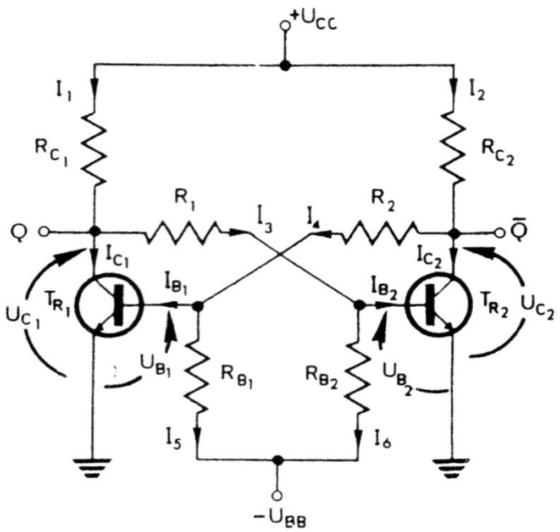

Figure 7.2 Basic flip-flop circuit.

$$U_{C_2} = 0;\; U_{B_2} = 0,\; I_2 = \frac{U_{CC}}{R_C}\;\; I_4 = \frac{U_{BB}}{R_1 + R_B} \tag{7.2}$$

also

$$I_{C_2} = I_2 - I_4 = \frac{U_{CC}}{R_C} - \frac{U_{BB}}{R_1 + R_B} \tag{7.3}$$

For T_{R_2} to be at saturation I_{B_2} must be

$$I_{B_2} > \frac{I_{C_1}}{h_{FE}(\min)} \tag{7.4}$$

The current I_3 and I_6 are

$$I_3 \sim \frac{U_{CC}}{R_1 + R_C},\; I_6 = \frac{U_{BB}}{R_B} \tag{7.5}$$

so that:

$$I_{B_2} = I_3 - I_6 = \frac{U_{CC}}{R_1 + R_C} - \frac{U_{BB}}{R_B} \tag{7.6}$$

The collector of T_{R_2} voltage is

$$U_{C_1} = U_{CC} - I_3 \cdot R_C \tag{7.7}$$

For T_{R_1} to be at cutoff:

$$U_{BE1} =< \left| U_{BE}\ \text{off} \right| \tag{7.8}$$

where U_{BE} off is the minimal forward bias assuring that T_{R_1} cutoff will not be affected by noise. It is given by:

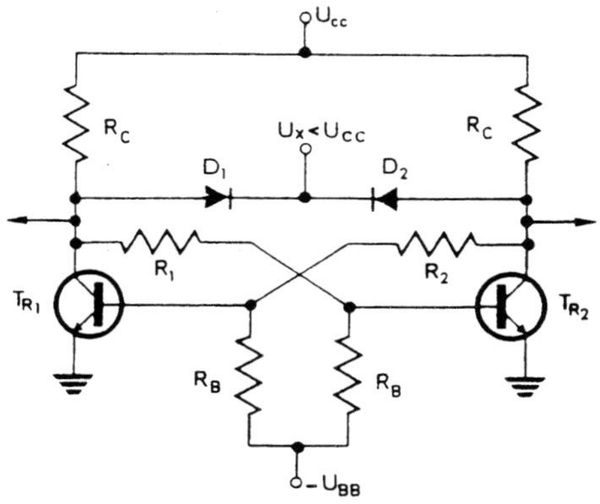

Figure 7.3 A flip-flop (FF) circuit with clamping diodes.

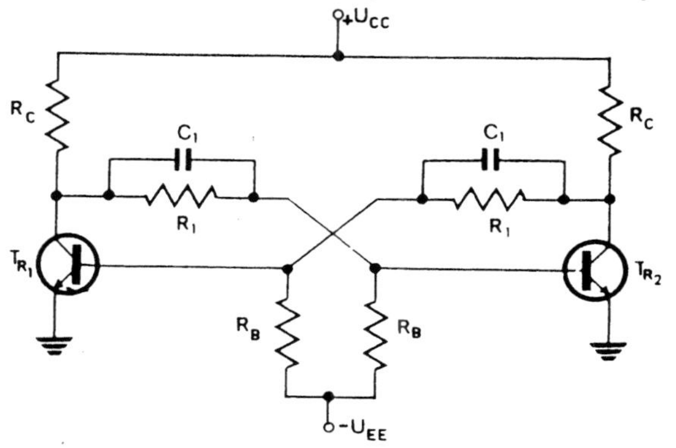

Figure 7.4 Flip-flop with speed-up capacitors.

$$U_{BB} = \frac{R_1}{R_1 + R_B} < U_{BE} \text{ off} \tag{7.9}$$

The flip-flop is generally connected to other circuits at its output and therefore loaded; this load current reduces the collector voltage of the cutoff transistor. This might reduce the base voltage of the saturated transistor below what is needed for saturation. A voltage lower than that defined as logic 1 will be interpreted as logic 0; to avoid that, R_C should be lower than R_1 plus the load resistance to assure closed loop gain higher than 1 in the transition from saturation to cutoff; and R and R_C should have the relation:

$$R_1 < h_{fe} \cdot Rc \tag{7.10}$$

In Figure 7.3 the output voltage is close to U_x, and the base current of the saturated transistor is constant. When T_{R_1} is at cutoff, its collector voltage is increased. When it gets to U_x, D_1 starts conducting and fixes the output voltage at U_x.

The steady states may be changed by an external trigger. Transition time between states can be speeded up by capacitors as seen in Figure 7.4. The value of C_1 may be determined by:

$$R_1 C_1 = R_B C \text{ in} \tag{7.11}$$

where Cin is the input capacitance of the active element. The time needed for capacitor charging is recovery time and the minimum time between two trigger pulses — resolving time (transition time + recovery time) — is given by:

$$\tau \sim C_1 \cdot \frac{R_1 \cdot R_B}{R_1 + R_B} \tag{7.12}$$

The maximum frequency is

$$F_{max} = \frac{1}{2\tau} = \frac{R_1 + R_B}{2 R_1 R_B C_1} \tag{7.13}$$

A flip-flop can be realized by two NOR gates as seen in Figure 7.5. If S = R = O, then $P = \overline{S+Q} = \overline{O+Q} = \overline{Q}$, $Q = \overline{R+P} = \overline{O+P} = \overline{P}$. If S = R = 1, then $P = \overline{Q+1} = \overline{1} = 0$, $Q = \overline{P+1} = \overline{1} = 0$, an impossible state since $P = \overline{Q}$ and $Q = \overline{P}$.

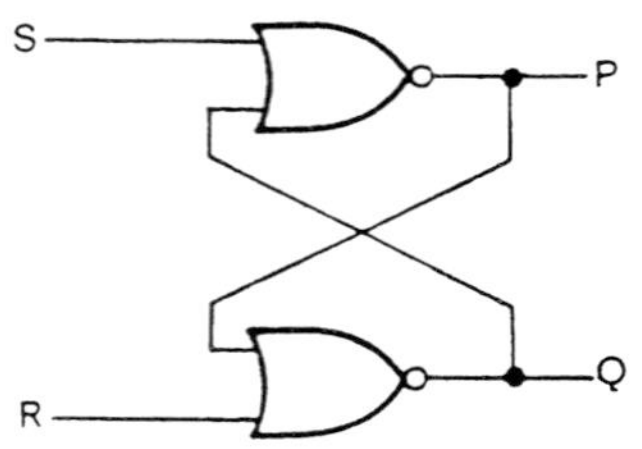

Figure 7.5 Set-reset flip-flop (SRFF) as two NOR gates.

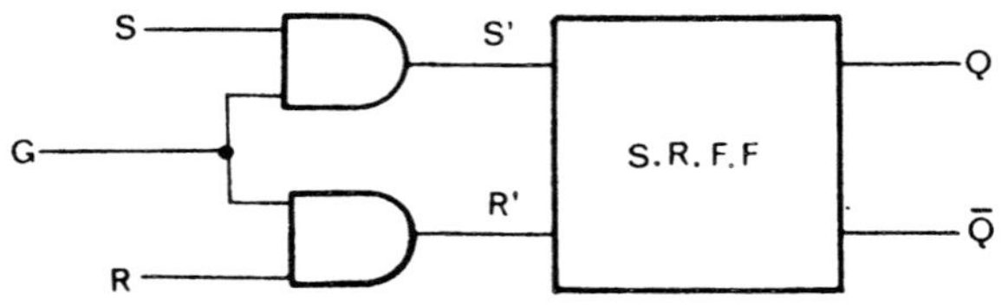

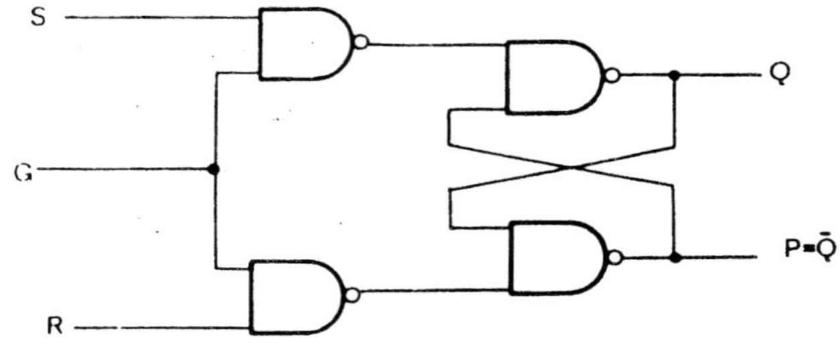

Figure 7.6 Clocked SRFF.

The SR flip-flop can be used as a memory element. A pair of input triggers: set S = 1, R = 0, and reset S = 0, R = 1 will leave the SRFF at its original state, i.e., remembering its earlier state. The various states are given in the following:

R	S	Previous state Q_n	New state Q_{n+1}
0	0	0	0
0	0	1	1
0	1	0	1
0	1	1	1
1	0	0	0
1	0	1	0
1	1	0	Unknown
1	1	1	Unknown

In a clocked flip-flop, the input pulses are synchronized by a clock (or a gate) as seen in Figure 7.6. As long as G = 0 the FF stays in the same state.

7.2.2 MEMORY CIRCUITS

Memory is a circuit element used in digital circuits for storing information. A general model of a memory circuit element is shown in Figure 7.7(a). Conceptually, it is a collection of 2^n addressable storage registers, each of which contains m bits. Associated with each storage register, which is called a memory location, is a unique memory address. As shown in Figure 7.7(a), the address of the first memory location is 0, that of the second memory location is 1, and so forth up to the last memory location which has an address of $2^n - 1$. With the specification of an n-bit address at the address inputs, the contents of any of the 2^n memory locations can be accessed directly, i.e., randomly, without the need to sequentially traverse the preceding locations to get to the specified location. For this reason, this type of memory is commonly called a random access memory. The m-bit data are transferred to and from a memory location through m-bidirectional (input/output) data lines of the memory unit.

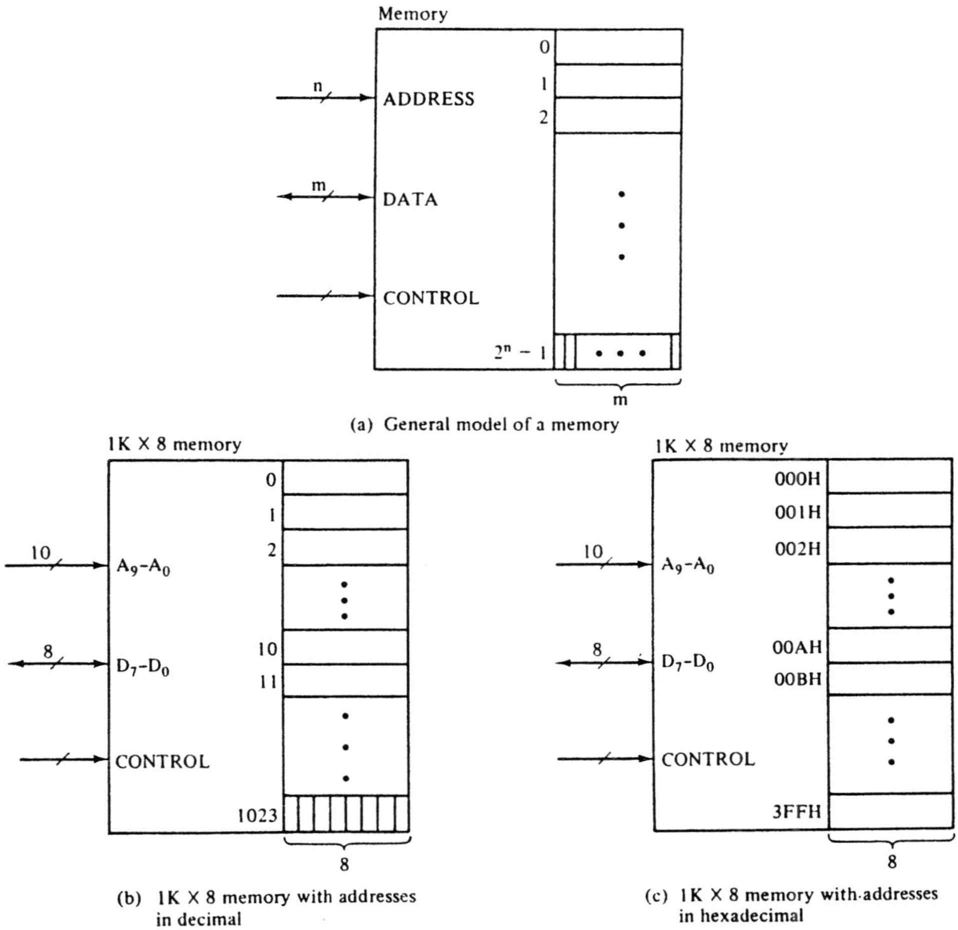

Figure 7.7 Models of memories. (a) General model of memory; (b) 1K × 8 memory with addresses in decimal; (c) 1K × 8 memory with addresses in hexadecimal.

For a read operation, data is retrieved from one of the memory locations by specifying an n-bit address at the address inputs, and applying appropriate control signals at the control inputs to cause the contents to be read from the specified location. After a time equal to the access time of the memory, the m-bit contents of that memory location are available on the data outputs. For a write operation, data are stored into one of the memory locations by specifying an n-bit address at the address inputs and applying the m-bit data to be stored at the data inputs. At the same time appropriate control signals are applied at the control inputs to cause the m-bit data to be stored at the specified memory location.

The capacity of a memory unit is characterized by the number of memory locations that it contains and the number of bits per memory location. The capacity of a memory unit can be determined from the number of its address input lines and the number of its data lines. With a 10-bit address, one can generate unique addresses for up to $2^{10} = 1024$ different memory locations. A memory unit with n address lines and m data lines has a capacity of $2^n \times m$ b.

A read-only memory (ROM) is a random access memory in which, under normal operation the data stored in each memory location can be read by a read operation, but cannot be altered by a write operation.

For a read operation, the n-bit address is specified at the address inputs, and the chip-enable (CE) input is made true (L), and the write-enable (WE) input is made false (H). After a time equal to the access time of the memory, the m-bit contents of the specified location become available at the data outputs. For a write operation, the n-bit address is specified at the address inputs. Additionally, the m-bit data to be stored are applied at the data inputs. Also, the CE and WE inputs are made true (L). Then after a time equal to the access time of the memory, the data are stored in the specified memory location.

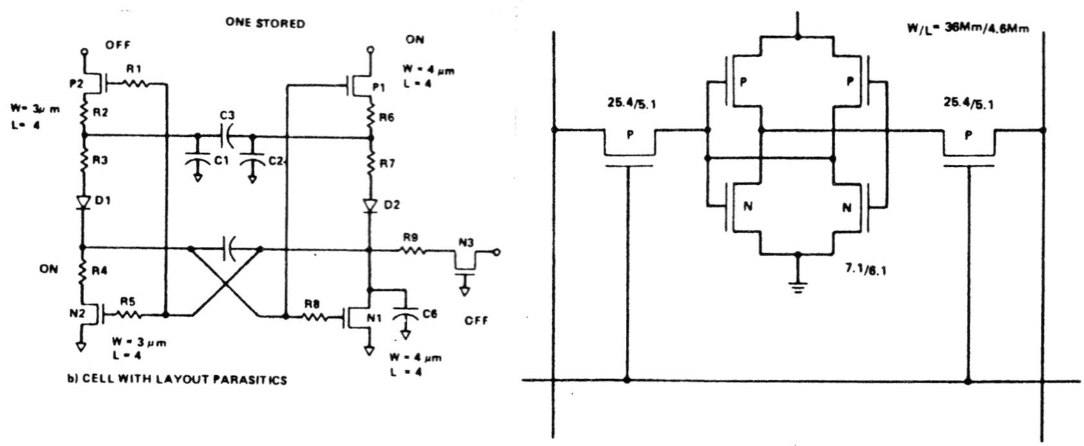

Figure 7.8 Examples of CMOS memory cells. (a) Circuit diagram of RCA TA 11246 5T memory cell design; (b) circuit diagram of RCA TA 11495 6T memory cell design. (From Brown, A. T. et al., *IEEE Trans. Nucl. Sci.*, NS33, 1484, 198X. With permission.)

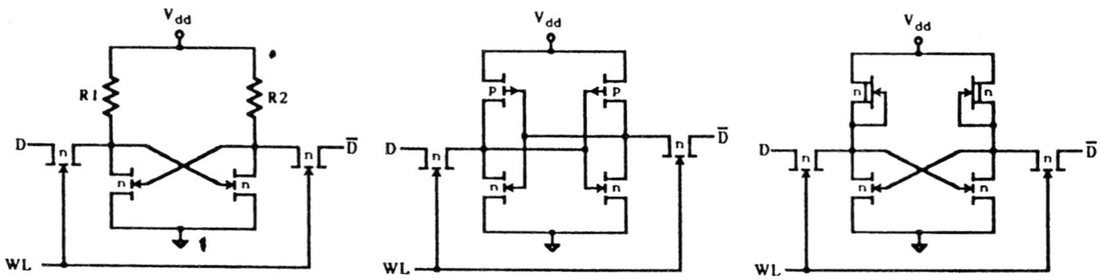

Figure 7.9 Circuit for three SRAM cells. (a) McDonnell Douglas resistive load E-JFET; (b) McDonnell Douglas complementary E-JFET; (c) Texas Instruments depletion load MESFET. (From Weatherford, T. R. et al., *IEEE Trans. Nucl. Sci.*, NS33(6) 1519, 1986. With permission.)

Read-write memory can be classified as either static random access memory (RAM) or dynamic RAM. A static RAM is a read-write memory in which data bits are stored in flip-flop storage elements. With this type of storage, the data bits retain their values as long as the memory is supplied with power. On the other hand, a dynamic RAM is a read-write memory in which data bits are stored as charges on capacitors. A dynamic RAM requires periodic refresh operations to retain the values of the stored data bits.

Two examples of complementary metal oxide semiconductors (CMOS) memory cells are shown in Figure 7.8.

Gallium arsenide integrated circuit (IC) technology is becoming increasingly important for circuits intended for spaceborne and military applications. GaAs circuits are faster and more temperature and radiation tolerant than similar silicon circuits, which makes them more suitable for harsh environments.[1]

Three different types of static random access memory (SRAM) cells are shown in Figure 7.9: the McDonnell Douglas resistive and complementary E-JFET SRAM cells, and the Texas Instrument depletion load MESFET SRAM cell.

The high speed, low power, and insensitivity to total ionizing dose and dose rate of GaAs memories make them promising candidates for military and space applications. Simulations and experimental evidence show that unhardened GaAs ICs have a single event sensitivity comparable to silicon devices of equivalent feature size.[2]

Memory cell local write times are an important measure of the impact of hardening approaches on device performance. For these evaluations, write times are defined as the shortest time the word line must be raised in order to write a “1” and then “0” into the cell. The word line and data bus voltages depend on the memory design.

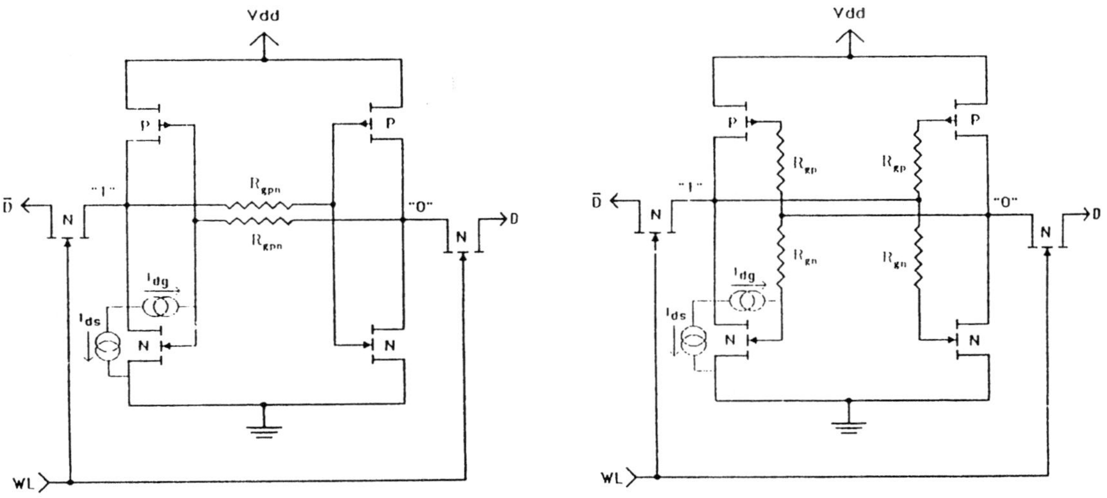

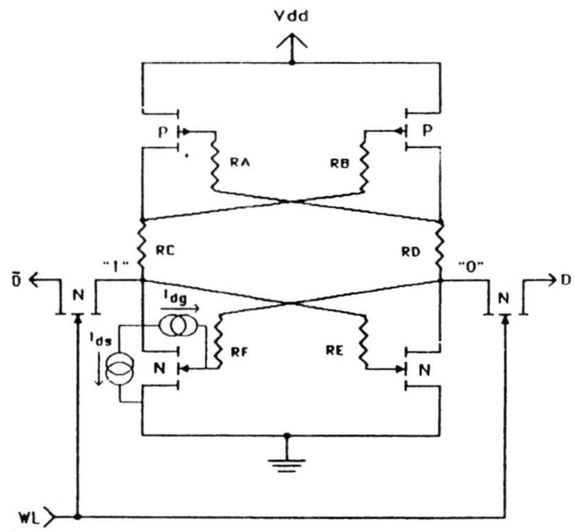

Figure 7.10 Complementary E-JFET cell incorporating (A) a two-resistor decoupling hardening approach; (B) a four-resistor decoupling hardening approach; and (C) a drain decoupling approach. (From Weatherford, T. R. et al., *IEEE Trans. Nucl. Sci.*, NS33(6), 1590, 1986. With permission.)

Published work has accurately predicted critical charge for the complementary E-JFET memory cell and suggested two approaches for decoupling memory cell with resistors. A two-resistor decouping approach decreases gate-to-drain critical charge and noise margin while a four-resistor approach improves gate-to-drain critical charge for large values of resistance (>500 K Ω).

Local write time analyses of these approaches predict that write time increases of approximately 5 nsec are required to reduce SEU soft error rate to 2E-6 errors per bit-day. A resistive-decoupling approach in GaAs memory technology degrades memory local write times by approximately the same amount as in silicon memories because resistive decoupling is essentially a time-discrimination technique that makes cells too slow to be written by single event pulses.

Response of the complementary E-JFET cell has been analyzed in terms of noise margin as a function of decoupling resistance. Figures 7.10(A-C) show three resistive decoupling approaches investigated: two-resistor approach with one common resistor for the gates of each inverter in the cell (10A), a four-resistor approach with a resistor for the gate of each device (10B), and a drain resistor approach (10C).

For an 0.8-V supply, the memory cell noise margin is not affected significantly by either the two- or four-resistor decoupling approach. A 1.0-V supply significantly increases gate leakage of the ON n-channel device. In the two-resistor approach, the voltage drop of the leakage current through the decoupling resistor decreases the gate voltage on both the OFF p-device and the ON n-device.[2]

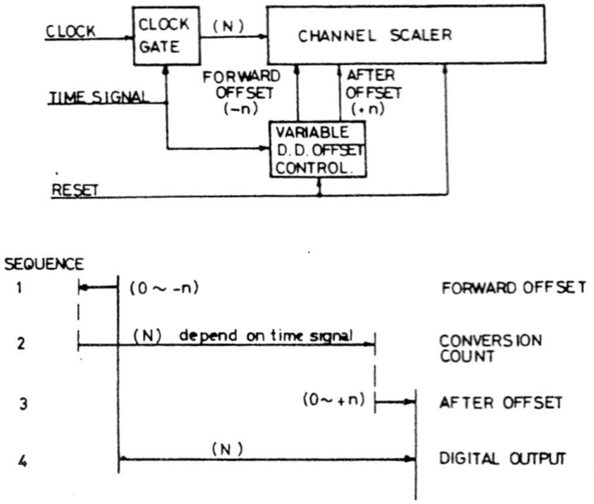

Figure 7.11 Block diagram and counting sequence of the variable digital-digital offset (VDDO) method. (From Kinbara, S., *Nucl. Instrum. & Methods,* 143, 267, 1977. With permission.)

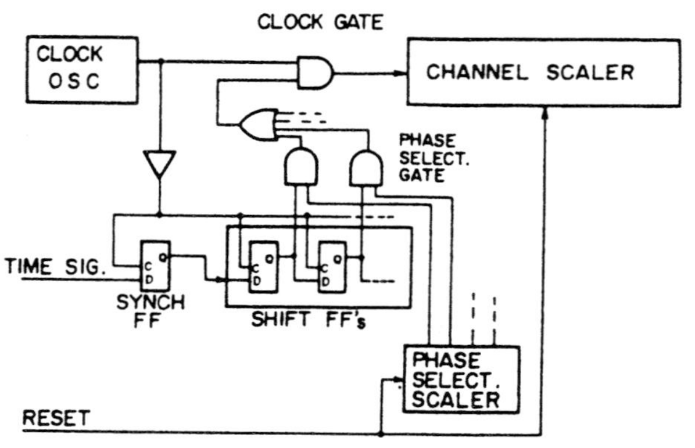

Figure 7.12 Block diagram of the variable phase gate (VPG) method. (From Kinbara, S., *Nucl. Instrum. & Methods,* 143, 267, 1977. With permission.)

7.3 SINGLE-PARAMETER MULTICHANNEL ANALYZERS

7.3.1 EARLY MODELS

The differential nonlinearity of a Wilkinson-type analog-to-digital converter (ADC) is dominated by the imbalance of even-numbered periods caused by the action of interference resulting from operation of a channel scaler.[3] To improve this situation Kinbara[3] has tested a new method which allows such action of interference to be dispersed. Due to the desire to increase the speed of the multichannel analyzer (MCA), the Wilkinson-type ADC was replaced by successive approximation ADC due to the differential nonlinearity problems which limits the ADC speed. To overcome this problem Kinbara tried to improve the odd-even imbalance by distributing it over many channels as done in the sliding-scale technique. The method is illustrated in Figure 7.11. The channel scaler is offset to a negative value (–n) before counting the block pulses and offset afterward by +n counts. The value of n is incorporated on successive input pulses and ranges over preselected amounts of 2, 4, or 8. This method is called variable digital-digital offset (VDDO).

Another method is to control the after offset at the clock gate as shown in Figure 7.12. It is also possible to place delayed elements between the clock gate and the channel scaler. In this method, interference takes place at the end of the time signal when the synchronizing circuit closes. The averaging effects occur because the channel scaler is reversed in phase on successive signal pulses.

A 4096-channel pulse height analyzer was described by Byrd and Goosey[4] that featured automatic differential dead time correction of spectra at input rates up to 5×10^4/s. A block diagram of the ADC is shown in Figure 7.13. The linear signal from the pulse amplifier was delayed for 2 μsec while the undelayed signal was inspected for appropriate amplitude limits and absence of pileup. If the requirements were met, a conversion command was generated in the control logic.

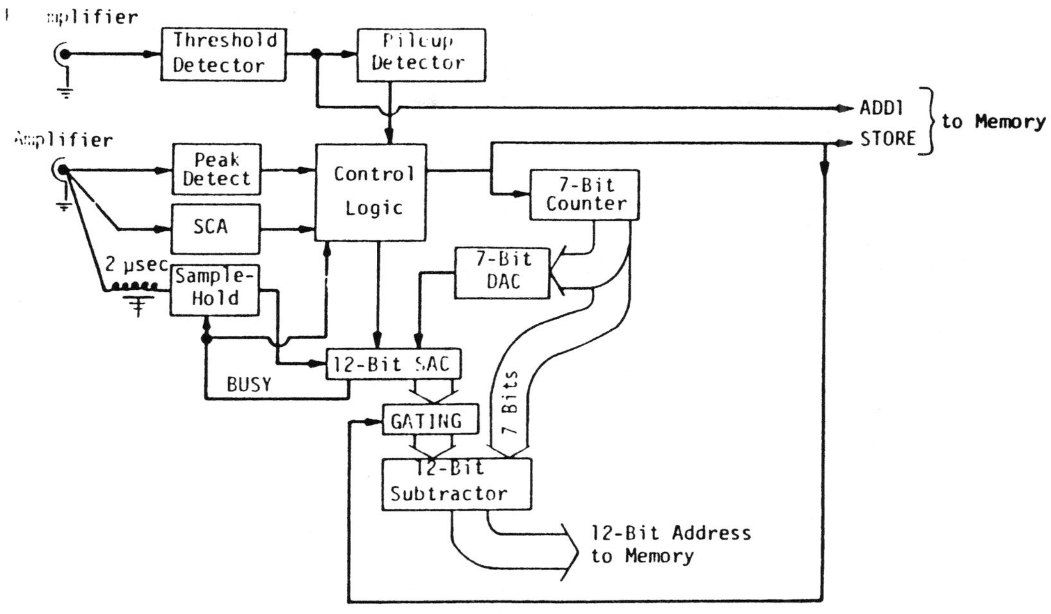

Figure 7.13 Block diagram of ADC. (From Byrd, I. S. and Goosey, M. H., *Nucl. Instrum. & Methods,* 158, 565, 1979. With permission.)

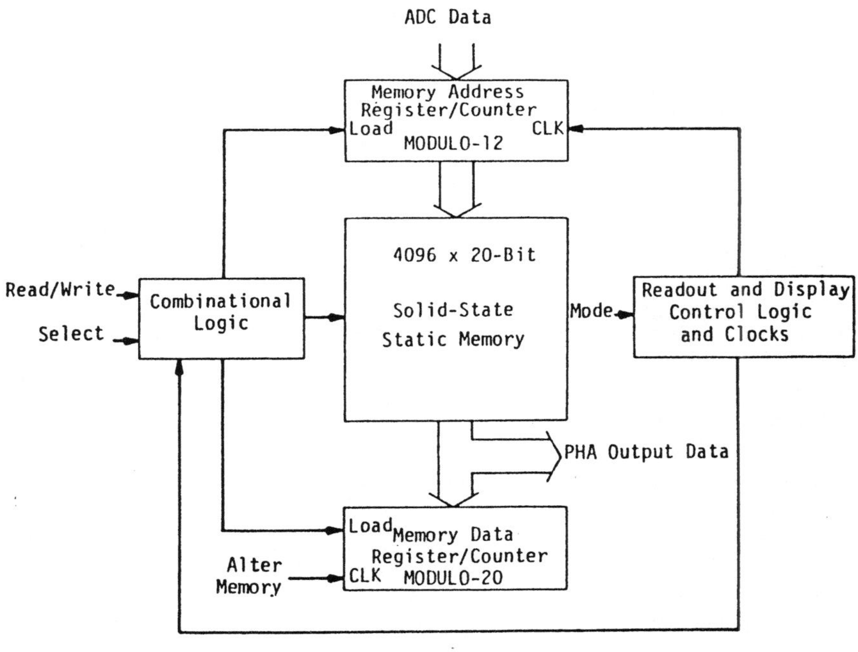

Figure 7.14 Block diagram of memory. (From Byrd, I. S. and Goosey, M. H., *Nucl. Instrum. & Methods,* 158, 565, 1979. With permission.)

ADD1 pulses to the memory unit were generated by a threshold detector circuit operating directly on the preamplifier output, which issued a standardized pulse for every event over the noise level. Automatic differential dead-time correction was achieved. The memory is shown in Figure 7.14. The pulse-height analysis (PHA) memory section is controlled with a pulse asynchronous sequential logic circuit (Figure 7.15). A small size low-cost 128–256 channel multichannel analyzer based on an SDK85 microcomputer

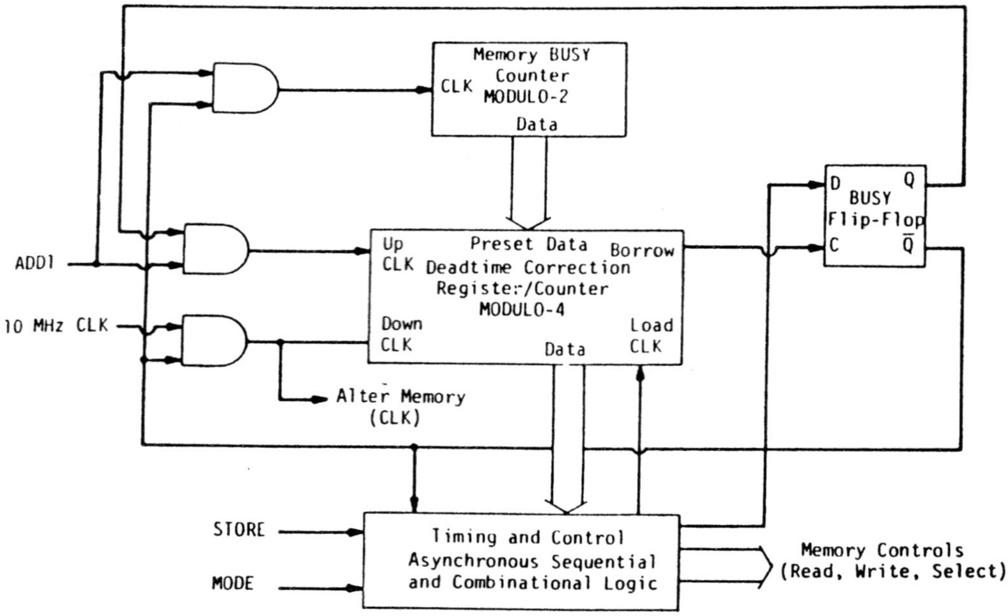

Figure 7.15 Memory control diagram. (From Byrd, I. S. and Goosey, M. H., *Nucl. Instrum. & Methods,* 158, 565, 1979. With permission.)

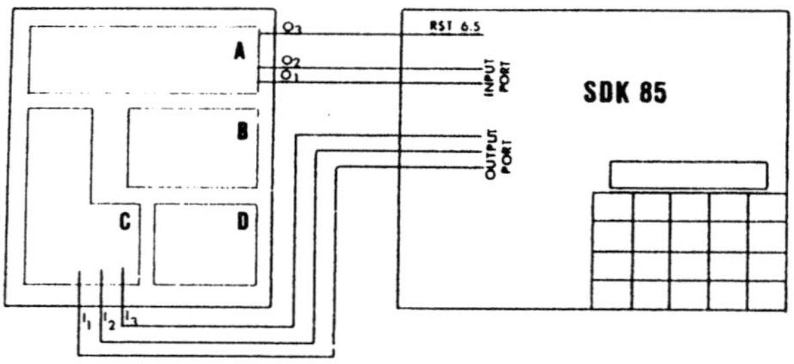

Figure 7.16 Analog board connected to the SDK85. (From Frago, F. A. F., *IEEE Trans. Nucl. Sci.,* NS29, 129, 1982. With permission.)

board using a minimum of external low-cost circuitry was described by Fraga.[5] The circuitry consisted only of a quad comparator, three operational amplifiers, three integrated analog switches, and very few discrete components. The pulse amplitude conversion was carried out fully by software. The hardware diagrams are shown in Figures 7.16 and 7.17 The SDK85 microcomputer is shown in Figure 7.16 with the analog signal circuit connected to the I/O ports.

Output lines to control the display digital-to-analog converters (DACs) and the six input lines needed to use an optional external keyboard are not shown. All input-output signals of the analog circuit were transistor-to-transistor logic (TTL) compatible so that the board was directly connected to the SDK85. The analog circuit is shown in more detail in Figure 7.17 and besides the two ADCs for X-Y output it is the only circuitry external to the SDK85.

This circuit is implemented with only six standard low-cost integrated circuits, one field effect transistor (FET), and a few resistors. Block A is the comparator, B the stretcher, C the analog gates, and D the constant current source.

The use of a microprocessor (µP) in a multichannel analyzer (MCA) allows the inclusion of features that would be impractical to implement in hard-wired instruments. The comparative slowness of software signal handling becomes unimportant in this application when use is made of direct memory access (DMA) techniques for data acquisition; the dead time of the MCA is then almost solely determined by the analog-to-digital converter (ADC) and, at the usual count rates of nuclear or X-ray spectroscopy. Data processing may be accomplished on-line or while other spectra are accumulating.

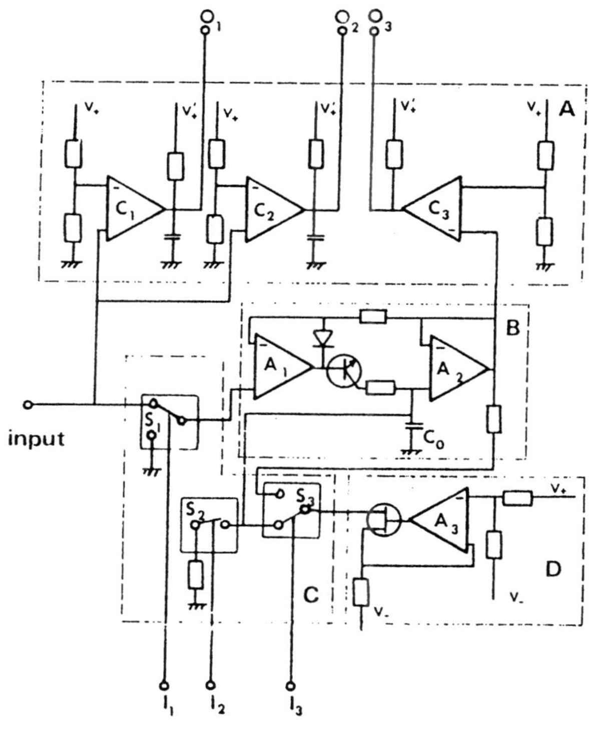

Figure 7.17 Analog circuit diagram. Only seven ICs, a transistor, a FET, and a few resistors are used. (From Frago, F. A. F., *IEEE Trans. Nucl. Sci.*, NS29, 129, 1982. With permission.)

A microprocessor-controlled MCA, in which direct memory access techniques were used to allow fast acquisition rates, was presented by Santos et al.[6] The hardware was assembled around an Intel Corporation 8086 central processing unit. Operating commands to the MCA were given through front panel switches and through a keyboard. After a valid signal was acknowledged at the ADC input, the instrument performed the following sequence of operations: (1) the input gate was closed and the ADC was ordered by the control logic to start conversion, (2) the resulting binary number was interpreted as a memory address and the contents of this address were incremented by one, (3) the control logic reset the ADC and reopened the input gate. The system is shown in Figure 7.18.

A megachannel pulse-height analysis system using a PDP-8/E computer and two moving-head disk memories has been developed by Camp et al.[7] The system had a storage capacity of 2^{20} memory locations, was capable of processing 1100 events per second, and provided on-line sorting and disk storage. An X- or Y-pulse-height spectrum in coincidence with one or several arbitrary pulse-height windows could be assembled in core for scope display and spectral analysis within 2 to 20 s. Reconstruction of a complete X- or Y-pulse-height spectrum required about 3 min.

The megachannel analyzer was used for on-line sorting of coincidence data with on-line data storage taking place on a moving-head disk. The basic system included a 32,000, 12-b per word, PDP-8/E computer; two moving-head disks, one with a removable cartridge; two ADCs; scope display with function box controls; line printer; and teletype.

The magnetic disk cartridges were divided into 406 tracks, each having 4096 words (12 b); and one track was accessible to a movable read/write head. One disk provided 2^{20} locations (bins) in an XY array, each capable of storing 2^{18} events, while the other drive provided storage for programs and data files. Because the time required to update a track is 90 msec, it was important that the disk be used efficiently for both the data acquisition program and the data recovery and display program.

An interface for transferring the contents of the memory of a Nuclear Data 2200 analyzer to a Digital Equipment Corporation PDP 10 computer, and its associated device handling software were described by Kieser et al.[8] Data were stored in the memory as 20-b binary words; on output to the BRPE high-speed paper tape punch driver module, the data control unit translated these data into binary coded decimal (BCD) representation and transmitted each data word as six serial 4-b bytes, each corresponding to a BCD character. The BCD data were sent from the data control unit to a 4-b parallel DTL register in the BRPE driver module. In the device controller, the data were taken character by character (4 b at a time) from

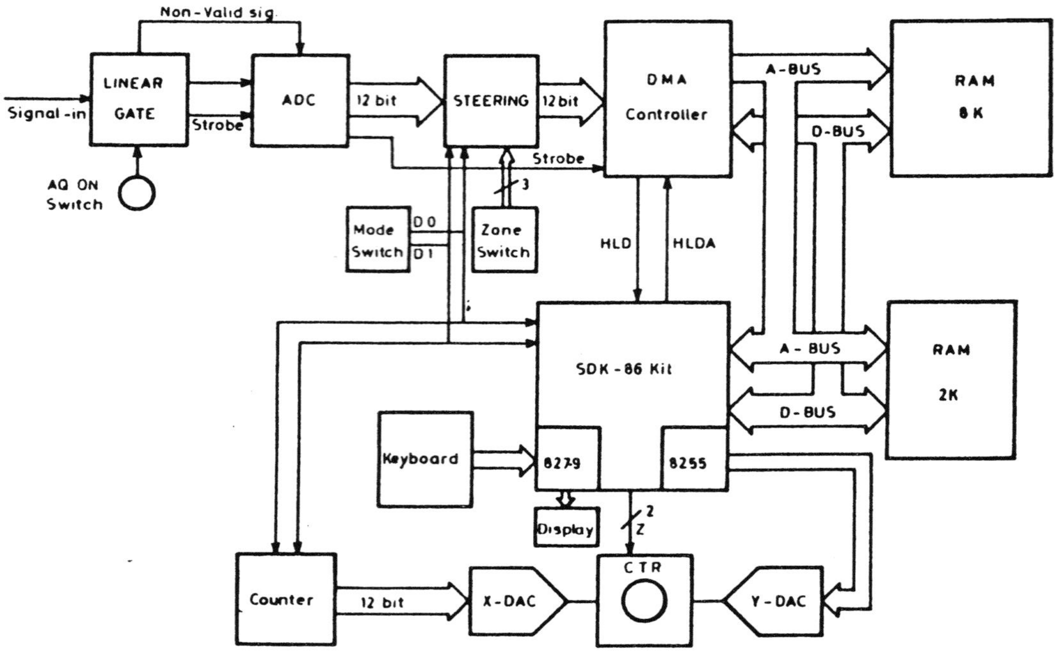

Figure 7.18 General block diagram of the microprocessor controlled multichannel analyzer. (From Santos, D. M. et al., *Nucl. Instrum. & Methods,* 196, 435, 1982. With permission.)

the line receivers and a complete BCD data word was assembled in a 24-b shift register. When the six data characters were received, the BRPE driver module was cycled through the transmission of the two special characters. When the computer (CPU) was ready, the data was strobed through gated level shifters, onto the I/O bus and into the PDP 10 memory.

A block diagram of the data transfer path and the control logic is given in Figure 7.19. A two-fold initialization was necessary for data transfer. The mode switch on the data control unit of the ND2200 must be set to read in/out which made the first character of the first word available to the BRPE driver module data register and sends a signal (CMMD) to the device controller to indicate that the first word is ready. Second, the device service routine must issue the command to start transmission (TRANST). On receipt of both CMMD and TRANST, the device controller issued a ready-to-receive pulse (PSYNC) to the BRPE driver module. The word cycle was controlled by the eight-step counter II which counted the load pulses from counter I.

An interface circuit was developed and constructed by Mildebrath and Cipolla[9] to read out the memory contents of printing-type multichannel analyzers into a time-shared remote computer to which communication is established via any EIA standard RS-232 compatible interface device.

The interconnection is indicated in the overall circuit schematic of Figure 7.20. The data words in the analyzer consisted of a series of decimal numbers up to six digits in length with leading zeros suppressed. The following grounding signal lines were used by the interface circuit: one each for the numbers 0 through 9, a tabulate (TAB) line, and a carriage return (CR) line.

The major components of the interface circuit were a 4-b encoder and a universal asynchronous receiver transmitter (UART). Besides the basic system there were three control elements which were essential to the operation of the circuit. The first was the strobe-delay controller, consisting of a series of three Schmitt triggers, which delays and reshapes the strobe pulse. The second was the LF generator, consisting of another series of three Schmitt triggers, which generated a separate LF pulse derived from the CR pulse. The third was the clock circuit for the UART. It was necessary to slow down the send rate of the analyzer so that the computer could accept the data. This was accomplished at the analyzer by changing the capacitor in the readout oscillator.

Because the interface was essentially receiver independent, no handshaking or any two-way communication was needed and there was little problem with adaptability to most computer systems.

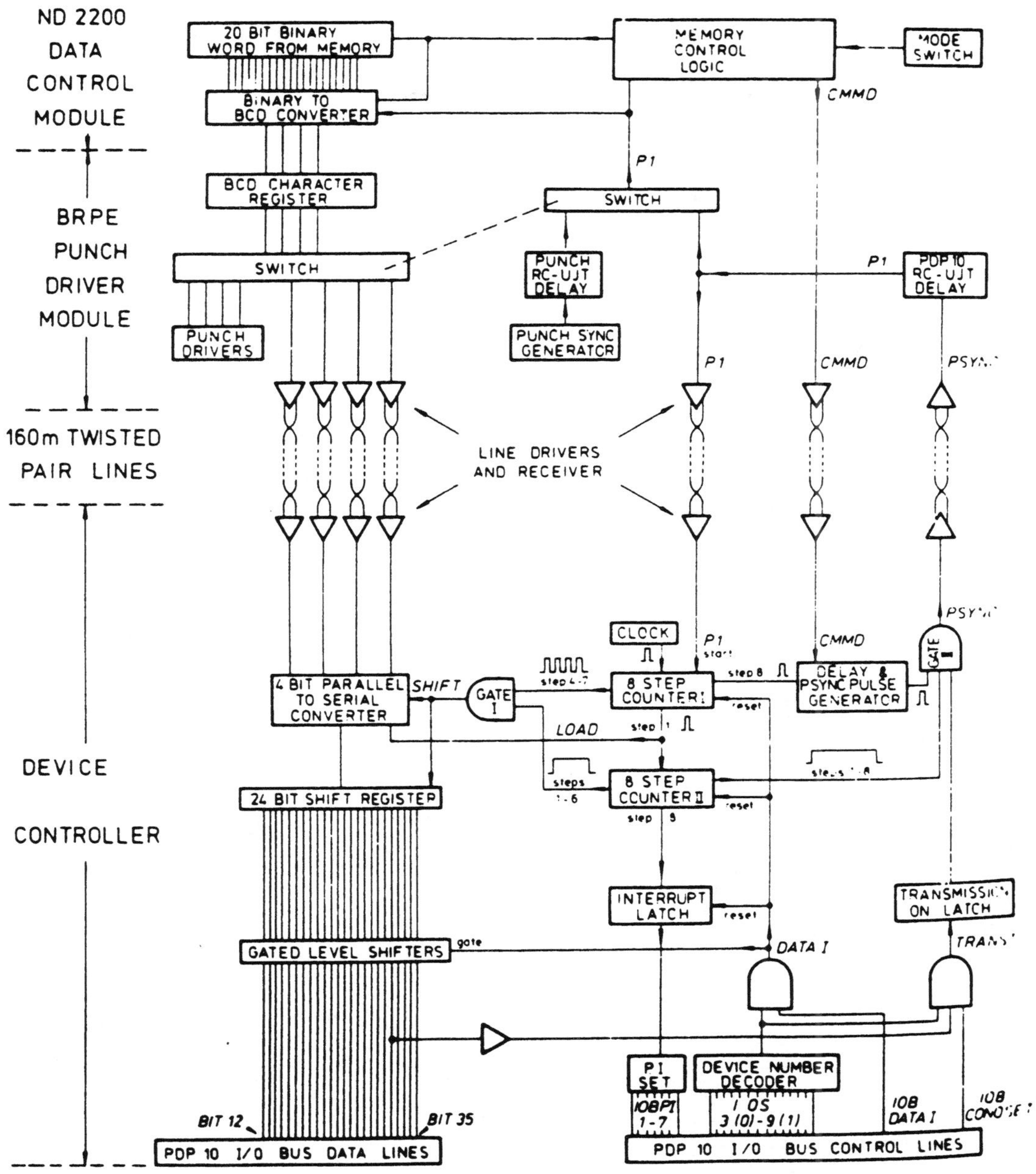

Figure 7.19 Block diagram of the ND2200 to PDP 10 interface. Standard DEC notation is used to describe the connections to the I/O bus. (From Kieser, W. E. et al., *Nucl. Instrum. & Methods,* 164, 587, 1979. With permission.)

7.3.2 CAMAC-AIDED MCA

Customer-made MCAs using the minicomputers, which are utilized for data accumulation as well as data processing, have generally less capability of data acquisition rate in accordance with the limit of operating speed of the computers, and need more complex operation.[10] The CAMAC standard is suitable for most computer-utilized experimental systems for the sake of its flexibility, cost benefit, and maintainability. As various kinds of the CAMAC products are supplied from various vendors, various instrumentation systems are easily constructed and modified by the users.

A CAMAC-composed MCA is realized by using dedicated CAMAC modules, which have combined features such as high data acquisition rate and easy operation as the commercially available ones, and such high flexibility and free choice as the customer-made ones.

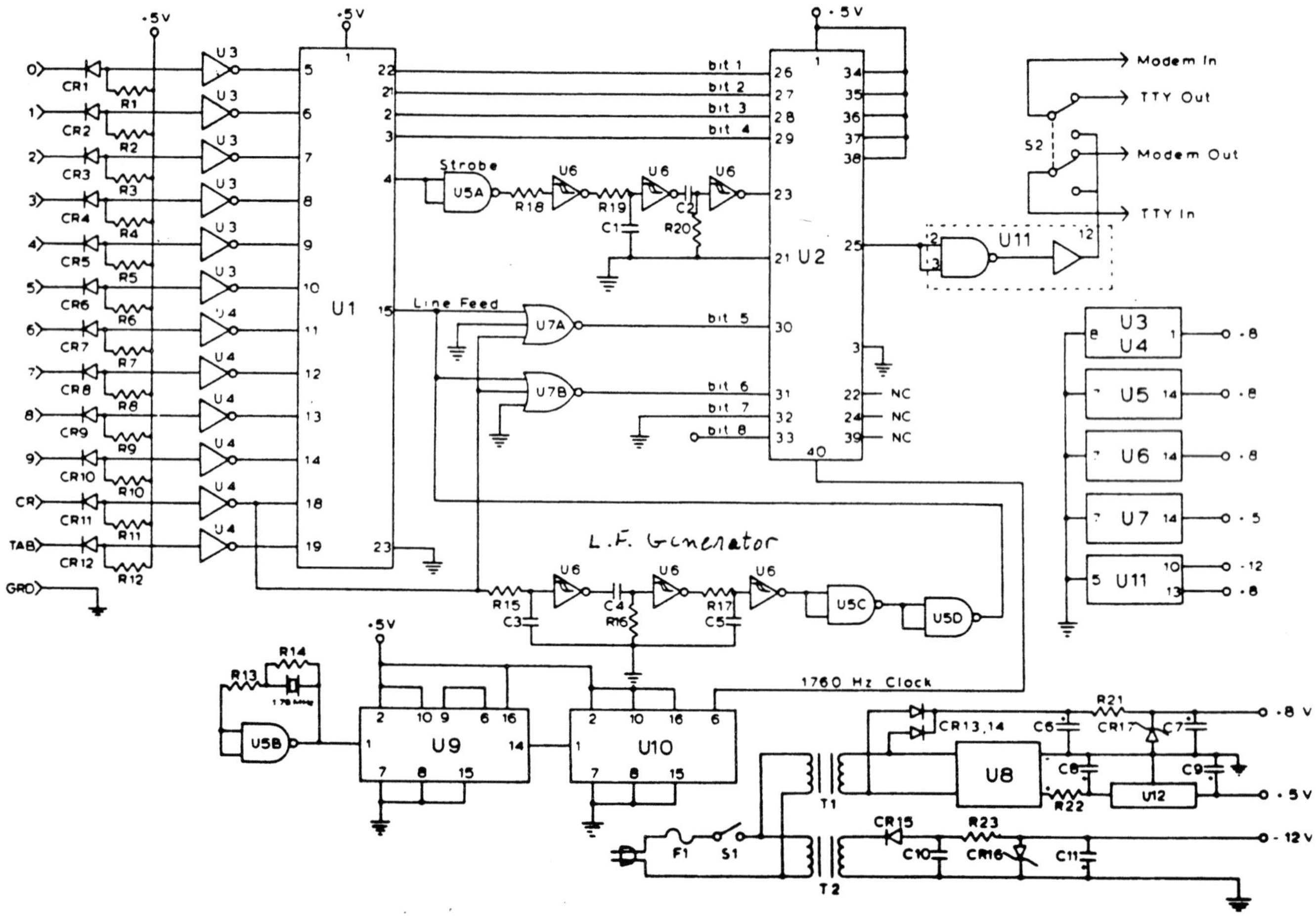

Figure 7.20 Schematic circuit diagram of the interface unit including the power supplies. U1 is the HD-0165 four-bit decimal-to-binary encoder, and U2 is the AY-3-1014A Universal asynchronous receiver/transmitter (UART). (From Mildebrath, M. and Cipolla, S., *Nucl. Instrum. & Methods*, 180, 615, 1981. With permission.)

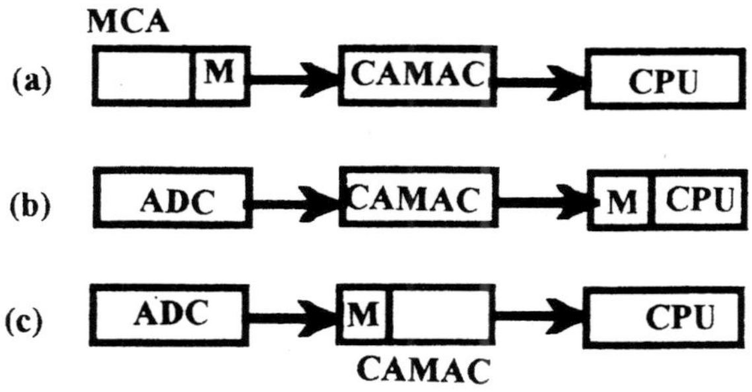

Figure 7.21 (a-c) Three kinds of system configuration of a CAMAC-type MCA. (From Kumahara, T. et al., *IEEE Trans. Nucl. Sci.*, NS26, 743, 1979. With permission.)

There are three different methods for CAMAC-based MCA as shown in Figure 7.21. A hard-wired MCA combined with an analogue-to-digital converter (ADC) and a data accumulation memory (M) is connected to a computer (CPU) through CAMAC interfaces in the upper drawing (a). This system has features of fast data acquisition and easy operation for the data acquisition but expensive cost because of needing double independent devices. The ADC and the CPU connected by the CAMAC interface as shown in the middle of the Figure 7.21(b) have the features of lower cost and higher flexibility but slower data acquisition in accordance with the limit of the CPU operation speed.

The system in Figure 7.21(c) has a fast data acquisition by direct connection of the ADC and the M, high flexibility by using CAMAC modules, moderate cost because of ability of wide use, and easy operation by adapting the front panel controls.[10]

A pulse-height analyzer with 1,048,576 channels has been developed by Tenten et al.[11] using a magnetic drum as a storage device. CAMAC instrumentation including a special-purpose drum controller was used for interfacing to a PDP-11 computer. One million channels cannot be fulfilled economically with ferrite core memories because of their high cost. Two methods have been used in the past. One is known as buffer-tape technique: the incoming events are converted in ADCs and then written onto a magnetic tape after some appropriate form of blocking by means of a core memory. The tapes are then processed on a large central computer facility. The number of channels sorted in one pass is determined only by the size of the available core of the central computer.[12] The second method is called an on-line method: the incoming events are processed by a real-time sorting program in a computer, which builds up the spectrum on a connected secondary storage device like a drum or a disk. The increment function is performed by software; hardware is used to transfer blocks of data between the computer memory and the disk or drum.[12] A rotating magnetic drum was used by Tenten et al. as the memory of the multichannel analyzer. The capacity of the drum was 1,024 tracks with 1,024 words each (1,048,576 channels). The drum was organized into a word structure: any single word on the drum could be read or written without disturbing the adjacent words. The drum was controlled by a special-purpose controller, which has been built as a CAMAC module, interfaced to a PDP-11 computer via a crate system controller.

Due to the rotation of the magnetic surface, an updated word, which was previously read, could be written on the same spot only after one full revolution. Random access was only possible with respect to tracks, because words were distributed along the track. The selection of a specific track without regarding the position within the track, however, could be done very fast. Because the drum had a head per track design, track switching was done within a few microseconds.

This was the principle of the megachannel analyzer (Figure 7.22): when a word had been accessed in one track, the next word could be accessed in any other track within one word time. Therefore all incoming events had to be sorted according to their position within a track and not to their track number for best efficiency of the system. The sorting is done by software in the PDP-11. The position of an entry within the list determined the word address within the appropriate track on the drum. The updated words on one revolution were written back into their proper locations during the next revolution of the drum. This was done by the drum controller (Figure 7.23) via direct memory address (DMA) access under the supervision of the PDP-11 software.

A data acquisition system for recording multiparameter digital data into a large memory array at over 2.5 MHz was described by Hendricks et al.[13] The system consisted of a MOSTEK MK8600 2048K × 24-b memory system, I/O ports to various external devices including the CAMAC dataway, a memory incrementer/adder, and a daisy chain of experiment-specific modules which calculated the memory address to be incremented. The design of the daisy chain permitted multiple modules and provided for

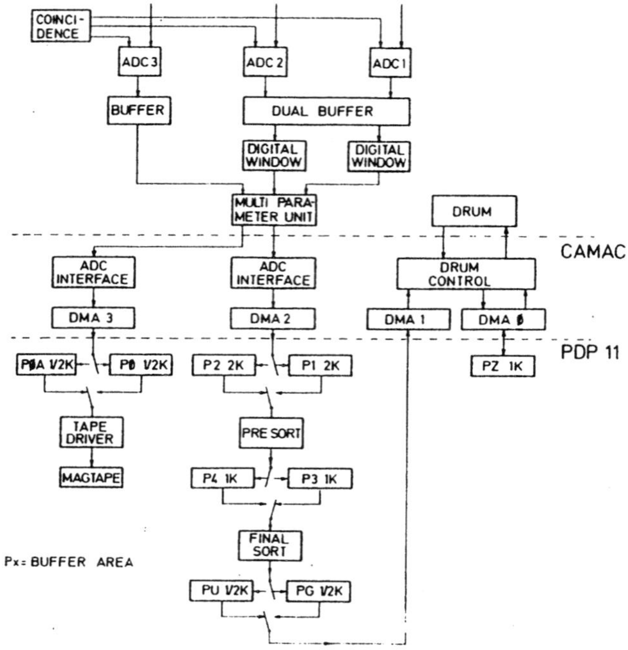

Figure 7.22 Megachannel analyzer. (From Tenten, W. et al., *IEEE Trans. Nucl. Sci.*, NS22, 529, 1975. With permission.)

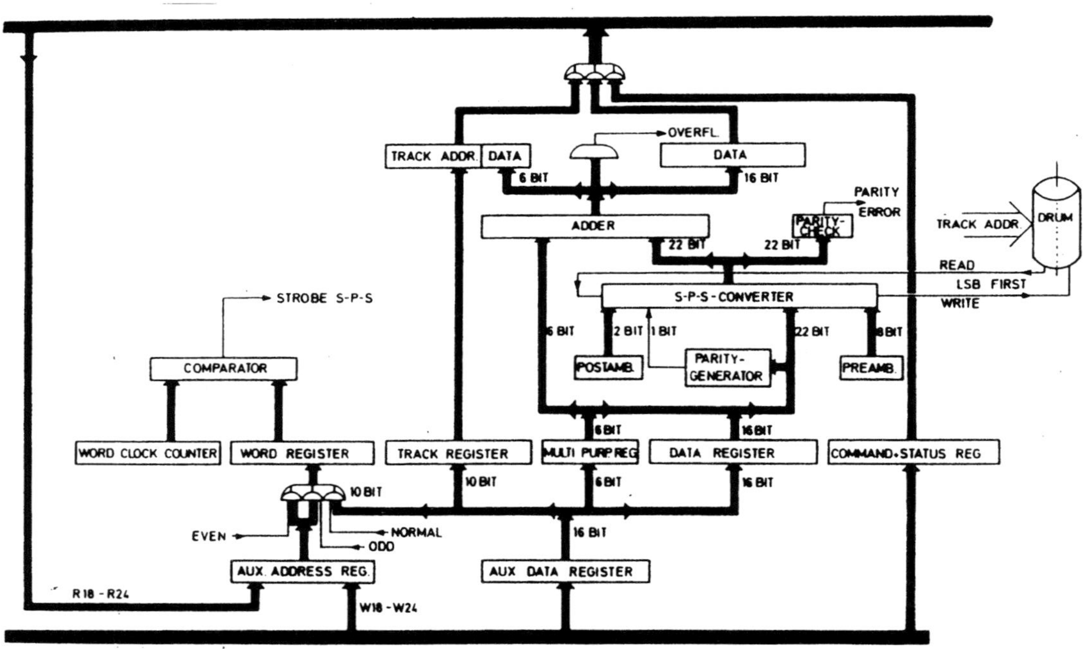

Figure 7.23 Megachannel analyzer drum controller block diagram. (From Tenten, W. et al., *IEEE Trans. Nucl. Sci.*, NS22, 529, 1975. With permission.)

each modification as experimental needs change. The system had been designed for use in multiparameter, multichannel analysis of high-speed data gathered by position-sensitive detectors at conventional and synchrotron X-ray sources as well as for fixed energy and time-of-flight diffraction at continuous and pulsed neutron sources. Modules developed included a buffer for two-dimensional position-sensitive detectors, a mapper for high-speed coordinate transformations, a buffered time-of-flight clock, a time correlator for synchronized diffraction experiments, and a display unit for data bus diagnostics.

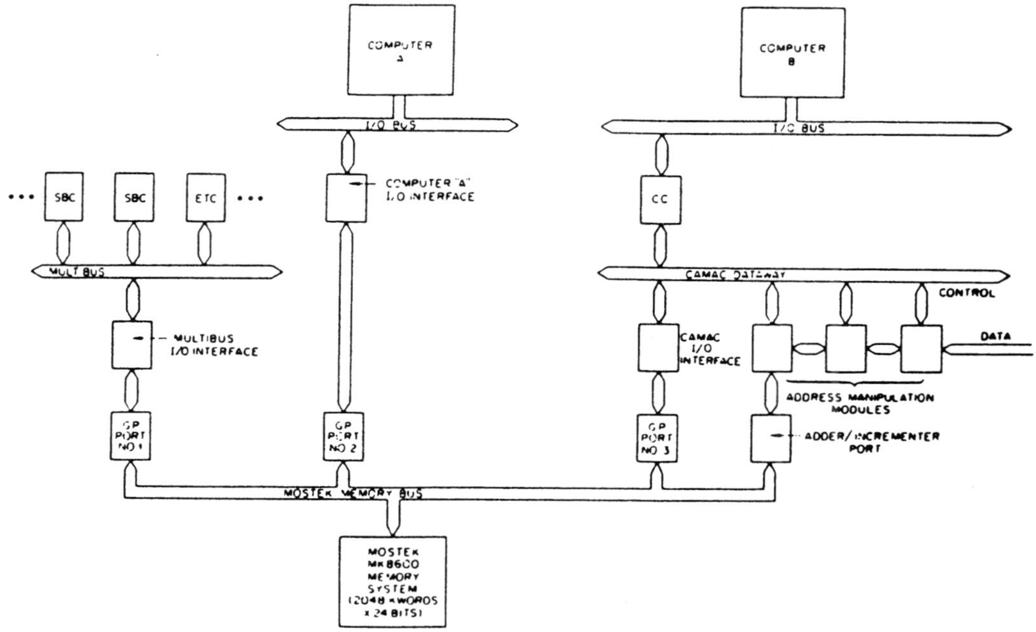

Figure 7.24 Functional block diagram of the data acquisition system. The interface to the memory consists of three general-purpose (GP) ports and an adder/incrementer port. Computer-specific interfaces connect the GP ports to single board microcomputers (SBCs) via multibus, directly to a host computer via its I/O bus, or to the CAMAC dataway and then to the host computer via the crate controller (CC). Experiment-specific CAMAC modules manipulate the address to be incremented prior to feeding it to the adder/incrementer port. (From Hendricks, R. W. et al., *Nucl. Instrum. & Methods,* 201, 261, 1982. With permission.)

When data rates over 1 MHz are anticipated, a CAMAC data acquisition interface system with its 1 μsec dataway cycle time is impractical.[13] On the other hand, data processing with TTL and with readily available inexpensive commercial memory systems at rates well in excess of 3 MHz is easily achievable.

A solution is to perform data acquisition on a separate, faster bus than the CAMAC dataway, but to continue to use CAMAC for lower speed module control.[14] In most data acquisition systems, the flow of data from the experiment to the recording system is unidirectional. However, it is usually essential that external devices (e.g., a computer) have bidirectional access to the memory but at a lower priority than the experiment.

To achieve maximum throughput and flexibility, Hendricks et al.[13] have concluded that the experiment (which determines the memory location to be incremented) and any other memory accessing devices should be on separate buses. A schematic configuration is shown in Figure 7.24. It consists of a large capacity 24-b memory system for which they have developed a four-port interface which contains three general-purpose I/O ports and a high-speed adder/incrementer port.

A multichannel analyzer composed of a dedicated CAMAC memory system has been developed by Kumahara et al.[10] It was combined with the flexibility of a computer and high performances of the hard-wired multichannel analyzer.

The data accumulation was performed by dedicated CAMAC modules: MCA timer, memory controller, 4-kiloword (kW) memory, and a CPT driver, which were directly connected to each other to get high-data acquisition rate.

The memory controller had the following operation modes: single pulse-height analysis (PHA), a dual pulse-height analysis (2 PARA, PHA), multichannel scaling (MCS), repeated multichannel scaling (MCSR), and continuous data storage (list). The memory controller block diagram is shown in Figure 7.25.

The memory cycle time in the accumulation mode was 2.8 μsec for loading data, incrementing the data, and storing the data in a certain channel. The 4-kW memory was extended to 16,384 channels by adding another module. The block diagram is shown in Figure 7.26. A modular MCA system using the dedicated CAMAC memory has been realized with the intelligent crate controller, and the system configuration is shown in Figure 7.27. In the system, the data acquisition is performed in the upper part

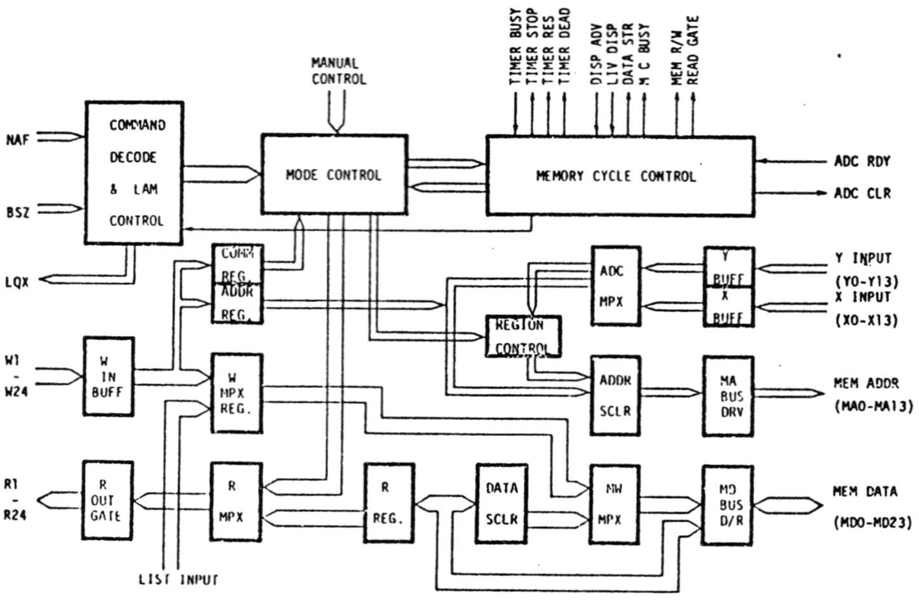

Figure 7.25 Block diagram of a memory controller. (From Kumahara, T. et al., *IEEE Trans. Nucl. Sci.,* NS26, 743, 1979. With permission.)

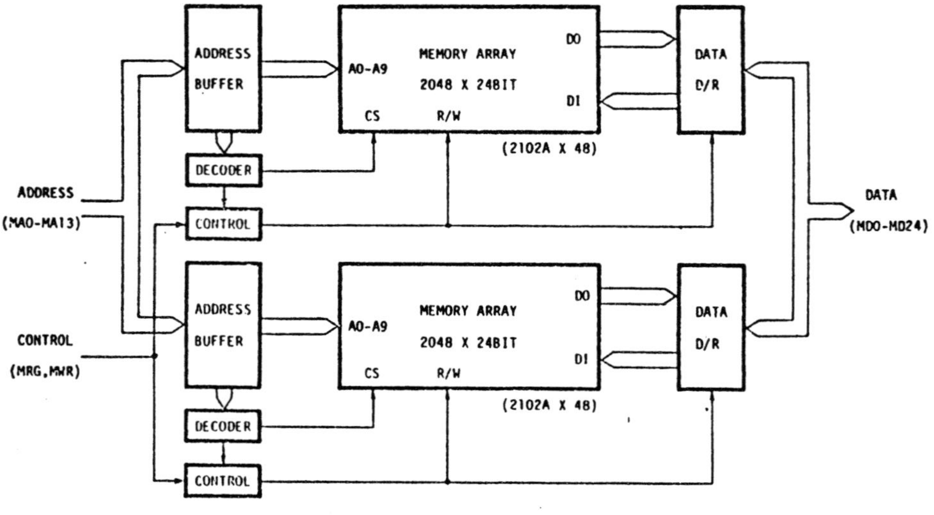

Figure 7.26 Block diagram of a 4-kW memory. (From Kumahara, T. et al., *IEEE Trans. Nucl. Sci.,* NS26, 743, 1979. With permission.)

which is composed of the MCA timer; the memory controller, the 4-kW memory, and the CRT driver with direct connections to each other. After the accumulation, the data were transferred to readout modules under control of the crate controller with a microcomputer as the communication adaptor.

A fast MCA capable of sampling at a rate of 10^7 s^{-1} has been developed by Berry et al.[16] The instrument is based on an 8-b parallel encoding converter ADC reading into a fast histogramming RAM system giving 256 channels of 64 K count capacity.

The MCA is based on the TRW model TDC 1007J 8-b parallel or flash encoding ADC and a histogramming memory. The 8-b output of the ADC is used to select one of 256 memory addresses. The contents of a selected address are automatically incremented by one, thus providing the histogramming feature. Control and readout of the fast MCA is through the CAMAC highway. Data from the memory

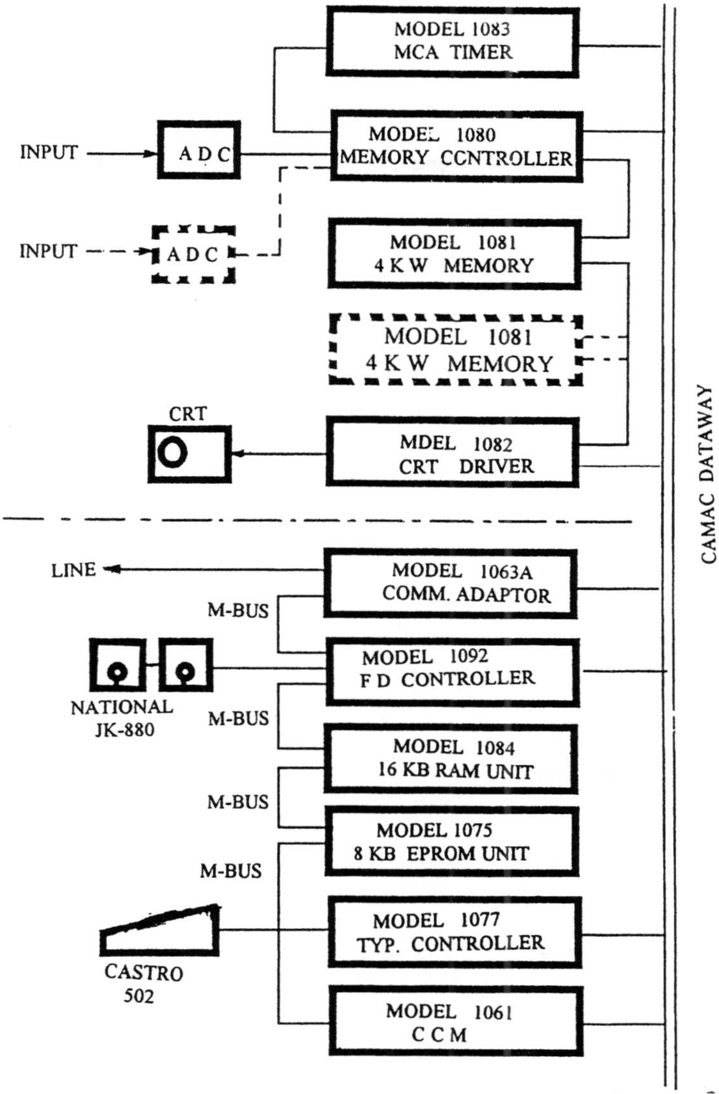

Figure 7.27 System configuration of a CAMAC composed MCA using a dedicated CAMAC memory. (From Kumahara, T. et al., *IEEE Trans. Nucl. Sci.,* NS26, 743, 1979. With permission.)

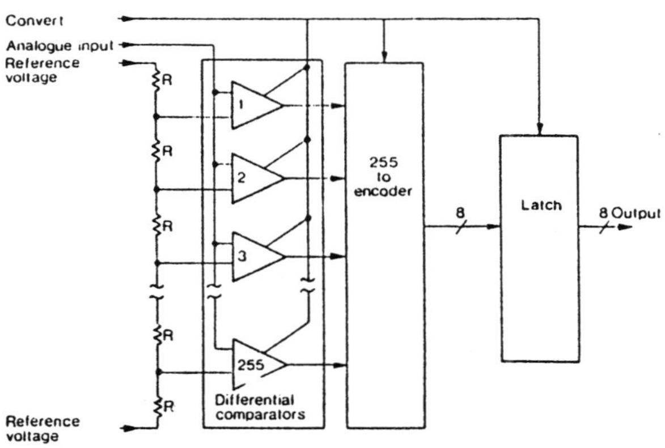

Figure 7.28 An 8-b parallel encoding ADC. (From Berry, A. et al., *Nucl. Instrum. & Methods,* 201, 241, 1982. With permission.)

can be read out during data acquisition on a memory cycle stealing basis. This will typically use 1 in a 100 memory cycles, giving a less than 1% decrease in data acquisition rate.

Figure 7.28 shows the principle of operation of the 8-b flash ADC. A reference voltage feeds a 256-element resistive voltage divider chain. Each tap of the divider is connected to an input of one of 255 discriminators providing an equal increment between successive discriminators. The other input of all

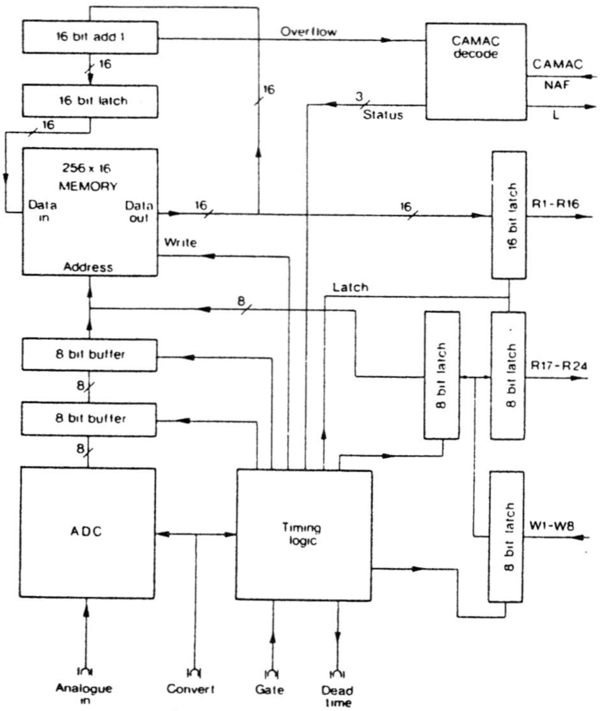

Figure 7.29 Block diagram of MCA. (From Berry, A. et al., *Nucl. Instrum. & Methods,* 201, 241, 1982. With permission.)

discriminators is fed by the input analog signal. The parallel output of all discriminators is encoded to form an 8-b word. Such an ADC gives the fastest possible conversion, but the number of elements (discriminators) increases geometrically with resolution. The TDC 1007J has a nominal conversion time of 30 nsec.

Since the MCA is not self-triggered, the CONVERT signal has to be generated externally and synchronized with the analog input signal. Operation of the unit is controlled by the CAMAC status register.

A block diagram of the complete instrument is shown in Figure 7.29. A data acquisition cycle is started by a CONVERT pulse, which clocks the output of the ADC into a two-word deep first-in-first-out memory (FIFO) consisting of two 8-b latches. This FIFO forms a derandomizing buffer, to enable memory access from CAMAC, thus permitting memory cycle stealing for readout during data acquisition.

The contents of a memory location addressed by the output of the ADC is applied to a 16-b adder, which increments the contents by one and passes the incremented value to a 16-b latch. Its output remains steady during the next operation in which the modified content is written back into memory. The adder circuit operates asynchronously to give maximum speed. The memory output also feeds a latch for CAMAC readout. The memory is based on the Fairchild 93422 static 256- × 4-b memory elements. The memory contents can overflow when a channel count reaches 65,535 (all 16-b set); the next count is zero. Alternatively the conversion process can be stopped when the first channel reaches 65,535. Data acquisition can be gated by "enable gate" command (status bit 3) and applying a gate signal to a front panel connector. The memory can be cleared either under CAMAC control, or by applying a signal to a front panel connector.

The standard deviation σ of the channel counts is a measure of differential nonlinearity of the ADC and is given by:[16]

$$\sigma = \left(\frac{1}{256} \sum_{i=1}^{256} \left(N_i - \overline{N} \right)^2 \right)^{1/2} \tag{14}$$

where N_i is count in the ith channel and N is mean number of counts per channel = $^1/_{256}\ \Sigma_{i=1}^{256}\ N_i$.

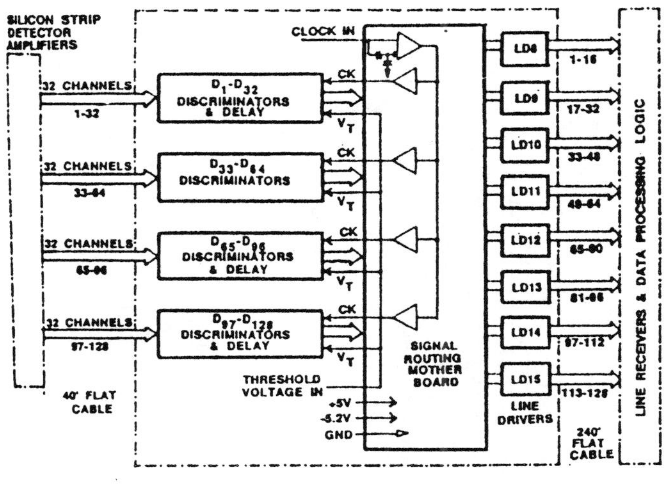

Figure 7.30 Block diagram of a 128-channel module. (From Turko, B. T. et al., *IEEE Trans. Nucl. Sci.*, 39, 758, 1992. With permission.)

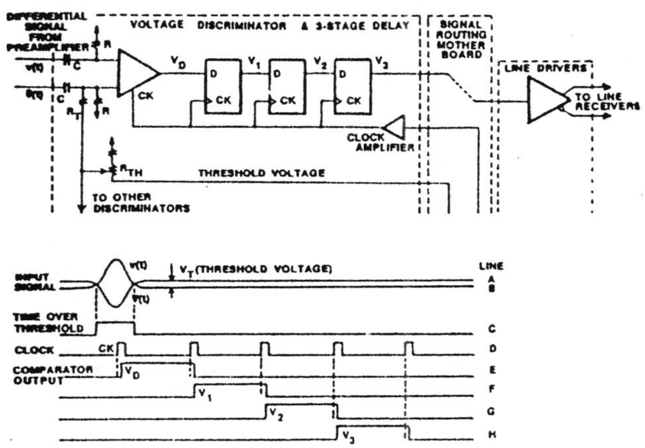

Figure 7.31 Block diagram of a single discriminator channel comprising a fast clocked discriminator, a three-state clock delay, and a line driver (top); time diagram (bottom). (From Turko, B. T. et al., *IEEE Trans. Nucl. Sci.*, 39, 758, 1992. With permission.)

A multichannel discriminator system for silicon strip detector readout was described by Turko et al.[17] The block diagram of a 128-channel subsystem is shown in Figure 7.30. Three 128-channel assemblies can be housed in a standard CAMAC crate shell (i.e., without the dataway board and power supply). A block diagram of an individual discriminator channel and its timing is shown in Figure 7.31. One half of a fast voltage comparator (HCMP96870A) and one half of a hex D-type ECL flip-flop (MC10176) are used; the remaining halves are used for the next channel. Input signals are AC coupled to optimize the signal shape and block the output DC offset of the preamplifiers. An RC of 75 nsec was found to be optimal. Threshold voltage is applied to the inverting comparator input through a resistor (R_t). The offset of the two comparators in the same package may vary by as much as several millivolts. Comparators were thus preselected in order to make the 32 discriminators on a board more uniform in threshold.

A differential input signal (Figure 7.31, lines A and B) would produce a square output (line C) were the comparator continuously enabled (time-over-threshold pulse). Since the comparator is strobed by the clock pulses (line D), the comparator output is latched by the trailing edge of the clock pulse (provided that the clock falls within the time-over-threshold period). The comparator output remains latched (line E) until the next strobe pulse. If (as shown) there is no signal exceeding the threshold at the comparator input at that point, the output returns to the low level. The three D-type flip-flops are clocked simultaneously. Since the flip-flop output follows the D input level at the leading edge of the clock pulse, the output V_1 changes state at the second clock pulse (line F). Each subsequent clock pulse shifts the event one step farther. The output V_3 (line H) appears after a delay of three clock periods and gets routed through the motherboard to the proper line driver.

A detailed circuit diagram of a two-discriminator channel is shown in Figure 7.32. Additional 15 dual channels are located on a single PC board. The discriminators are served by 16 clock amplifiers,

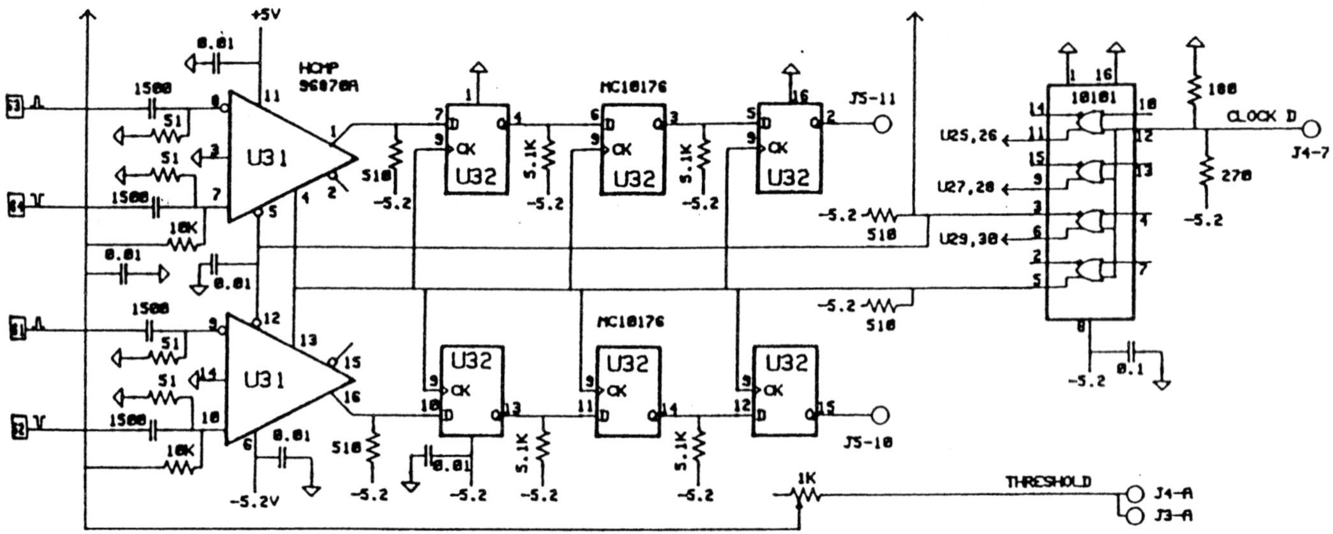

Figure 7.32 Circuit diagram of a dual-clocked discriminator channel with a three-stage delay. One each dual comparator and ECL hex D-type latch are required along with one gate of a good line driver. (From Turko, B. T. et al., *IEEE Trans. Nucl. Sci.*, 39, 758, 1992. With permission.)

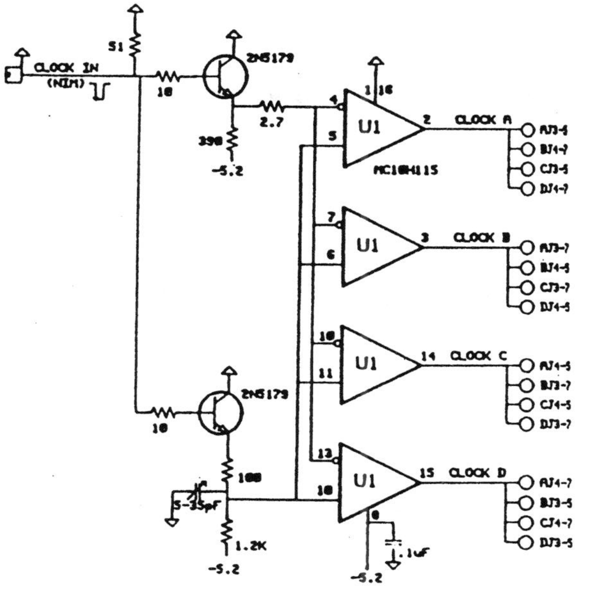

Figure 7.33 Clock fan-out circuit on the motherboard includes a 50-Ω receiver/shaper and a quad ECL driver (one section for each 32-channel discriminator board). (From Turko, B. T. et al., *IEEE Trans. Nucl. Sci.,* 39, 758, 1992. With permission.)

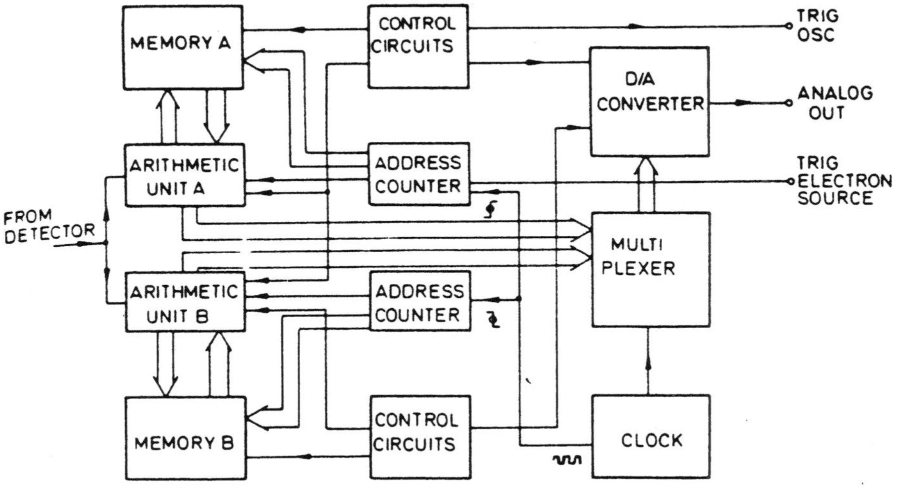

Figure 7.34 Block diagram of the multichannel analyzer. (From Norell, K. E. and Baltzer, P., *Nucl. Instrum. & Methods,* 223, 122, 1984. With permission.)

distributed evenly on the board to ensure uniform propagation delay of clock pulses. Only one quad driver is shown. The circuit on the motherboard (Figure 7.33) shapes the input nuclear instrument module (NIM) clock pulse into an emitter-coupled logic (ECL) signal distributed to the four 32-channel discriminator boards. The clock pulse width can be adjusted by a variable capacitor, C2. External DC threshold voltage can be connected through a LEMO connector to the discriminators. Also, power is connected at two –5.2-V points, one at +5 and at ground points.[17]

A high-duty cycle multichannel analyzer has been designed by Norell and Baltzer.[18] The memory capacity is 64K counts. The number of channels is 8192 with a time resolution of 100 nsec. The block diagram of the multichannel analyzer is shown in Figure 7.34. It consists of clock circuits, multiplier D/A converter, and two identical memories. Each memory has its own address counter, control circuits, and arithmetic unit. The number of channels (a channel is defined as a memory cell that stores the total number of counts for a given flight time) of a memory is 4096, which gives a total number of 8192 channels.

The even-numbered channels (0, 2, 4, ..., 8190) of the multichannel analyzer are located in memory A; and the odd-numbered channels (1, 3, 5 ..., 8191), in memory B. The memory channels are addressed sequentially by the address counter. During an address cycle the arithmetic unit works in two different modes, a data acquisition mode and an arithmetic mode; and the two arithmetic units A and B are always in opposite modes. By the construction involving two memories with associated circuits (address counter, control circuits, and arithmetic unit), the dead time in electron detection when shifting between two adjacent channels is zero. The clock frequency is 5 MHz, which gives an address cycle time for a given channel of 200 nsec. The data acquisition time of each memory channel is 100 nsec.

The address counter is triggered by the clock signal which also starts the data acquisition of the arithmetic unit. When a channel is addressed, data are read out from the memory to the arithmetic unit. If there is a new count present at the input, after the data acquisition time of 100 nsec, it will be added to the previous number of counts in the arithmetic unit. The new number is then written back into the memory. As can be seen in the block diagram, memory A is triggered by the positive-going edge of the clock signal while memory B is triggered by the negative-going edge.

A positive-going edge of the clock signal starts the data acquisition of arithmetic unit A, and the content of the addressed channel is read out simultaneously. During the second half of the clock period, while data of memory A is processed in the arithmetic unit and written back into the memory, the data acquisition is continued in arithmetic unit B.[18]

Figure 7.35 shows the circuit diagram of memory A, arithmetic unit A, address counter, and control circuits. The corresponding circuit diagram of memory B with associated circuits is identical to Figure 7.35 excluding IC4b, IC6a and b; IC7a, IC9a and b; and IC8a and TR1.

The arithmetic unit consists of the circuits IC10c and d, IC5, and IC2. IC1 is the address counter, and the memory is IC21. The other circuits in the diagram are to be referred to as control circuits.

The memory is built around Intel 2147 H-1, a 4096- × 1-b static RAM. The 2147 H-1 is directly TTL-compatible and has an access time of 35 nsec. Each memory consists of 16 packages, which gives the capacity of 64K counts. The address counter IC1 (3 × 74S163) provides 12 parallel address bits. When a channel is addressed, data is read out from the RAM into a latch and clocked through the latch to the adder. If there is a new count present at the least significant bit (LSB) input of the adder, it will be added to the previous number of counts and the new number is written into the RAM by the write-pulse, $\overline{WE}$. If there is no new count, the old (previous) number is written back into the RAM. The data hold time of 2147 H-1 is minimum 10 nsec and the latch IC2 prevents input data to the adder from disappearing with the write-pulse; 4 × 74S175 with a maximum propagation delay time of 17 nsec are used as the latch.[18]

The latch is clocked by the positive-going edge of the output pulse of a monostable multivibrator, IC3a. IC3a is triggered by the positive-going edge of the output pulse of another monostable multivibrator, IC3b, which is triggered by the clock pulse to the address counter. The latter multivibrator is used as a delay to compensate for the access time of the RAM and to generate the write-pulse, as will be described below. The outputs of the latch are connected to the A inputs of the adder IC5 (4 × 74S83). The LSB input of the adder is connected to the arithmetic unit, and the other 15 B inputs are connected to ground. The add time of two 16-b words is typically 30 nsec.[18]

The detector pulse is fed to a NAND gate, IC10a, followed by an S-R latch, IC10c and d. The detector pulse sets the latch if IC10a is enabled. IC10a is enabled by clock pulses through IC12a, which is a NAND gate used as an inverter. The data acquisition time of a channel is determined by the length of the enable pulse, which is set to 100 nsec.

Figure 7.36 shows the circuit diagram of the clock and output circuits. The clock pulses are generated by a 10.00 MHz crystal oscillator. The 10.00 MHz frequency is divided by 2 in the flip-flop IC17a. The function described above is obtained with S3 in the count position, as it is in the circuit diagram. The readout operation will be described below. The 5-MHz clock signal to the address counters is fed through the NAND gates IC19a and c. The address counter 74S163 can only be triggered by positive-going edges, but the address counter of memory B must be triggered by the negative-going edges of the clock signal as described above. This is accomplished by the inverter IC12d, which gives the inverted clock signal to the address counter of memory B. The pulses taken from B in the circuit diagram of Figure 7.36 generate the enable pulses to the arithmetic units through the NAND gate IC12a.[18]

A programmable multichannel correlation module (PMCCM) was designed by Arnone and Brunson.[19] The correlation circuit consists mainly of a shift register of finite length that is clocked at a fixed rate. The object is to create a time window using the shift register, thereby allowing correlated neutron pulses to be contained within this window. The correlation circuit records the number of neutron pulses in the

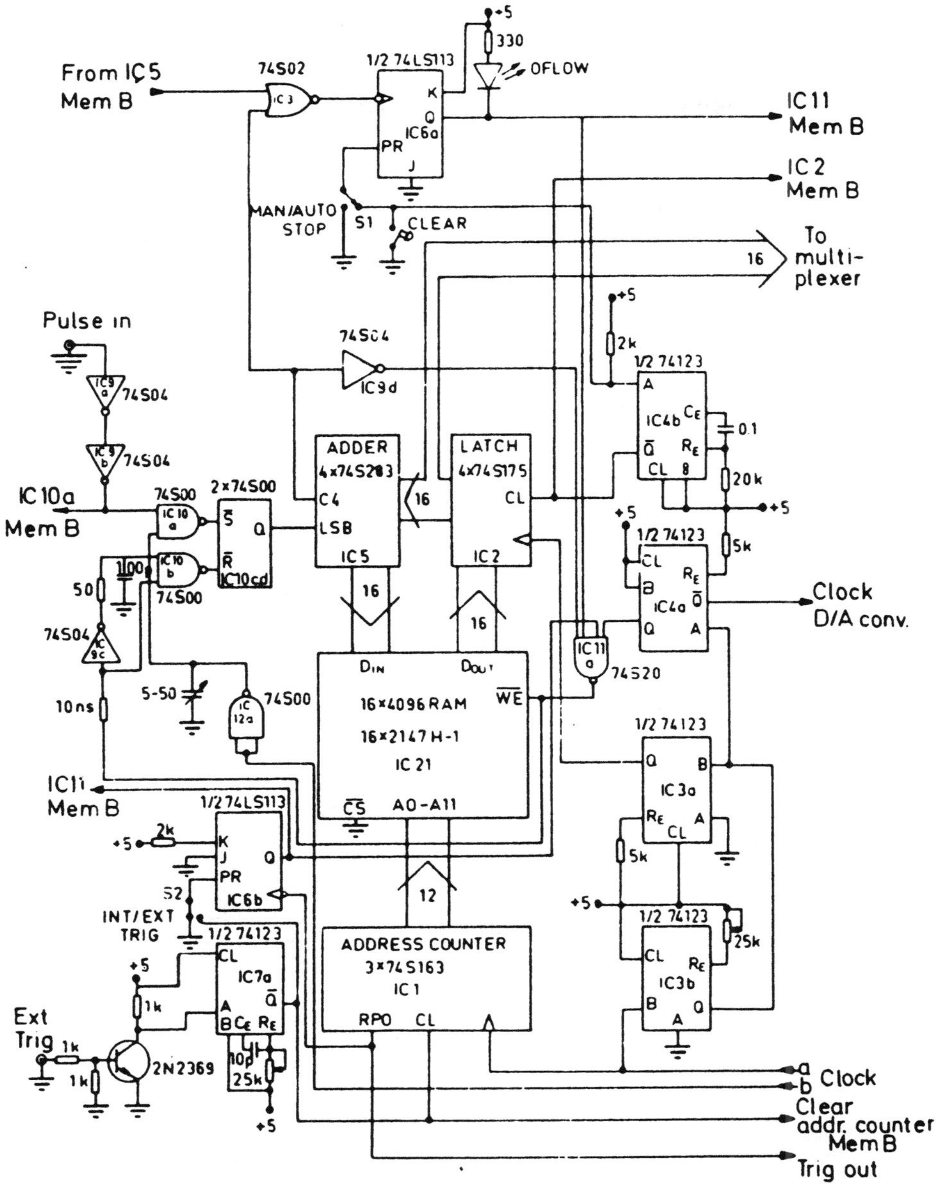

Figure 7.35 Circuit diagram of the memory, arithmetic unit, address counter, and control circuits. (From Norell, K. E. and Baltzer, P., *Nucl. Instrum. & Methods,* 223, 122, 1984. With permission.)

shift register when a neutron pulse exits the shift register. The correlation circuit also records the contents of the shift register at a periodic rate that is not related to the neutron pulse stream. To lower the overall dead time, a separate shift register was used for each detector channel and by minimizing the time that each channel input is disabled from processing input pulses. This dead-time is 12 nsec. per shift register clock.

The PMCCM comprises the following components and their operations (see Figure 7.37).

The oscillator generates the overall PMCCM system timing and drives the programmable clock circuit and the programmable background read-clock circuit.

The programmable clock which generates the shift register clock signal, can be programmed to be 2 MHz, 1 MHz, 500 kHz, or 250 kHz, thus providing a clock period or shift-register bin time of 500 nsec 1, and 2, or 4 µsec, respectively.

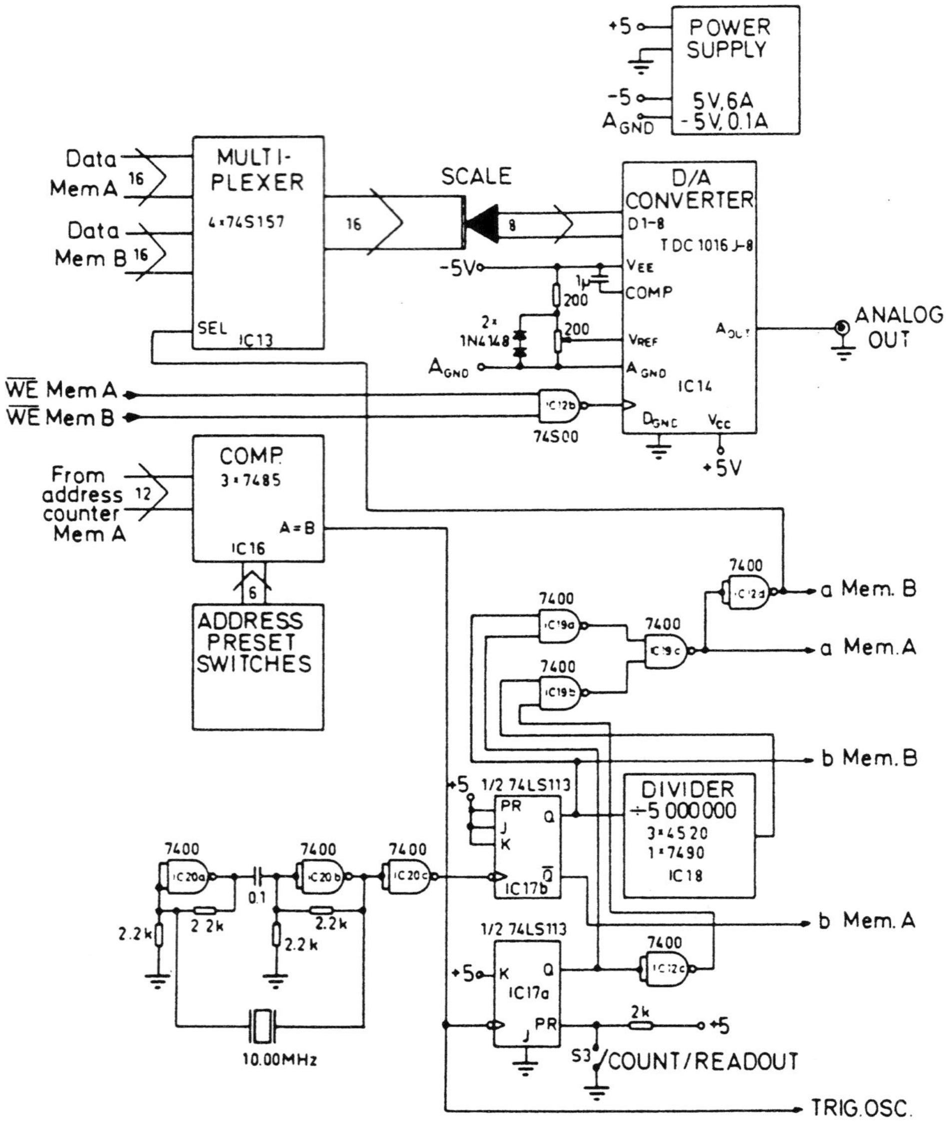

Figure 7.36 Clock and output circuits. (From Norell, K. E. and Baltzer, P., *Nucl. Instrum. & Methods,* 223, 122, 1984. With permission.)

The input synchronizer latches pulses (hits) that occur on any of the 15 inputs of the PMCCM module. Only one pulse per channel can be latched every clock period. The inputs to the PMCCM can be disabled via the CAMAC interface.

The summing programmable read-only memory (PROM) basically sums the number of input channels hit during the current clock period and produces a 4-b binary value (data byte) from 0 to 15. The output of this circuit is sent to the shift register and the up/down counter via the adder.

The programmable shift register (Figure 7.38) shifts the data byte received for each clock period. The shift-register length is programmable from the range of 1 to 1024 bits. The shift register bit time is based on the setting for the programmable clock, which ranges from 500 nsec to 4 μsec. The shift register consists of a standard 1024 by 4 static RAM chip, with address lines that are driven by a 10-b binary counter.

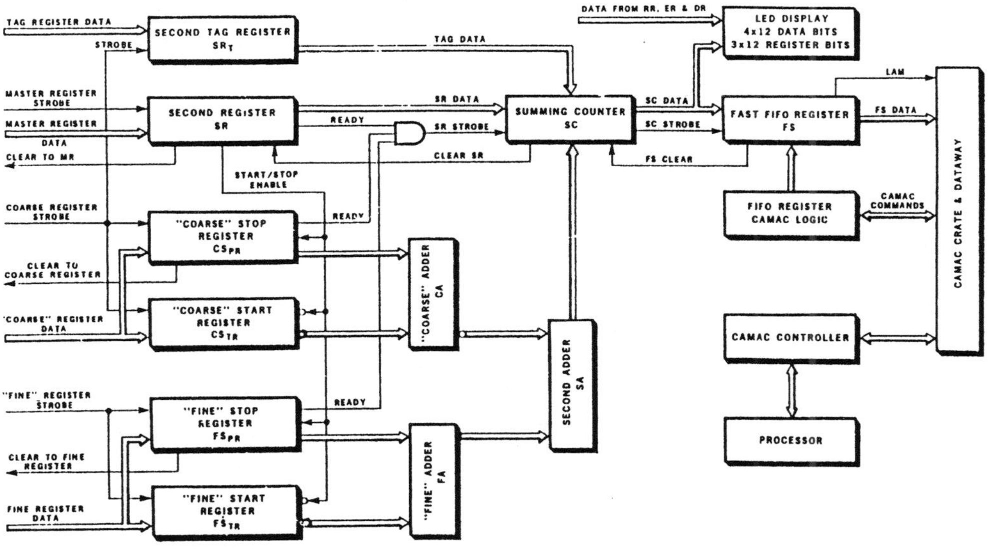

Figure 7.37 Block diagram of the PMCCM. (From Arnone, G. J. and Brunson, G. S., *IEEE Trans. Nucl. Sci.*, 37, 403, 1990. With permission.)

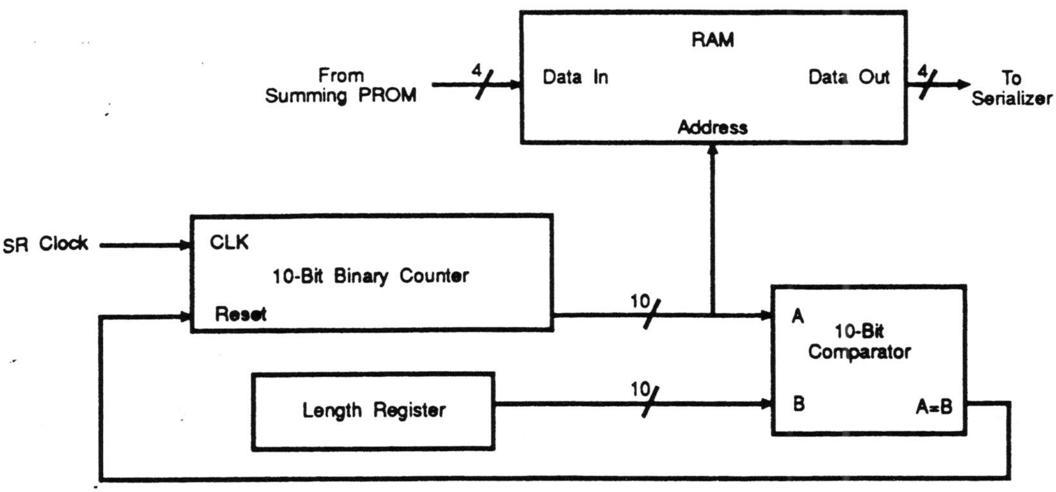

Figure 7.38 Block diagram of the programmable shift register. (From Arnone, G. J. and Brunson, G. S., *IEEE Trans. Nucl. Sci.*, 37, 403, 1990. With permission.)

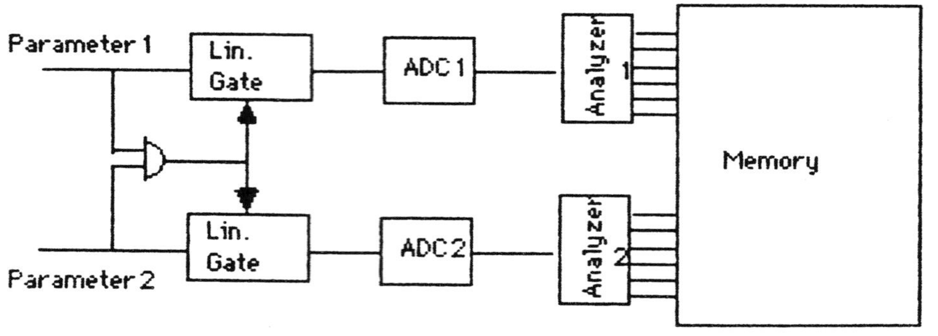

Figure 7.39 Two-parameter analyzer.

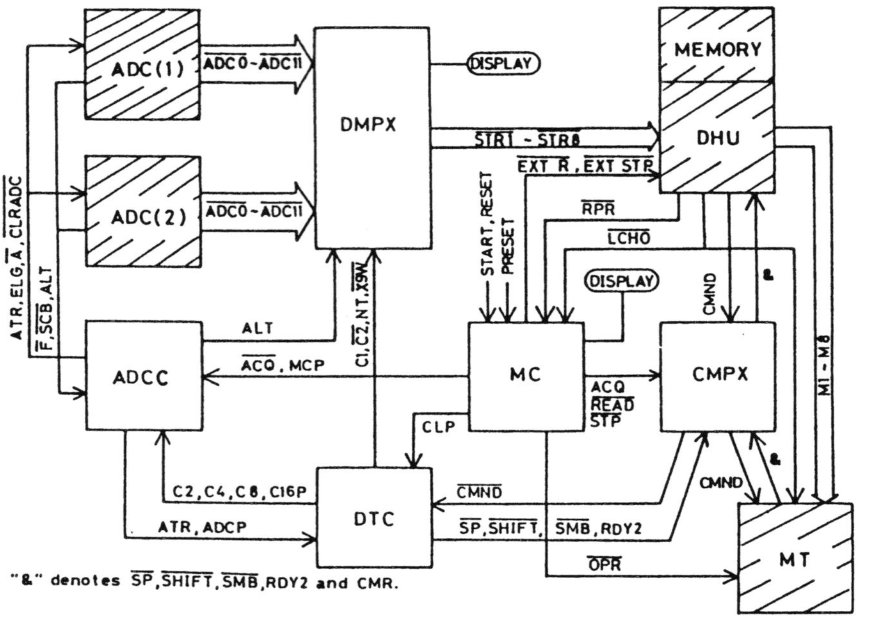

Figure 7.40 Block diagram of the two-parameter multichannel analyzer system. The hatched area is the ND-2200 analyzer. (From Toyoshima, K., *Nucl. Instrum. Methods,* 138, 557, 1976. With permission.)

During one SR clock period, the carrier memory location is read out and loaded into the serializer; then the data byte from the summing PROM is loaded into the same memory location. The up/down counter stores the sum of data bytes that are contained in the shift register.

The 5- to 32-channel decoder is used to decode the up/down counter value and route a pulse out of the appropriate scaler channel signal line to the selected Lecroy 32-channel scaler module.

Both the signal and background scalers are Lecroy Model 4434 CAMAC scaler modules. Each of these units contains 32 separate 30-MHz scalers that are 24 b deep. The PMCCM 32-channel outputs are based to the 32-channel inputs of Lecroy 4434. Only one of the scaler units is enabled at a time.

The INT RD\EXT RD switch, located on the front panel, determines how the up/down counter value will be sent to the signal scaler. The status of this switch can also be obtained via the CAMAC interface. If this switch is in the INT RD mode, the up/down counter value will be read out each time a down count is sent out of the serializer.

The background clock circuitry controls the background read synchronization. The up/down counter to the background scaler is read out at a programmed rate.

The CAMAC interface circuitry is used to interface the PMCCM with the CAMAC bus, which allows the CAMAC controller (under software control) to program the PMCCM and read its status.[19]

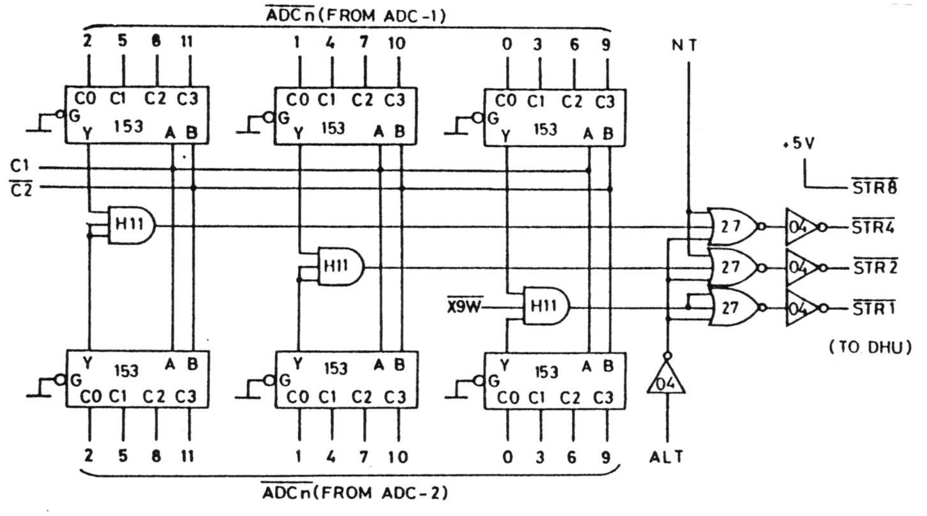

Figure 7.41 Data multiplexer logic. (From Toyoshima, K., *Nucl. Instrum. Methods,* 138, 557, 1976. With permission.)

7.4 MULTIPARAMETERS MULTICHANNEL ANALYZERS

Sometimes the recording of more than one parameter is needed with the correlation between the two parameters. Such measurement requires a separate ADC to each parameter as shown in Figure 7.39. Each ADC directs the data to another region in the memory. The recording is done only when there are two inputs. Addition of a parameter is analogous to an addition of a dimension to the memory; thus if the number of channels needed for one parameter recording is N, two parameters recording requires N^2 channels. An alternative to a very large memory for this case can be recording some of the data on magnetic tape. A three-dimensional image on the cathode ray tube (CRT) is obtained by shifting the dots so that diagonal lines are obtained. The brightness can also be changed for this purpose.

Two conventional multichannel pulse-height analyzers were adopted by Toyoshima[20] for 4096 × 4096 two-parameter measurement. A block diagram of the modified system is shown in Figure 7.40. The binary data of the two ADCs are transferred to the memory register in ND2200 through the data multiplexer (DMPX) by exploiting the READ IN/OUT mode of the data handling unit (DHU) of the ND2200. Sequentially timed control signals for execution of the transfer process are provided by the data transfer control logic (DTC) in communication with the DHU. This process was initiated by an $\overline{\text{ADCP}}$ signal which was generated by the ADC control (ADCC) logic when the ADCs have analyzed a pair of input pulses. Data acquisitions in core memory and tape recordings were alternated by the mode control logic (MC), which governed the control signal multiplexer (CMPX), the ADCC, and the magnetic tape unit of the ND2200.

When both ADCs are set, $\overline{\text{F}}$ signals from them come down and $\overline{\text{SCB}}$ signals go up, which are received by the ADCC, resulting in the ADCP signal going up. This signal turns into the RDY2 pulse through an OR gate in the DTC, which is sent to the DHU through the CMPX. As the DHU is set for the READ IN/OUT mode, it responds with the CMND pulse. Then the flip-flop (SN74279) of the DTC is set and the oscillator in it is released, pulses of which are counted by two SN7493As. The outputs of the binary counter are connected to several gates in the DTC, and the control signals necessary for data transfer are provided in sequence. That is, the ATR(1) signal (transfer command for ADC-1) goes up and the output data of the ADC-1 appear on the lines to the DMPX (Figure 7.41). By signals C1, $\overline{\text{C2}}$ (select signals for the DMPX) and $\overline{\text{SP}}$, $\overline{\text{SHIFT}}$ (which cause data shift on the memory register), 12 b of data are transferred to the memory register in four steps, each carrying 3 b of them. After the memory register has acquired the data of the ADC-1, the X7 signal takes place, triggering a univibrator in the DTC, and signal $\overline{\text{SMB}}$ comes down and initiates a store cycle in the ND2200. Signal $\overline{\text{SMB}}$ also activates the RDY2 signal line from the DTC, and the data of the ADC-2 are read into the memory in the same way as the ADC-1 data are moved, except for activations of signals ATR(2) and C16P. Signal C16P triggers $\overline{\text{SMB}}$ and $\overline{\text{CLRADC}}$, and the latter clears both ADCs. This process is completed within 120 μsec after the ADCs are set.

Dual-parameter time and energy spectrometry in positron annihilation was discussed by Sen and MacKenzie.[21] The Doppler broadening of the annihilation gamma-rays was measured by a closed coaxial Ge(Li) detector of 9.5% efficiency and recorded in the 4096-channel X-ADC. The timing discriminator was of a conventional type with two 463 coincidence first (C.F.) discriminators, a 437 A time-to-amplitude converter (TAC1) and a 4096-channel Y-ADC. Large scintillators, 5×5 cm^2, of NE104 were used with RCA 8575 tubes to combine efficiency with acceptable time resolution.

Both of the ADCs of the two-parameter analyzer operated in the coincidence mode with the gating signal provided by a coincidence between the Ge(Li) and start detectors. The 473 C.F. discriminator was set in the slow rise-time reject mode, giving almost 100% coincidence efficiency with a resolving time of less than 8 nsec. The shape of the spectrum going to the Y-ADC was basically symmetrical since both C.F. discriminators accepted the same energy spectra. Coincidence gating of the two-parameter analyzer was used to reduce chance coincidences, but on the debit side this resulted in the nonrecording of the large majority of TAC1 signals.

Berube et al.[22] described a multiparameter data acquisition, display, and on-line analysis system. It enabled analyzing multiparameter data during acquisition. Input from eight physical devices, independently or in coincidence, could be acquired and stored in 16 configurations in increment or list mode. During acquisition the data could be analyzed on-line by user-generated microprograms and stored in 16 additional independent configurations. This capability was achieved through the use of a high-speed synchronous bus, distributed processing (seven microcomputers and microprocessors), and overlapping memory cycles.

Figure 7.42 shows the multiparameter acquisition (MPA) hardware modules and their interconnections. The lower part of the diagram is the ND6600 product base; the upper part is the MPA display, acquisition, and processing subsystem.

The MPA hardware consisted of the following: display processor, ADC interface, data buffer, list processor, acquisition interface, and function control box. This system used the powerful high-speed bipolar microprocessor board (MIP). For acquisition and on-line analysis the MAI and LIP boards were attached to the MIPs.

A fast, versatile single dual-parameter multichannel analyzer was described by Imperiale.[23] The analyzer has a count capacity of $2^{32} - 1$ counts/channel, a minimum dead time of 200 nsec, and a maximum conversion gain of 2^{20} channels; the acquired data are transferred from the instrumentation to the computer under interrupt control; and several interrupt possibilities are foreseen for the sake of flexibility.

The MCA characteristics are as follows:

1. The instrument must be fully programmable.
2. The spectrum must be updated in real time, dead time ranging from 200 nsec to 2 μsec. This updating phase must be independent of either the electrical pulse amplitude from the detector or the adopted record system.
3. One- and two-parameter measurements must be allowed, the one-parameter measurement resolution ranging from 8 to 18 b, and the two-parameter measurement maximum conversion gain being 2^{20} channels.
4. The instrument must be allowed to interact on-line with different computing systems, the choice issuing from the investment and from the real necessities of the utilizer. Real-time graphic display must not interrupt either the outstanding spectrum acquisition or the outstanding spectrum probable processing.

The instrumentation design starts from the following considerations.[23]

The key to high performance in the multiprocessing system is allowing all of the processors to run concurrently in their own private environments. For this to occur, each functional module must contain its own CPU, memory, and I/O resources.

Even though the CPU in the acquisition module directly controlled the spectrum acquisition and updating phase, it would be slower than dedicated electronics. If the electronics were autonomous in this phase and moreover at a relatively low cost, they would resolve either the problem of the acquisition speed or that of the whole system efficiency.

Consider the acquisition module (AM) as a module furnished with different units: (1) a work memory which stores the spectrum updated in real time, (2) a first-in-first-out (FIFO) memory which can exchange spectrum data (channel address) with the external world in the course of the spectrum acquisition, (3) a control unit which superintends the various acquisition phases, and (4) an interface unit which allows the acquisition module to exchange information with the computer.

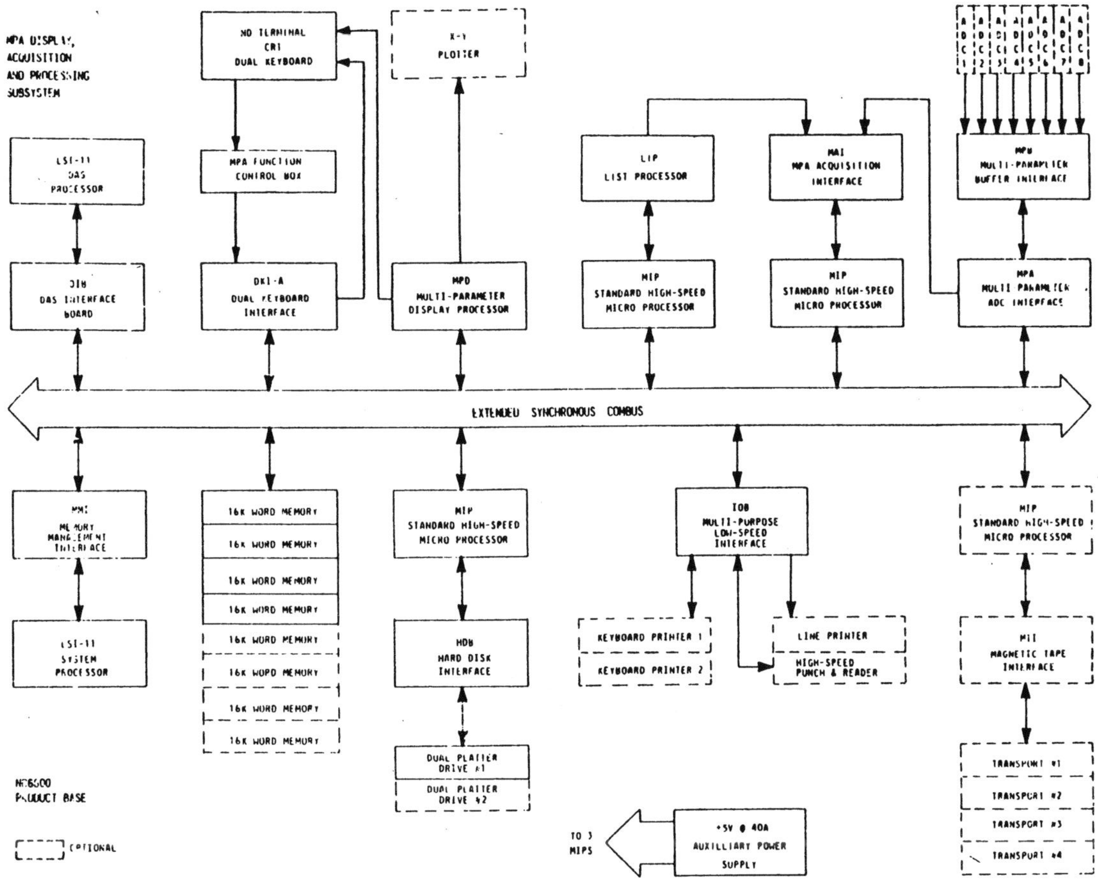

Figure 7.42 ND6660 multiparameter system (MPA) configuration. (From Berube, R. et al., *IEEE Trans. Nucl. Sci.*, NS26, 750, 1979. With permission.)

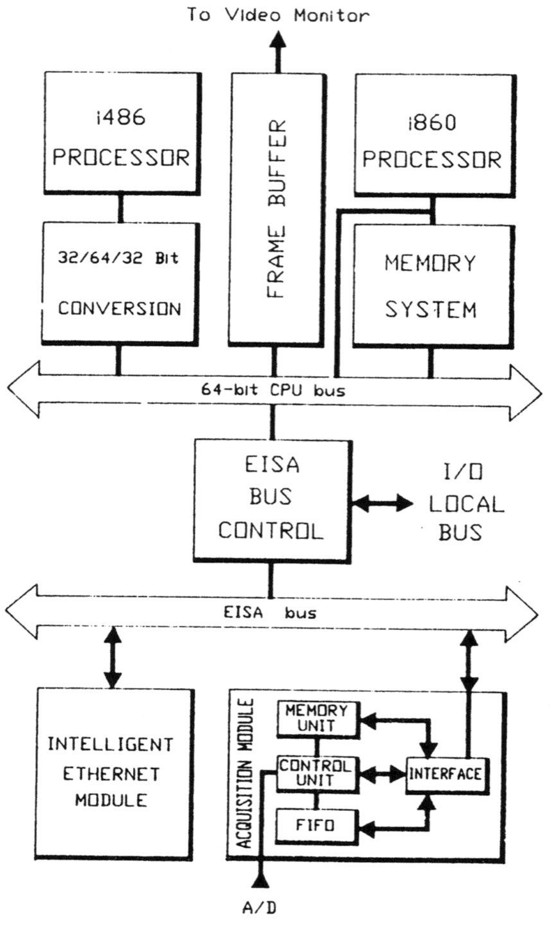

Figure 7.43 Acquisition module interfaced with a personal computer constructed around a Hauppauge 4860 motherboard. (From Imperiale, C., *IEEE Trans. Nucl. Sci.*, 39, 283, 1992. With permission.)

The AM is interfaced with both a powerful personal computer system constructed around a 33-MHz Hauppauge 4860 motherboard (Figure 7.43) and a Sun Microsystems sparkstation 2GT. The first system is an i486/i860 dual-processor system incorporating the fast 32-b EISA I/O bus (33 Mbyte/s). Coupled with its 12 kbyte on-chip cache memory, the i860 RISC processor can execute up to 25 million instructions per second, and is used to enhance both three-dimensional imaging and scientific calculations.

Both the AM and an EISA intelligent Ethernet controller are connected to the system by two EISA connectors. The Ethernet controller allows the user to store and process experimental data locally and to realize communication with the host computer either to analyze the experimental data off-line or to store them. The resulting system is fast and compact, and moreover has a contained cost. It complies with the specified aims when performing the acquisition and processing of spectra only; otherwise (gated clinical studies, digital filtering, etc.) it is necessary to increase the degree of parallelism.

It is convenient to manage the autonomously acquired measurement points, grouping them in measurement blocks, especially for long measurement times. The block dimension depends on either the parameter number or the precision required in the measurement of each parameter, i.e., the ADC bit number. The block number depends on the used work memory.

By choosing four 256K memory modules, the two- and one-parameter measurements of practical interest can be executed. The two most significant bits in the address word select the memory chip; the other 18 b select the word in the work memory.

The ADC conversion speed must increase while decreasing the instrumentation dead time; then flash-type ADCs and correlated techniques must be used. Analogously, a fast pseudorandom transformation compatible with the short dead time of the instrumentation must be realized while using the associative techniques.

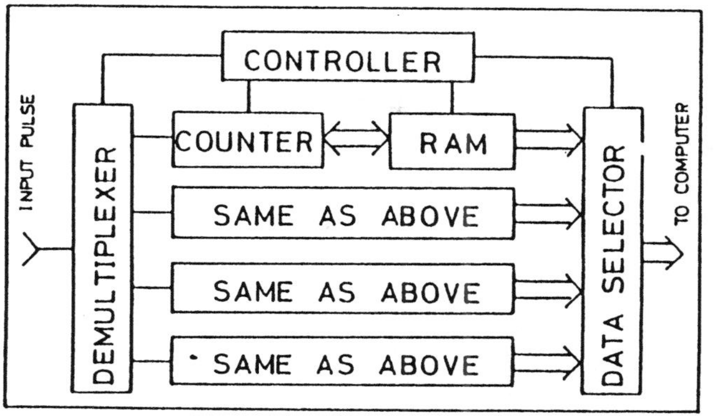

Figure 7.44 Four parallel groups of counter-RAM assemblies. From Okayasu, T. et al., *IEEE Trans. Nucl. Sci.,* 34, 115, 1987. With permission.)

The complete data acquisition process passes through a fast autonomous acquisition phase (FAPH) and a computer-controlled phase (CPH). The CPH initializes the AM by menu, and transfers the updated spectrum data to the host, either at the end of each block or at the end of the measurement; the host processes stores, etc. these data. During the FAPH, the spectrum is autonomously acquired in real time at the maximum frequency compatible with the programmed dead time; a prefixed number of acquired data may be sent to the system graphic section under interrupt control. Then a block can be read (1) at the end of the measurement, (2) at the end of the acquisition of each block and before the beginning of the next one, or (3) in the course of the block acquisition.[23]

7.5 MULTISCALER

Multiscaling is measurement of the time distribution of the pulses. The use of a multichannel analyzer as a multiscaler is done by using the content register as a scaler for given time periods. At the end of each period the accumulated number of counts is transferred to one channel and the register is cleared. In the next period the register counts and the number of counts is transferred to the next channel and so on. This way a decay curve or any other time-dependent curve is obtained.

A fast multichannel scaler achieving the minimum dwell time of 50 nsec was described by Okayasu et al.[24] The dead time due to memory cycles was eliminated by four-phase operation of four parallel groups of counter RAMs. The MCS had 4000 channels in total. Differential nonlinearity was less than 0.4%. In design of the RAM section, they have preferred fast NMOS RAMS, μPD2149D-2 (cycle time = 35 nsec, 4 b × 1024 address) instead of ECL RAMs from the viewpoint of the number of chips required, power consumption, and noise immunity.

In the repetitive mode of operation, i.e., averager mode, the signal processing per channel consists of (1) loading the accumulated data from the RAM into the counter, (2) counting the input pulses starting from the accumulated data, and (3) storing the new data into the RAM. The total time required for one cycle is 35 nsec (READ) + (dwell time) + 35 nsec (WRITE) + (delay of TTLs). This time is approximately from 150 to 200 nsec when the dwell time is 50 nsec. In order to eliminate the dead time due to these cycles, they prepared four parallel groups of counter-RAM assemblies as is shown in Figure 7.44. Input pulses must be distributed properly to be counted in one of four groups of Counter-RAM assemblies, according to the arrival time of the pulse. An edge-triggered demultiplexer was designed to avoid simultaneous counting of one pulse in the adjacent two counters. In the circuit of Figure 7.45(a), only one selector is active at a time and other three selectors are disabled. At every edge of the dwell time clock, position of an active selector is shifted sequentially by the SELECT CONTROL levels (a HIGH level SELECT CONTROL enables a selector LOW level SELECT CONTROLs disable three selectors). Figure 7.45(b) shows the circuit diagram of a selector. After the SELECT-CONTROL signal to a D-FF is made HIGH, the positive edge of a first-input pulse (indicated as P1 in Figure 7.45[c]) raises the Q output of the D-FF, which opens the AND gate to pass the pulse to the counter through a one-shot pulse shaper. The use of an edge-triggered D-FF prevents the splitting of one pulse into two channels. See Figure 7.45(c). Pulse P3 may be split into pulse p3 and pulse p3′ if the edge triggering is not used. Dead time exists around the channel boundary (see Figure 7.45[d]) due to the recovery time from the CLEAR removal or due to the insufficient pulse width of the AND output to trigger the one shot shaper (tw in Figure 7.45(d) is the minimum pulse width required for triggering).

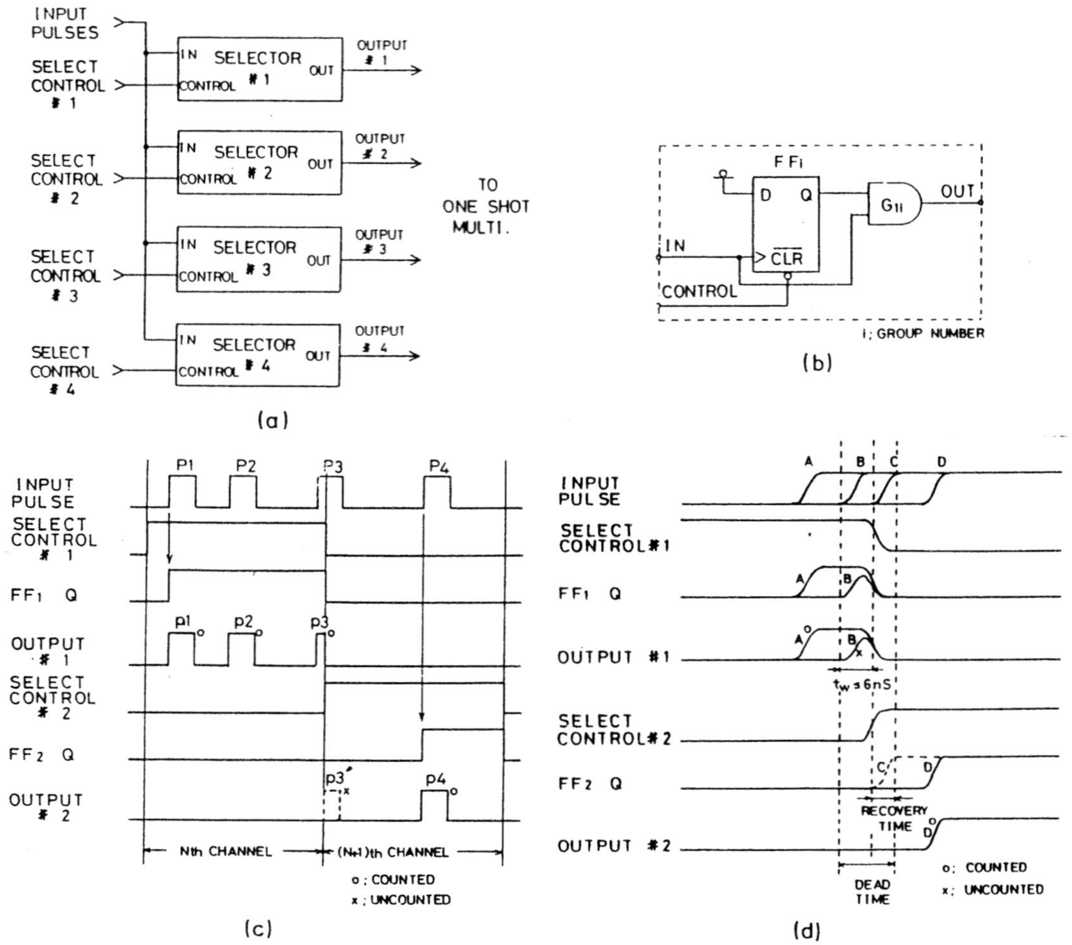

Figure 7.45 Edge-triggered demultiplexer. (a) Block diagram of the demultiplexer; (b) circuit diagram of the SELECTOR unit; (c) waveforms in the demultiplexer; (d) enlarged presentation of the waveform to illustrate the dead time. Input pulse width = 20 nsec. (From Okayasu, T. et al., *IEEE Trans. Nucl. Sci.,* 34, 115, 1987. With permission.)

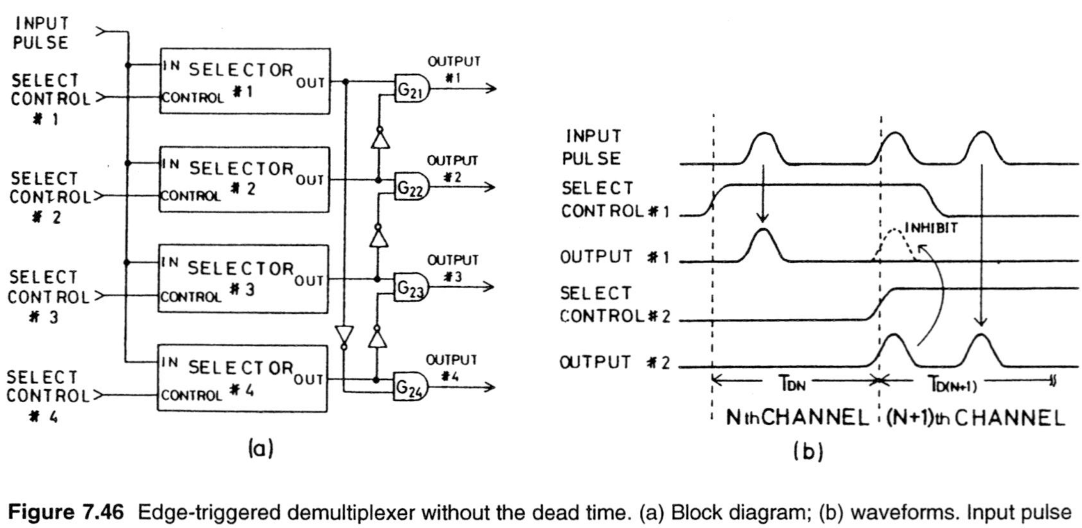

Figure 7.46 Edge-triggered demultiplexer without the dead time. (a) Block diagram; (b) waveforms. Input pulse width = 20 nsec. (From Okayasu, T. et al., *IEEE Trans. Nucl. Sci.,* 34, 115, 1987. With permission.)

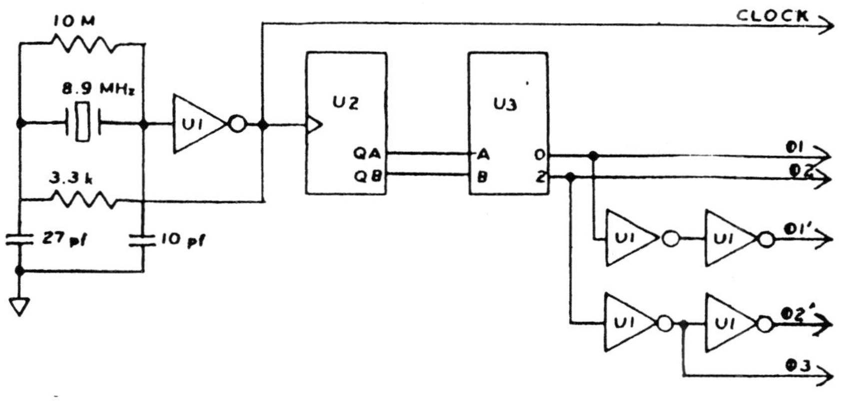

Figure 7.47 Clock generator circuitry for the multichannel scaler. (From Murphy, D. M., *Nucl. Instrum. & Methods,* A236, 349, 1985. With permission.)

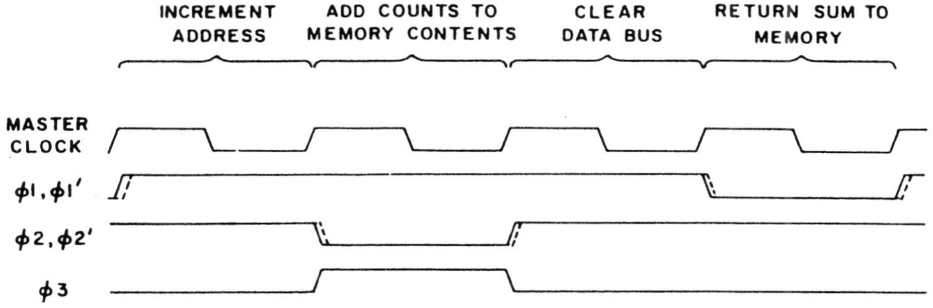

Figure 7.48 Timing of clock lines used in the multichannel scaler. (From Murphy, D. M., *Nucl. Instrum. & Methods,* A236, 349, 1985. With permission.)

Further improvement of the circuit was made to achieve zero dead time. The control signals for time quantizing are duplicated at the channel boundary to ensure the pulse capture (see the waveforms of SELECT-CONTROL 1 and 2 in Figure 7.46[b]). If an input pulse arriving around the channel boundary is captured by both selectors, the pulse from the preceding selector is inhibited at an AND gate G2i (i = 1, 2, 3, or 4) and only the pulse from the succeeding selector is sent to the counter. If a pulse arrives slightly before or after the channel boundary, it is captured by one selector.[24]

A multichannel scaler (MCS) has been constructed by Murphy[25] which is capable of running as fast as 250 nsec per channel. It has four major sections: a lock generator, addition logic and memory, bookkeeping logic to keep track of the number of scans completed and so forth, and readout memory.

Figure 7.47 shows the clock generator. A crystal oscillator provides the clocking for the scaler. Since a four-phase clock controls the addition logic, the dwell time is four times the clock period. Other dwell times would require different crystal, resistor, and capacitor values from those shown in Figure 7.48. The various clock phases generated are shown in Figure 7.48; $\phi 1'$ and $\phi 2'$ are slightly delayed versions of $\phi 1$ and $\phi 2$, respectively.

The heart of the MCS is the addition logic which adds incoming counts to those already accumulated in a given channel. An abbreviated schematic of this section is shown in Figure 7.49. It consists of a counter (U7–9), memory (U14–15), arithmetic logic units (ALUs) with look ahead carry (U18–22, and a latch (U23–24). The operation may be understood by considering the four clock phases in succession. The address is incremented by a rising edge on $\phi 1$ advancing the counter at the start of the first clock phase. The memory, ALU, and latch are disabled or ignored during the first phase. The address has settled by the beginning of the second phase, when the memory output is enabled. During this phase, the ALU adds the number of counts present at its input to the memory contents. The count input must therefore be stable during this phase. The origin of the count input depends on the application. In a typical use, the input would be provided by a high-speed counter followed by a latch, with both reset by $\phi 2$. A

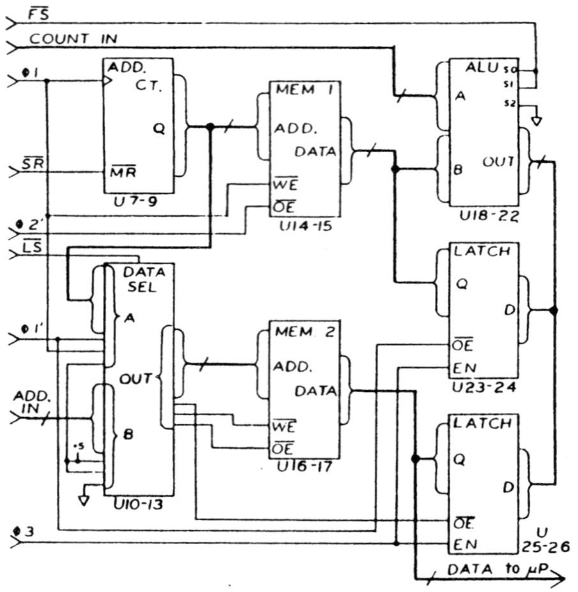

Figure 7.49 Addition and memory circuitry to accumulate counts in successive channels; 01 through 03 are generated by the circuit shown in Figure 7.48. FS. LS, and SR refer to first scan, last scan, and scan reset, respectively. They are generated by the circuit shown in Figure 7.50. (From Murphy, D. M., *Nucl. Instrum. & Methods,* A236, 349, 1985. With permission.)

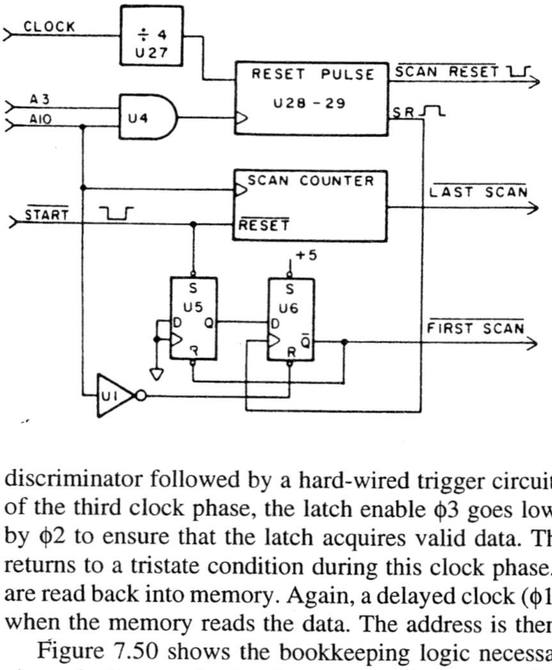

Figure 7.50 Bookkeeping circuitry to keep track of scans completed. A_3 and A_{10} are address lines from the output of U7-9 in Figure 7.49. (From Murphy, D. M., *Nucl. Instrum. & Methods,* A236, 349, 1985. With permission.)

discriminator followed by a hard-wired trigger circuit was used to generate the count input. At the start of the third clock phase, the latch enable $\phi 3$ goes low to latch the sum. The memory output is disabled by $\phi 2$ to ensure that the latch acquires valid data. The data bus between the memory, ALU, and latch returns to a tristate condition during this clock phase. During the fourth clock phase, the latch contents are read back into memory. Again, a delayed clock ($\phi 1'$) is used to ensure that the latch output is still valid when the memory reads the data. The address is then incremented to start the next channel.

Figure 7.50 shows the bookkeeping logic necessary. After the MCS has scanned through all of its channels, the counter for the number of scans completed is incremented. The channel address counter is reset to zero by SR, and the flip-flops indicating the first or last scan of a measurement are changed if appropriate. A dead time the length of eight channels is used to perform these functions. Allowing this much time allows the use of low power CMOS logic for these bookkeeping functions. Pull-up resistors not shown in the figures are used between CMOS and TTL logic.[25]

7.6 MULTIPLEXER

When data of a large number of detectors have to be analyzed, a multiplexer is used. Two general methods are available to measure the charges from a large number (N) of counters. Method A uses N sample-and-hold channels, each equipped with its own ADC; method B uses N sample-and-hold channels sharing a single ADC. The advantages for method A include:[26] conversion times for the ADCs are less critical, hence cheaper ADCs can be used; since all the analog-to-digital conversion can be done in parallel, demand on the hold time stability of the sample and holds is much less stringent; and no analog multiplexing is necessary.

The disadvantages for this method are that it is relatively costly due to ADC costs and requires digital multiplexing in the computer readout.

The advantages for method B are that it saves (N-1) ADCs (and in the case when N is large, this saving is significant); and because only a single ADC is involved, one can afford to use a high-quality ADC. The disadvantages are that it requires analog multiplexing and has a much more stringent requirement on hold-time stability.

An eight-channel sample-and-hold unit, serially addressable with multiplexed output, was described by Chang and Larsen.[26] The unit accepted clipped inputs of 5 nsec (nominal) FWHM; the positive overshoot of the clipped pulse was rejected. The outputs were held to ±1 mV for ≥1 msec, and were linear to ±1% from 30 mV to 1 V. The output sensitivity is 10 mV/pC of input charge.

Because of lower cost, the second method — that of using a single ADC for N sample-and-hold channels — was chosen by Chang and Larsen. Each sample-and-hold channel accepted clipped inputs of 5 nsec (nominal) FWHM. The unit was specifically designed to reject the positive portion of the clipped pulse. The input signal was sampled and held to ±1 mV over a time interval of ≥1 msec. During this hold time, the analog outputs were multiplexed through an eight-channel FET multiplexer which was serially addressable through an 8-bit shift register. The system block diagram is shown in Figure 7.51.

A simplified circuit schematic is shown in Figure 7.52. Fast clipped signals about 3 to 15 nsec wide and up to 1-V peak amplitude passed through a buffer with approximately ×1.2 gain. After the buffer was a fast diode gate, the driver circuit of which was transformer coupled and could accept gates up to 50 nsec wide. This was followed by another stage with a gain of about 7. After this there was a current source driving the sampling capacitor C1 through diode D_1. The charge was isolated by D_1 so that the waveform at the input of A_1 was a pulse with a decay time constant of about 18 μsec.

The area of the stretched pulse represents the integral of the input signal. This signal passed through an amplifier A_1, the output of which was coupled and restored; and an integrator A_2, which was restored by a separate clamp. This combination of clamped capacitors avoids the need for DC coupling of the stretched signal.

A_1 had a voltage gain of 2 and lowered the drive impedance to effectively provide a high current (or charge) gain at the input of A_2. The coupling capacitor, C_2, was restored to ground by clamp S1 after each operation. The second integrator, A_2, provided an output pulse which was flat to ±1 mV for ≥1 msec. The FET clamp, Q_1, restored the integrating capacitor, C_3, when a measurement was not in progress. This held the input of A_2 essentially to 0 V. The hold time was determined by the timing resistors and capacitors of the one shots.

Groom and Marshall[27] have used a sampling digitizer employing the sample-and-hold technique in photomultiplier experiments. This method has been found to introduce dispersions in the result due to the nonideal stretcher. As the number of detectors increase, the stretcher droop becomes an increasing source of error. However, only one ADC is required independent of the number of detectors.

A mixer-route technique of summing the outputs of shaping amplifiers and then routing the data into a multichannel analyzer was described by Bayer and Borsuk.[28] The major disadvantage of this method is that the output includes the quadrature sum of the noise from each input which can degrade the total resolution of a three-detector experiment by as much as 35%. Additionally, the mixer router must reject any coincident signals by gating the storage device off or by using a linear gate in the multiplexed signal path, resulting in a low-count rate capability. This method requires only one ADC to digitize the output and is an inexpensive and simplistic means to multiplex spectroscopy data when high resolution is not mandatory.[29] Systems that use linear gating in each input signal path offer greatly improved energy resolution, compared to mixer routers, while effectively adding the detector efficiencies.[29]

A nuclear spectroscopy multiplexer was described by Renner et al.[29] that can multiplex up to 14 linear signals within the range of ±10 V. The multiplexer gating was provided by inputs on the rear panel which had a common adjustable threshold. When the multiplexer was used with a multidetector array, the resulting system resolution was the efficiency-weighted average of the resolution of the separate detectors. The apparent system efficiency was the sum of the separate detector efficiencies. Only one channel was open at any time and there were no priority channels. Isolation from any off channel was 65 dB, and the noise contribution of all nonselected channels was essentially negligible.

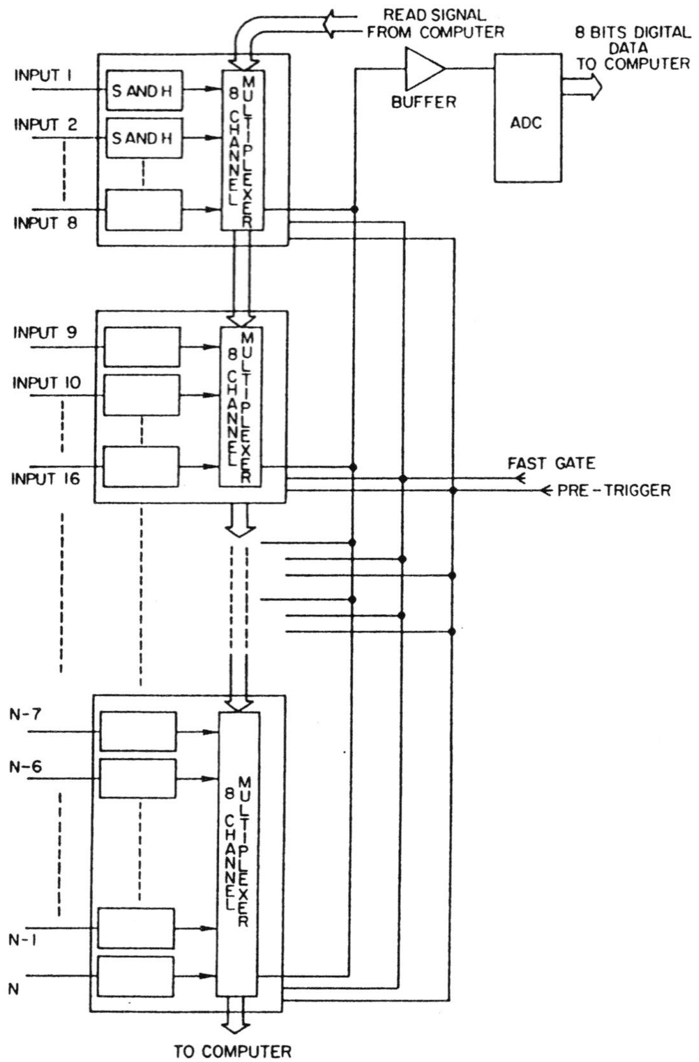

Figure 7.51 System block diagram. (From Chang, A. K. and Larsen, R. S., *IEEE Trans. Nucl. Sci.,* NS20(1), 216, 1973. With permission.)

The 14-channel multiplexer offered a simple means of effectively adding the detector efficiencies in a multidetector array while obtaining an energy resolution approximately equal to the average of the energy resolution of the detectors.

A system block diagram is shown in Figure 7.53. The multiplexer contained 14 discriminators, one for each channel, which monitored the bipolar output of shaping amplifiers or other gating inputs, thereby determining which linear channel has useful information. A front panel potentiometer labeled DISC was used to vary the threshold voltage which was common to all discriminators. The discriminator threshold voltage range can be set depending on the noise input level to the multiplexer which was a function of the detector-preamplifier-amplifier combination. The discriminator threshold adjustment and busy indicator light-emitting diode (LED) offered a unique and simple means to adjust the noise threshold. With only background radiation striking the detectors, the discriminator threshold could be lowered until the busy indicator began to glow. At this point, the threshold was set at the optimum value.

The discriminator was buffered through low-power TTL inverters to two 8-line to 3-line decoders (shown as decoder no. 1). As a bipolar input signal exceeded the discriminator threshold setting, the respective line on the decoder went low and the address for the channel appeared at the decoder output.

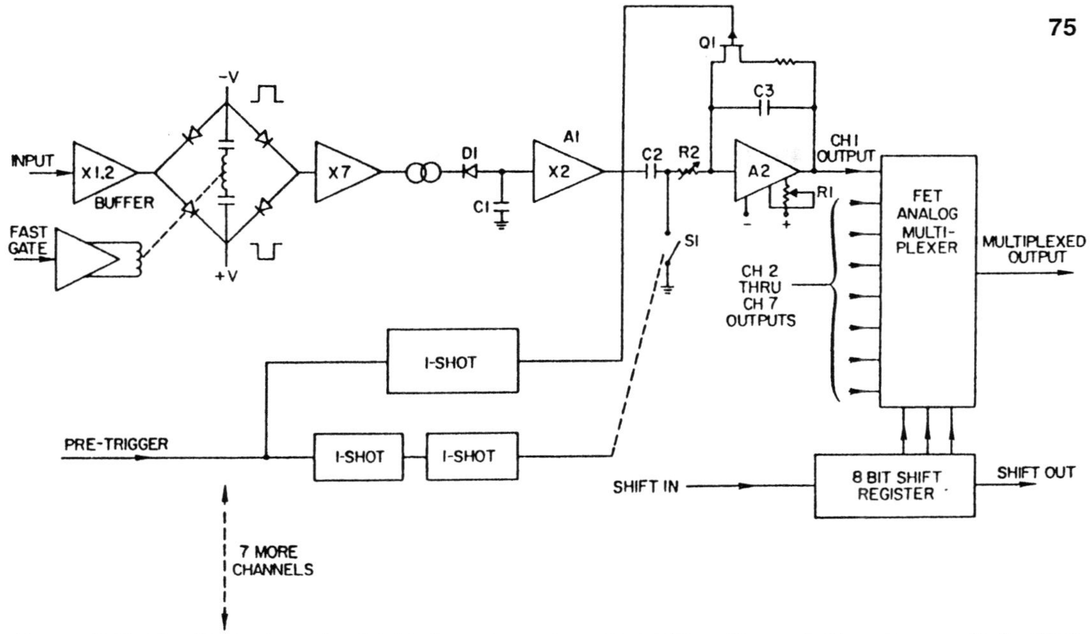

Figure 7.52 Simplified schematic of eight-channel sample and hold. (From Chang, A. K. and Larsen, R. S., *IEEE Trans. Nucl. Sci.,* NS20(1), 216, 1973. With permission.)

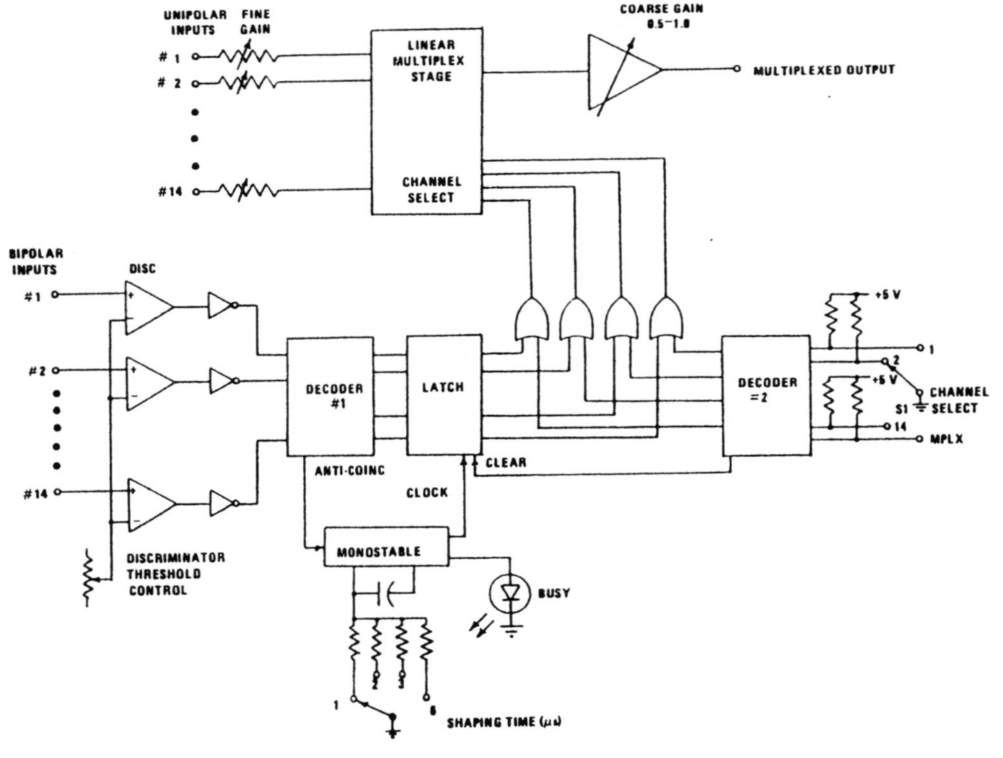

Figure 7.53 Block diagram of the 14-channel multiplexer. (From Renner, K. W. et al., *IEEE Trans. Nucl. Sci.,* NS25, 62, 1978. With permission.)

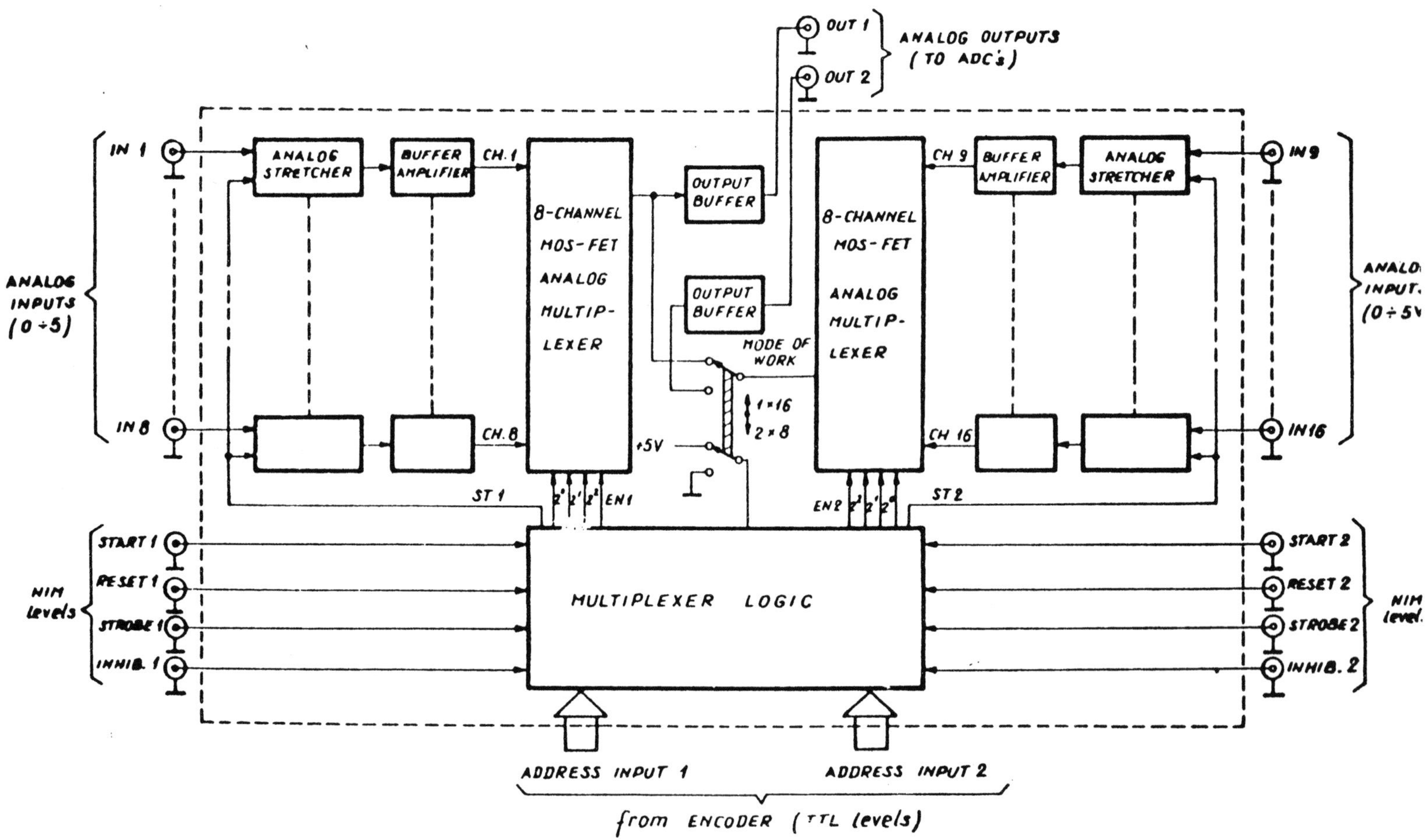

Figure 7.54 Block diagram of the analog store and multiplexer unit. (From Brossard, M. and Kulka, Z., *Nucl. Instrum. & Methods,* 160, 357, 1979. With permission.)

Simultaneously, the anticoincidence signal from the decoder triggered the monostable multivibrator which latched the decoded address into a clocked D-type flip-flop.

The width of the monostable pulse was set to approximately seven times the shaping time constant (τ from 1 to 6 μsec, selected by a jumper on the printed circuit board). The anticoincidence signal from the decoders changed states only after the inputs to all discriminators were below the threshold, and then one exceeds the threshold. This ensured that the multiplexer did not switch to a channel which might be past the peaking time but still above the discriminator threshold level.

A 16-channel analog store and multiplexer unit (ASMU) was described by Brossard and Kulka.[30] The unit enabled storing and selection of analog information which was then digitally encoded by a single ADC. A block diagram of the ASMU is shown in Figure 7.54. It had 16 analog channels, each of them consisting of a pulse stretcher with storage capacitor as well as a buffer amplifier with high-input impedance. The outputs of the analog channels were connected to two 8-input MOS-FET-integrated multiplexers.

Input analog pulses were fed directly to each stretcher from a corresponding amplifier and detector. In order to eliminate the possibility of storing undesirable analog signals, all FET clamps in pulse stretchers were ON in the quiescent state. The FET clamps must be switched OFF before analog pulses arrive. Each analog channel accepted positive pulses of amplitudes up to 5 V and duration of 200 nsec. The input resistance was 50 Ω and DC coupling has been used. The multiplexed output pulses had amplitudes up to about 5 V with a pedestal near 30 mV and variable duration of 5 μsec.

REFERENCES

1. **Brown, A. T. et al.,** *IEEE Trans. Nucl. Sci.,* NS33(6), 1519, 1986.
2. **Weatherford, T. R. et al.,** *IEEE Trans. Nucl. Sci.,* NS33(6), 1590, 1986.
3. **Kinbara, S.,** *Nucl. Instrum. & Methods,* 143, 267, 1977.
4. **Byrd, I. S. and Goosey, M. H.,** *Nucl. Instrum. & Methods,* 158, 565, 1979.
5. **Frago, F. A. F.,** *IEEE Trans. Nucl. Sci.,* NS29, 129, 1982.
6. **Santos, D. M. et al.,** *Nucl. Instrum. & Methods,* 196, 435, 1982.
7. **Ruhter, W. D. et al.,** *IEEE Trans. Nucl. Sci.,* NS24, 440, 1977.
8. **Kieser, W. E. et al.,** *Nucl. Instrum. & Methods,* 164, 587, 1979.
9. **Mildebrath, M. and Cipolla, S.,** *Nucl. Instrum. & Methods,* 180, 615, 1981.
10. **Kumahara, T. et al.,** *IEEE Trans. Nucl. Sci.,* NS26, 743, 1979.
11. **Tenten, W. et al.,** *IEEE Trans. Nucl. Sci.,* NS22, 529, 1975.
12. **Murray, G. L. and Macefield, B. E. F.,** *Nucl. Instrum. & Methods,* 51, 229, 1967.
13. **Hendricks, R. W. et al.,** *Nucl. Instrum. & Methods,* 201, 261, 1982.
14. **Machen, D. R.,** LASL-7683-MS, 1979.
15. **Berry, A. et al.,** *Nucl. Instrum. & Methods,* 201, 241, 1982.
16. **Berry, A. et al.,** *Nucl. Instrum. & Methods,* 201, 241, 1982.
17. **Turko, B. T. et al.,** *IEEE Trans. Nucl. Sci.,* 39, 758, 1992.
18. **Norell, K. E. and Baltzer, P.,** *Nucl. Instrum. & Methods,* 223, 122, 1984.
19. **Arnone, G. J. and Brunson, G. S.,** *IEEE Trans. Nucl. Sci.,* 37, 403, 1990.
20. **Toyoshima, K.,** *Nucl. Instrum. & Methods,* 138, 557, 1976.
21. **Sen, P. and MacKenzie, I. K.,** *Nucl. Instrum. & Methods,* 141, 293, 1977.
22. **Berube, R. et al.,** *IEEE Trans. Nucl. Sci.,* NS26, 750, 1979.
23. **Imperiale, C.,** *IEEE Trans. Nucl. Sci.,* 39, 283, 1992.
24. **Okayasu, T. et al.,** *IEEE Trans. Nucl. Sci.,* 34, 115, 1987.
25. **Murphy, D. M.,** *Nucl. Instrum. & Methods,* A236, 349, 1985.
26. **Chang, A. K. and Larsen, R. S.,** *IEEE Trans. Nucl. Sci.,* NS20(1), 216, 1973.
27. **Groom, D. E. and Marshall, J. H.,** *Rev. Sci. Instrum.,* 33, 1249, 1962.
28. **Bayer, R. and Borsuk, S.,** *Nucl. Instrum. & Methods,* 133, 185, 1976.
29. **Renner, K. W. et al.,** *IEEE Trans. Nucl. Sci.,* NS25, 62, 1978.
30. **Brossard, M. and Kulka, Z.,** *Nucl. Instrum. & Methods,* 160, 357, 1979.
31. **Brusker, G. J. et al.,** *IEEE Trans. Nucl. Sci.,* NS33, 1484, 1986.

Chapter 8

Constant-Fraction Timing Discriminators

Constant-fraction timing is a method used to overcome the problem of timing dependence on pulse height. Due to variation in pulse height the time of crossing a certain voltage level varies widely. To overcome this problem a fixed small portion of the pulse is sampled and timing is determined according to this portion. The timing variation due to pulse-height variation (walk) is reduced considerably.

Several methods are used to reduce the walk in timing circuits. The best one is the constant-fraction method. The detector pulse is delayed (Figure 8.1a) and a part of the pulse before the delay is inversed and attenuated (Figure 8.1b). The sum of the two gives a pulse of shape (Figure 8.1c) which crosses zero and provides an output pulse (Figure 8.1d). This method is independent of pulse height and thus free of walk.

Another method[1] uses a tunnel diode (TD) to provide a trigger. The bias depends on the input pulse so that always a trigger is obtained at a constant fraction of the pulse. In this method an integration is carried out instead by inversion and attenuation. The TD is triggered independently of pulse rise time and thus free of walk.

It is important to find the pulse fraction that will give the smallest error. In the case of a scintillator, the rise time to the trigger level is about 1/4 nsec. The load resistor (50 Ω), anode, and pick-off circuit capacitor also have a time constant of ~1/4 nsec. It was found that the pulse fraction that will give minimum and time fluctuation is about 10%,[2] for the NE213 scintillator and an XP1040 PM tube. For NaI(T1) and 56 AVP it is less than 1%.[3]

In solid-state detector use, the timing depends on the detector because of charge collection time which is dependent on the initial charge distribution. A surface barrier (SB) is the fastest solid-state detector. The collection time in an SB detector is ~1 nsec. In a Ge(Li) detector the collection time is up to 100 nsec; the time uncertainty can be FWHM = 30 nsec.

Great improvements of timing have been achieved, both in leading-edge and fast zero-crossover methods. The leading-edge method, however, suffered from walk in a wide dynamic range; and consequently the zero-crossover method, which has the principle advantage of less time walk, was preferred.

A modified fast zero crossing, the so-called constant fraction of pulse-height trigger, was found to be comparable or even better than the leading-edge timing when a wide dynamic range was employed. This constant-fraction pulse-height trigger method is essentially a fast zero crossing from the triggering point of view. The constant-fraction discrimination method is demonstrated in Figure 8.2.

It consists of delaying the original pulse by a time interval t_d, then amplifying it by a gain K, and inverting it. Finally the inverted pulse is summed with the original input pulse. The resulting waveform is applied to a zero detector, to mark the crossover time, t_o. From Figure 8.2:

$$t_o = \frac{K}{K-1} t_d \tag{8.1}$$

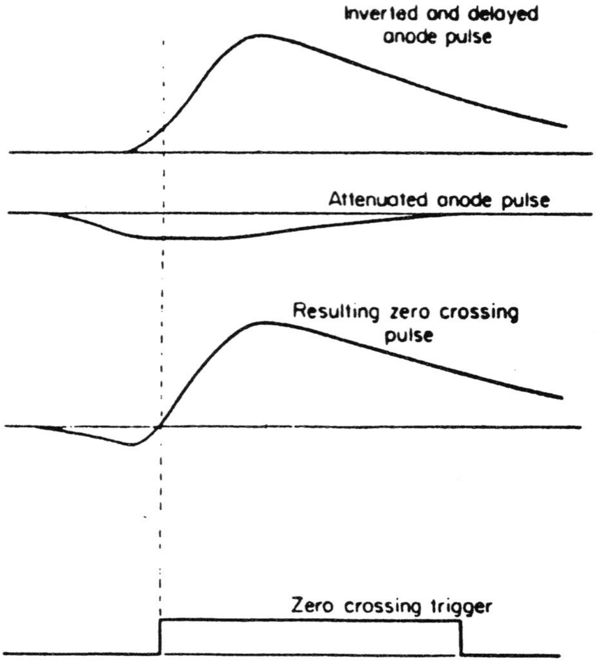

Figure 8.1 Constant fraction of height pick off.

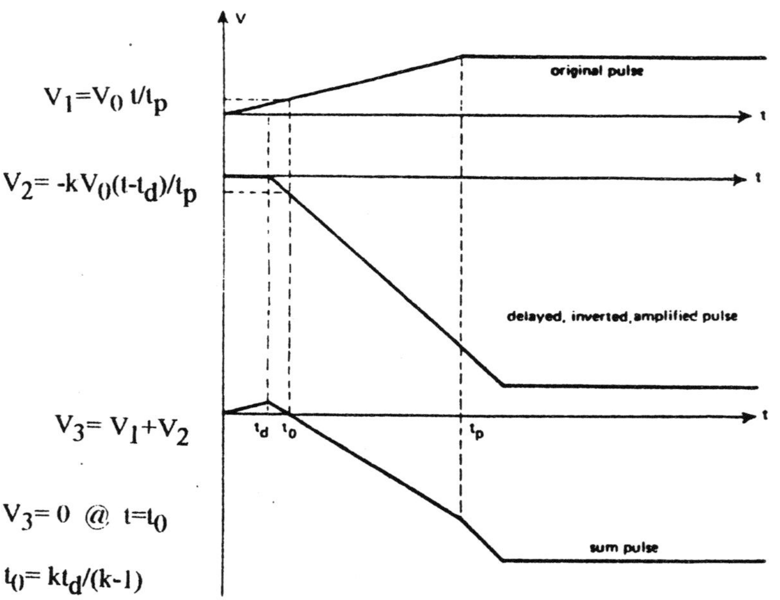

Figure 8.2 Constant-fraction discrimination.

The independence of the amplitude V_0 of the original pulse and of its rise time t_0 is thus provided, the walk being eliminated. This is true when:

$$t_d < \frac{K-1}{K} t_p \tag{8.2}$$

For this purpose the delay time should be small and the gain large.

In a design of a constant-fraction pulse-height trigger Gedcke and McDonald[4] used the attenuation-subtraction technique to produce a bipolar pulse with a zero-crossing phase point corresponding to the

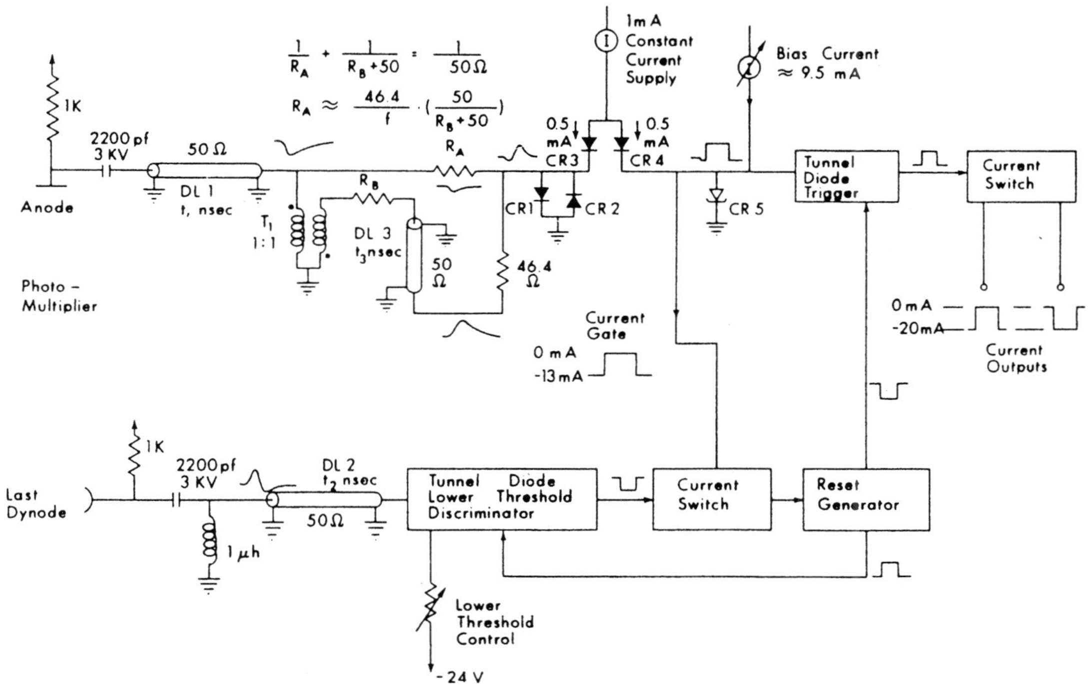

Figure 8.3 A block diagram of the constant fraction of pulse-height trigger. (From Gedeke, D. A. and McDonald, W. J., *Nucl. Instrum. & Methods,* 58, 253, 1968. With permission.)

desired fraction-trigger level on the leading edge of the anode current pulse. The principle is illustrated in Figure 8.1. If a 20% fractional-triggering level is required, the prompt anode current pulse is attenuated to 20% of its initial amplitude, and it is delayed by a time interval t_3 and inverted. The delay time t_3 is such that the 20% phase point on the delayed inverted pulse lines up with the maximum amplitude of the attenuated pulse. The two phases are added to produce a bipolar pulse. The attenuated pulse cancels the delayed-inverter pulse at the 20% phase point. This zero-crossing pulse is presented to a true zero-crossing discriminator which will trigger at a time defining the 20%, fractional-triggering level on the original anode current pulse.

A block diagram of the constant fraction of pulse-height trigger is shown in Figure 8.3. The input anode pulse is negative. R_A feeds the attenuated version of the pulse to the adding point CR_1, CR_2. The transformer T_1 inverts the pulse and the DL_3 delay line causes the delay (t_3). The desired zero-crossing fraction f is determined by R_A and R_B. (R_B is chosen to match the impedance at DL_1 to 50 Ω). $1/R_A + \{1/(R_B + 50)\} = 1/50$; $R_A \simeq (46.4/f)\ \{50/(R_B + 50)\}\Omega$.

The remainder of the circuit is a true zero-crossing discriminator. CR_3 and CR_4 provide a fast current limiter stage. It limits the pulse applied to tunnel diode CR_5 to 0.5 mA, and it also decouples CR_5 from the input. The tunnel diode CR_5 is the zero-crossing detecting element.

The current output of the last dynode is clipped with 1 μH inductance to limit the length of the positive phase of the pulse to about 25 nsec. This signal passes through DL_2 and is used to trigger the lower threshold discriminator. The output of the lower level discriminator triggers a current switch which removes the gating current from CR_5 and activates the reset generator.

A fast timing circuit suitable for zero crossing and a constant-fraction pulse-height trigger constructed with matched pairs of emitter-coupled integrator circuits and a tunnel diode were described by Cho and Helgesson,[5] the former for the negative leading pulse recognition and the latter for the true zero-crossing detection. A current pedestal was generated internally to assure true zero-crossing detection without external bias signals. Zero-level biasing compensation was found to be necessary for slow zero-crossing detection.

The principle of the circuit is shown in Figure 8.4. In its initial state, the tunnel diode is at a low-voltage state with the current somewhere below a half of I_p. When a negative leading part of a bipolar pulse arrives and triggers the current switch, the tunnel diode is set to its peak current (voltage) by a supply of current

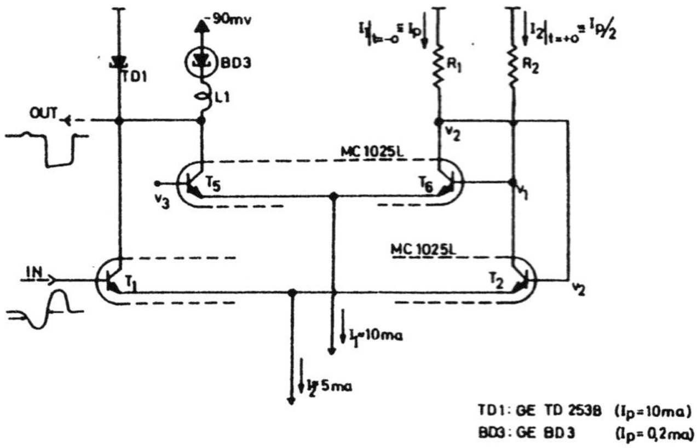

Figure 8.4 A simplified circuit diagram of the zero-crossing detector. (From Cho, Z. H. and Helgesson, U., *IEEE Trans. Nucl. Sci.*, NS18(1), 108, 1971. With permission.)

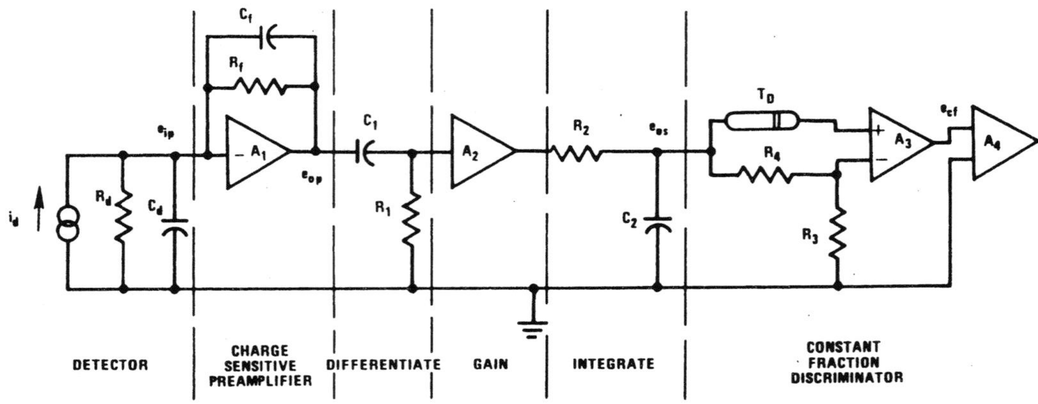

Figure 8.5 Simplified circuit model of the timing system. (From Bedwell, M. O. and Paulus, T. J., *IEEE Trans. Nucl. Sci.*, NS23, 234, 1976. With permission.)

I_p. When this state is reached, a small positive current will switch the tunnel diode to its high-voltage state. In this way, a fast forward switching of the tunnel diode is achieved at the moment of positive zero crossing. This operation is achieved in practice by using two emitter-coupled pairs as is shown in Figure 8.4.

Assume that the transistors T_1 and T_2 are initially so biased that T_1 is on while T_2 is off. Since the currents I_1 and I_2 are adjusted in such a way that I_1 is close to the tunnel diode peak current I_p ($\simeq$10 mA) while I_2 is $I_p/2$, then the voltage developed across R_2 is smaller (or more positive) than the fixed bias voltage at T_5, i.e., V_3 and consequently T_6 is on while T_5 is off. This, in turn, in its initial state, automatically determines the voltage V_2 at T_2 base and sets the negative leading bipolar input pulse integral discrimination level.

A constant-fraction timing system having good time derivation properties over a wide dynamic range of signals was described by Bedwell and Paulus[6] Leading-edge walk has been significantly reduced by arming the constant-fraction zero-crossing circuit with a discriminator that was separated from the energy discriminator. A simplified circuit model for the timing system is shown in Figure 8.5.

Using standard circuit analysis techniques, the transfer function from the detector current to the constant-fraction-shaped signal from A_3 is given by:

$$\frac{E_{cf}(s)}{I_d(s)} = \frac{E_{op}(s)}{I_d(s)} \times \frac{E_{os}(s)}{E_{op}(s)} \times \frac{E_{cf}(s)}{E_{os}(s)} \tag{8.3}$$

For the circuit model, this transfer function becomes:

$$\frac{E_{cf}(s)}{I_d(s)} \simeq \left(\frac{-R_p}{1+sR_pC_{f1}}\right)\cdot\left(\frac{A_2sR_1C_1}{(1+sR_1C_1)(1+sR_2C_2)}\right)\left(e^{-sT_D}-f\right)\left(A_3\right) \tag{8.4}$$

where

$$R_p = R_f \| R_d(1+A_1) \tag{8.5}$$

$$C_{f1} = C_f + C_d/(1+A_1) \tag{8.6}$$

$$f = R_3/(R_3+R_4) \tag{8.7}$$

and the approximation sign in Equation 8.4 results from assuming $|A_1| \gg 1$. The transfer function in Equation 8.4 can be simplified under most operating conditions.

Incorporating some assumptions[6] into Equation 8.4 and solving for $E_{cf}(s)$ yields:

$$E_{cf}(s) \simeq \frac{I_d(s)}{sC_{f1}}\left(f-e^{-sT_D}\right)A_2A_3 \tag{8.8}$$

The total charge collected in the detector for a particular event is equal to the integral with respect to time of the current pulse, I_d, from 0 to ∞. Since division by s in the s-domain corresponds to integration in the time domain, Equation 8.8 can be rewritten as:

$$E_{cf}(s) \simeq \frac{Q(s)}{C_{f1}}\left(f-e^{-sT_D}\right)A_2A_3 \tag{8.9}$$

The inverse Laplace transformation of Equation 8.9 is

$$E_{cf}(t) = \left[\frac{fq(t)}{C_{f1}}U(t) - \frac{q(t-T_D)}{C_{f1}}U(t-T_D)\right]A_2A_3 \tag{8.10}$$

The values of C_{f1}, f, and T_D in Equation 8.10 are fixed in a given timing system. Consequently, the variations in charge collection time in the detector produce similar variations in the rise time of the input signals to be the constant-fraction circuit.

A block diagram of a typical second generation constant-fraction timing discriminator (CFTD) is shown in Figure 8.6. Neglecting the propagation delays of the active circuit components, the circuit operates as follows. The input signal starts at time equal zero and is applied to the constant-fraction circuitry and to the leading-edge (LE) discriminator simultaneously. If the input signal exceeds the LE threshold, the LE discriminator produces an output pulse which crosses the logic threshold of the zero-crossing AND gate at the time t_{LE}. The constant-fraction bipolar signal crosses the logic threshold of the zero-crossing gate at time t_{cf}, which for ARC timing with linear input signals, is given by:[6]

$$t_{cf} = T_D/(1-f) \tag{8.11}$$

If the output pulse from the LE discriminator satisfies the condition:

$$t_{LE} > t_{cf} \tag{8.12}$$

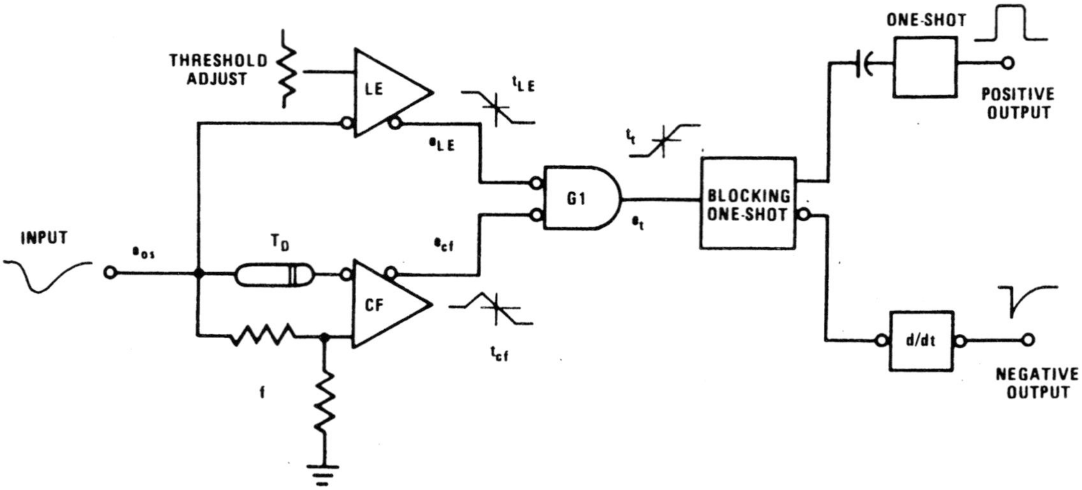

Figure 8.6 Block diagram of a second generation constant-fraction discriminator. (From Bedwell, M. O. and Paulus, T. J., *IEEE Trans. Nucl. Sci.,* NS23, 234, 1976. With permission.)

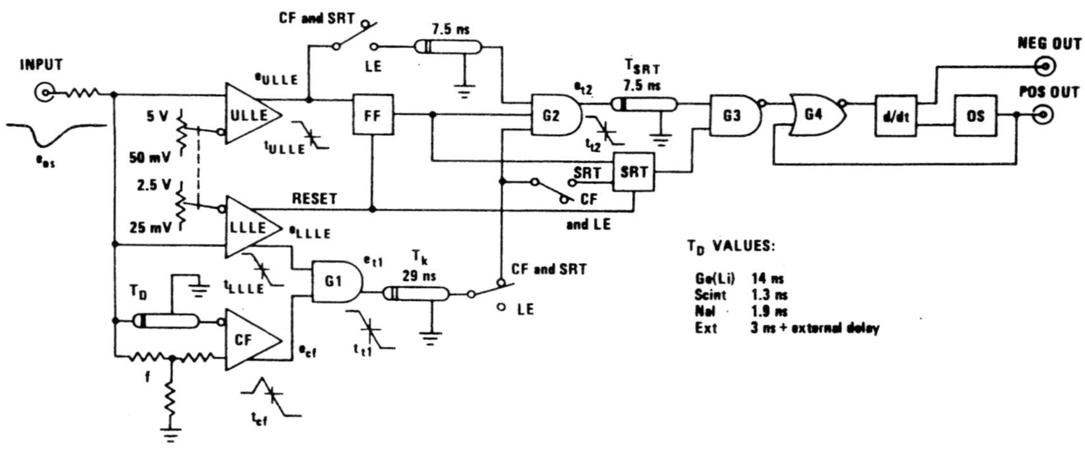

Figure 8.7 Simplified block diagram of the third generation constant-fraction discriminator discussed in this chapter. (From Bedwell, M. O. and Paulus, T. J., *IEEE Trans. Nucl. Sci.,* NS23, 234, 1976. With permission.)

then the time t_t of occurrence of the logic signal at the output of the zero-crossing AND gate is related to the beginning of the input signal by Equation 8.11, where $t_t = t_{cf}$.

Major sources of timing error in the CFTD shown in Figure 8.6 occur for input signals with amplitudes near the LE discriminator threshold and for input signals with exceptionally long rise times. The time at which the input signal actually crosses the LE discriminator threshold can be such that t_{LE} does not satisfy Equation 12. When t_{LE} is greater than t_{cf}, the timing signal at time t_t is based on t_{LE} and not on t_{cf}. This produces leading-edge timing from the CFTD. The problem of walk due to leading-edge timing is particularly severe with Ge(Li) detectors, since their output signals can have a wide range of rise times.

A third generation CFTD is shown in Figure 8.7. This CFTD differs from the second generation unit in several important ways. First, there are two leading-edge discriminators. The lower level leading-edge (LLLE) discriminator arms the zero-crossing AND gate G1. The upper level leading-edge (ULLE) discriminator determines whether the input signal is of sufficient amplitude to produce an output pulse from the instrument. The lower level threshold is internally set at one half the upper level threshold. Consequently, signals that satisfy the UL threshold requirement exceed the LL threshold by a factor of

2. This greatly reduces leading-edge walk in the timing signal e_{t1}, at the output of G1. The signal e_{t1} is free of leading-edge walk if the time t_{LLLE} at which the LLLE output signal occurs fulfills the inequality:[6]

$$t_{LLLE} < t_{cf} \tag{8.13}$$

A second important aspect of this CFTD is the addition of the fixed delay T_k in Figure 8.7. This delay provides additional time for the input signal to cross the UL threshold. Leading-edge walk is not introduced by the ULLE signal at the upper level coincidence gate, G2, if the time at which the ULLE signal occurs fulfills the inequality:

$$t_{ULLE} < \left(t_{t1} + T_k\right) \tag{8.14}$$

A third important difference is the addition of an optional slow rise-time (SRT) reject circuit. The SRT evaluation circuit determines whether Equation 8.14 is satisfied. If not, the SRT evaluation circuit blocks the SRT coincidence gate, G3, for the timing signal e_{t2}. The delay T_{SRT} is necessary to allow the SRT evaluation circuit sufficient time to form a decision.

The two important delays in the CFTD shown in Figure 8.7 are the shaping delay, T_D, and the additional delay, T_k. The value of T_k is selected so that the SRT circuit can reject all signals which contain leading-edge walk. Assume a linear input signal of the form:

$$e_{os}(t) = \frac{E}{T_r}\, tU(t) - \frac{E}{T_r}\left(t - T_r\right)U\left(t - T_r\right) \tag{8.15}$$

where E is the maximum value, T_r is the 0 to 100% rise time, and E/T_r is the slope of the linear input signal. At time t_{ULLE} the value of e_{os} is E_{UL}. Solving Equation 8.15 for t_{ULLE} yields:

$$t_{ULLE} = E_{UL}\left(T_r/E\right) \tag{8.16}$$

Since the LLLE threshold, E_{LL}, is set at $E_{UL}/2$, the value of t_{LLLE} is

$$t_{LLLE} = \left(E_{UL}/2\right)\left(T_r/E\right) \tag{8.17}$$

Assuming an input signal of the form:

$$e(t) = E\left(\frac{t}{T_r}\right)^2 U(t) - E\left(\frac{t^2 - T_r^2}{T_r^2}\right)U\left(t - T_r\right) \tag{8.18}$$

The SRT circuit can eliminate all timing signals resulting from leading-edge walk if:[6]

$$T_k < \left[\sqrt{2} - 1\right]\left[T_D/\left(1 - \sqrt{f}\right)\right] \tag{8.19}$$

A constant-fraction discriminator (CFD) with improved stability and walk characteristics was described by McGervey et al.[7] Generally CFD improves time and walk resolution over a wide range of input signal amplitudes by producing an output signal at a fixed time after the input pulse reaches a predetermined fraction of its amplitude. The basic technique is to convert the negative input anode signal into a zero-crossing signal by mixing a portion of the input with a delayed and inverted portion of the input.

The CFD proposed by McGervey et al. was based on a low-power fast logic system. It was based on the comparison technique of Maier and Sperr[8] who produced the crossover signal by feeding an attenuated portion of the input to one side of a fast differential amplifier and a delayed portion to the other side. High-

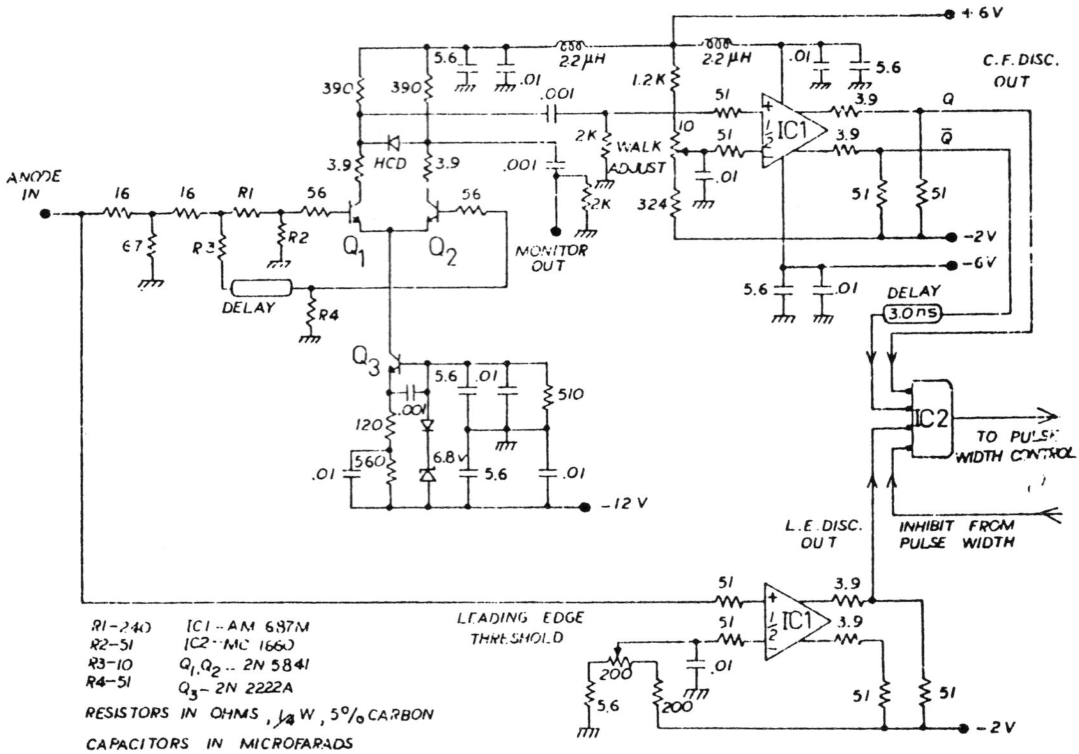

Figure 8.8 Front end of constant-fraction discriminator. (From McGervey, J. D. et al., *Nucl. Instrum. & Methods,* 143, 435, 1977. With permission.)

speed emitter coupled logic (ECL) comparators then sample the resulting bipolar pulse. The McGervey et al. circuit differs from that of Maier and Sperr in that MECL-III ICs were used throughout and a wide bandwidth discrete differential amplifier was used at the input stage. The front end of the circuit is shown in Figure 8.8.

The anode input was connected to two distinct circuits, the constant-fraction discriminator and a leading-edge discriminator. The leading-edge discriminator was one half of a dual high-speed ECL comparator (Advanced Micro Devices AM 687) with an input offset voltage of ±2.0 mV on a range of –3.3 to 2.7 V. The threshold was variable from –28 mV to –1.0 V. The output of this stage was an ECL signal with the width of the anode signal below the threshold. The anode signal also fed the constant-fraction stage of the circuit. The signal was attenuated by a factor of two to keep the signals from saturating the stage. The signal was then fed to both inputs of the differential amplifier, the one input attenuated by R_1 and R_2, the other input delayed and attenuated by R_3 and R_4; R_4 also terminated the delay line and had to be equal to R_2 to maintain DC balance. R_1 and R_3 were then chosen to yield the desired fraction of the pulse height, and to maintain the proper fraction of the pulse height and the proper input impedance. The differential amplifier Q_1, Q_2, Q_3 did a bipolar subtraction of the signals, with the crossover point a function of the attenuation ratio and the delay line. One output of the amplifier was AC coupled to the front panel to serve as a monitor point, while the other output was AC coupled to the other half of the AM 687 comparator.

Figure 8.9 shows the timing of the discriminator pulses as they reach the coincidence circuit (MECL 1660). The leading-edge discriminator must reach the coincidence first (C.F.) and will do so because of the delay line and cumulative transit times of the C.F. stage. The complementary outputs of the constant-fraction discriminator are brought to the coincidence with $\overline{Q}$ output delayed by 3.0 nsec. This forms a leading-edge differentiation, yielding a 3.0-nsec pulse out of the stage and inhibiting the circuit for the duration of the C.F. output pulse. The final input of the fourth input gate is fed by an inhibit from the pulse-width control circuit to prevent the gate firing until the output pulse has been completed.

A constant-fraction discriminator designed by Mÿllÿla,[9] using the fastest ECL series (MECL-III) is shown in Figure 8.10. The circuit is based on the technique whereby timing occurs at the instant when

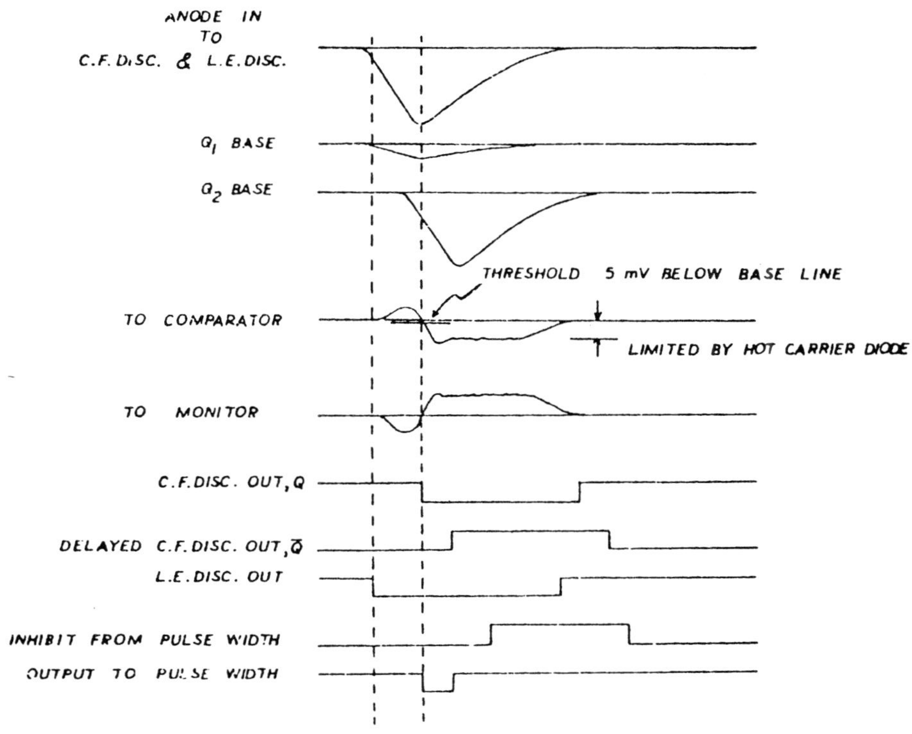

Figure 8.9 Timing of pulses in constant-fraction discriminator. (From McGervey, J. D. et al., *Nucl. Instrum. & Methods,* 143, 435, 1977. With permission.)

the pulse reaches an amplitude which is a predetermined fraction of the value of the pulse amplitude after a constant delay t_d from the beginning of the pulse. The optimum values of the fraction and the delay depend on the properties of the photomultiplier pulses. A fraction (f) of 0.4 and a delay (t_d) of (1 – f) times the pulse rise time were chosen on the basis of a theoretical analysis.

The comparator circuit MC1651 has many advantages compared with the triple differential amplifiers. It needs no level shifter to ECL voltage levels at its input, and it has no tendency to oscillate because of hysteresis. The walk can easily be controlled by the potentiometer P1 and the delay t_d, and the fraction can be selected by the attenuator in the inverting input. The timing instant is determined from the pulse by first attenuating it by the factor f and then inverting and adding the result to the original pulse delayed by the time t_d. The zero crossover of this pulse constitutes the timing moment. The other half of the MC1651 rejects noise pulses and pulses of insufficient amplitude.

A zero-crossing discrimination technique was used by Gal and Bibok[10] for constant-fraction timing. It utilized a simple method that made walkless timing possible, independent of delay and fraction settings. Duration discrimination was used for the zero-crossing detection timing, combining it with an amplitude discrimination that did not cause any walk.

The method is shown in Figure 8.11. The bipolar signal was coupled to a comparator biased to zero level. This comparator was triggered by true events and noise as well. The output of the zero-crossing comparator was fed to an integrator. The amplitude of the output of this integrator was proportional to the zero-crossing comparator output pulse width. In the case of a true signal, the amplitude of the integrator output was higher than if it were triggered by noise. This integrator output signal was fed to a second comparator, which selected signals by their amplitude, i.e., their pulse duration. As seen in Figure 8.11 the duration discrimination is performed on the second lobe of the bipolar signal. To avoid spurious output pulses due to pileup, the integrator output has to be restored to zero level by the negative edge of the first comparator.

A constant-fraction discriminator having excellent timing characteristics over a wide dynamic range of signals has been designed and tested by Bedwell and Paulus.[11] To achieve maximum timing performance, the most critical block of circuitry has been implemented using hybrid microcircuit technology.

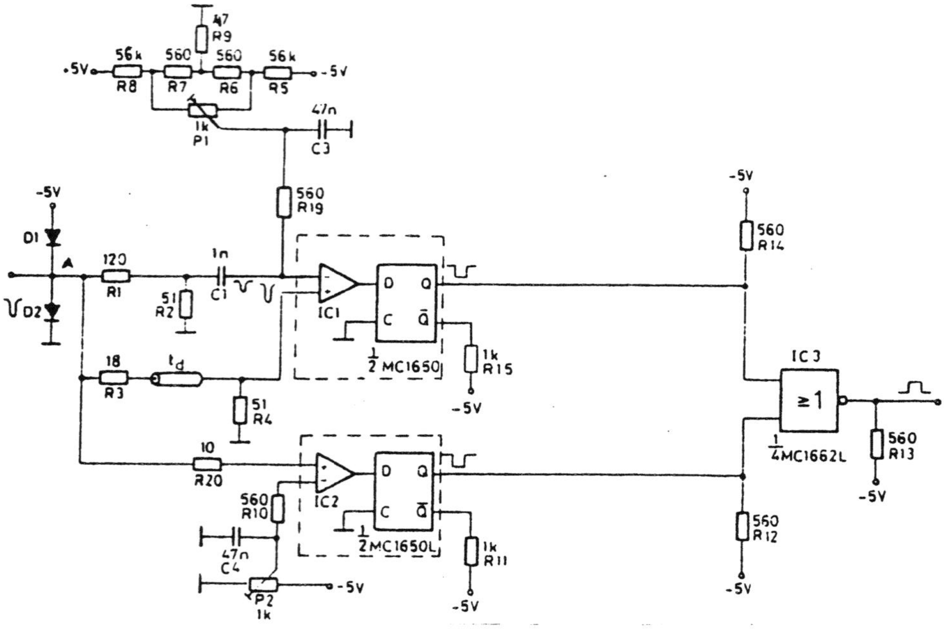

Figure 8.10 Schematic diagram of the constant-fraction timing discriminator. (From Mÿllÿla, R., *Nucl. Instrum. & Methods,* 148, 267, 1978. With permission.)

This block included the leading-edge arming discriminator, the constant-fraction shaping circuitry and the zero-crossing pick-off. The basic constant fraction building element indicated above operates to a minimum threshold of –30 mV and can accept input signals to –10 V without overload. Amplitude-dependent time walk was less than ±150 psec for a 200:1 dynamic range of input signals (t_r = 1 nsec). The count rate capability of this constant fraction hybrid exceeds 100 MHz.

A functional representation of a constant-fraction trigger is shown in Figure 8.12. In the constant-fraction method the input signal to the circuit was delayed and a fraction of the undelayed pulse was subtracted from it. A bipolar pulse was generated and its zero-crossing was detected and used to produce an output logic pulse.

The left-hand side of Figure 8.13 is a simplified block diagram of the hybridized constant-fraction element used. A differential transformer technique was used to passively form the constant-fraction bipolar signal. Particular attention has been given to the design of the associated thick film resistive network to ensure proper impedance matching. This network sets a constant-fraction attenuation factor of f = 0.2. The related 50-Ω shaping delay must be provided externally. The passively derived bipolar timing signal is amplified by the fast differential constant-fraction amplifier (CFA) and applied to the input of gate G1, which detects the zero crossing. Walk adjustment was accomplished by varying the voltage applied to one input terminal of the CFA. The fast leading-edge discriminator (LLED) provided the leading-edge arming signal to the zero-crossing gate G1.

The technique of passively forming the constant-fraction bipolar signal has three primary advantages over most active techniques.[11] First, the bipolar pulse is formed without the noise at zero crossing that is normally introduced by active components. Second, the upper bandwidth limitation of the differential transformer in the pulse-forming network exceeds 350 MHz. Third, the dynamic range limitation of the passive pulse-shaping network far exceeds that of most active devices.

Figure 8.13 in all is a simplified block diagram of one channel of the four-channel, 100-MHz updating constant-fraction discriminator. The circuit operates as follows.[11] A 0 to –10 V amplitude input signal starts at time zero and is applied simultaneously to the constant-fraction (CF) circuitry and to the leading-edge discriminator (LLED) circuitry. The LLED has an adjustable threshold, ranging from –30 mV to –1 V, which determines the minimum input signal amplitude required to produce an output pulse from the instrument. If the input signal exceeds the LLED threshold, that comparator produces an output pulse which arms the zero-crossing gate G1 at time t_{LE}. The passively formed constant-fraction bipolar timing signal is

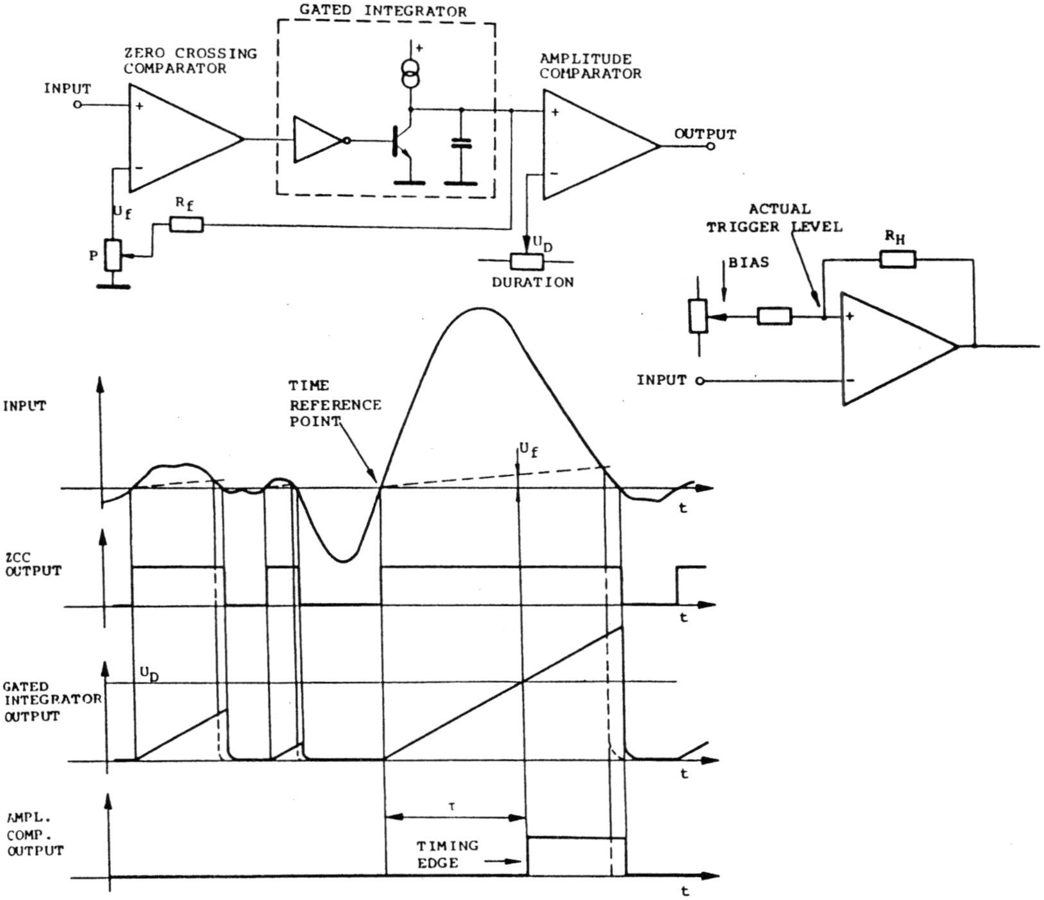

Figure 8.11 Block diagram and simplified waveforms of zero-crossing discrimination. (From Gal, J. and Bibok, G., *Nucl. Instrum. & Methods,* 163, 535, 1979. With permission.)

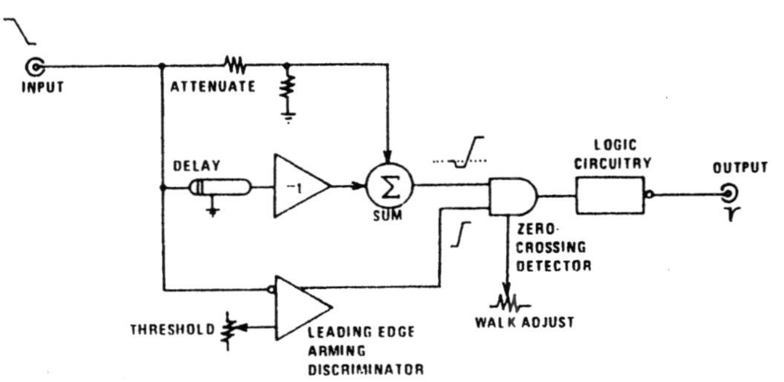

Figure 8.12 Functional representation of a constant-fraction trigger. (From Bedwell, M. O. and Paulus, T. J., *IEEE Trans. Nucl. Sci.,* NS25, 86, 1978. With permission.)

amplified by the constant-fraction amplifier (CFA) and crosses the logic threshold at the zero-crossing gate at time t_{CE}. The time t_{CF} is determined by the selection of the external constant-fraction shaping delay, $t_{D(Ext)}$, subject to the rise times and pulse shapes of the input signals. A fixed internal delay of 0.6 to 0.7 nsec must be added to $t_{D(Ext)}$ to obtain the total constant fraction shaping delay:

$$t_{D(Tot)} \simeq t_{D(Ext)} + 0.65 \text{ nsec} \tag{8.20}$$

The timing logic signal from gate G1 triggers a fast one-shot, comprised of an ECL-type D master-slave flip-flop, FF1, and amplifier A_2. This one-shot produces a reshaped logic pulse with a fixed 2-nsec width

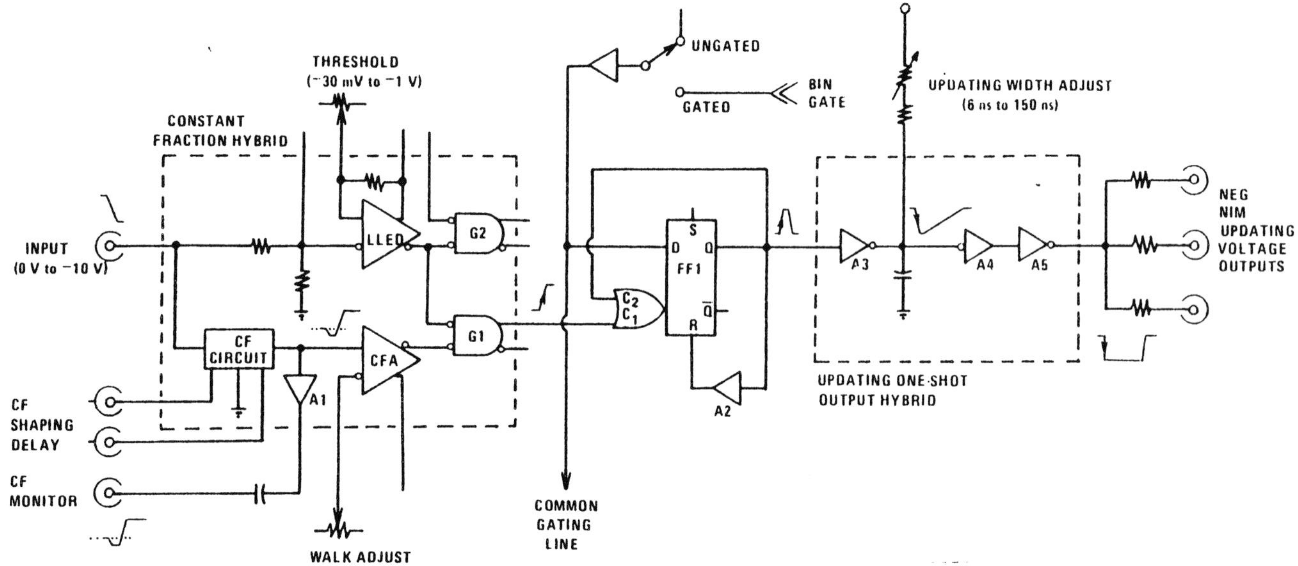

Figure 8.13 Simplified block diagram of one channel of the four-channel 100-MHz updating constant-fraction discriminator. (From Bedwell, M. O. and Paulus, T. J., *IEEE Trans. Nucl. Sci.*, NS25, 86, 1978. With permission.)

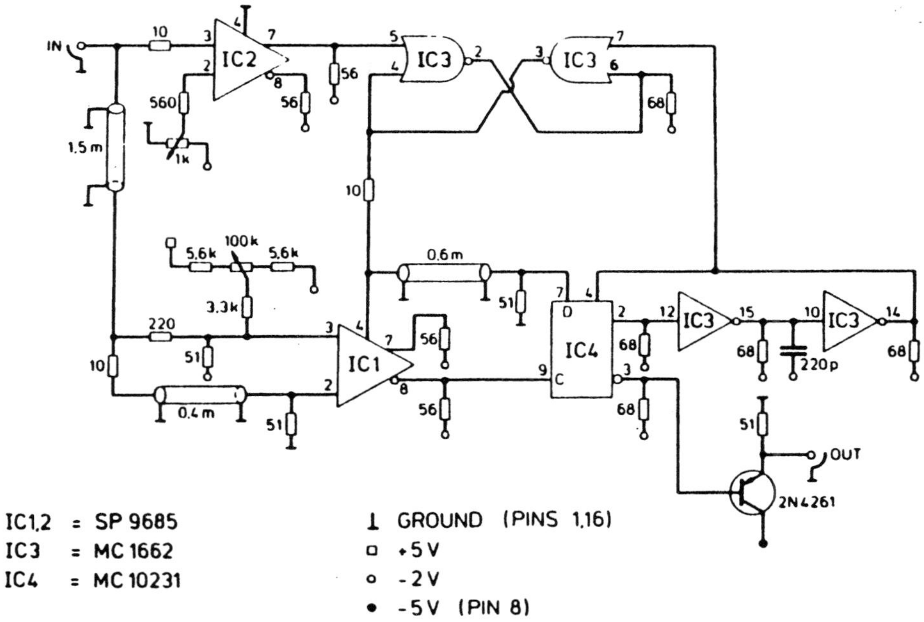

Figure 8.14 Circuit diagram of the latched constant-fraction discriminator. (From Rytsola, K., *Nucl. Instrum. & Methods,* 199, 491, 1982. With permission.)

for each signal from gate G1. The reshaped signal drives a hybridized, updating one-shot and output circuit, comprised of a capacitor and amplifiers A_3 through A_5. Each 2-nsec wide signal from FF1 causes A_3 to charge the updating capacitor to its maximum amplitude. At the end of the 2-nsec interval, the updating capacitor begins to discharge through the series resistor and to update width potentiometer toward its quiescent voltage level. Elements A_4 and A_5 amplify and shape the signal on the updating capacitor to a fast voltage output signal, capable of simultaneously driving three 50-Ω loads with negative nuclear instruments module (NIM) standard fast logic pulses.

A constant-fraction differential discriminator has been designed and tested by Bedwell and Paulus.[12] To achieve maximum timing performance, the block of circuitry including the leading-edge arming discriminator, the constant-fraction shaping circuitry, and the zero-crossing pick-off, has been implemented using hybrid microcircuit technology.

The instrument can be operated either as an integral discriminator or as a fast timing single-channel analyzer (SCA) for the anode signals from photomultiplier tubes (PMTs). The amplitude-dependent time walk of the discriminator was less than ±150 psec for 1-nsec rise-time input signals from –30 mV to –10 V in the integral mode (–30 mV to –5 V, differential mode).

A constant-fraction discriminator with a latched timing comparator was described by Rytsola.[13] The design improved the temperature stability of the device while providing excellent timing characteristics in a simple construction. The time resolution of a complete fast/slow coincidence system was 65 psec, measured with a ^{60}Co gamma source. The CF discriminator was used with RCA C31024 photomultipliers. The anode pulse was as fast as the fastest ECL logic.

The circuit diagram of the discriminator is shown in Figure 8.14. The CF-timed pulse was formed in the usual way by resistive attenuation, cable delay, and summing and inverting at the inputs of comparator. A very fast comparator, Plessey SP9685, was used both in the timing and the lower level discrimination channel. In this comparator type a latch function is provided for a sample and hold operation. The output from the low-level (LL) channel was brought to the latch input of the timing comparator so that the comparator was activated only when the lower level setting had been exceeded. This completely eliminates the ringing of the comparator and makes its output extremely clean. The input pulse in the timing channel was delayed by a separate 7-nsec delay to allow time for forming the latch pulse and for the timing comparator to open.

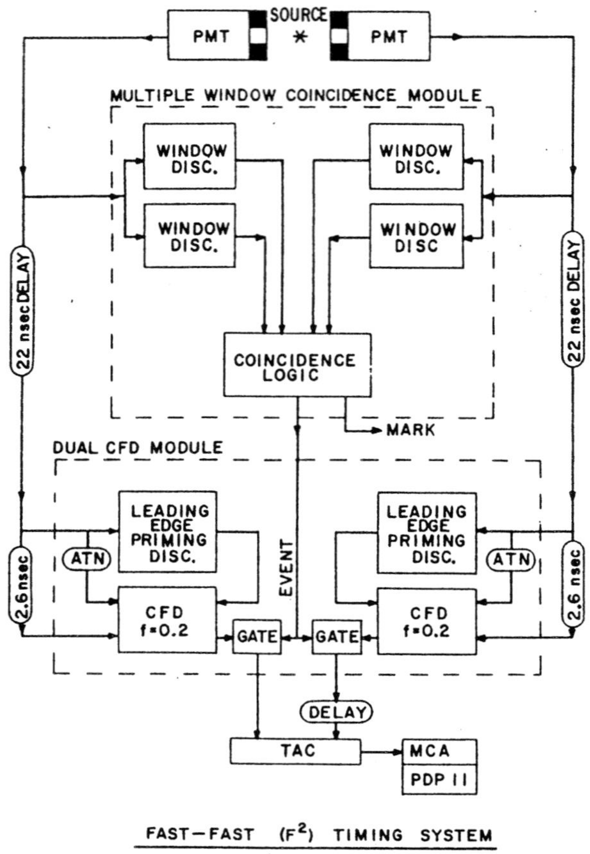

Figure 8.15 Functional diagram of the system consisting of source, two scintillators, two photomultiplier tubes, the MWCM, the DCCFD, TAC, MCA, and PDP11. (From Hardy, W. H., II and Lynn, K. G., *IEEE Trans. Nucl. Sci.*, NS23, 229, 1976. With permission.)

The input pulse from the photomultiplier and the output of the comparator are so fast that a flip-flop was necessary in the LL channel for walk-free operation. Two ECL-III gates (MC1660) were used. MC1023 was fast enough and used as the summing element for the two channels.

A dual-channel constant-fraction timing discriminator incorporating multiple energy windows and coincidence logic has been developed by Hardy and Lynn.[14] A prototype system employing the fast-fast (F^2) approach is shown in Figure 8.15. The system consists of two quad-width NIM modules: a multiple window coincidence module (MWCM) and a dual-channel constant-fraction discriminator (DCCFD). The prototype has been constructed primarily for positron annihilation lifetime studies.

The MWCM contained eight discriminators connected to form four windows, two in each channel. Each discriminator consisted of one element of MECL 10102 gates. A window was formed by requiring that the lower threshold fires while the upper threshold does not. Only a single level of amplification has been provided in order to minimize propagation delays.

Each channel of the DCCFD contained a leading-edge priming discriminator consisting of an MECL 10216 triple differential amplifier (Figure 8.16), and a constant-fraction discriminator consisting of two stages of MECL 10216. A fraction of 0.2 was chosen on the basis of the work of Gedeke and McDonald.[4] The delay in the CFD was initially chosen to be equal to the rise time of the photomultiplier (PMT) output pulse from 20% to maximum.

A different method was used by Maier and Sperr[8] to obtain a constant-fraction trigger (see Figure 8.17). The input pulse (C_1) is split into two parts. One part (A) is attenuated and applied to the inverting input of a fast differential discriminator. The other part (B) is delayed and applied to the noninverting input of the same discriminator. The output voltage of the discriminator determined the difference of the input voltages. The output pulse (AB) crosses the threshold at the following gate at V_{BB} (–1.2 V). If the voltages at the input are equal, crossing the threshold supplies the timing information. Timing (corresponding to any fraction) must be done at the occurrence of the pulse maximum; thus the attenuated pulse must be delayed to

$$t_{delay} = t_{rise}(1 - \text{fraction}) \tag{8.21}$$

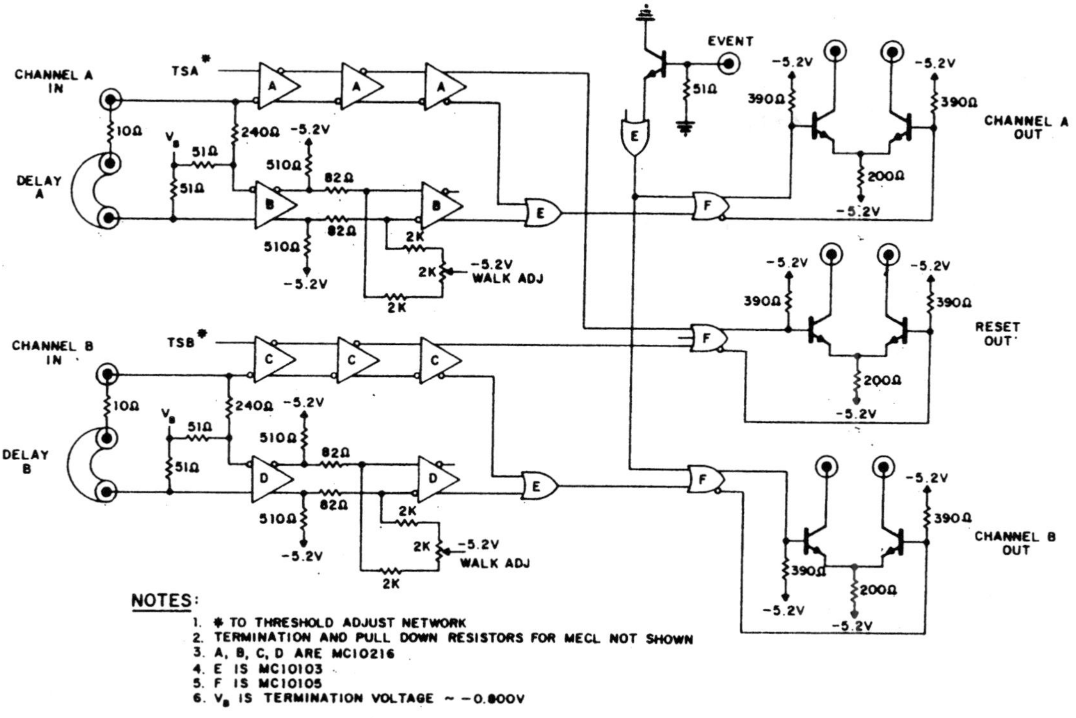

Figure 8.16 Circuit diagram of dual-channel constant-fraction discriminator. (From Hardy, W. H., II and Lynn, K. G., *IEEE Trans. Nucl. Sci.,* NS23, 229, 1976. With permission.)

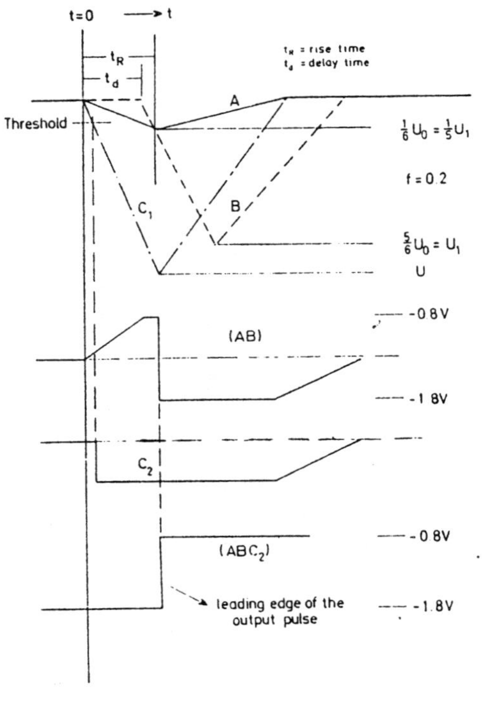

Figure 8.17 Pulse shapes at the specified points in the electronic circuit of Figure 18. (From Maier, M. R. and Sperr, P., *Nucl. Instrum. & Methods,* 87, 13, 1970. With permission.)

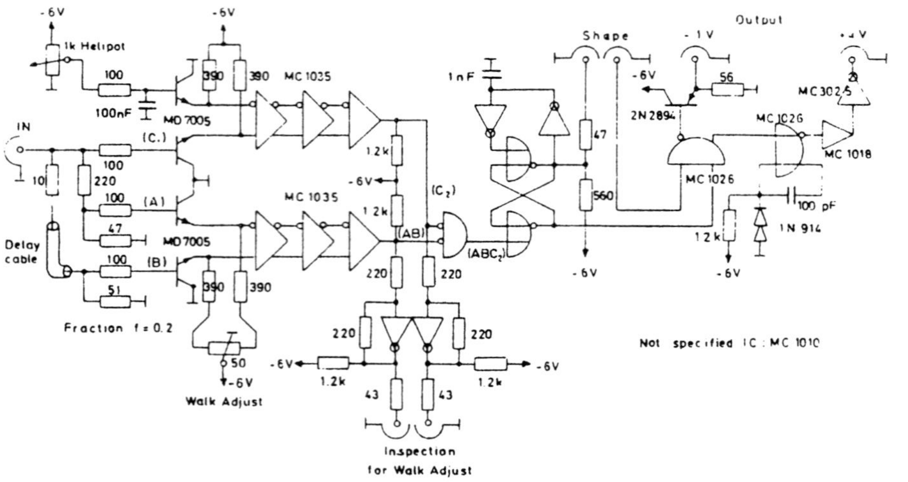

Figure 8.18 Electronic circuit of the constant-fraction trigger. (The point where the pulses of Figure 8.17 can be observed are labeled by bracketed letters.) (From Maier, M. R. and Sperr, P., *Nucl. Instrum. & Methods,* 87, 13, 1970. With permission.)

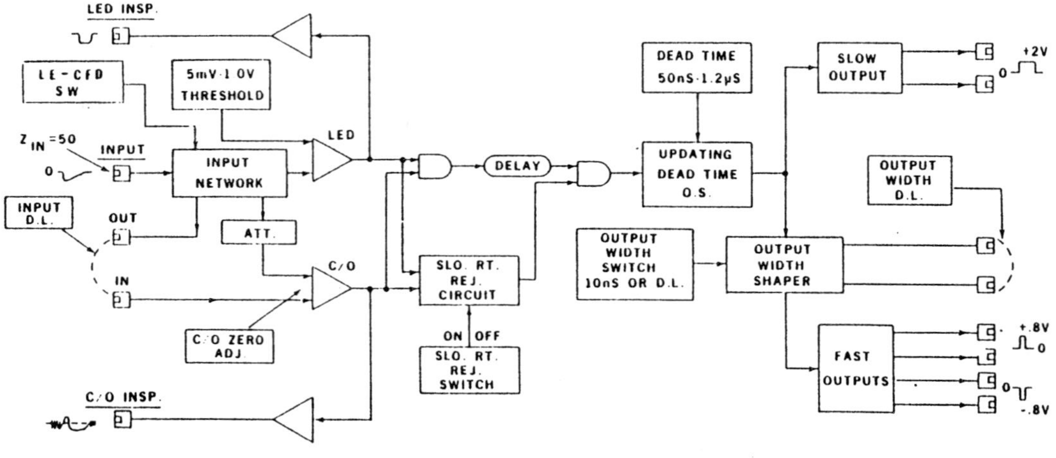

Figure 8.19 Block diagram of the new constant-fraction trigger. (From Maier, M. R. and Landis, D. A., *Nucl. Instrum. & Methods,* 117, 245, 1974. With permission.)

The circuit is shown in Figure 8.18. The fast discriminator is made of three cascade differential amplifiers MC1035. The input pulse is split by a resistor network. The delay time is a 50-Ω resistor. The resistance in the path of delayed pulse is $f \cdot Z_0$. The output pulse AB crosses V_{BB} when the particular selected fraction of the maximum amplitude occurs in the delayed pulse.

A later version of the constant-fraction designed by Maier and Landis[15] is shown in Figure 8.19. A circuit was included for the suppression of pulses having slow rise times. It removed the tails on peaks in the time spectra associated with low-rising pulses in germanium detectors. This slow rise-time suppression circuit inhibited the output of the unit when the output signal of the leading-edge discriminator (LED) integrated circuit (IC) occurred after the signal of the CFD IC.

A combination of a snap-off diode and a constant-fraction circuit was proposed by Elscint.[16] The snap-off diode integration is based on the properties of a special silicon diode. In a conventional tunnel diode the response to a bipolar pulse results in current flowing in only one direction when the diode is biased in the forward direction. Switching from the conducting to nonconducting state occurs more or less at the crossover time. Usually, the charge accumulation is as small as possible, being an error cause. In a snap-

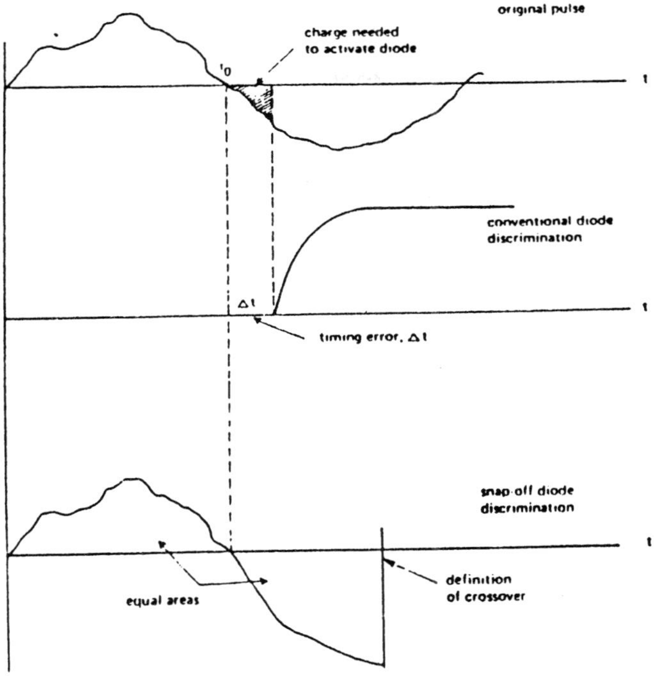

Figure 8.20 Tunnel diode and snap-off diode discrimination. (Reprinted with permission from Elscint Snap-Off Timing Discriminator STD-N-1 Manual.)

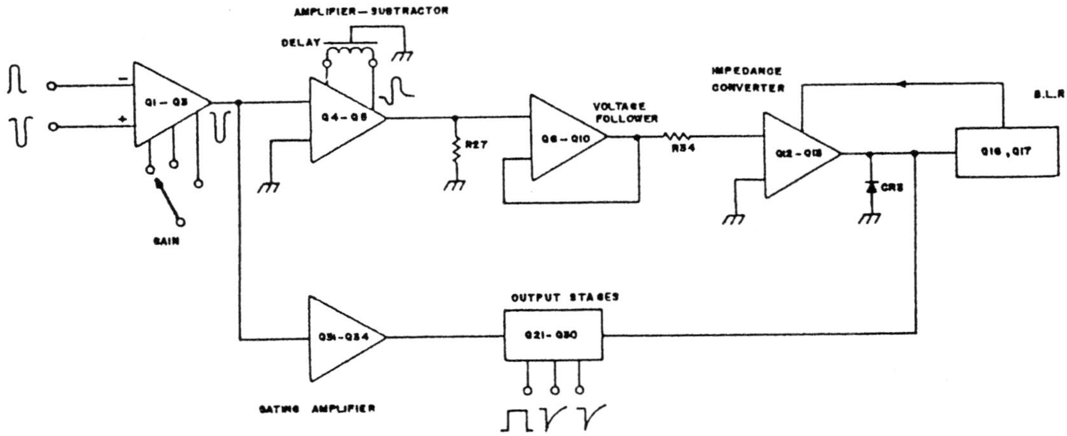

Figure 8.21 Block diagram of snap-off constant-fraction system. (Reprinted with permission from Elscint Snap-Off Timing Discriminator STD-N-1 Manual.)

off diode, on the contrary, the reversal of the current does not force the diode immediately to the nonconducting state.

A rapid change of state occurs only when the charge accumulated during the forward half cycle is drained off during the backward half cycle. When this is accomplished, an abrupt switch-off of the current occurs within picoseconds. In this way, the accumulation of charge (when the positive charge is not too large) is used to delay the switching, this moment being used as a time reference (Figure 8.20). Figure 8.21 shows the block diagram of the snap-off constant-fraction system. The input signal is amplified by A_1 and delayed by an external delay line. It is amplified then again and subtracted from the original undelayed pulse in A_2 in order to obtain a zero crossover at an instant corresponding to a constant fraction of the pulse height. The output of the amplifier-subtractor is fed to the voltage follower A_3, which reduces the output impedance and converts it through R34 into a current. This current is finally fed from the high-impedance source A_4 into the snap-off diode CR3. The voltage across the snap-off diode is suitably

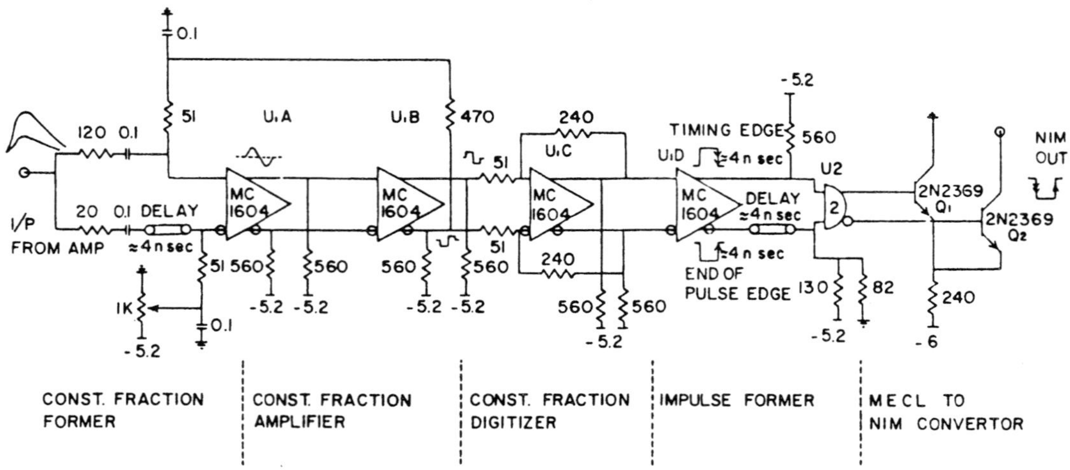

Figure 8.22 The constant-fraction discriminator unit. (From Bryman, D. et al., *Nucl. Instrum. & Methods,* 141, 573, 1977. With permission.)

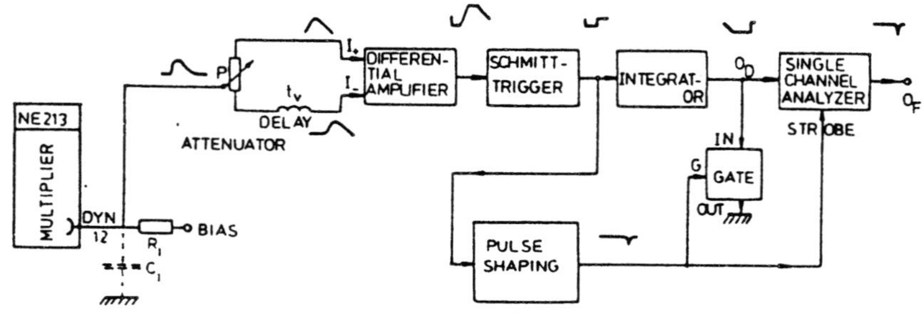

Figure 8.23 Block diagram of the pulse shape discrimination circuit. (From Plischke, P. et al., *Nucl. Instrum. & Methods,* 136, 579, 1976. With permission.)

shaped by A_5, and gated by A_6, in coincidence with the input signal. BLR is a baseline restorer, providing equal currents through A_4 in the absence of pulses.

A constant-fraction discriminator used by Bryman et al.[17] is illustrated schematically in Figure 8.22. The positive pulses were passed into U1A via two branches. The first branch passed the signal, attenuated, to the noninverting input; and the second branch passed the signal via a delay line to the inverting input of U1A. These signals were mixed in U1A and amplified by U1B. U1C is a Schmitt trigger, the output of which, after further amplification by U1D, was applied to the pulse former U1D via a delay line. The output of U2 which was an MECL level was converted into a standard NIM pulse by the output transistors Q_1 and Q_2.

A constant-fraction circuit was used by Plischke et al.[18] to design a pulse shape discriminator for n-gamma separation with NE213. The advantages were stability at high counting rates and energy independence over a wide range. The system block diagram is shown in Figure 8.23. The output pulse of diode 12 of the PM tube was assumed to be composed of two exponential components:

$$I(t) = I_o\left(A_1 e^{-t/t_1} + A_2 e^{-t/t_2}\right) \tag{8.22}$$

Integration of I(t) with time constant $t_i = R_iC_i$ (see Figure 8.23) yielded a voltage phase over resistor R_i:

$$U_i(t) = I_o R_i\left[\frac{A_1 t_1}{t_1 - t_i}\left(e^{-t/t_1} - e^{-t/t_i}\right) + \frac{A_2 t_2}{t_2 - t_i}\left(e^{-t/t_2} - e^{-t/t_i}\right)\right] \tag{8.23}$$

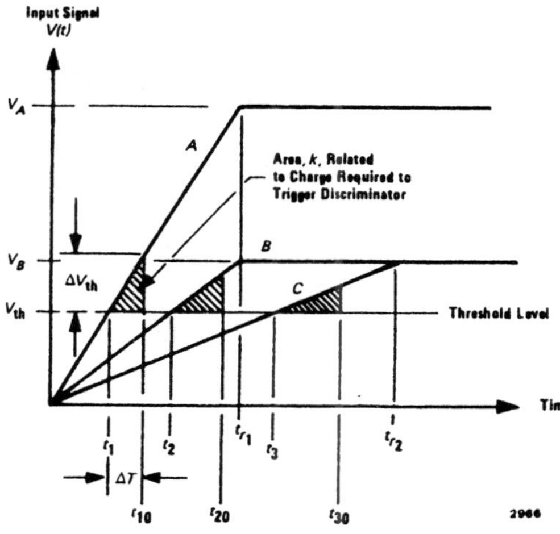

Figure 8.24 Walk in a leading-edge discriminator due to amplitude and rise-time variations and charge sensitivity. (From Paulus, T. J., *IEEE Trans. Nucl. Sci.*, NS32, 1242, 1985. With permission.)

The pulse was turned into a bipolar pulse for the zero-crossing time measurement. A simple differentiation with an RC network ($t_d = R_dC_d$) yielded the following bipolar pulse:

$$U_d(t) = I_oR_i \sum_{v=1}^{R} \frac{A_v t_o}{(t_i - t_v)} \times \left[\frac{t_d}{t_d - t_i} e^{-t/t_i} - \frac{t_d}{t_d - t_v} e^{-t/t_v} - \frac{t_i}{t_d - t_i} e^{-t/t_d} \right] \qquad (8.24)$$

The zero-crossing time depends on A and t_v.

For the constant-fraction timing the signal $U_i(t)$ was split into two pulses by means of a signal distribution with an adjustable distribution ratio P. The pulses $PU_i(t)$ and $U_i(t - t_v)$, with t being a constant delay time, were fed into a differential amplifier. The amplifier output pulses were

$$u_a(t) = \begin{cases} PU_i(t) & t \le t_v \\ PU_i(t) - U_i(t - t_v) & t \ge t_v \end{cases} \qquad (8.25)$$

Practical values found to be P = 1.3, t_i = 0.5 μsec, and t_v = 0.35 μsec.

Paulus[19] reviewed time pick-off techniques comparing leading-edge timing and constant-fraction timing. There are three important sources of error in time pick-off measurements: walk, drift, and jitter. Walk is the time movement of the output pulse relative to the input pulse due to variations in the shape and amplitude of the input pulse. Drift is the timing error introduced by component aging and by temperature variations. Jitter is the time uncertainty that is caused by noise and by statistical fluctuations of the signals from the detector.

A leading-edge discriminator is the simplest means of deriving a time pick-off signal. Using a comparator, a logic signal is produced when the input signal crosses a fixed threshold level. A disadvantage of this method is that the time of occurrence of the output pulse is a function of the amplitude and rise time of the input signal.

Time walk of a leading-edge comparator due to input signal variations is shown in Figure 8.24. Signals A and B have the same rise time but different amplitudes. Signal A crosses the threshold, V_{th}, at time t_1 while signal C crosses the threshold at time t_2. The difference in threshold-crossing time causes the output logic pulse from the leading-edge discriminator to walk along the time axis. Walk is most pronounced for signals with amplitudes that only slightly exceed the threshold level.

Time walk due to charge sensitivity is also illustrated in Figure 8.24. After an input signal crosses the discriminator threshold level, a small additional amount of charge is required to actually trigger the discriminating element. Times t_{10}, t_{20}, and t_{30} are the times when the output signals actually occur, relative to times t_1, t_2, and t_3. Timing errors introduced by charge sensitivity are greater for signals with longer rise times and for signals with smaller pulse amplitudes above threshold. Due to charge sensitivity signal

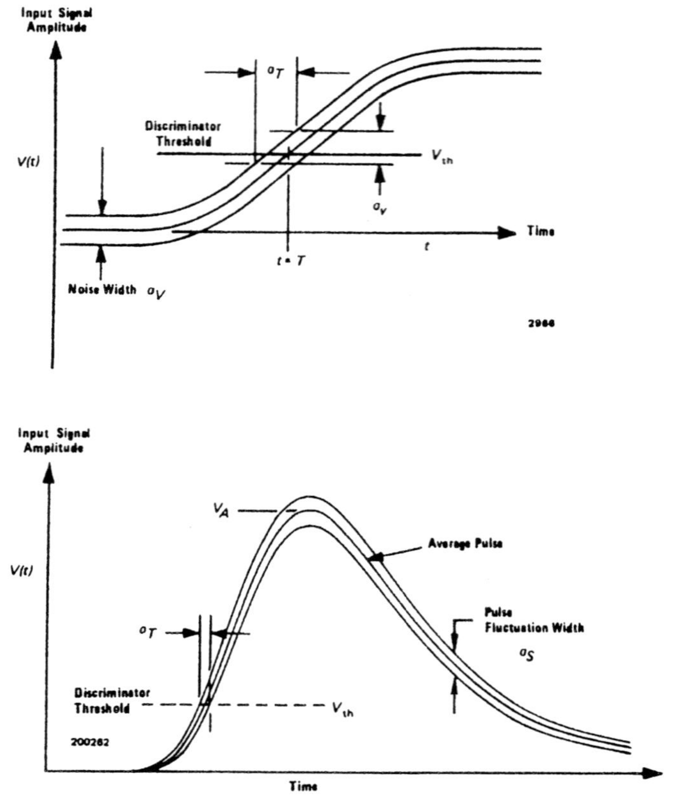

Figure 8.25 Time jitter in a leading-edge discriminator due to (a) noise on the input signal and (b) statistical pulse-height variations. (From Paulus, T. J., *IEEE Trans. Nucl. Sci.*, NS32, 1242, 1985. With permission.)

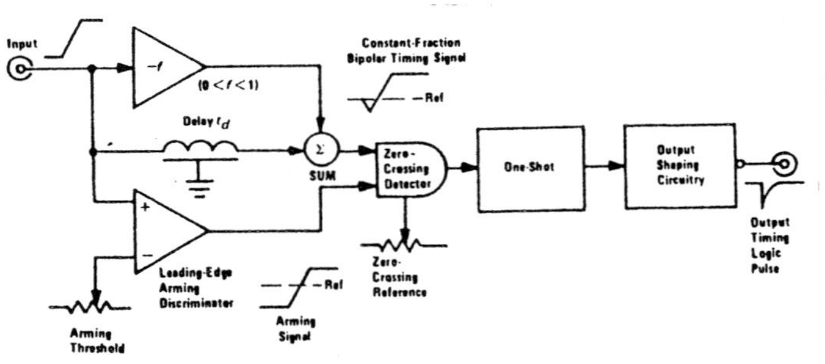

Figure 8.26 Functional representation of a constant-fraction trigger. (From Paulus, T. J., *IEEE Trans. Nucl. Sci.*, NS32, 1242, 1985. With permission.)

A has an effective threshold of $V_{th} + \Delta V_{th}$. Two sources of timing uncertainty are illustrated in Figure 8.25 for an ideal leading-edge discriminator. The noise-induced root-mean-square (RMS) uncertainty, σ_T, in threshold-crossing time for the leading-edge discriminator is given (with reasonable accuracy) by the triangle rule as:[19]

$$\sigma_T = \sigma_V / dV(t)dt \,\big|_{t=T} \tag{8.26}$$

where the input signal V(t), is assumed to be approximately linear in the region of threshold crossing. σ_V is the RMS value of noise amplitude.

Figure 8.26 shows a functional diagram of a constant-fraction discriminator (CFD). The input signal is delayed and a fraction of the undelayed input is subtracted from it to produce a bipolar pulse. The zero crossing is detected and used to produce an output logic pulse. The use of a leading-edge arming discriminator provides energy selection capability and prevents the sensitive zero-crossing device from triggering on the noise of the constant-fraction baseline.

With the constant-fraction technique, walk due to rise time and amplitude variations of the input signal is minimized by proper selection of the shaping delay, t_d. Jitter is minimized by proper selection of the attenuation factor, f.

Two cases must be considered in determining the zero-crossing time of the constant-fraction bipolar signal. The first case is for true-constant-fraction (TCF) timing, and the second is for the amplitude-and-rise-time-compensated (ARC) timing.[19]

In the TCF case the time of zero crossing occurs while the attenuated input signal is at its full amplitude. Figure 8.27 illustrates the signal formation in an ideal CFD for TCF timing with linear input signals. The amplitude independence of the zero-crossing time is depicted for input signals A and B which have the same rise time but different amplitudes. From signals B and C the zero-crossing time is seen to be dependent on the rise time of the input signal. For linear input signals that begin at time zero the constant-fraction zero-crossing time for the TCF case, T_{TCF}, is

$$T_{TCF} = t_d + ft_r \tag{8.27}$$

where t_r is the input signal rise time.

Two criteria for t_d must be observed to ensure TCF timing. The shaping delay must be selected so that:

$$t_d > t_r(1-f) \tag{8.28}$$

This constraint ensures that the zero-crossing time occurs after the attenuated linear input signal has reached its maximum amplitude.

The second case to be considered is ARC timing, when the time of zero crossover occurs before the attenuated input signal has reached its maximum pulse height. This condition eliminates the rise-time dependence of the zero-crossing time that limits the application of the TCF technique. Figure 8.28 illustrates the signal formation in an ideal CFD for ARC timing with linear input signals. The amplitude independence of the ARC zero-crossing time is depicted for input signals B and C, which have the same amplitude but different rise times. For linear input signals that begin at time zero the zero-crossing time T_{ARC} is[19]

$$T_{ARC} = t_d/(1-f) \tag{8.29}$$

One criteria for t_d to ensure ARC timing is

$$t_d < t_{r(min)}(1-f) \tag{8.30}$$

where $t_{r(min)}$ is the minimum expected rise time for any input signal. This constraint ensures ARC timing for all linear input signals with rise times greater than $t_{r(min)}$ regardless of input pulse width.

In ARC timing the fraction of the input pulse height at which the time pick-off signal is generated is not constant. The effective triggering fraction for each input pulse is related to the attenuation fraction by the input signal rise time. For linear input signals the effective ARC-timing triggering fraction is

$$f_{ARC(eff)} \simeq ft_d/t_r(1-f) \tag{8.31}$$

which is always less than f.

For TCF timing:

$$\sigma_{T(TCF)} \simeq \sigma_V\left(1+f^2\right)^{1/2}/\left(V_A t_r\right) \tag{8.32}$$

For ARC timing:

$$\sigma_{T(ARC)} \simeq \sigma_V\left(1+f^2\right)^{1/2}/\left(V_A(1-f)/t_r\right) \tag{8.33}$$

where V_A is the input signal amplitude and t_r is the input signal rise time.

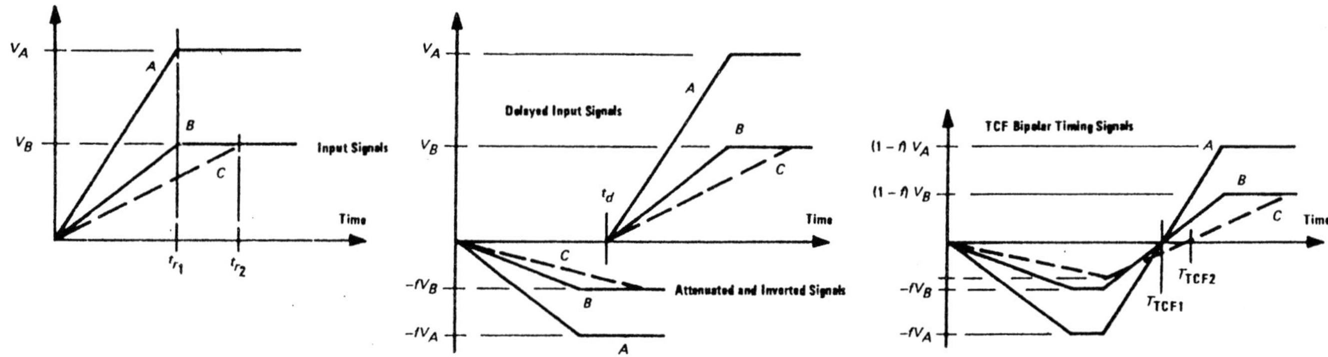

Figure 8.27 Signal formation in a constant-fraction discriminator for TCF timing. (From Paulus, T. J., *IEEE Trans. Nucl. Sci.*, NS32, 1242, 1985. With permission.)

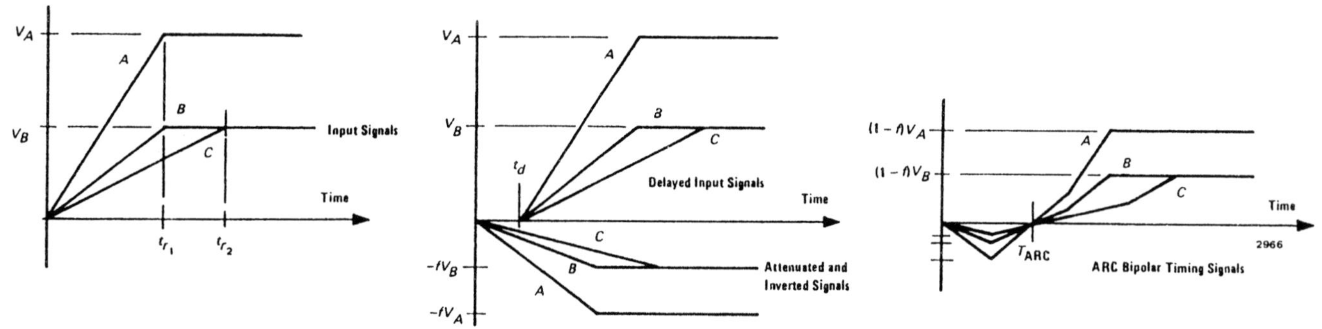

Figure 8.28 Signal formation in a constant-fraction discriminator for ARC timing. (From Paulus, T. J., *IEEE Trans. Nucl. Sci.*, NS32, 1242, 1985. With permission.)

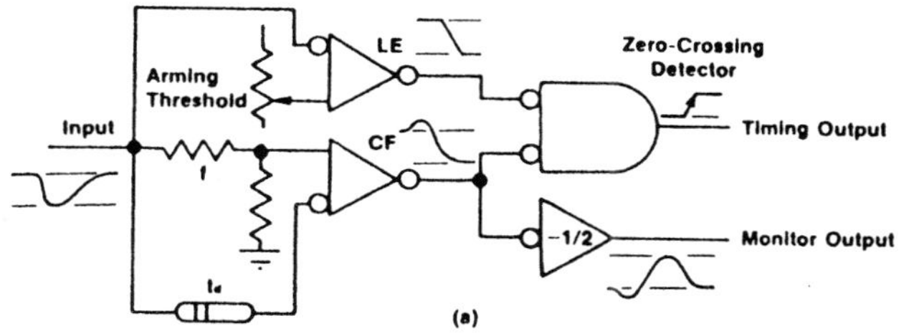

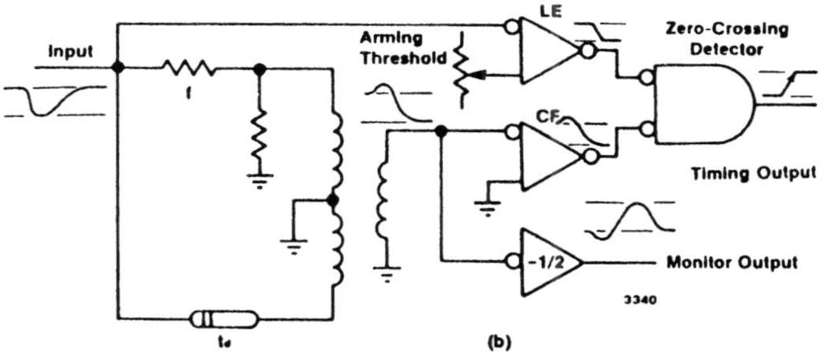

Figure 8.29 Two general circuits used in modern constant-fraction discriminators. (From Paulus, T. J., *IEEE Trans. Nucl. Sci.*, NS32, 1242, 1985. With permission.)

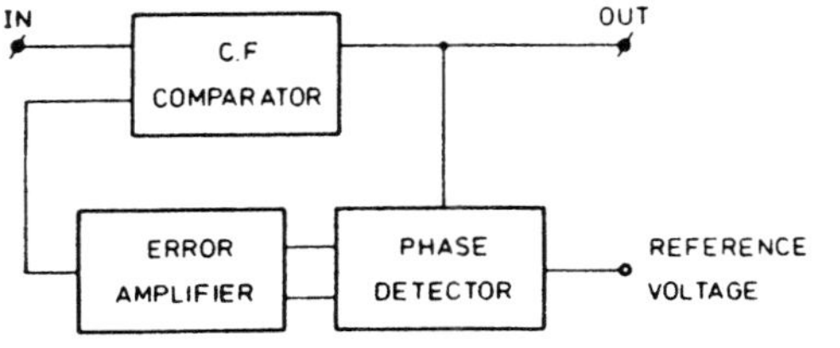

Figure 8.30 Block diagram of the stabilization of triggering point by means of feedback loop principle. (From Bialkowski, J. and Schoeps, W., *Nucl. Instrum. & Methods*, 228, 110, 1984. With permission.)

Under identical input signal and noise conditions and for the same attenuation fraction, the jitter is usually worse for ARC timing than it is for TCF timing due to the more shallow slope of the ARC zero-crossing signal. Under identical input and noise conditions, the jitter is usually worse for TCF timing than for leading-edge timing by the factor $(1 + f^2)^{1/2}$. However, TCF timing virtually eliminates time jitter due to statistical amplitude variations of the signals from the detector so that TCF timing often results in less overall jitter.[19]

Two general circuits are used in modern CFDs. A block diagram of one type is shown in Figure 8.29(a). The upper comparator is a leading-edge discriminator whose output arms the zero-crossing detector. The constant-fraction signal is formed actively in the input differential stage of the lower comparator. The monitor signal is taken at the output of the constant-fraction comparator which is typically an ECL logic level having an 800-mV signal range. Thus the monitor signal is clamped at about 400 mV peak-to-peak. The lowest threshold setting is approximately 5 mV and is determined by the characteristics of the leading-edge comparator.

The block diagram of a second and fundamentally different input circuit is shown in Figure 8.29(b). In this circuit, the constant-fraction signal is formed passively in a differential transformer. The bandwidth of the transformer can be very high, >400 MHz. The monitor output is a close approximation of the actual constant-fraction signal since it is picked off at the input to the constant-fraction comparator. The arming and zero-crossing detector circuits are the same as in Figure 8.29(a). The minimum threshold is about 30 mV due to the attenuation in the transformer resulting from maintaining a 50-Ω impedance level.

A constant-fraction discriminator equipped with an automatic controlled triggering point based on a feedback loop controller principle was described by Bialkowski and Schoeps.[20] The circuit can accept at the input DC (±150 mV) and low-frequency interference (±100 mV ≥ 400 Hz) as well as high-counting rates of change (0 to 10 MHz of 10-nsec FWHM, 2-V pulses) without degradation of timing performance. The trigger was designed for signal rates up to 100 MHz and provides walk equal to ±100 psec over a dynamic range of 400.

Stabilization of the triggering point based on a feedback loop controller is shown in Figure 8.30. The stabilization level is determined by a reference voltage and does not depend on input signal asymmetry

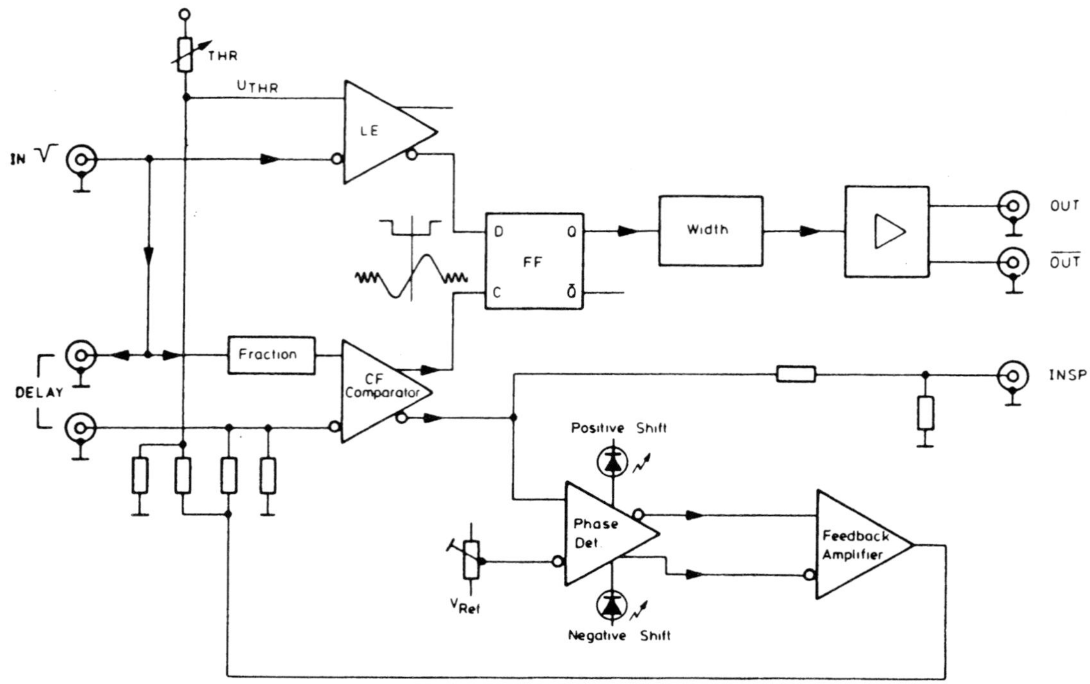

Figure 8.31 Block diagram of the constant-fraction trigger. (From Bialkowski, J. and Schoeps, W., *Nucl. Instrum. & Methods,* 228, 110, 1984. With permission.)

and frequency. This is because of symmetrical processing of the input signal by means of a differential amplifier. An interference signal at the input causes an asymmetry at the differential amplifier output and produces a correction signal from the feedback amplifier to the CF comparator. This also explains why the stability level does not depend on the input frequency.

The circuit description is based on the block diagram shown in Figure 8.31. The constant-fraction discriminator is equipped with a leading-edge (LE) discriminator for energy selection, for which output is presented at the data input D of a fast flip-flop (FF). This FF is triggered at the clock input C by the output of the constant-fraction comparator. In such a mode of triggering of the FF, the LE threshold does not influence the triggering point, which is determined only by crossing the V_{BB} level of the C input by the output pulse of the CF comparator.

The stabilization of the triggering point is ensured by a feedback loop controller which is composed of a high-gain CF comparator; a stable voltage reference; a fast differential amplifier (phase detector); and a feedback amplifier, which corrects the triggering point at the input of the CF comparator. The correction voltage is also added to the threshold of the LE discriminator. Normal operation of the circuit is displayed by equal light intensities of two LEDs. This indicates that in actual conditions of AC and DC interferences at the input the optimal triggering point is ensured by the loop stabilization system working in the linear range. When only the upper or lower diode is lit, the CF comparator input is out of the linear range by a positive or negative voltage, respectively.[20]

An important requirement on the time pickup circuit of the drift chamber is its capability to pick up the timing signal from the heavy core ionization among many background delta-rays which might distort the timing signal of core ionization. A constant-fraction discriminator (CFD) was chosen as the best compromise for dealing with the wide dynamic range of the signals produced by the high-energy heavy ions. Particles with Z = 6 at 1.3 GeV/A produce approximately one energetic delta-ray (whose range is longer than 5 mm) per path length in the chamber. If this delta-ray is on the side of the track nearest the anode, a leading-edge discriminator would trigger on it and not on the heavy ion. To avoid this kind of mistake, the circuit should pick up only the largest pulses during the maximum drift time of ~250 nsec. A schematic diagram of the circuit is shown in Figure 8.32 with pulse shapes at several points in the circuit expected from light ions such as carbon.[21]

A monolithic, 2-μm complementary metal oxide semiconductor (CMOS) constant-fraction discriminator (CFD) was developed by Binkley et al.[22] for use with moderate time resolution (about 4-nsec

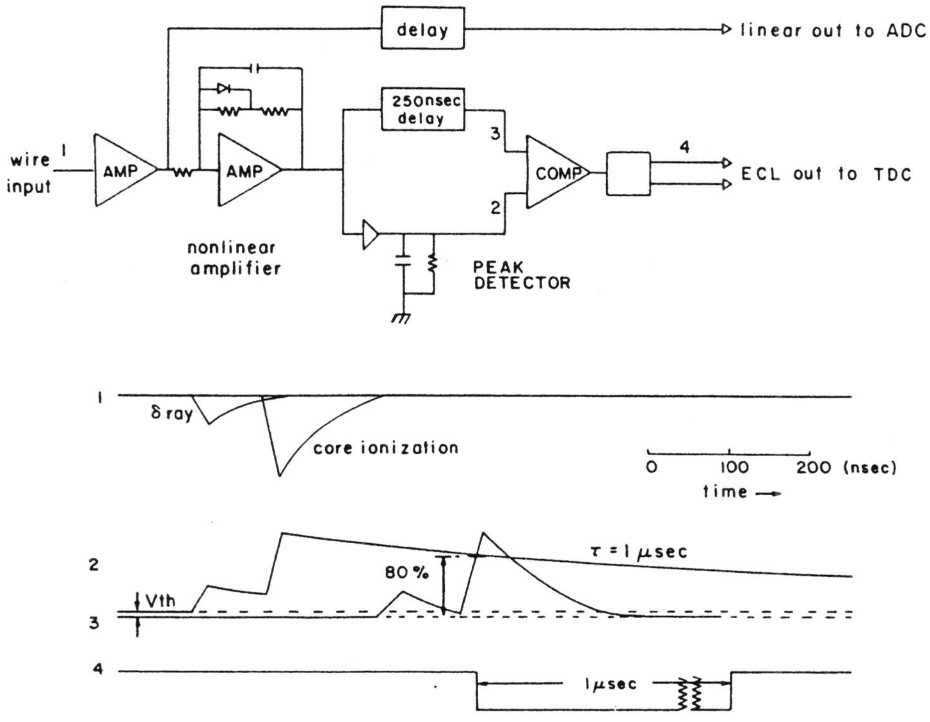

Figure 8.32 Functional block diagram of the front end. Pulse shapes at several points in the circuit are also shown as expected for light ions such as carbon. The circuit picks up the timing signal from the largest pulse inside a 250-nsec time window with a fraction of 80%. (From Kobayashi, T. et al., *Nucl. Instrum. & Methods,* A254, 284, 1987. With permission.)

FWHM) scintillation detectors such as BGO. The circuit was fabricated through the metal oxide-semiconductor implementation service (MOSIS) using a double-metal, double-polysilicon digital process.

The monolithic CMOS CFD consists of an arming comparator, a constant-fraction comparator, and a coincidence gate working from the comparator outputs. No external components for the comparators are provided on the monolithic CMOS CFD. A schematic diagram of the CFD with its external components is shown in Figure 8.33. Additionally, a separate general-purpose comparator is shown which was included on the CMOS CFD for separate evaluation. Figure 8.34 contains a schematic diagram of the constant-fraction and arming comparators, as well as the coincidence gate that make up the CFD.

The external arming-comparator hysteresis (positive feedback) network in Figure 8.33 is included to prevent arming-comparator oscillation and noise triggering. Additionally, an external arming-comparator threshold network is included to set the lower limit on the amplitude of signals that will arm the CFD and result in a timing output.

External constant-fraction attenuation and delay networks are included (Figure 8.33 to set the operating fraction and delay parameters for the CFD. Additionally, an external walk adjustment is used to correct for the constant-fraction comparator input-offset voltage. The walk adjustment can be adjusted beyond the comparator input-offset voltage to minimize CFD walk over a specific input signal amplitude range. Finally, external networks are also included (Figure 8.33) to develop bias voltages for the n- and p-channel comparator output cascade networks.

A single-stage, continuous-time comparator topology was chosen for the CFD-arming and constant-fraction comparators (Figure 8.34). Positive feedback, although frequently used in clocked CMOS comparators to minimize response time, was not employed here because the comparators are applied in a continuous-time application where the comparator voltage decisions occur at arbitrary points in time.

A monolithic constant-fraction discriminator was develop by Tanaka et al.[23] with a bipolar ASIC technology from NEC. In order to pick out timing information, the following techniques were used: (1) leading-edge discriminator, (2) the clipping stub technique, and (3) the constant-fraction technique. The first method has a penalty on time walk The second method requires use of a cable delay, so that it is

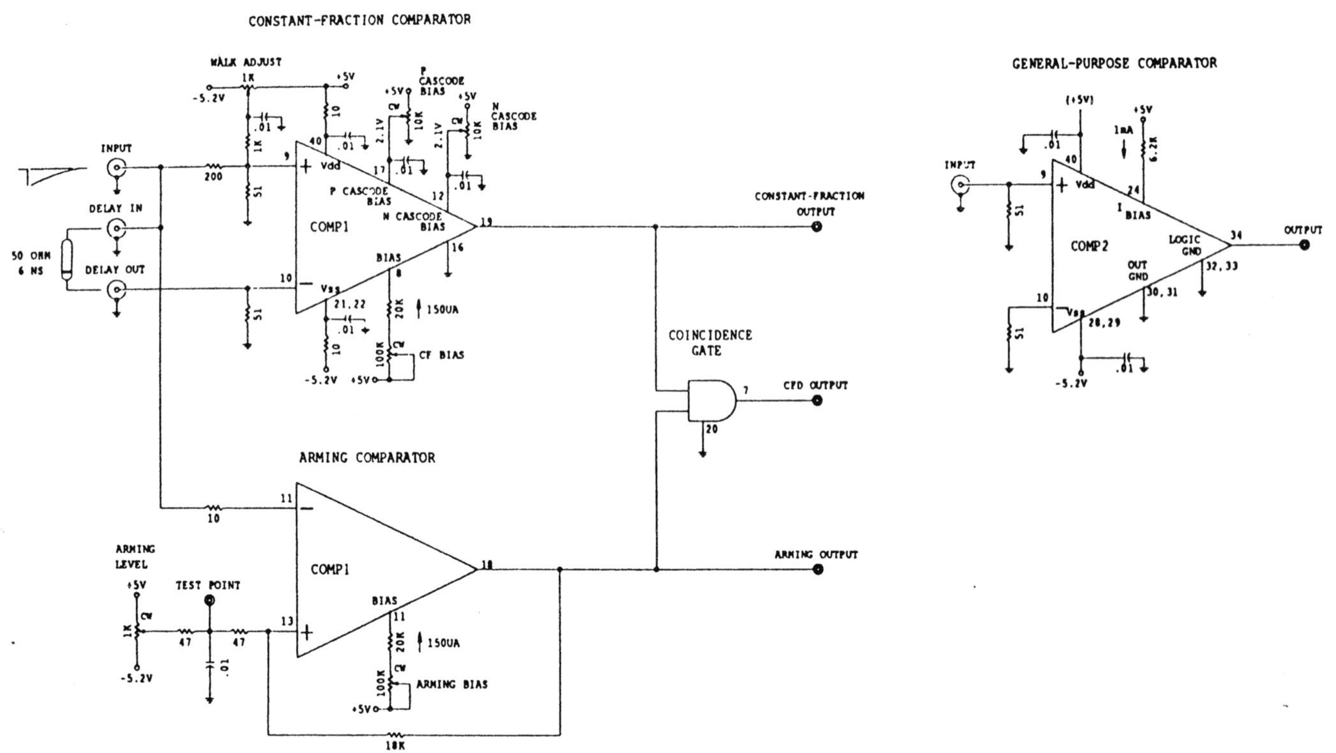

Figure 8.33 Schematic diagram of the CFD including external circuits. (From Binkley, D. M. et al., *IEEE Trans. Nucl. Sci.*, 38, 1754, 1991. With permission.)

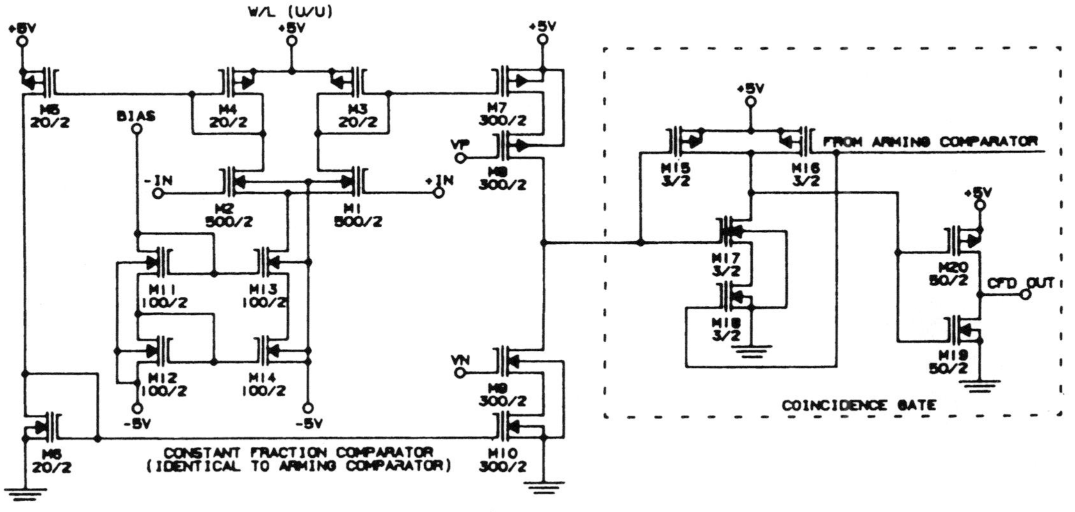

Figure 8.34 Schematic diagram of the CFD comparators and coincidence gate. (From Binkley, D. M. et al., *IEEE Trans. Nucl. Sci.,* 38, 1754, 1991. With permission.)

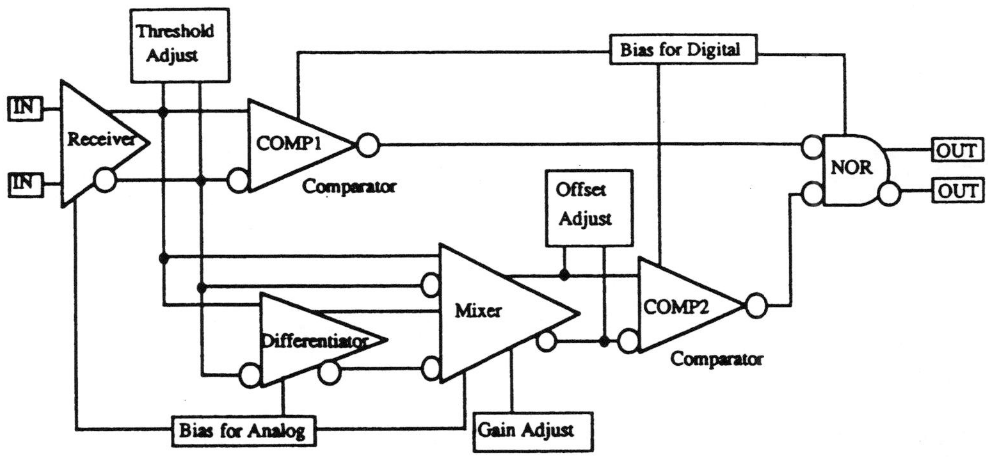

Figure 8.35 Block diagram of the CFD IC. (From Kinbara, S. et al., *Nucl. Instrum. & Methods,* 67, 261, 1969. With permission.)

difficult to achieve stability and high-density integration. The constant-fraction discriminator (CFD) has been used because of its negligible time walk.[24]

The CFD maintains a constant-timing edge since it is not affected by the amplitude of the input pulse. In order to achieve a constant-timing edge, it is customary to use an attenuator and a delay. The input signal is split. One part is delayed, and the other is attenuated. After these two signals are remixed, a comparator detects the zero-cross points. The timing of the zero-cross point is not influenced by the amplitude of the input signal. The delay has a large temperature coefficient, and the component is not suited for the high-density integration technique due to their mechanical constraints.

Therefore, the CFD IC was designed. Its block diagram is shown in Figure 8.35, and its timing chart is shown in Figure 8.36. A significant feature of this design is that the use of a delay and an attenuator is avoided. The input signal is divided into two parts. One part is fed into a leading-edge discriminator (COMP1). The other one is differentiated and mixed with the original signal. The other comparator (COMP2) picks up the zero-cross point of the mixed signal. At the last stage the two signals from the comparators are logically ANDed, and the output signal is differential ECL.

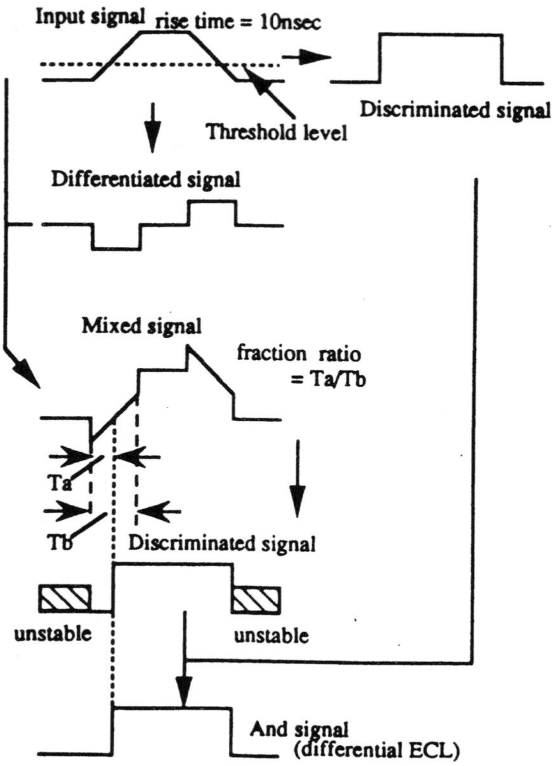

Figure 8.36 Timing chart of the CFD IC. (From Kinbara, S. et al., *Nucl. Instrum. & Methods,* 67, 261, 1969. With permission.)

The use of a high slew rate circuit is advantageous, because the time fluctuation which is caused by the voltage noise decreases. In the CFD IC, the most important feature is the speed of the comparator. If that comparator can detect a zero-cross point completely, the fluctuation of the timing edge of the output signal decreases to almost zero.

REFERENCES

1. **Karlson, L.,** *Nucl. Instrum. & Methods,* 100, 193, 1972.
2. **McDonald, W. J. and Gedcke, D. A.,** *Nucl. Instrum. & Methods,* 55, 1, 1967.
3. **Braunsfurth, J. and Korner, H. J.,** *Nucl. Instrum. & Methods,* 34, 202, 1965.
4. **Gedeke, D. A. and McDonald, W. J.,** *Nucl. Instrum. & Methods,* 58, 253, 1968.
5. **Cho, Z. H. and Helgeson, U.,** *IEEE Trans. Nucl. Sci.,* NS18(1), 108, 1971.
6. **Bedwell, M. O. and Paulus, T. J.,** *IEEE Trans. Nucl. Sci.,* NS23, 234, 1976.
7. **McGervey, J. D. et al.,** *Nucl. Instrum. & Methods,* 143, 435, 1977.
8. **Maier, M. R. and Sperr, P.,** *Nucl. Instrum. & Methods,* 87, 13, 1970.
9. **Mÿllÿla, R.,** *Nucl. Instrum. & Methods,* 148, 267, 1978.
10. **Gal, J. and Bibok, G.,** *Nucl. Instrum. & Methods,* 163, 535, 1979.
11. **Bedwell, M. O. and Paulus, T. J.,** *IEEE Trans. Nucl. Sci.,* NS25, 86, 1978.
12. **Bedwell, M. O. and Paulus, T. J.,** *IEEE Trans. Nucl. Sci.,* NS26, 423, 1979.
13. **Rytsola, K.,** *Nucl. Instrum. & Methods,* 199, 491, 1982.
14. **Hardy, W. H ., II and Lynn, K. G.,** *IEEE Trans. Nucl. Sci.,* NS23, 229, 1976.
15. **Maier, M. R. and Landis, D. A.,** *Nucl. Instrum. & Methods,* 117, 245, 1974.
16. Elscint Snap-Off Timing Discriminator STD-N-1 Manual.
17. **Bryman, D. et al.,** *Nucl. Instrum. & Methods,* 141, 573, 1977.
18. **Plischke, P. et al.,** *Nucl. Instrum. & Methods,* 136, 579, 1976.
19. **Paulus, T. J.,** *IEEE Trans. Nucl. Sci.,* NS32, 1242, 1985.
20. **Bialkowski, J. and Schoeps, W.,** *Nucl. Instrum. & Methods,* 228, 110, 1984.
21. **Kobayashi, T. et al.,** *Nucl. Instrum. & Methods,* A254, 284, 1987.
22. **Binkley, D. M. et al.,** *IEEE Trans. Nucl. Sci.,* 38, 1754, 1991.
23. **Tanaka, M. et al.,** *IEEE Trans. Nucl. Sci.,* 39, 1321, 1992.
24. **Kinbara, S. et al.,** *Nucl. Instrum. & Methods,* 67, 261, 1969.

Chapter 9

Pulse Shape Discriminators

CONTENTS

9.1 Introduction 107
9.2 Zero Crossing 107
9.3 Other Methods 113
References 130

9.1 INTRODUCTION

Pulse shape discriminators (PSD) are generally used to distinguish between pulses created by different kinds of radiation. The method is based on the fact that sometimes different kinds of radiation have different mechanisms of interaction with the detector; thus pulses of different shapes (different rise time or fall time) are created. The most common use is the distinction between neutron and gamma-ray pulse counting. The phenomenon has been found in the three major kinds of detectors: gas filled, scintillators, and solid state.

There are two main methods used to obtain pulse shape discriminations. The first is pulse integration (leading-edge discrimination [LED]), and the second is zero-crossing discrimination (ZCD). The pulse shape discrimination must be independent of pulse amplitude. One of the first systems designed by Brooks[1] is shown in Figure 9.1. The system is composed of two integrators of different integration times and a circuit that gives the difference between the two. When a scintillation detector is used, it is possible to obtain one pulse from the dynode and the other from the anode which are inverted to each other. The dynode integration time is long and that of the anode is short. The diodes are used as pulse stretchers. The fast pulse overshoot is canceled by the resistors. The slow reaction pulse charges C_1 but not C_2, and the dynode pulse is larger than that of the anode; thus the combination gives a positive pulse whenever a neutron is present and a negative pulse when there are no neutrons.

In the ZCD method, use is made of the fact that after double differentiation zero-crossing time is independent of amplitude of the original pulse but on its rise time. Integration of the detector pulse provides a pulse of rise time relative to the scintillation pulse and fall time thus different for neutron and gamma. Two discriminators are combined; one gives the start signal by leading-edge discriminator and the second gives the end by zero crossing. The time difference is translated by a time-to-amplitude converter (TAC) to pulse height (Figure 9.2).

9.2 ZERO CROSSING

A zero-crossover pulse shape discriminator was described by Taylor and Kalyana.[2] The circuit is shown in Figure 9.3. The fast anode pulse is divided into three pulses: noise, and upper and lower level discriminators. The combined pulse $A \cdot B \cdot \overline{C}$ supplies a start pulse to a TAC circuit and operates a crossover control gate. The crossover detector produces an output timed from the crossover of the input bipolar pulse which is used to produce a stop pulse to the TAC.

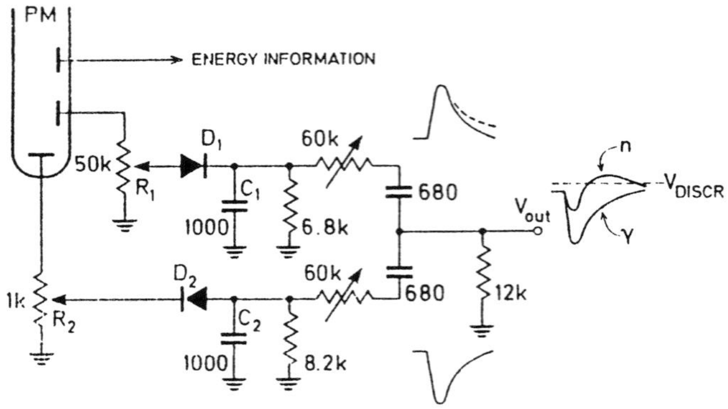

Figure 9.1 Pulse shape discriminator. (From Brooks, F. D., *Nucl. Instrum. & Methods,* 4, 151, 1959.)

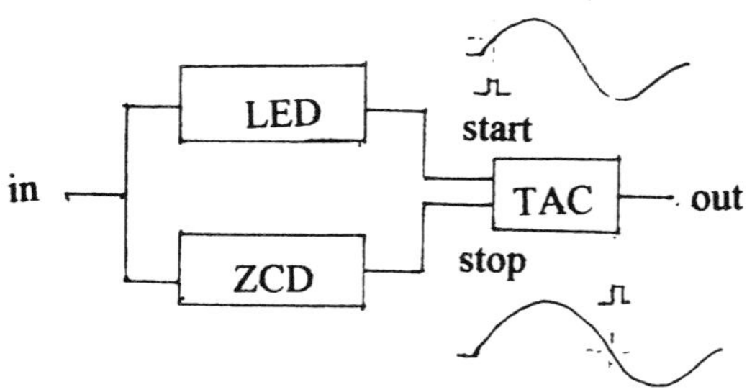

Figure 9.2 LED and ZCD pulse shape discriminator.

The crossover detector operates as follows: the dynode pulse is amplitude limited at the inputs by a back-to-back diode limiter which, however, preserves the crossover characteristics. An emitter follower Q_2 feeds the signal to a tunnel diode via a grounded base transistor. Another emitter follower is used to provide an output test point for inspection of the crossover signal. The tunnel diode is thus triggered at the crossover and is reset when the reverse bias current is restored.

Pulse shape discrimination systems were investigated by Winyard and McBeth[3] using the zero-crossing technique. The figure of merit (gamma-neutron peak separation divided by the sum of full-width half maximum of the two peaks) of several scintillators was investigated. Some results are given in Table 9.1.

A block diagram of the system used to determine the lower energy limit of PSD in NE213 is shown in Figure 9.4.

Improved PSD zero crossing discrimination was described by Rendic et al.[4] The system is shown in Figure 9.5. The pulse current was supposed to be composed of two exponents

$$I = I_{of} e^{-t/\tau_f} + I_{os} e^{-t/\tau_s} \tag{9.1}$$

The time difference t measured by the TAC is proportional to:[3]

$$t \sim \tau_s \ln\left[(Q_s/Q)\left(1 - e^{-\tau_d/\tau_s}\right)(\tau_{rc}/\tau_d)\right] \tag{9.2}$$

where τ_f, τ_s are the time constants of the fast and slow components, Q_s is the slow component pulse charge, Q is total pulse charge, τ_d is the delay-line time constant, and τ_{rc} is the time constant of the RC-differentiating network.

A simple PSD was designed by Sperr et al.[5] It was based on the zero-crossing method which had the advantage of large dynamic range >1:100. The block diagram of the system is shown in Figure 9.6. The anode pulse was integrated and differentiated in a preamplifier A_1 so that the zero-crossing point of the output pulse was determined by the fall time of the input pulse. The shaped signal was then fed to a high-gain limiting amplifier A_2. The separation of the zero-crossing points corresponding to gammas or neutrons was greatly enhanced as shown in Figure 9.7. The overlap of the amplifier signal and the strobe pulse is formed in a gate G. If the strobe signal is placed between the crossing points of gamma and

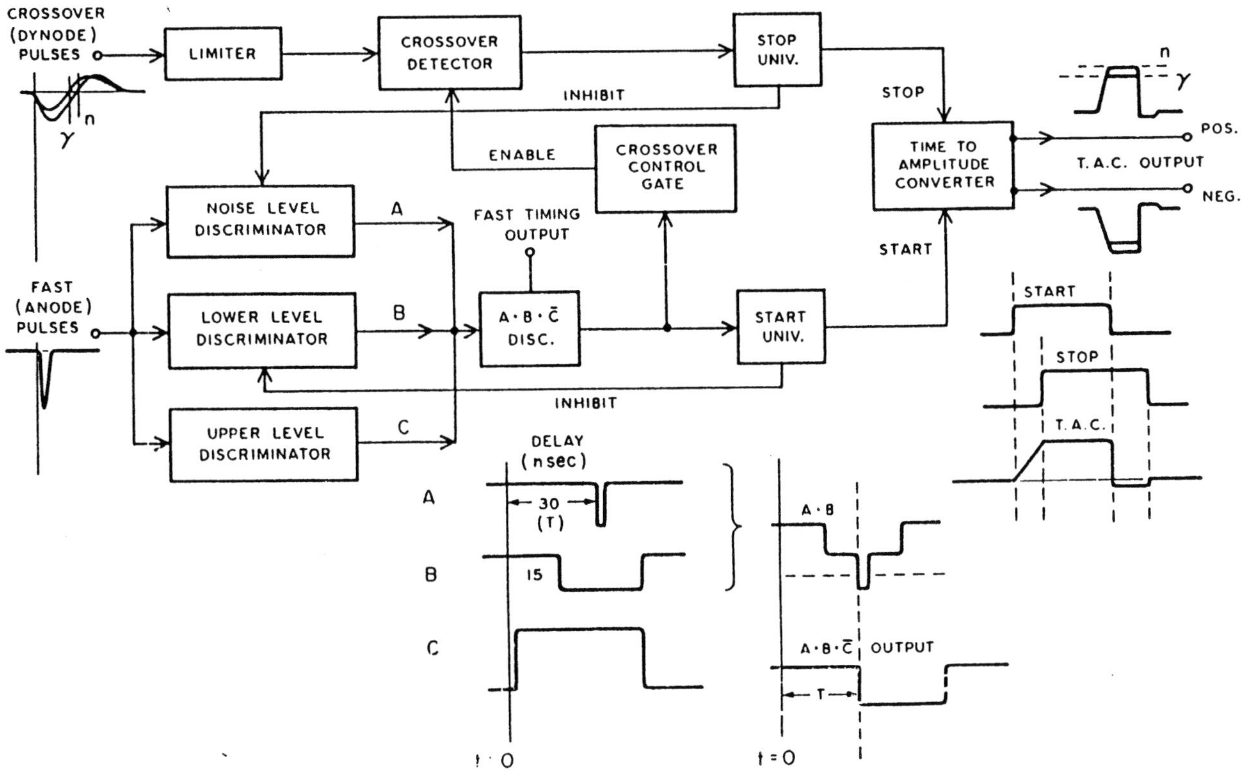

Figure 9.3 Block diagram of the pulse shape discriminator circuit. (From Taylor, I. J. and Kalyana, J., *Nucl. & Instrum.* Methods, 88, 267, 1970. With permission.)

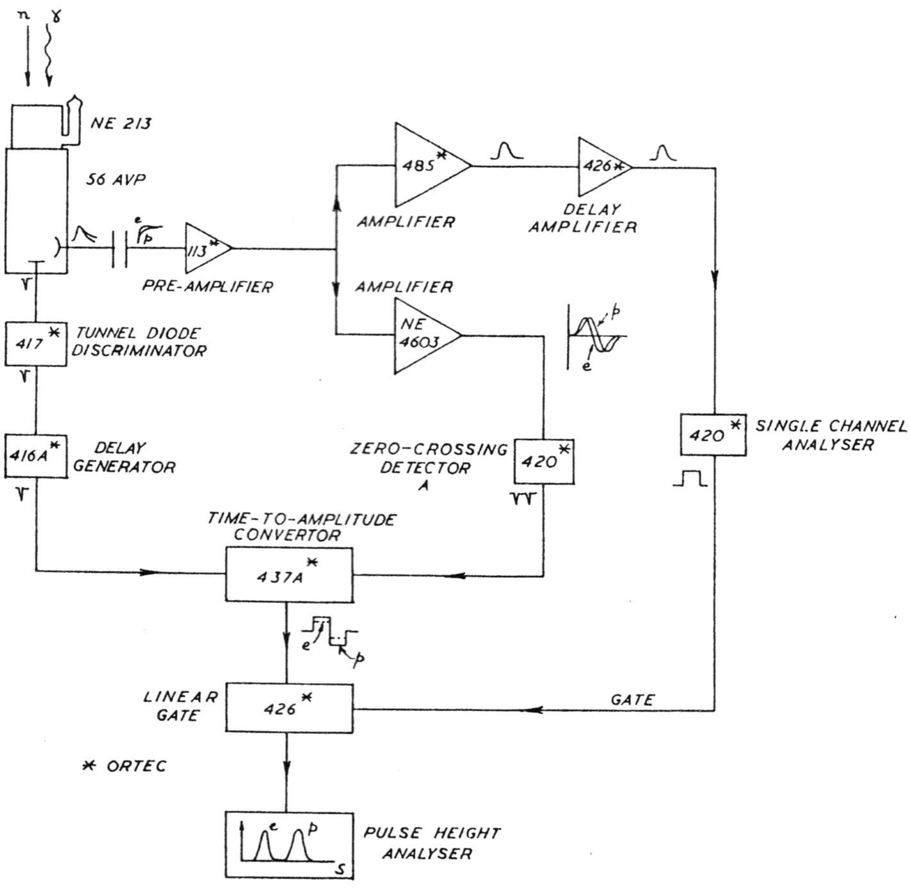

Figure 9.4 Block diagram of the electronics employed in the determination of the lower energy limit to PSD in NE213. (From Winyard, R. A. and McBeth, G. W., *Nucl. & Instrum. Methods,* 98, 525, 1972. With permission.)

Table 1 Figures of Merit for Neutron, Gamma Discrimination in Liquid Scintillators.

Scintillator	Solvent	$M_{n,\gamma}$[a]	
Pilot PSD Fluid	1 Toluene	2.7	3.3
Pilot PSD Fluid	2 Xylene	2.7	3.5
NE211	Xylene	1.5	2.4
NE213	Xylene	2.5	3.1
NE216	Xylene	1.9	2.0
NE218	Mixture of trimethylbenzenes	1.5	1.8
NE220	Dioxan	1.7	2.2
NE223	Decalin	1.4	1.5
NE224	Pseudo-cumene	2.1	2.8
NE230	Deuterated benzene	1.5	2.2
NE231	Benzene	1.4	1.8

[a] 0.05 to 7 MeV: 0.9 to 1.1 MeV.

From Winyard, R.A. and McBeth, G.W., *Nucl. Instrum. Methods,* 98, 525, 1972. With permission.

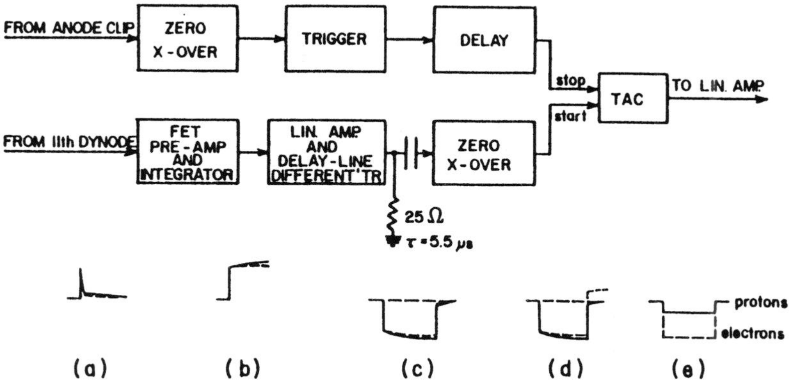

Figure 9.5 Wave forms (a–e) and a block diagram of the pulse shape discrimination system (PSD). (From Rendic, D. et al., *Nucl. Instrum. & Methods,* 99, 189, 1972. With permission.)

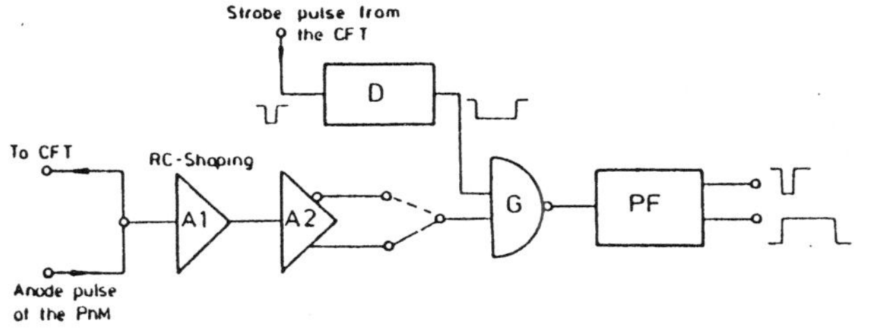

Figure 9.6 Block diagram of the PSD circuit. (From Sperr, P. et al., *Nucl. Instrum. & Methods,* 116, 55, 1974. With permission.)

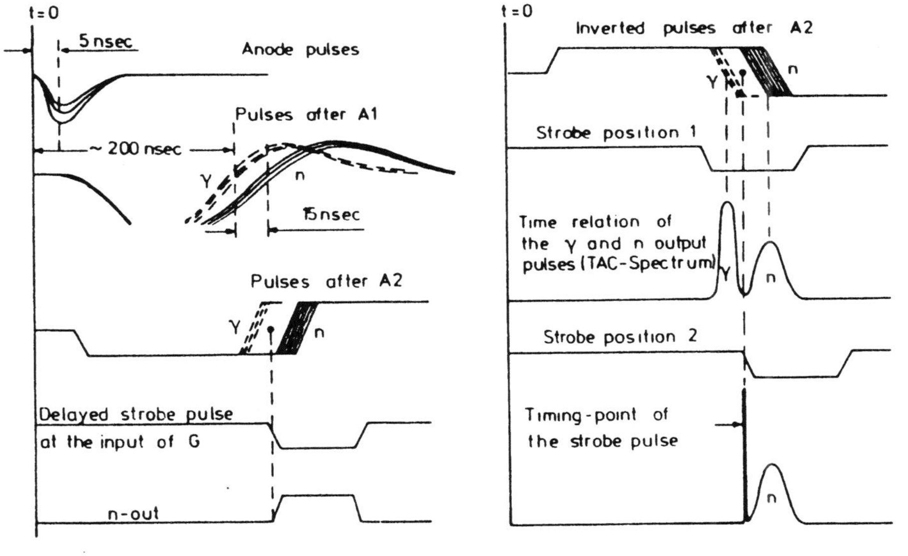

Figure 9.7 Time relationship between the pulses at various points of the PSD circuit. (a) Normal operation; (b) during adjustment. (From Sperr, P. et al., *Nucl. Instrum. & Methods,* 116, 55, 1974. With permission.)

neutron pulses, only pulses caused by neutrons will produce an output pulse. In Figure 9.6 the strobe signal was triggered by the output pulse of the constant-fraction trigger (CFT) which was delayed and stretched in D.

A fast pulse shape discriminator circuit was described by Betz et al.[6] The circuit provided the pulse shape (PS) information within 300 nsec and allowed the gate the fast-timing signal with the logic neutron or gamma information. The block diagram (Figure 9.8) shows the essential parts of the neutron-gamma

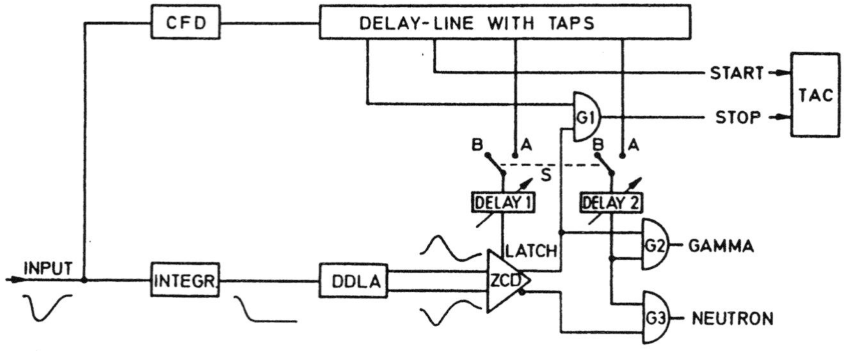

Figure 9.8 Block diagram of the neutron-gamma discriminator. (From Betz, P. et al., *Nucl. Instrum. & Methods,* 119, 199, 1974. With permission.)

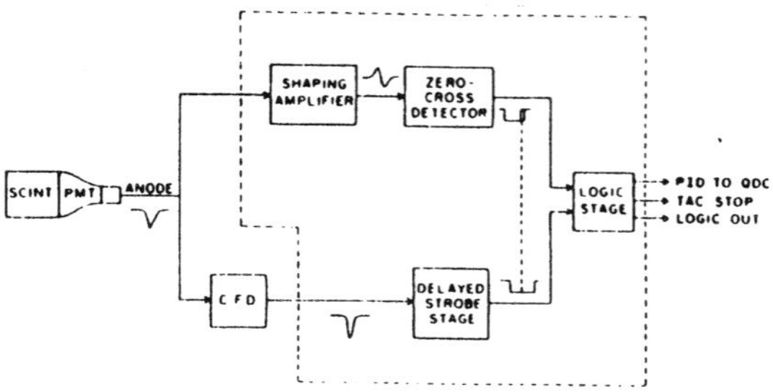

Figure 9.9 Block diagram of the PSD circuit. The part in the dashed box is contained in the module being described and tested (From Pai, S. et al., *Nucl. Instrum. & Methods,* A278, 749, 1989. With permission.)

discriminator: (1) constant-fraction discriminator (CFD); (2) integrator (INTEGR); (3) double delay-line amplifier (DDLA); (4) zero-crossing detector (ZCD) with latch; and (5) gating network (G1, G2, G3).

The anode pulse of the photomultiplier is split at the input and simultaneously fed into the CFD and the integrator. The input load is 50 Ω. The decay time of the output signal is long compared to the rise time, which is given by the width of the anode pulse. In the subsequent DDLA this signal is clipped twice. As the rise time of the integrator output pulse depends on whether it is generated by a neutron or a gamma, the zero-crossing points of the respective DDLA signals appear at different times, as well as the related edges of the ZCD outputs. With switch S in position B, a time spectrum of both gammas and neutrons can be taken using the signals START (from the CFD) and STOP (from the ZCD) as inputs for a time-to-amplitude converter (TAC). With switch S in position A, the outputs of the ZCD are locked in their current logical states, as long as a signal is supplied to the latch input. If the leading edge of that signal appears between the distinct zero-crossing points of gammas and neutrons from the DDLA, the ZCD will trigger for a gamma but not for a neutron.

The delayed CFD output is supplied to the inputs of gate G2 and G3, which according to the state of the ZCD provide the output pulses in the case of a gamma or a neutron, respectively.

A versatile electronic pulse shape discriminator was designed by Pai et al.[7] It is based on the zero-crossing technique for determining the time dependence of the scintillations that are produced by different species of particles. The circuit determines the time difference between the initial rise of an anode pulse and the time at which a bipolar pulse, produced by a shaping circuit, crosses zero. This time difference is sensitive to the proportion of decay components in the light output but is insensitive to the amplitude of the pulse.

A block diagram of the pulse shape discrimination circuit is given in Figure 9.9. The required input signals are an anode signal from a photomultiplier tube and a nuclear instrument module (NIM) standard negative pulse from an external constant-fraction discriminator (CFD). The anode signal is first passed through a shaping amplifier. The output of this stage is ready for gating after passing through a zero-crossing detector. The signal from the CFD is delayed, reshaped, and used as a strobe. The overlap of the zero-crossing detector output and the strobe is obtained in the logic stage.

There are three output signals which are produced as shown in Figure 9.9. One is a NIM standard negative pulse which can be used to stop either a TAC or a time-to-digital converter. Another pulse, the particle identification signal (PID), is meant to be fed to a charge-to-digital converter (QDC). Both of these output pulses contain equivalent information about the shape of the anode pulse. The third output pulse is a NIM logic pulse which does not contain pulse shape information but merely signifies that an output pulse has occurred within the timing window. This pulse may be useful for certain gating requirements.

A more detailed version of the circuit is presented in Figure 9.10. This version is designed to minimize the number of components. The anode signal is processed by the pulse-shaping amplifier to generate a bipolar signal. The zero-crossing point of the output pulse is determined by the time structure of the input pulse. The pulse-shaping amplifier utilizes a wideband video amplifier (Signetics NE592), configured as a bandpass filter. It allows for a gain which is adjustable from 0 to 400 and for an adjustable pass band. Hence, the filter network components can be designed to give the best results for each type of scintillator with its characteristic decay times. The output of the NE592 is capacitively coupled to the zero-crossing detector in order to block the DC component in the output of the video amplifier, thereby ensuring zero-crossover detection. The complementary output of the video amplifier is buffered through a transistor and brought out to a monitoring point on the front panel through a coupling capacitor.

The zero-crossing detector is an ultrafast comparator (Plessey SP9680). The reference level for this comparison can be adjusted by a knob on the front panel.

The strobe from the CFD is fed to the emitter-coupled logic (ECL) delay stage, after a level translation by a transistor circuit to generate an ECL compatible signal. The timing of this signal is modified using a combination of two monostable multivibrators. Only those comparator transitions which occur during the time window are allowed to produce the three output signals. The output of the comparator is gated by the strobe in the logic stage using a set-reset flip-flop. One of the outputs of this flip-flop is the PID to QDC output, shown in Figure 9.10. The other flip-flop is used to trigger a monostable multivibrator which generates the LOGIC pulse. Finally, the outputs of the monostable and the later flip-flop are OR gated together and used to generate TACSTOP. The three output pulses are then level shifted to produce NIM compatible pulses.[7]

A pulse shape discriminator based on the zero-crossing principle was described by Bialkowski et al.[8] Due to DC negative feedback loops stabilizing the shaping amplifier and the zero-crossing discriminator, the working of the circuit was not affected by the high-counting rate and the temperature variations. The pileup rejection circuit built into the discriminator improves the quality of the neutron-gamma separation at high-counting rates.

The block scheme of the neutron-gamma pulse shape discriminator is presented in Figure 9.11. The circuit is composed of linear and logic parts. In the linear part of the discriminator, the anode pulse of the photomultiplier is integrated and differentiated in the input amplifier so that the zero-crossing point of the output pulse is determined by the fall time of the input pulse. Afterward the signal is sent to the fast zero-crossing comparator to distinguish a difference between the neutron and gamma pulses.

The logic circuitry triggered by the strobe pulse from the external constant-fraction discriminator consists of a pileup detector and delay selection and time reference circuits. The timing information about the zero-crossing point is standardized by means of the "neutron + gamma FF" flip-flop. This flip-flop is triggered by the strobe signal via the delay section circuit and reset at the zero-crossing discriminator. Two outputs of the neutron + gamma FF circuit as well as the outputs of the delay selection and the pileup detection circuits are used to produce gamma + neutron, neutron and gamma signals. These signals are standardized for TTL (0.5 μsec duration) and fast NIM (10- to 500-nsec) pulses. The zero-crossing discriminator was designed using the fast AM685 comparator operated in the linear range.

9.3 OTHER METHODS

A pulse shape discriminator was described by White[9,10] whose operating principle was based on the comparison of two weighted time integrals of the detector signal. The method used was that of charge comparison. A simplified block diagram with waveforms is shown in Figure 9.12. The signal was integrated in a pure integrator for a time T_1 (25 nsec) and then held; a second integrator opened coincidentally with the first but continued integrating for T_2 (400 nsec). The outputs of the integrators were so weighted that when they were compared, the output of the comparator indicated a neutron, if the output of Integrator 2 exceeded that of Integrator 1 and a gamma-ray if the reverse was true.

This can be represented by the following. If:

$$K_1Q_1 > K_2(Q_1 + Q_2) \tag{9.3}$$

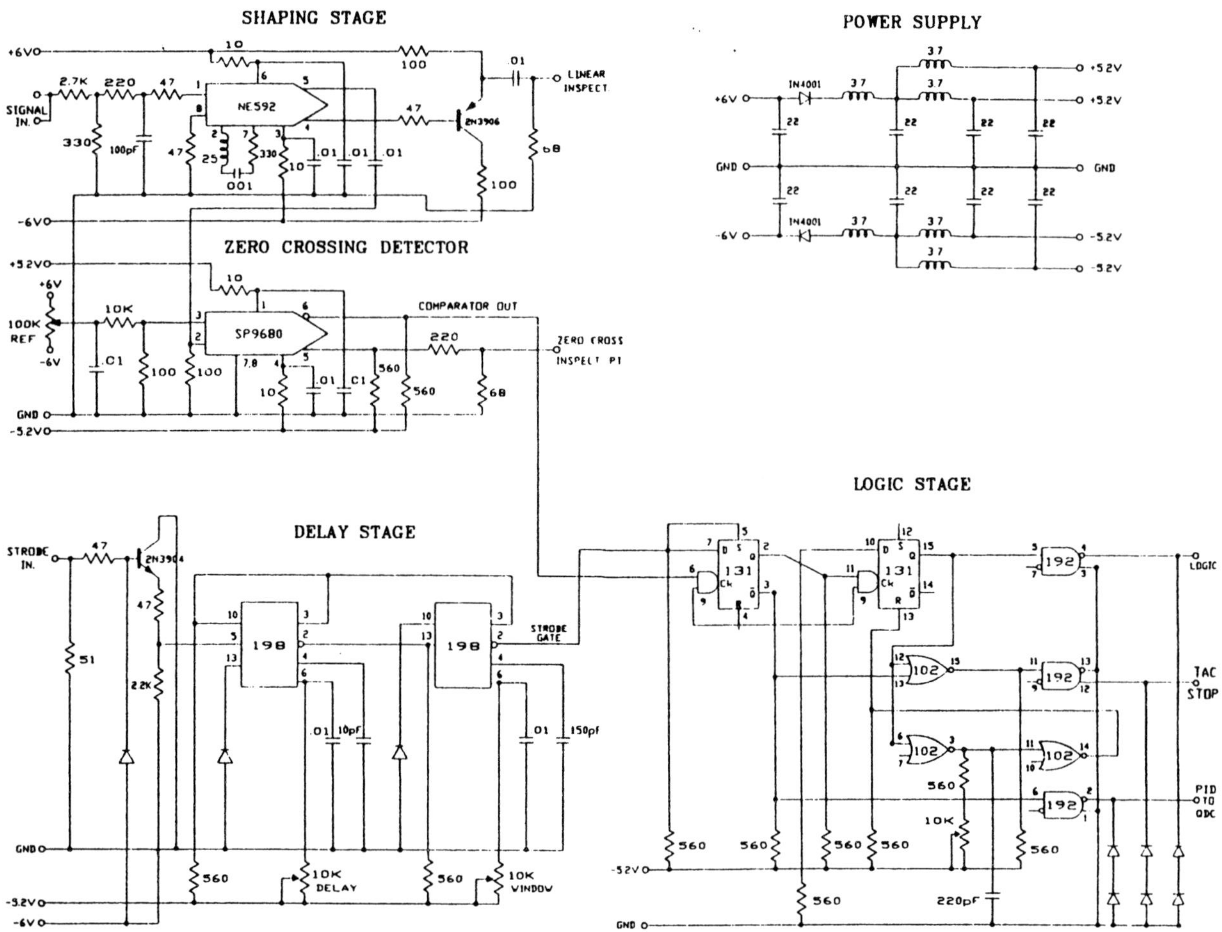

Figure 9.10 Circuit diagram of the version of the pulse shape discriminator which used the minimum number of components. Unless otherwise indicated, the values for the capacitors are given in microfarad, for the resistors in ohm and for the inductors in microhenry. Each unlabeled diode is a 1N914. Several ECL components are labeled by the last three digits of their identification number, e.g., 102 means MC10H102. (From Pai, S. et al., *Nucl. Instrum. & Methods,* A278, 749, 1989. With permission.)

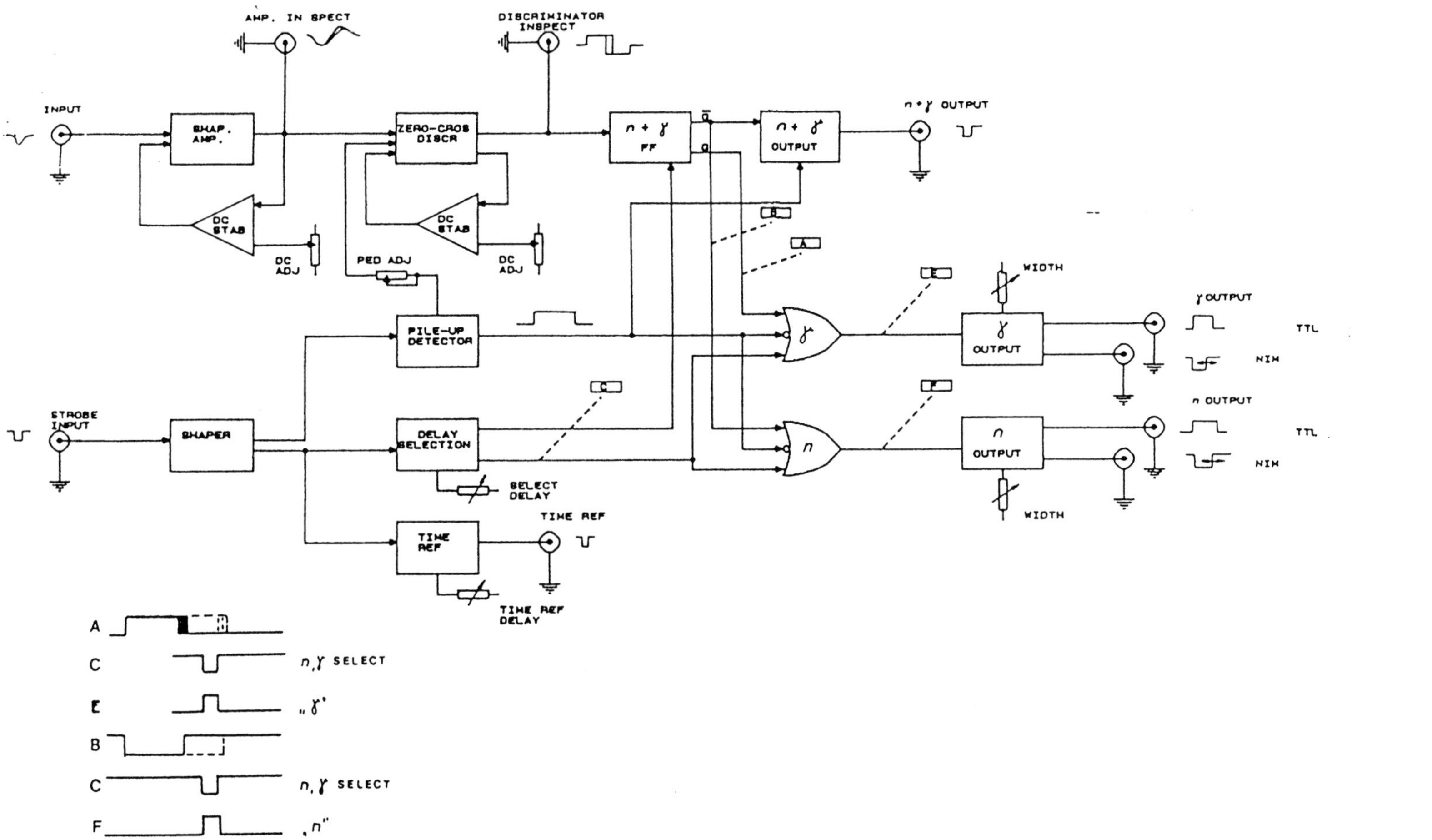

Figure 9.11 Block scheme of the neutron-gamma pulse shape discrimination circuit. (From Bialkowski, J. et al., *Nucl. Instrum. & Methods,* A275, 322, 1989. With permission.)

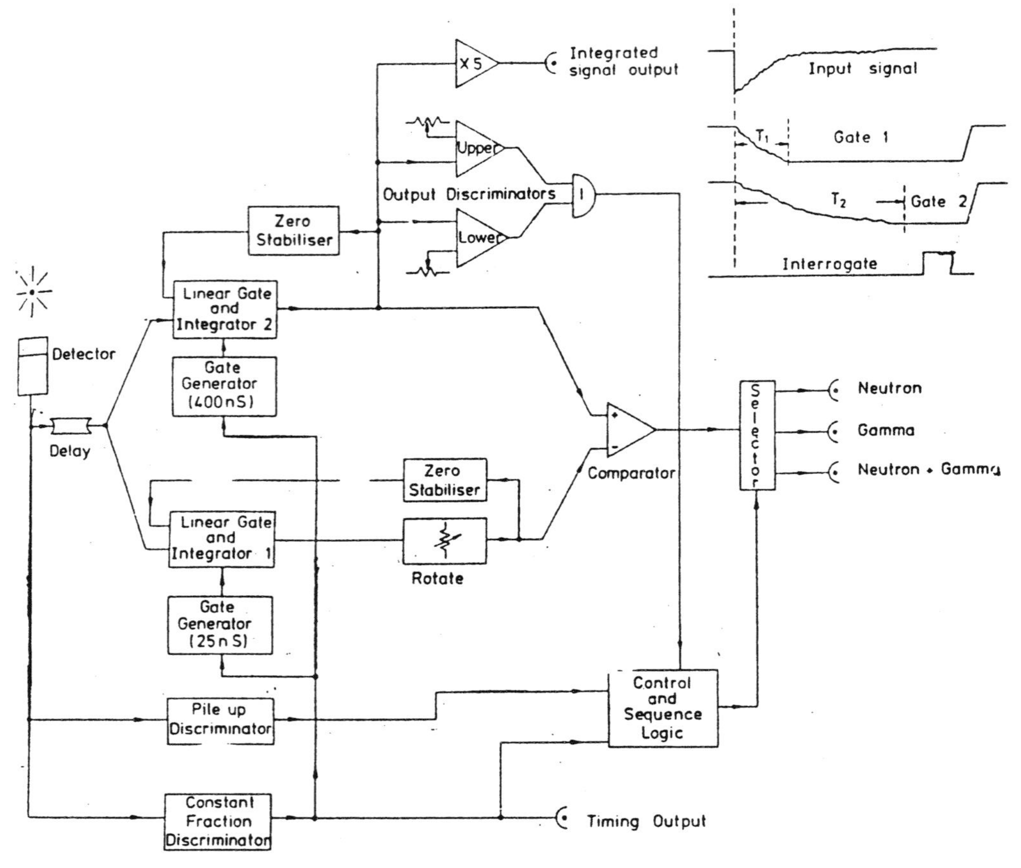

Figure 9.12 Simplified block diagram with waveforms. (From White, G., in *Proc. 2nd Ispra Nucl. Elec. Symp.* EUR 5370, 1975, p. 447.)

then a gamma-ray is indicated and if:

$$K_1Q_1 < K_2(Q_1 + Q_2) \tag{9.4}$$

a neutron is indicated, where Q_1 is the charge collected in the time T_1, $Q_1 + Q_2$ is the charge collected in the time T_2, and K_1 and K_2 are the normalizing weights.

The two gate time generators are the control and sequence logic; these were initiated by a constant-fraction discriminator which was triggered from the input signal.

The signal to the gated integrators was delayed by a short cable until the gate was fully open. The gate of the first integrator closed after about 25 nsec (T_1). After 400 nsec, the second gated integrating time (T_2), the state of the comparator was interrogated by time control and sequence logic and a neutron or gamma-ray was then indicated. If the pileup discriminator detects a second event within t_2, the neutron and gamma outputs are inhibited. Illustration of the system output is shown in Figure 9.13.

An application of antiwalk discriminators was made by Emmer[21] in pulse shape discriminators for separating different mass charge particles and gamma-rays in inorganic and organic scintillators. The circuit shown in Figure 9.14 has been used to separate gamma-rays from alpha particles in NaI scintillators. Since the crossover time of a double differentiated pulse depends on the pulse shape, pulse shape discrimination can be accomplished by a time coincidence between the crossover time and the delayed leading edge. The circuit shown in Figure 9.14 is identical in operation to the fast coincidence circuit with the exception that the back edge of a delay trigger pair (T1 and T2) was adjusted to be in time coincidence with the shaped antiwalk discriminator current pulse. Either the gamma or the alpha pulses can be selected by proper adjustment of the delay trigger pair.

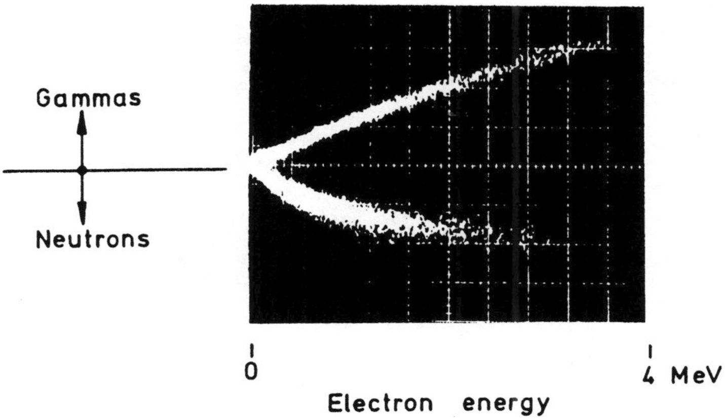

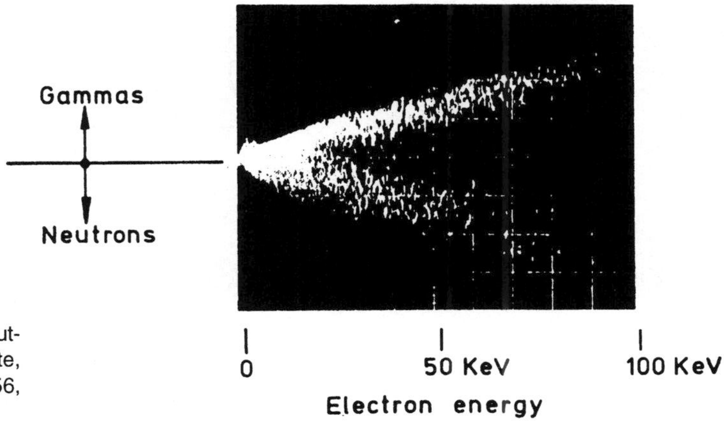

Figure 9.13 Oscilloscope display output. (From Adams, J. M. and White, G., *Nucl. Instrum. & Methods,* 156, 459, 1978. With permission.)

ORTEC[11] developed a pulse shape analyzer which measures the time during which the input waveform falls from 90% of its peak amplitude to 10% of its peak amplitude, providing the peak amplitude is sufficient to trigger the input discriminator. The input pulse fall time is converted into a linear output pulse, with an amplitude range of 0 to 10 V corresponding to the selected time range. The total range for the time measurement is 200 nsec to 10 μsec. The input discriminator must be triggered by an input pulse in order for its fall time to be measured.

The ORTEC 458 pulse shape analyzer consists of three basic functions: an input and compression circuitry, a dual constant-fraction discriminator, and a time-to-pulse-height converter. These three basic functions lead to the generation of a linear output. A single-channel analyzer is also included for logical analysis of the linear output amplitudes.

The dual constant-fraction discriminator generates two time markers. The first time marker is generated at a point on the trailing edge of the unipolar input pulse when the pulse has decayed to approximately 90% of its peak amplitude. The second marker is generated when the input pulse has decayed to approximately 10% of its peak amplitude.

The time-to-pulse-height converter then measures the time between the 90 and 10% markers and stores the time information on a capacitor. The voltage on the capacitor is strobed, shaped, and routed to the linear output. This voltage is in the range of 0 to 10 V, independent of the input amplitude from the linear amplifier.

Three different types of low-power pulse shape discriminators have been designed and built by Wilson et al.[12] for applications in space experiments. The techniques were a zero-cross PSD, a decay-time PSD, and a rise-time PSD. They were used for investigation of dynamic range, jitter due to photon statistics, and NaI-CsI mixture separation (in Phoswich).

In the zero-cross PSD, the width of the positive lobe of a bipolar-shaped detector pulse was used to distinguish a NaI pulse from a CsI pulse (or mixture). A diagram of the circuit is shown in Figure 9.15. The leading edge of the pulse was detected by the comparator with a positive threshold. There was a time

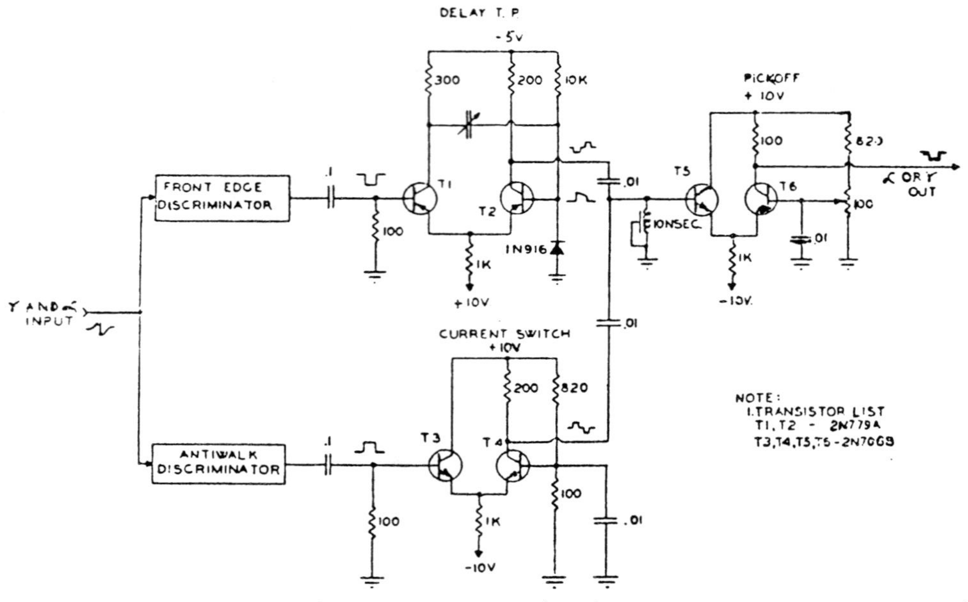

Figure 9.14 Pulse shape discriminator.[21] (From Emmer, T. L., in NAS-NRC Pub. 1184, Monterey, 1953.)

walk which depends on the pulse amplitude. A first-order correction to the time walk was made using a negative threshold on the negative lobe of the bipolar pulse.

A timing threshold generator by a one-shot was used to distinguish a NaI zero-cross time from a CsI mixture. The usual technique was to perform a time to pulse-height conversion and then discriminate against a CsI pulse shape with a pulse-height discriminator. The veto signals from the PSD and level discriminators were connected to timing multivibrators and logic gates so that all of the critical timing could be adjusted and verified to work over a wide range of pulse heights.

The decay-time PSD circuit operated by subtracting a generated NaI pulse from the composite NaI and CsI input pulse from the detector. This difference was proportional to the amount of CsI present. A sampled comparator operating on the subtracted signal triggered a veto multivibrator if the CsI level was above a threshold. The circuit also contained an active baseline clamp which made it unnecessary to use a differentiator in front of the PHA.

A block diagram of the circuit is shown in Figure 9.16. A unipolar pulse integrated as little as possible, consistent with amplifier slew rate and linearity restrictions, and was used as the input to the discriminator. In one branch of the circuit, this pulse crossed an early threshold which triggered a 1.8-μsec timing multivibrator. In another branch of the circuit, the pulse was delayed by 0.2 μsec, buffered, and then capacitor coupled to a clamp and then to a buffer amplifier. The clamp held the signal at ground level until the threshold was actuated. At this time (0.1 μsec before the pulse), the clamp opened and the signal came through at the baseline that had been present 0.1 μsec before the pulse. The output of the buffer amplifier was further integrated, delayed, and fed to the PHA.

In another circuit branch the pulse was fed to a delay line differentiator and cut off on the trailing edge so that the full time can be regulated by an RC time constant. The output pulse (generated NaI) had an amplitude equal to that of the input pulse and the shape of NaI pulse.

The generated NaI pulse and the input pulse were of opposite polarity. They were summed in a resistor network, and the difference signal is amplified (CsI amplifier). The CsI output was then integrated by a clamped RC network. The capacitor charge was sampled at 1.8 μsec. If a positive threshold (indicating the presence of a CsI component) were crossed, then the sampled output would trigger a multivibrator which was the veto signal.

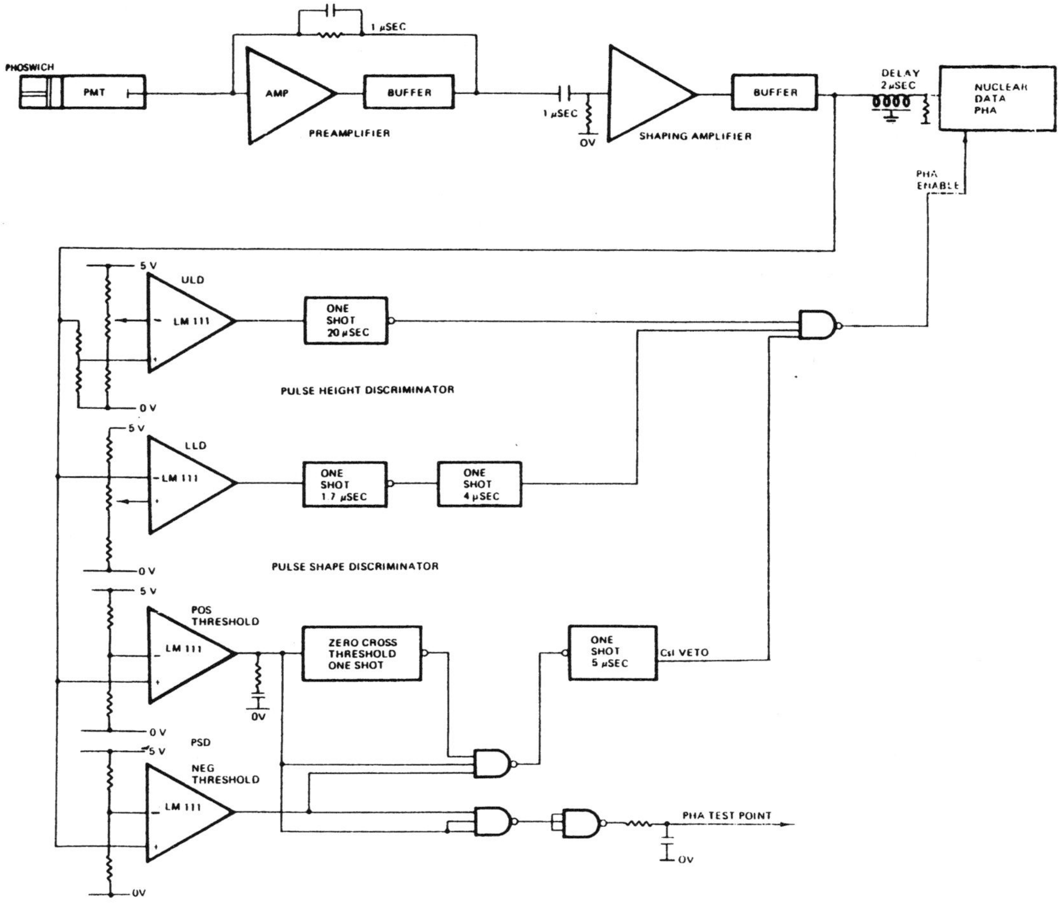

Figure 9.15 Zero-cross PSD system. (From Wilson, S., *IEEE Trans. Nucl. Sci.*, NS19(1), 512, 1972. With permission.)

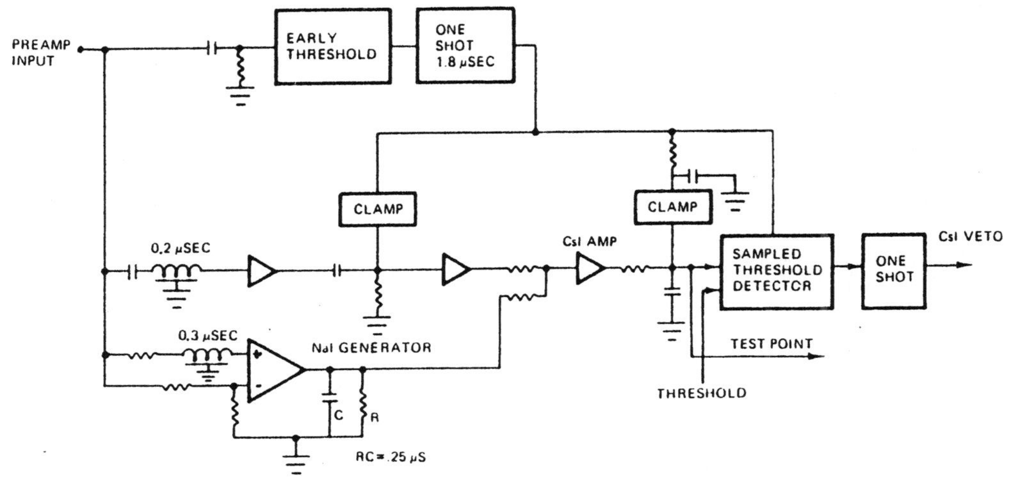

Figure 9.16 Decay-time PSD. (From Bradley, A. E. and Willes, E. H., *IEEE Trans. Nucl. Sci.*, NS20(1), 80, 1973. With permission.)

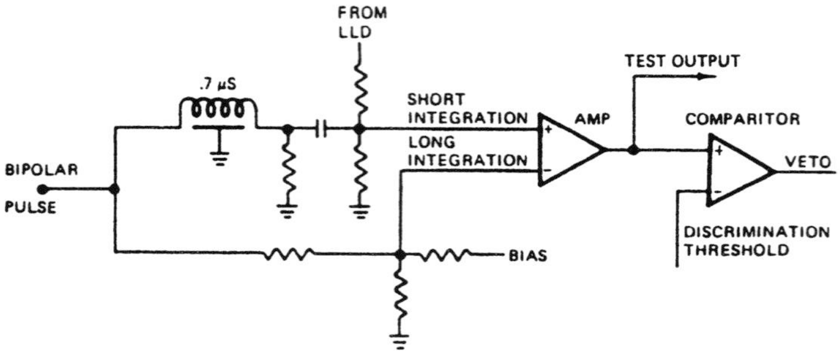

Figure 9.17 Rise-time PSD. (From Wilson, S., *IEEE Trans. Nucl. Sci.*, NS19(1), 512, 1972. With permission.)

The rise-time circuit took the output of the charge integrating preamplifier which integrated the signal up to a time t_2 and compared it with the integral at time t_1; i.e., the circuit measures:

$$C = \int_0^{t_2} \text{PMT dt} - x\int_0^{t_1} \text{PMT dt} \quad (9.5)$$

where the integrals are the output of the preamplifier, x is a constant which comes from the theory, and C is the amount of the CsI component. This technique has been used for solid-state detectors.[12]

A circuit which accomplishes the above is shown in Figure 9.17. The integral up to t_2 is simply the PMT amplifier output. The integral up to t_1 is obtained from the RC network. It is delayed by 0.7 μsec so that the two integrals can be subtracted. The pulse heights are adjusted so that complete cancellation occurs at time t_2 for a NaI pulse only. Cancellation will not occur for a CsI pulse.

The design of a rise-time discriminator circuit (RTD) was described by Bradley and Willes[13] with associated anticoincidence/coincidence (AC/C) logic and its application to the counting of low-activity tritium samples in gas proportional counter systems.

The RTD circuit utilized the property that gamma photons and high-energy micromesons produce ionization tracks in the proportional counter that were typically longer and had lower specific charge density than those tracks produced by low-energy (tritium) beta particles. Thus, these events had longer charge collection times and result in slower signal rise times at the output of a charge-sensitive (integrating) preamplifier.

The RTD circuit differentiated the wideband preamplifier signal which produced a pulse with amplitude proportional to the signal rise time. The derivative was then peak-detected, stretched, delayed, and inverted. An amplitude comparison was then made between the stretched derivative and the linear input signal using a linear summation amplifier. The purpose of the comparison was to prevent signal amplitude information from entering into the generation of the rise-time gate. A rise-time gate was generated and routed to the AC/C logic where it functioned as one of three conditioning inputs. In this system, an event qualified as a true event when the following three conditions were met: (1) the event was not in coincidence with a guard event; (2) the event had a fast rise time which triggered the RTD; and (3) the amplitude of the linear signal fell within the energy window as defined by a single-channel pulse-height analyzer. A simplified block diagram of the tritium counter system is shown in Figure 9.18.

The rise-time discriminator (RTD) required the unshaped preamplifier pulse for satisfactory operation. The RTD produced an output gate pulse whenever the input signal rise time was less than a preselected value. The RTD gate functioned as an enabling signal for the system logic.

A simplified block diagram of the rise-time discriminator (RTD) and associated anticoincidence/coincidence (AC/C) logic is shown in Figure 9.19.

The FAST INPUT signal from the proportional counter preamplifier was coupled to an input buffer which had constant 7.5 K-Ω input impedance for either positive or negative input signals. The buffer was followed by a fast amplifier. The FAST AMPLIFIER output was split into two parallel channels, designated as the differentiator (D) and reference (R) channels, which produced signals for summation in the sum amplifier. Additionally, the FAST AMPLIFIER output drove a threshold (LLD) circuit which was set to discriminate against low-level noise. Signals above a preset threshold generated an ENABLE logic signal.

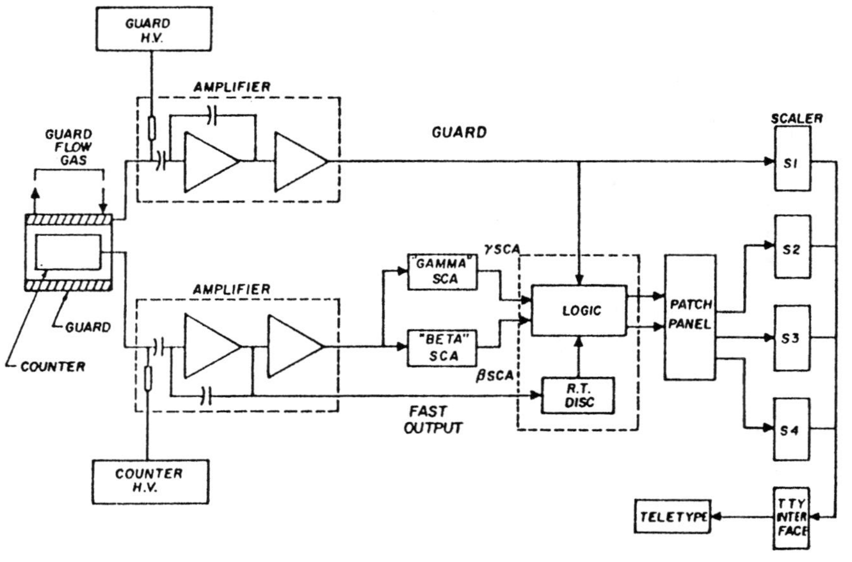

Figure 9.18 A block diagram of the low-level tritium counter system. The counters are enclosed in a lead shield. (From Bradley, A. E. and Willes, E. H., *IEEE Trans. Nucl. Sci.,* NS20(1), 80, 1973. With permission.)

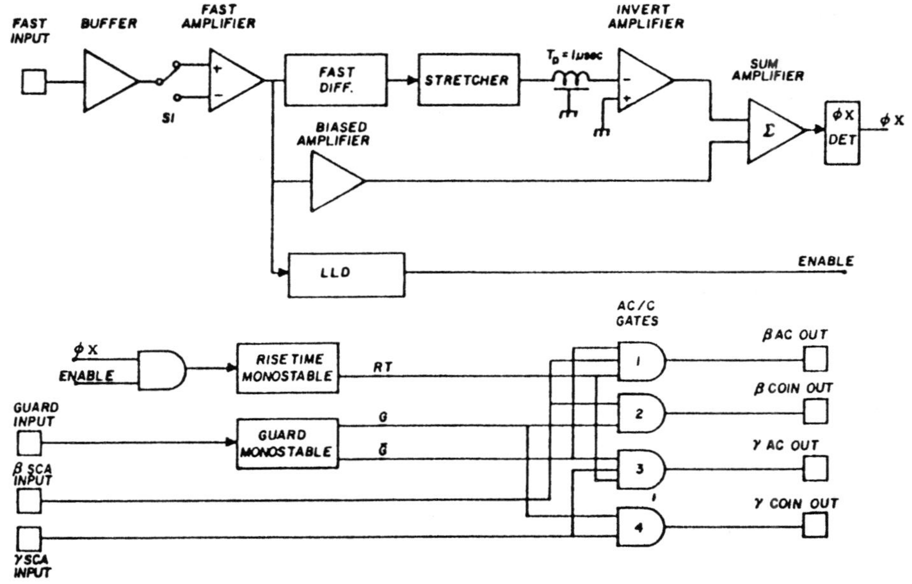

Figure 9.19 A block diagram showing the organization of the rise-time discriminator and the associated anticoincidence/coincidence logic. All circuitry is incorporated on a single printed circuit board packaged on a double-width NIM. (From Bradley, A. E. and Willes, E. H., *IEEE Trans. Nucl. Sci.,* NS20(1), 80, 1973. With permission.)

The sum amplifier algebraically added the two inputs and produced an output signal waveform characterized by a fast-rise negative-going signal (from the biased amplifier) followed in 1 μsec by a positive-going signal (from the stretcher). This composite waveform was fed to the zero-crossing detector, φX DET. If the positive going edge exceeded ~0 V, the φX DET generated a gate signal (φX) which could be shown to depend only on signal rise time and to be independent of input signal amplitude within the dynamic range (50:1) of the circuitry.

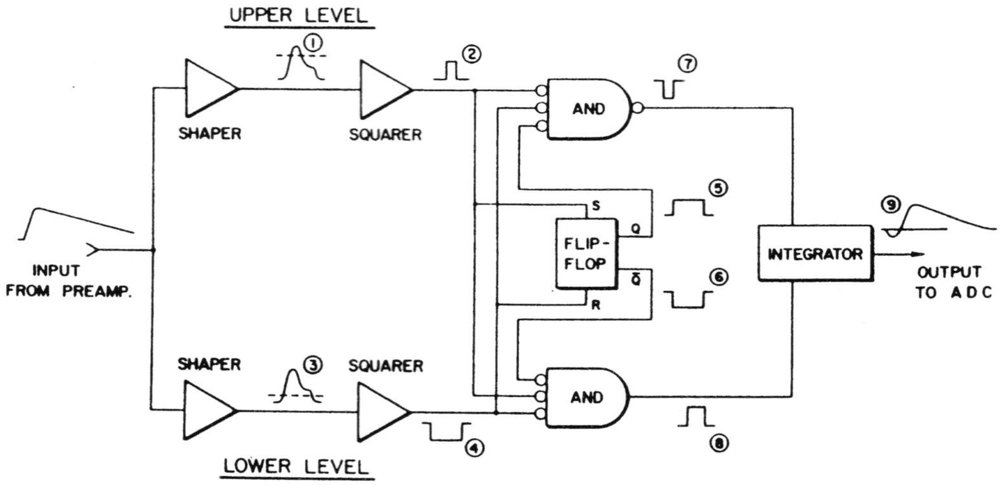

Figure 9.20 Block diagram of pulse shape analyzer. Shapers recover current pulse (by differentiating preamplifier signal) and offset it from baseline. Portion of pulse above baseline (dashed) is squared. AND gates and flip-flop in combination with integrator produce output signal whose amplitude varies linearly with difference between fall- and rise times of current pulse as shown in Figure 9.21. (From Sherman, I. S. and Strauss, M. G., *Nucl. Instrum. & Methods,* 117, 285, 1974. With permission.)

The position of a gamma-ray interaction in a true coaxial Ge(Li) detector was investigated by Sherman and Strauss.[14] The angular position of interaction was obtained by noting the sector in which a gamma-ray was detected. The radial distance was determined from the difference between the rise and fall times of the recovered current pulse obtained from the p-core.

A simplified block diagram of the pulse shape analyzer is shown in Figure 9.20. The two shapers differentiate the integrated pulse from the preamplifier to recover the current pulse. Each shaper offsets the current pulse from the baseline by a different amount. Figure 9.21 shows representative waveforms in time sequence. Referring to the horizontal dashed lines which represent the baseline, it is seen that the upper level shaper introduces a fractional offset which is large and the lower level shaper a fractional offset which is small (waveform 3). Each squarer produces a rectangular pulse (waveforms 2 and 4) whose leading and trailing edges correspond to the times at which the current pulse (1 and 3) rises above and falls below the baseline. Since the squarers respond to baseline crossing and not to amplitude, the system is pulse-height or energy independent. Pulse 9 amplitude corresponds to $t_2 - t_1$, i.e., the fall time minus the rise time.

Piercy et al.[15] have developed pulse pair-detection techniques, which when used in conjunction with common pulse shape discrimination circuits can markedly improve their performance at high event rates. Performance at rates above 10^6 events per second has been improved by more than 100%.

The principle of the typical zero-crossing PSD is shown in Figure 9.22. The anode signal from the detector is split into a fast bipolar amplifier (FBA) and a constant-fraction discriminator (CFD). The CFD generates a fast, pulse-height independent start pulse for the time-to-pulse height converter (TPHC). The FBA converts the anode charge pulse into a bipolar pulse with a zero-crossing time that is related to the pulse shape of the input signal. The output of the ZCD is used to stop the TPHC producing a distribution of pulse heights which are then recorded in a multichannel analyzer (MCA). In circuits such as the one represented by the lower two signal paths in Figure 9.22, the separation of the neutron and gamma peaks is dependent on the neutron:gamma ratio, the total detector count rate and the dynamic range of the total spectrometer circuitry. As detector count rates rise much above the inverse of the FBA shaping time, the neutron and gamma peaks in the MCA will broaden and develop long tails. This degradation arises from several mechanisms, the most important of which are usually pulse pileup in the FBA and event confusion in the TPHC. As shown in the upper branch of Figure 9.22, a pulse-pair rejector (PPR) was introduced in the circuitry which generates an output pulse when the pulse in question is separated from the previous and succeeding pulses by the pulse-pair resolution, τ_{pp}. If τ_{pp} is greater than the width of the positive lobe of the FBA output pulse, then piled-up events, which shift the zero crossing of the signal, will be rejected

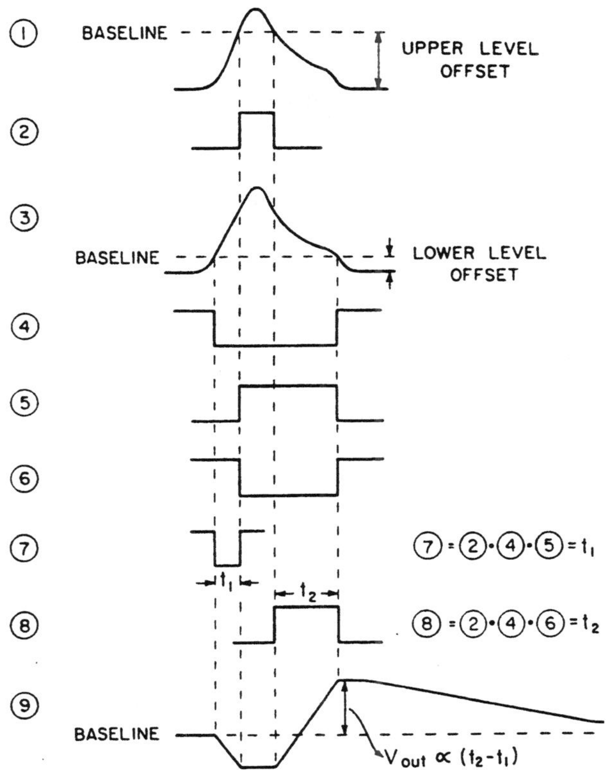

Figure 9.21 Pulse shape analyzer waveforms shown in time sequence. AND gates switch state only when all inputs are negative. Thus, pulse 7 (t_1) corresponds to rise time of current pulse (1 and 3) and pulse 8 (t_2) corresponds to fall time. Output is integral of negative t_1 and positive t_2 and therefore corresponds to fall time minus rise time of current pulse. (From Sherman, I. S. and Strauss, M. G., *Nucl. Instrum. & Methods,* 117, 285, 1974. With permission.)

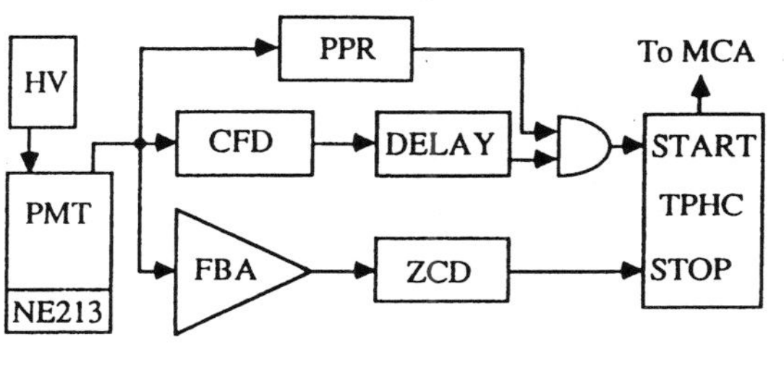

Figure 9.22 A schematic diagram of a typical pulse shape discrimination circuit with pulse-pair rejection. (From Piercy, R. B. et al., *IEEE Trans. Nucl. Sci.,* 34, 82, 1987. With permission.)

before starting the TPHC. Also, if τ_{pp} is longer than the ramp time of the TPHC, event confusion in the TPHC (which is caused by starts and stops from different events) will be eliminated.[15]

A fast module for pulse shape analysis was described by Annand.[16] A block diagram of the circuit is given in Figure 9.23 where those components not contained within the PSD module are horizontally shaded. The waveforms and relative timing of pulses at various points on the circuit are sketched in Figure 9.24 where the vertical axis labels, A–I, refer to the similarly labeled circuit points in Figure 9.23. Time zero is referenced to the leading edge of the analog output from the linear fan out (Figure 9.23). Anode pulses from a photomultiplier are fanned out (a passive splitter works equally well if the pulse height is not too small) and fed to a timing discriminator and the PSD module. The leading edge of the discriminator logic output (A, Figure 9.24) triggers the gate generators (C and D, Figure 9.24) for the charge integrators and also disables the clear line (G, Figure 9.24) which otherwise holds the integrator capacitors in a discharged state. For best performance walk-free timing is necessary. A constant fraction discriminator may be used, but a dual threshold type was preferred because it had a slightly lower propagation delay and was empirically found to give particle separation at least as good as a constant fraction type. After the fan out, the anode pulse is delayed (B, Figure 9.24) and fed into the PSD module, where it is equally split and passed to the charge integrators. Two 100-Ω resistors, connected to the virtual ground inputs in parallel, give an effective 50-Ω input impedance. The integrators consist of current-to-

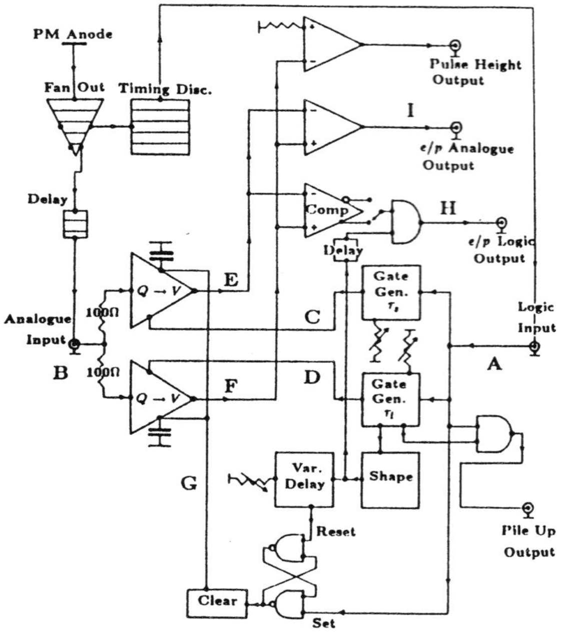

Figure 9.23 A simplified block diagram of the PSD circuit. Horizontally shaded components are external NIM modules. Q → V denotes a charge integrator; τ_1 and τ_s, denote long and short gate generators, respectively; and Comp denotes the fast voltage comparator. (From Annand, J. R. M., *Nucl. Instrum. & Methods,* A262, 371, 1987. With permission.)

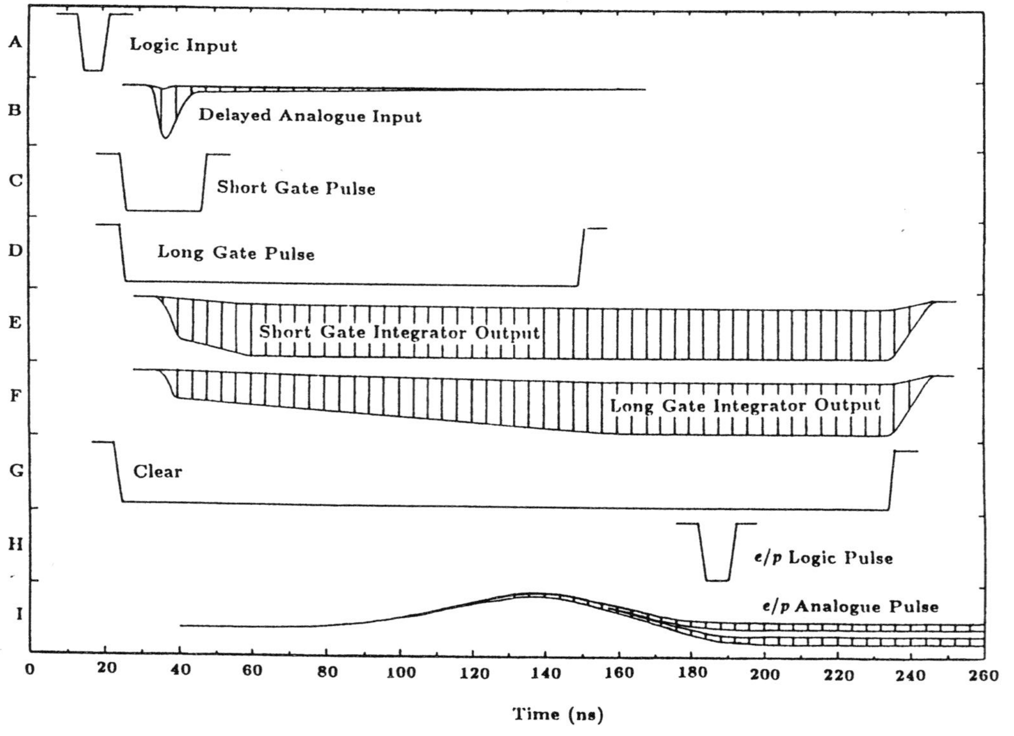

Figure 9.24 The relative timing and shapes of pulses at points on the PSD circuit labeled by letters A–I in Figure 9.23. Vertical shading denotes the amplitude range of analog pulses. (From Annand, J. R. M., *Nucl. Instrum. & Methods,* A262, 371, 1987. With permission.)

current converters which allow charge to flow for the duration of the gate pulse, and this integrated charge is stored on a capacitor. The resultant voltages are buffered and fed to a fast comparator, which provides a logic decision on particle type (e.g., recoil electron or proton); and to a fast difference amplifier, which produces an analog output (I, Figure 9.24) with a pulse height that is related to the effective scintillation decay time.[16]

The comparator used to make the decision on particle type is an ultra-fast type with a 2-nsec response time. It has Q and $\overline{Q}$ outputs to that switching between the two in the present application gave a logic 1 level signaling either a detected electron or proton. This logic output is shaped and timed (H, Figure 9.24) with respect to the trailing edge of the long gate pulse, delayed slightly to allow comparison to be made on the flat sections of the integrator outputs (E and F, Figure 9.24). Its leading edge falls at time $= \tau_1 + 60$ nsec so that with τ_1 set at 100 nsec, an electron/proton logic decision is produced in 160 nsec. The PSD analog output (I, Figure 9.24) proportional to the difference of the integrator voltages, is provided by a difference operational amplifier buffered to give 50-Ω output impedance.

A digital pulse shape discriminating unit for neutron-gamma discrimination with NE213 was described by Kumahara and Tominaga.[17] A block diagram of the digital pulse shape discriminating circuit is shown in Figure 9.25. The input signal, from a photomultiplier tube, is fed to an input buffer and the output DC level is stabilized at the front stage. A trigger pulse generated by a couple of lower level discriminator and constant-fraction discriminator sets a busy flip-flop in the RUN condition for opening a couple of input gate of gated integrators, and starts a couple of preset counter-to-counts crystal-controlled clock pulses for getting precise gating time of the integrators.

A coupling of gated integrators is provided with DC level stabilizers for integrating prompt part and total part of the input pulse separately. Both of the integrated pulses are digitized by a couple of flash-type A–D converters in fast speed. Two data corresponding to the prompt integration and the total integration of the input pulse are operated in division by a divider. The quotient is digitally compared to reference data for the discrimination between neutron and gamma-ray events. The output pulses corresponding to neutrons and gamma rays are accumulated in separate counters for event rate display and automatic control.

After an elapsing preset interval (10^{-1} to 10^4 s), an integrated single-chip computer reads the neutron and the gamma ray counts, then converts them in decimal value in terms of counts per s (cps), and writes out to the numerical displays, respectively. In case a large amount of the gamma counts change, the computer writes new reference data to the digital comparator.[17]

A charge-integrating ADC has been used by Heltsley et al.[18] to sample the intensity in two different time regions of a pulse and thus to sense the shape of the pulse. If one channel of a charge-integrating ADC (QDC) is used to integrate the entire anode pulse from a scintillation spectrometer and one to integrate either the early or (equivalently) the late part of it, then for a given total charge the second integral depends on the particle type. One must choose a suitable time t_d in the pulse to use as an effective dividing line between the fast and slow components when generating that second integral. Unfortunately, the value of t_d that maximizes the difference is not the value that minimizes the spread. Initially the gamma-ray pulse is bigger than the neutron pulse, but for a given total charge the pulses intersect and cross over at a later time.

The choice of which one to use is, therefore, based on secondary considerations. For use with a single QDC module, in which all channels receive the same gate pulse, a long delay (~300 nsec) is required if we use the early integral while Q total is being obtained in another channel. The total pulse and the tail of the pulse can be integrated simultaneously by delaying the total pulse before integration, but the delay is much shorter (~50 nsec).

The pulse shape discrimination system is illustrated in Figures 9.26 and 9.27. Only the anode signal was used, thus obviating the need for a preamplifier and a long delay line. The anode signal is divided into three identical signals, S_A, S_B, and S_C, by using a passive splitter. Two of these signals, S_A and S_B, are given different delays, A and B, and then sent to their respective QDC channel inputs. The third signal, S_C, is delayed and passed through a CFD to produce a logical flag to indicate that an event has occurred and to trigger the gate signal to the QDC.

The relative timing of these signals is displayed in Figure 9.27. The delay of the third anode signal, delay C plus the electronics delay, determines when the gate of the QDC opens. This delay t_d above, is made long enough to include the long tail of the anode signal S_A. Anode signal S_B is delayed a greater amount so that the gate essentially includes the entire pulse.[18]

A simple design of a pulse shape discrimination circuit for Phoswich scintillation detectors was described by Malkar and Kunte.[19] Phoswich arrangement of NaI(T1) and CsI(Na) scintillation crystals has been widely used for detection of cosmic X-rays in X-ray astronomy. The two scintillators have different decay times: 0.25 μsec for NaI(T1) and 0.65 μsec for CsI(Na). Pulse shape discrimination (PSD) is used to determine whether the scintillation occurred in NaI(T1) or CsI(Na).

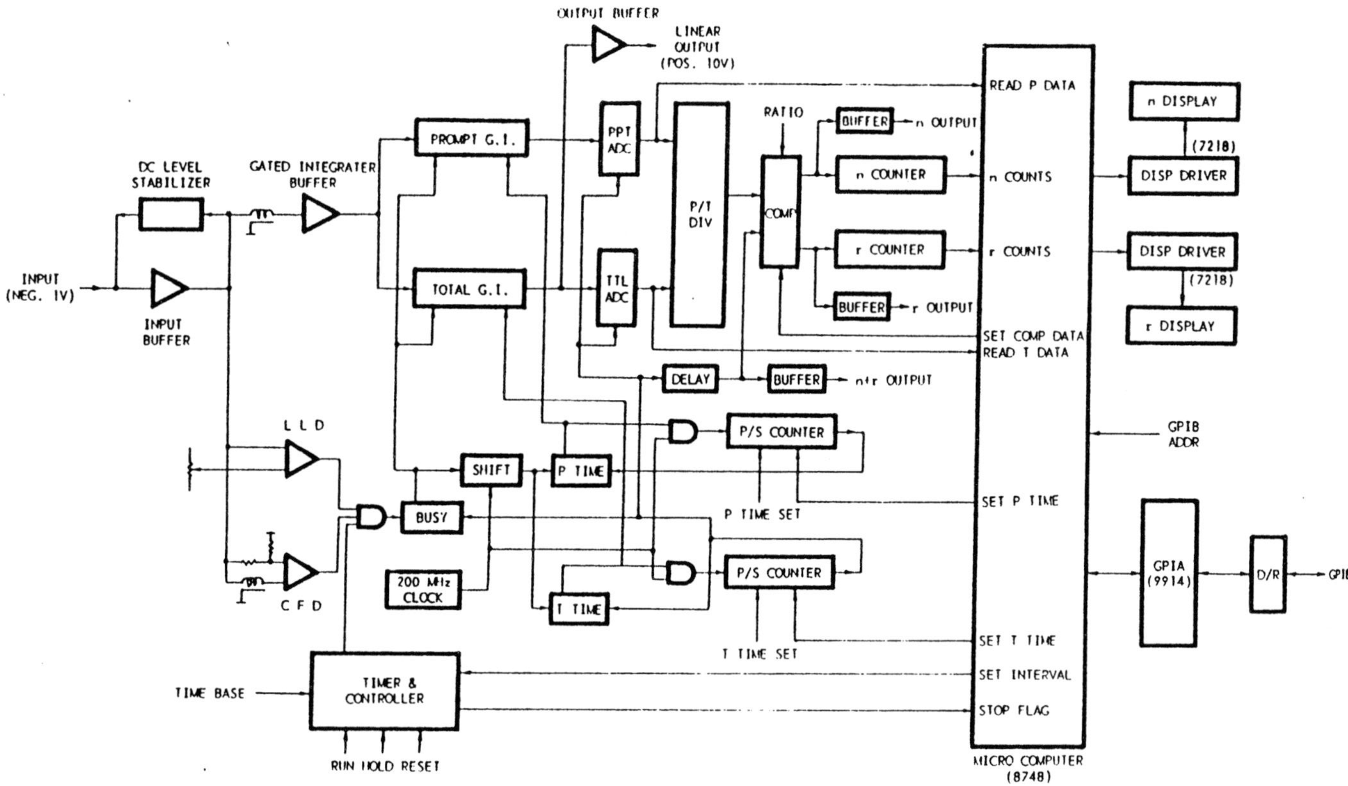

Figure 9.25 Block diagram of a digital pulse shape discriminating unit. (From Kumahara, T. and Tominaga, H., *IEEE Trans. Nucl. Sci.*, NS31, 451, 1984. With permission.)

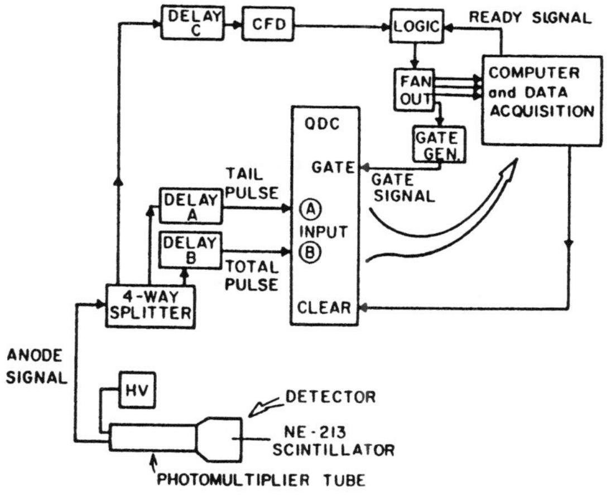

Figure 9.26 Schematic of the experimental setup using a QDC for pulse shape discrimination. (From Heltsley, J. H. et al., *Nucl. Instrum. & Methods,* A263, 441, 1988. With permission.)

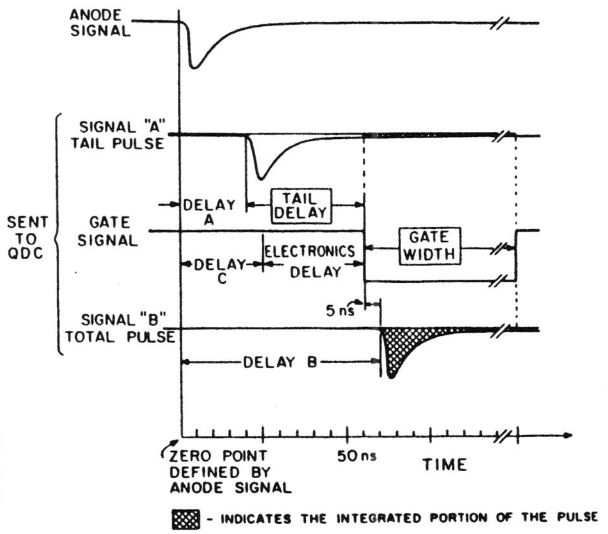

Figure 9.27 Display of the relative arrival times of signals at the QDC. (From Heltsley, J. H. et al., *Nucl. Instrum. & Methods,* A263, 441, 1988. With permission.)

The block diagram of the electronics for Phoswich is shown in Figure 9.28. The pulses which appear at the anode of the photomultiplier tube are fed to a preamplifier; later they are in turn fed to the linear postamplifier (LPA) and to a fast amplifier (FA) simultaneously. The LPA contains a double differentiating-integrating network with again a control amplifier. The FA contains single differentiation at the preamplifier and gain control system.

The LPA output is given to the level discriminators which are set for every energy interval 20 to 120 keV for X-rays. The events are analyzed within the above energy range if the same event has not occurred during the dead time of the previous event. This dead time includes busy time of height to time converter (HTC) circuit, PSD circuit, and time required for the transmission of encoded data. LPA and FA outputs are fed through linear gates HTC and PSD circuits, respectively. The analog values of pulse height and pulse shape of an event are digitized in 128 channels. Digital information of HTC and PSD circuits is given to the time base telemetry encoder. The above event analysis procedure is controlled by the logic control circuits.

The PSD circuit is shown in Figure 9.29. FA output is fed through a switch gate G2 to PSD which is controlled by logic circuits. If the pulse is of interest, the gate G2 is closed and the pulse appears at the input of a peak detector. The peak detector output appears at the condenser C1. The same pulse is also given to the noninverting input port of the comparator COMP 1 and an attenuated peak magnitude (say 37% of peak height) of the same pulse from peak detector is applied to the inverting input. The comparator COMP 1 output width W_1 (Figure 9.30) becomes a measure of decay time constant and rise time of the

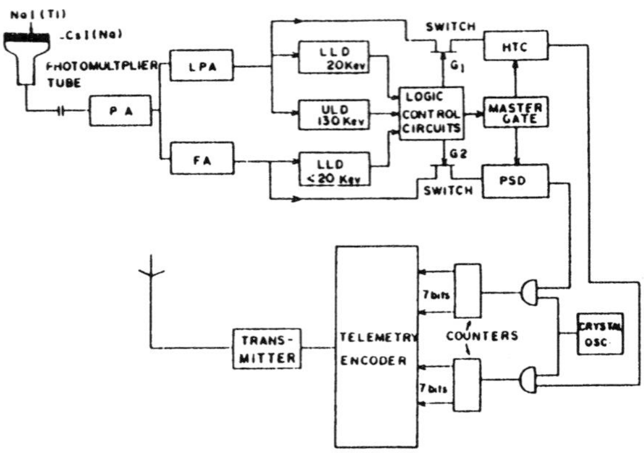

Figure 9.28 The Phoswich electronics block diagram. PA, preamplifier; FA, fast amplifier; LLD, lower level discriminator; HTC, height to time converter; G_1G_2, gate switches; LPA, linear postamplifier; OSC, oscillator; ULD, upper level discriminator; PSD, pulse shape discriminator. (From Malkar, J. P. and Kunte, P. K., *Nucl. Instrum. & Methods,* 202, 465, 1982. With permission.)

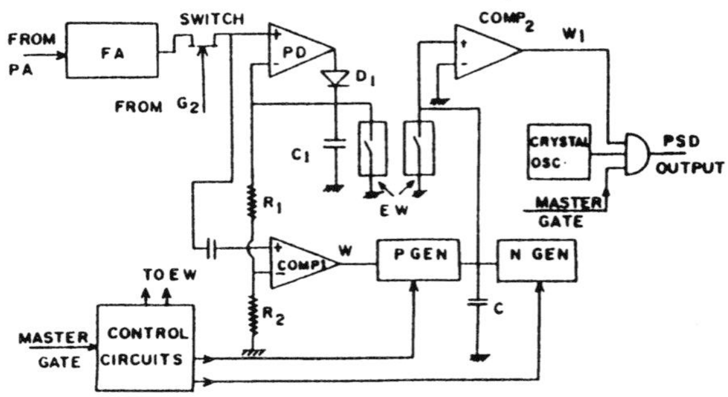

Figure 9.29 Pulse shape discriminator block diagram. PA, preamplifier; PD, peak detector; COMP, comparator; P GEN, positive current generator; OSC, oscillator; FA, fast amplifier; D_1, diode; EW, electronic switches; N GEN, negative current generator; and R_1R_2, resistors. (From Malkar, J. P. and Kunte, P. K., *Nucl. Instrum. & Methods,* 202, 465, 1982. With permission.)

same input pulse. The width W remains constant irrespective of the pulse height and varies only with the different decay time of the pulses. As the width W is not sufficient for measuring in terms of digital information through telemetry, it is further amplified in the time scale. The condenser C is charged linearly by a positive-current generator for the time equal to the width W and is made to discharge linearly by a negative-current generator which in effect can be stretched to a suitable length of time, namely, width W1. This gate width, W1, is proportional to the gate width W. Both the 500-kHz clock and the master gate generated by logic control circuits are and gated with gate W1 in order to obtain the digital information. During the busy time of the HTC and PSD circuits, switch gate G2 remains open and does not permit any pulse to appear at the input of PSD circuit. For all uninterested pulses the condenser C and C1 in Figure 9.29 are short circuited by electronic switches, EW, which are activated by logic control circuits.[19]

A simple pulse shape discriminator method for the Phoswitch counter was developed by Murakami et al.[20] The output signal from a photomultiplier is differentiated and integrated to form a triple-polar pulse. The time interval between the first and second zero crossings depends on the decay time of the scintillation light. The circuit selects signals from the proper scintillator by using this interval. By this method, the effect of time slewing in the low-energy region is significantly reduced. The pulse shape discriminator is composed of a shaping amplifier and a timing logic. In the shaping amplifier, the signal is differentiated and integrated to form the unipolar, bipolar, and triple polar pulses. The time interval between the first and second zero crossings in the triple polar pulse is measured by the timing logic circuits. Physically the PSD is composed of four hybrid ICs: three for the shaping amplifier (150 mW/channel/IC) and one for the timing logic (200 mW/channel). Sixteen PSDs and additional circuits for monitoring are arranged in two double-width NIM modules.

The triple polar shaping is realized by adding one CR differentiation to a bipolar shaping amplifier. A triple polar shaping amplifier needs three CR differentiations and N RC integration stages (CR^3-RC^N). In this shaping amplifier, the number of stages are CR^1-CR^4 for the unipolar, CR^2-RC^4 for the bipolar, and CR^3-RC^4 for the triple polar pulses.

There are two zero cross points in the triple polar waveform. The time interval between the first zero and second zero-crossing points depends on the rise time of the input pulse. The transfer function of the CR^3-RC^N pulse-shaping amplifier is

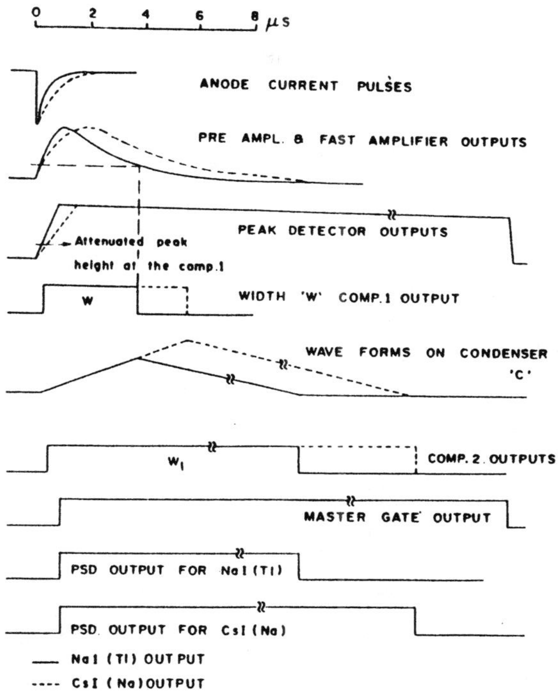

Figure 9.30 The waveforms of various elements of pulse shape discriminator. (From Malkar, J. P. and Kunte, P. K., *Nucl. Instrum. & Methods,* 202, 465, 1982. With permission.)

$$G(s) = \frac{s^3}{T^N (s + 1/T)^{N+3}} \tag{9.6}$$

where T is the RC time constant of the shaping amplifier. The input to the PSD is a signal through a charge-sensitive preamplifier, and takes the following shape:

$$V_i(s) = \frac{Q}{\tau} \frac{1}{s + 1/\tau} \frac{1}{sC_f} \tag{9.7}$$

where τ is the decay time constant of the input current pulse, which is the time constant of the scintillator. The signal at the output of the shaping amplifier in the PSD is given as:

$$V_o(s) = V_i(s) G(s)$$

$$= \frac{Q}{\tau} \frac{1}{s + 1/\tau} \frac{1}{sC_f} \frac{s^3}{T^N (s + 1/T)^{N+3}} \tag{9.8}$$

where Q is the total charge of the input signal. The inverse Laplace transformation of V(s) is given as:

$$\upsilon_o(t) = \frac{Q}{C_f} \frac{1}{\tau T^N N!} \left\{ \left(\frac{T}{\tau - T} \right)^2 \left(\frac{\tau T}{\tau - T} \right)^{N+1} N! \times \left[e^{-t/\tau} - e^{-t/T} \sum_{r=0}^{N} \frac{\left(\frac{\tau - T}{\tau T} t \right)^r}{r!} \right] \right. \tag{9.8}$$

$$\left. + \frac{\tau}{T - \tau} t^{N+1} e^{-t/T} \left[\left(\frac{1}{N+1} - \frac{1}{N+2} \right) \frac{t}{T} + \frac{\tau - 2T}{T - \tau} \frac{1}{N+1} \right] \right\}$$

The block diagram and the principle of operation of the PSD are shown in Figures 9.31 and 9.32. The first zero-crossing point is detected by the positive to negative zero-cross time pick-off. The second zero-crossing is detected by the negative to positive zero-cross time pick-off. The lower level discriminator in

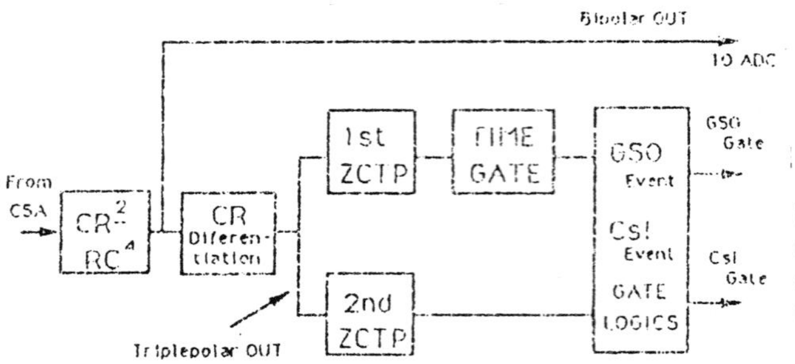

Figure 9.31 The block diagram of the PSD. CSA, a charge-sensitive preamplifier; ZCTP, zero-cross time pick-off discriminators. (From Murakami, H. et al., *IEEE Trans. Nucl. Sci.*, 39, 1316, 1992. With permission.)

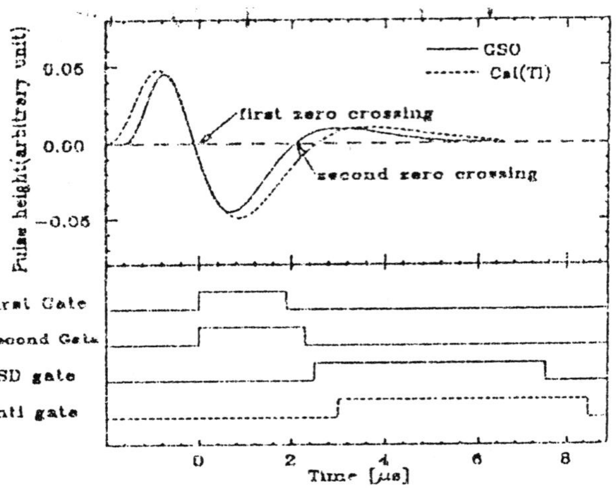

Figure 9.32 Principle of the pulse shape discriminator. The PSD pulse is issued only if the second zero-crossing point falls in the first gate and the second gate. (From Murakami, H. et al., *IEEE Trans. Nucl. Sci.*, 39, 1316, 1992. With permission.)

the first time pick-off circuit defines the minimum energy accepted. If the signal exceeds the discriminator level, two logic gates are generated at the first zero-crossing time. If the second zero-crossing point falls in between the two gates, the PSD system issues a pulse (the PSD gate) at the second zero-crossing timing. If the second zero-crossing point does not fall between the gates, the Antigate is issued.[20]

REFERENCES

1. **Brooks, F. D.,** *Nucl. Instrum. & Methods,* 4, 151, 1959.
2. **Taylor, I. J. and Kalyana, J.,** *Nucl. Instrum. & Methods,* 88, 267, 1970.
3. **Winyard, R. A. and McBeth, G. W.,** *Nucl. Instrum. & Methods,* 98, 525, 1972.
4. **Rendic, D. et al.,** *Nucl. Instrum. & Methods,* 99, 189, 1972.
5. **Sperr, P. et al.,** *Nucl. Instrum. & Methods,* 116, 55, 1974.
6. **Betz, P. et al.,** *Nucl. Instrum. & Methods,* 119, 199, 1974.
7. **Pai, S. et al.,** *Nucl. Instrum. & Methods,* A278, 749, 1989.
8. **Bialkowski, J. et al.,** *Nucl. Instrum. & Methods,* A275, 322, 1989.
9. **White, G.,** in *Proc. 2nd Ispra Nucl. Elec. Symp. EUR 5370,* 1975, p. 447.
10. **Adams, J. M. and White, G.,** *Nucl. Instrum. & Methods,* 156, 459, 1978.
11. ORTEC 458, Pulse Shape Analyzer Manual.
12. **Wilson, S.,** *IEEE Trans. Nucl. Sci.,* NS19(1), 512, 1972.
13. **Bradley, A. E. and Willes, E. H.,** *IEEE Trans. Nucl. Sci.,* NS20(1), 80, 1973.
14. **Sherman, I. S. and Strauss, M. G.,** *Nucl. Instrum. & Methods,* 117, 285, 1974.
15. **Piercy, R. B. et al.,** *IEEE Trans. Nucl. Sci.,* 34, 82, 1987.
16. **Annand, J. R. M.,** *Nucl. Instrum. & Methods,* A262, 371, 1987.
17. **Kumahara, T. and Tominaga, H.,** *IEEE Trans. Nucl. Sci.,* NS31, 451, 1984.
18. **Heltsley, J. H. et al.,** *Nucl. Instrum. & Methods,* A263, 441, 1988.
19. **Malkar, J. P. and Kunte, P. K.,** *Nucl. Instrum. & Methods,* 202, 465, 1982.
20. **Murakami, H. et al.,** *IEEE Trans. Nucl. Sci.,* 39, 1316, 1992.
21. **Emmer, T. L.,** in NAS-NRC Pub. 1184, Monterey, 1963, p. 112.

Chapter 10

Photomultiplier Tube Voltage Divider

The last three decades have brought about the development of fast photomultipliers and scintillators, as well as a better understanding of their timing properties. The bialkali photocathode with a quantum efficiency about 50% higher was introduced in photomultipliers. The high-gain dynode structure applied to RCA 8850 and C31024 photomultipliers strongly reduced the contribution of the electron multiplier to the total time jitter of these photomultipliers. A time resolution for single photoelectrons equal to 250 to 400 psec (FWHM) has been measured using light-emitting diode generators.[1]

The scintillation-counter time resolution is limited by the space-charge effect in photomultipliers. This effect causes a broadening of the output-current pulse, while the total charge of the pulse is still proportional to the integral of the light pulse from the scintillator. Hence, in the energy spectrum, there is no deviation of the photomultiplier response from linearity. The leading edge of the current pulse is not affected by the space charge, but the maximum of the pulse becomes nonlinear vs. energy.

One of the methods used to improve the dynamic range of the photomultiplier linearity is to take the output signal from the central part of the dynode chain.

The timing properties of the XP 2020 and RCA 8850 tubes were studied by Bengston and Moszynski[1] with several combinations of voltage chains. The C31024 was used for comparison with the high-current voltage chain recommended by the manufacturer to ensure good linearity of the output pulse.

Figure 10.1 presents the voltage chains used for RCA 8850. In the standard structure with the output from the anode, the last few diode stages operate at progressively increasing high voltage to ensure a high-speed response as well as a high-range linearity. The voltage between the first high-gain dynode and the photocathode is stabilized by means of zener diodes at 750-V potential. In the modified version of the voltage chain, the output pulse was picked up from the seventh dynode, but the next two stages were kept active (see Figure 10.1b). Thus the electrons leaving the output dynode were extracted in the subsequent stages.

Figure 10.2 presents the voltage chains used for the XP 2020 photomultipliers. The standard B′ voltage chain recommended by the manufacturer for timing applications and voltage chain A′ being an intermediate version between the linear one and the B′-type, both with the anode output, were studied. The latter voltage chain was found to be better at high gamma-ray energies, in timing with plastic scintillators. Figure 10.2(b) shows the modified version of the voltage chain, with the output signal taken from the ninth dynode.

In spite of the reduced number of stages in the modified version of the voltage chains used for both RCA 8850 and XP 2020 photomultipliers, it was rather easy to get an output pulse height of 1.5 to 2.0V for the range 300 to 1000 keV of energy transferred from gamma-quanta to the plastic scintillators. Note that the dynode output of the photomultipliers delivers a positive signal significantly faster than that from the anode[1]

A voltage divider should keep all the voltages across the dynodes as constant as possible, in order to achieve a stable photomultiplier (PM) gain vs. counting rate. A modification was made by Bonnet et al.[2] by adding: (1) parallel capacitors to cope with sudden high-peak anode currents, (2) additional power

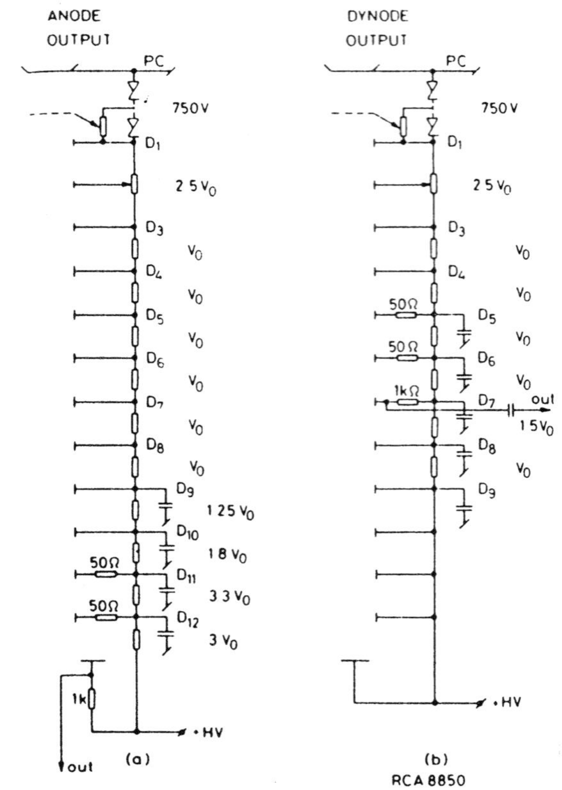

Figure 10.1 Voltage chains studied with the RCA 8850 photomultiplier. (From Bengtson, B. and Moszynski, M., *Nucl. Instrum. & Methods,* 204, 129, 1982.)

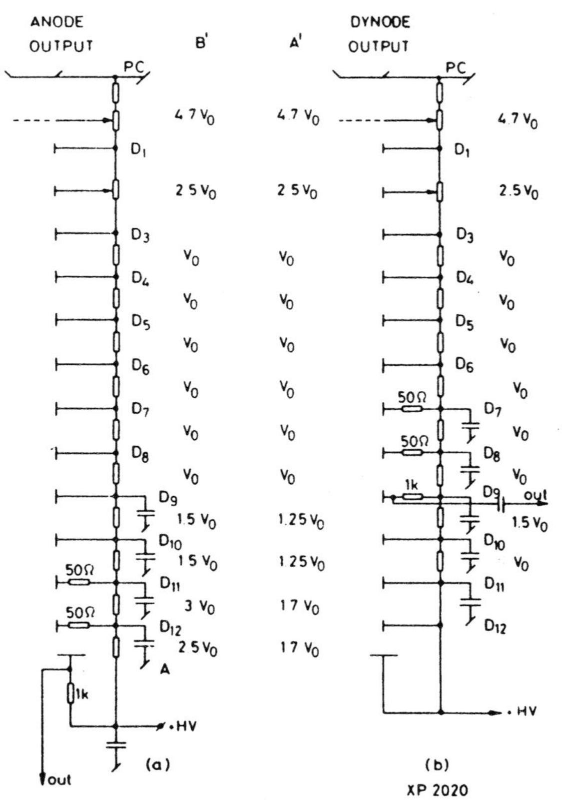

Figure 10.2 Voltage chains studied with XP 2020 photomultipliers. Note that for the anode output, two different voltage distribution, B′ and A′, were used. (From Bengtson, B. and Moszynski, M., *Nucl. Instrum. & Methods,* 204, 129, 1982. With permission.)

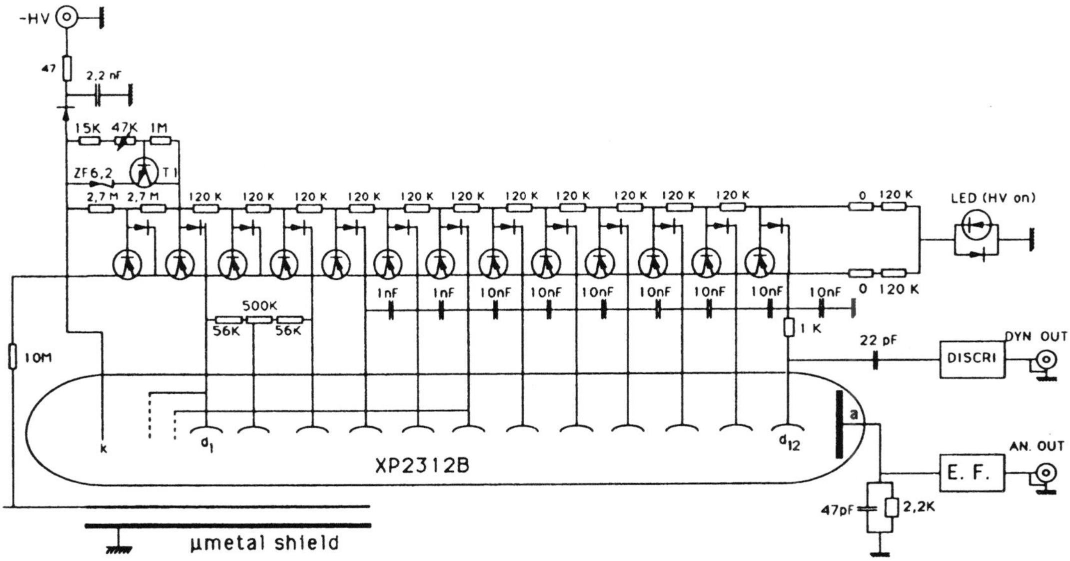

Figure 10.3 Schematic view of an XP 2312B photomultiplier base; all transistors are BF472, except T1 (BF471). T1 is used in conjunction with the ZF6.1 low-voltage zener diode to stabilize the photocathode-to-first dynode voltage; this voltage is adjustable for optimum resolution through the 47K potentiometer. The diodes preventing overbias at the base of the transistors are 1N4146. The dynode output is a logical one; the anode output is adapted to the 50-Ω transport line by an emitter follower (EF). (From Bonnet, L. et al., *Nucl. Instrum. & Methods,* A292, 343, 1990. With permission.)

supply or high-voltage zener dynodes, and (3) active dividers consisting of emitter-follower transistors. The layout, displayed in Figure 10.3 is designed for a 3-in. diameter, 12-stage Philips XP 2312B PM, incorporating the following particular features:

- Emitter-follower transistors are used to stabilize the voltages of all the dynodes. The resistive divider impedance is greatly decreased, by the beta or the h_{Fe} factor of the transistors, which is typically 50.
- Particular care is paid to the photocathode-to-first-dynode potential V_{kd1}, where the stability is controlled by a low-voltage zener diode ZF6 in series with a transistor BF471. The temperature dependence of these devices are opposite and matched in order to get a null dependence. V_{kd1} can be varied and optimized, e.g., vs. the energy resolution of a gamma-ray line, while keeping the total voltage constant.
- The PM mean current is decreased by charging the anode and the last dynode with a high impedance (typically 1 or 2 kΩ) which improves the stability by a factor of at least 20 with respect to the common 50 Ω by permitting the decrease of the current in the tube. This implies that an emitter follower has to be included in the voltage divider housing.

A simple gating system was described by Sené[3] for the Hamamatsu R1250 photomultiplier, for use in neutron time-of-flight work. The gain of the photomultiplier is reliably restored less than 0.5 µsec after the gating pulse.

A gating pulse is applied to the first focusing electrode (g_1) of the photomultiplier tube (PMT). Measurements of the gain of the PMT as a function of the voltage applied to g_1 indicated that for voltages greater than –80 V with respect to the photocathode the PMT was turned off completely. The circuit developed to gate all four PMTs in one of the neutron detectors is shown in Figure 10.4. A positive transistor-to-transistor logic (TTL) pulse derived from the linac trigger is presented to IC1, a photocoupler whose output triggers the monostable IC2. Adjustment of VR1 enables gating pulses from 0.1 to 1.0 µsec to be obtained. The output pulse from IC2 is presented to IC3 and IC4, dual metal oxide semiconductor (MOS) clock drivers, which provide high-output current drives, via a 15-Ω resistor to damp oscillations, to the gates of the four power MOSFETS. When turned on, the MOSFET drains, normally at +300 V,

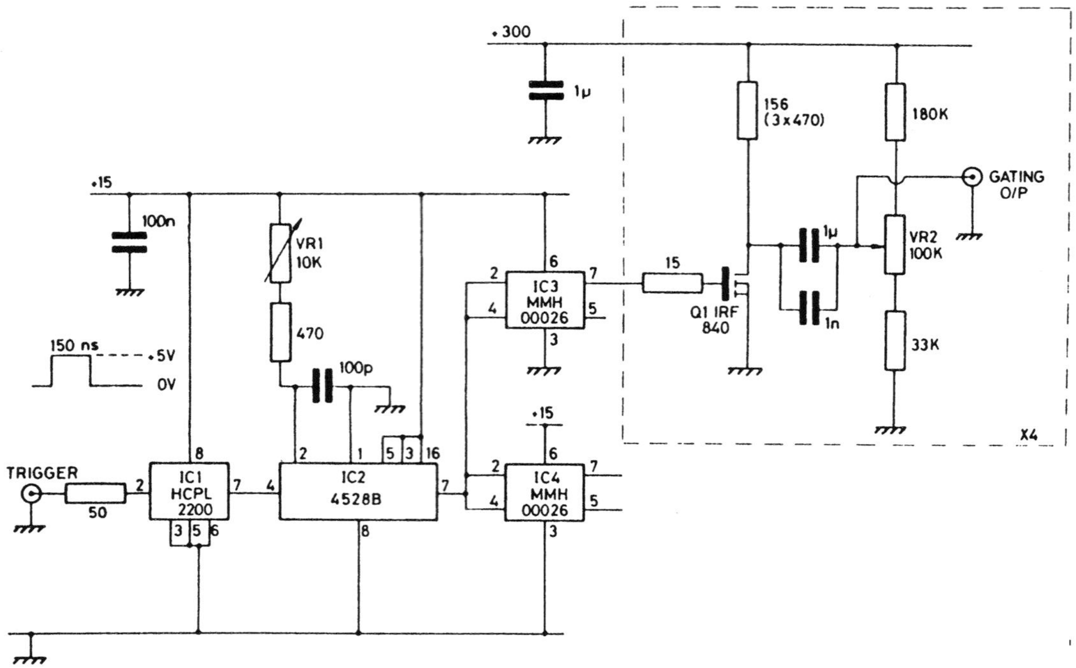

Figure 10.4 Gating circuitry for detector with four Hamamatsu R1250 photomultipliers. (From Sené, M. R., *Nucl. Instrum. & Methods,* A278, 503, 1989. With permission.)

go to ground. The 300-V negative going gating pulses are capacitively coupled to the gating outputs, normally held at voltages appropriate to the focusing electrodes of the PMTs by simple resistive dividers. The gating output is connected directly to the focusing electrode g_1 as shown in Figure 10.5, which is a diagram of the dynode chain used. Capacitive coupling of the dynodes d_1 and d_2 to ground, as shown, minimizes feedthrough of the gating pulse to the anode.

An adjustable voltage divider chain has been developed by Randors-Pehrson et al.[4] for use with 5-in. PM tubes in a neutron time-of-flight detector array. Best time response and pulse shape discrimination are obtained by maintaining high fixed gradients from the cathode to the third dynode. Best saturation characteristics are obtained by maintaining high fixed gradients for the last several dynodes and the anode. Gain adjustments are made by reducing the voltage gradients in the noncritical central dynodes while simultaneously reducing the total tube voltage such that all tubes in the array can be powered by a common supply.

As the energy of the incident neutron increases, the anode pulse height is no longer a linear function of the energy deposited in the scintillator. The shape of the anode pulse becomes distorted, and the pulse is no longer suitable as an input to a constant-fraction timing discriminator.

The range of linear anode response of the photomultiplier can be extended considerably by adopting a progressive bias resistor chain in which the voltage drop between adjacent dynodes increases with dynode number especially for the last few dynodes. Progressive bias resistor chains can be used to produce very large anode pulses without loss of linearity.

The divider chain designed for use with RCA 4522 and RCA 8854 tubes is shown in Figure 10.6. The essential feature of this chain is that as the ganged potentiometers R_1 and R_2 are adjusted, the overall gain of the tube is reduced by reducing only the voltages between the noncritical central dynodes.[5] The reduction is accomplished by shunting more of the bias current through R_2 and R_3 while simultaneously increasing the resistance of R_1. Using the gain-adjusting scheme, the voltage between the photocathode and dynode 3 remain fixed as does the voltage across the final stages from dynode 11 to the anode. It should be mentioned that the present design has some advantage over an alternative method[6] which employs zener diodes to maintain constant-voltage gradients at the photocathode and at the higher stages, namely, less sensitivity to high-voltage supply fluctuations.

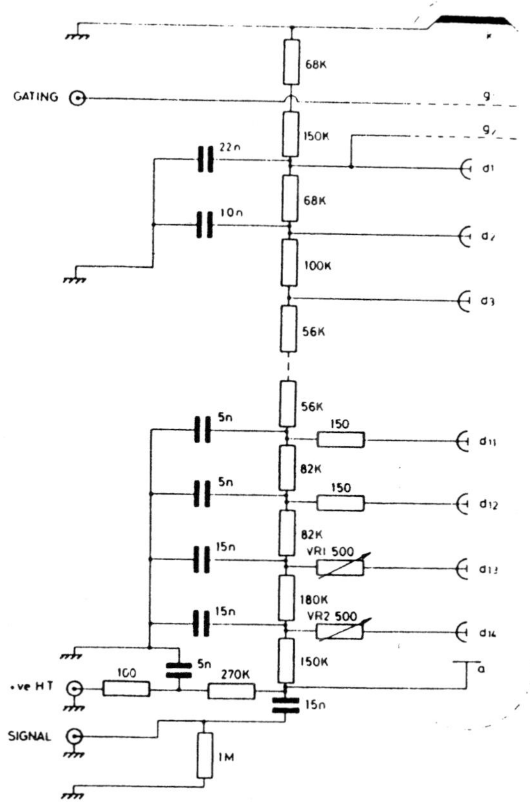

Figure 10.5 Dynode chain used with gating system. VR1 and VR2 are adjusted to give the cleanest output pulse. (From Sené, M. R., *Nucl. Instrum. & Methods,* A278, 503, 1989. With permission.)

A steep progression of the resistors between dynodes 11 and 14 is necessary to assure good linearity at large pulse heights. Capacitors are also connected to the final stages as usual. The 20-MΩ resistor R_4 is present as a safety measure since the outside surface of the RCA 4522 is at cathode potential.

The unusual feature of the circuit shown in Figure 10.7 is that both the timing and linear signals are derived from the anode, whereas the linear signal is normally obtained by capacitively coupling to one of the dynodes. Modern constant-fraction discriminators require a linear (nonsaturated) anode response for proper operation. It is simple to collect that signal on a capacitor to obtain an integrated charge for pulse height and pulse shape analysis. The fast pulse current transformer has been added to the circuit to eliminate line frequency pickup which had been present. A very wideband operational amplifier (Burr-Brown Model 3554) completes the circuitry in the tube base.

Fast timing with photomultiplier dynode pulses was discussed by de Vries and Kelling.[7] Timing was obtained with a time definition system using photomultiplier dynode pulses without any further shaping.

Dynode pulses are obtained from the circuit of the photomultiplier base of the XP 2020 (Q) shown in Figure 10.8. The dynode pulse pick-off part equals the circuit suggested by Bengston et al.[8] A zener stabilized first amplification step was chosen so that this critical part is not significantly influenced by adjustments of the total amplification of the tube or of the internal fraction. The pulse pick-off was attached as close as possible to the base connection pin, minimizing pulse distortion due to line inductances. In order to have a continuously and easily adjustable internal fraction, the voltage step between d8 and d9 was made variable.

Figure 10.9 shows a block diagram of the zero-crossover time discriminator. It contains three comparators: one for timing and two for upper and lower level energy discrimination. The timing comparator switches at the zero crossover of the input signal. As usual in constant-fraction timing, the reference voltage of this comparator will be called the walk adjust. The walk adjust serves as compensation for the small DC component at the input of the comparators. The outputs of the comparators are combined in an AND gate and this AND gate feeds the output circuit. The delay in the output of the timing comparator is set such that this signal arrives slightly later at the AND gate than the outputs of both level

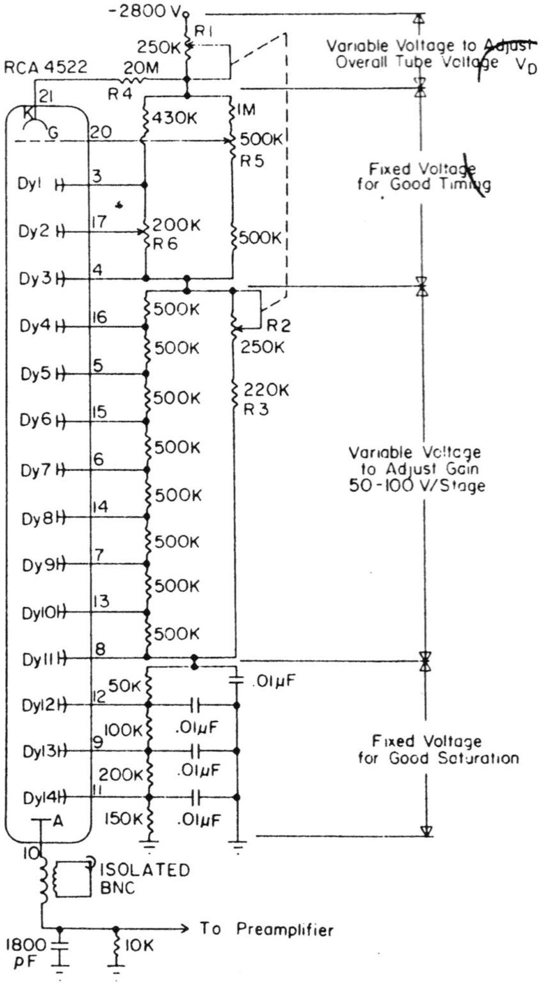

Figure 10.6 Proposed new voltage divider chain for RCA 4522 and RCA 8854 photomultiplier tubes. (From Catalog P001/E79 (EMI), GENCOM Inc., 80 Express Street, Plainview, NY 11803, 1979. With permission.)

discriminators. The output circuit contains a monostable with a reset time slightly longer than the input pulses, including any ringing. In this way, the discriminator is inactive for the duration of the reset time on arrival of a proper event. Figure 10.10 shows an elaborate diagram. The monostables directly after the level comparators account for the finite rise time of the output pulses of the AND gate. Omitting the monostables would cause leading-edge timing on pulses just exceeding the lower level. To save space and to have a discrete time delay, gates were used to define the reset time of the monostables.

Fast timing with photomultiplier dynodes pulses was described by de Vries and Kelling.[9] The dynode pulses are fed into a zero-crossover time discriminator with upper and lower level energy selections. Timing measurements were performed with the XP2020 (Q) photomultiplier and several scintillators.

Using dynode pulses, there are two advantages. The pulses have a smaller rise time than those from the anode and they are bipolar. The negative part of the dynode pulse represents the electrons arriving at the specific dynode and the larger positive part represents the electrons leaving the dynode after amplification. The smaller rise time of the pulse is due to the fact that a parasitic contribution to the anode pulse is introduced by electrons traveling toward the last dynode from the preceding one.

Thus, it is possible to obtain a timing signal directly from the zero crossover of the dynode pulse. Shaping pulses internally in a photomultiplier is comparable with constant-fraction shaping, but has fewer degrees of freedom. That is, if the internal fraction is defined as the ratio of the amplitude of the negative part and the positive part of the pulses, the internal fraction can be changed by changing the voltage between the dynode from which the signal is taken and the previous one. The internal shaping process has fewer degrees of freedom because the secondary emission factor of the dynodes is limited to a factor

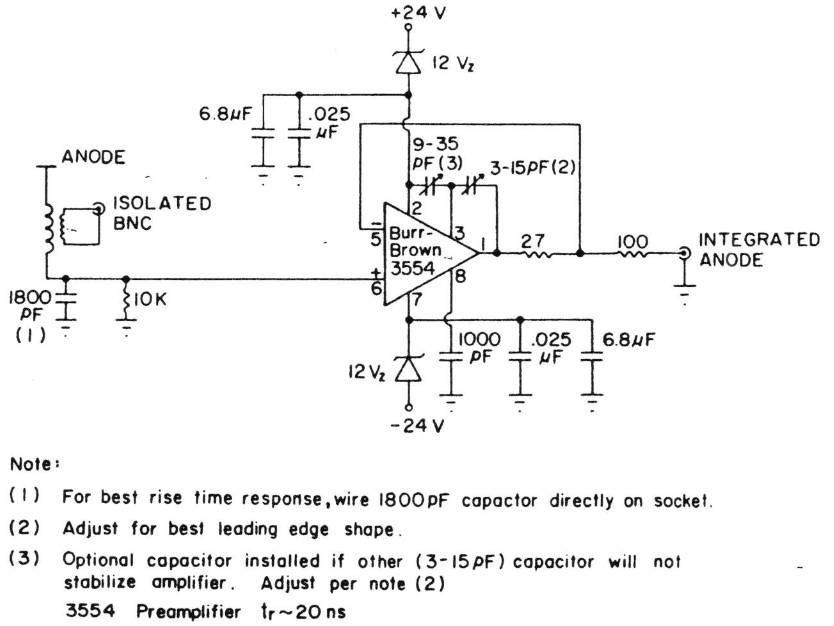

Figure 10.7 Preamplifier circuit used to obtain the integrated linear pulse-height signal from the photomultiplier anode. The output signals from the photomultiplier have a rise time of about 20 nsec. (From Catalog P001/E79 (EMI), GENCOM Inc., 80 Express Street, Plainview, NY 11803, 1979. With permission.)

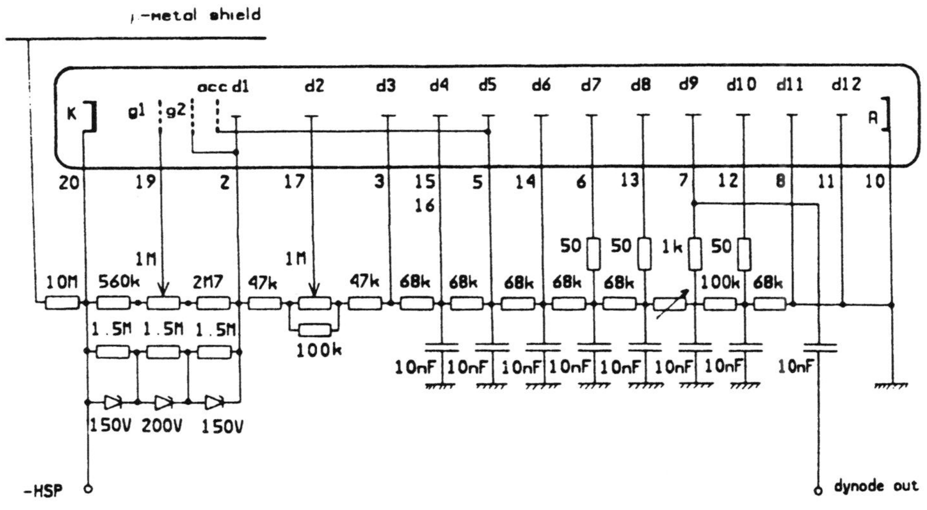

Figure 10.8 Photomultiplier base circuit. (From de Vries, J. and Kelling, F. E. T., *Nucl. Instrum. & Methods,* A262, 385, 1987. With permission.)

of about 10 and also because the time delay between the negative part and the positive part cannot be adjusted as in constant-fraction timing. This delay is of the order of 2 nsec, which is, however, close to the value required by the constant-fraction method for this type of pulse.[9]

Counting efficiency of the miniature photomultiplier RI635 was studied by Ohmori et al.[10] Two methods of the voltage stabilization have been used: (1) the emitter-follower type by npn transistors and (2) the booster type in which the constant dynode voltage is supplied from a separator power supply. The test was performed by using a light-emitting diode and 10-MeV protons from a tandem accelerator.

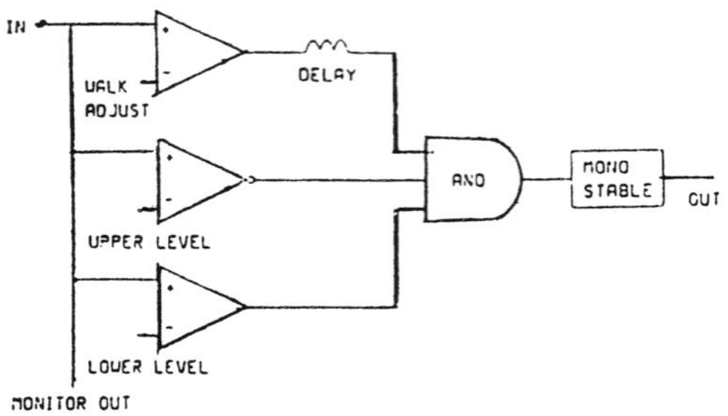

Figure 10.9 Block diagram of the time discriminator. (From Bengtson, B. and Moszinski, M., *Nucl. Instrum. & Methods,* 204, 129, 1982. With permission.)

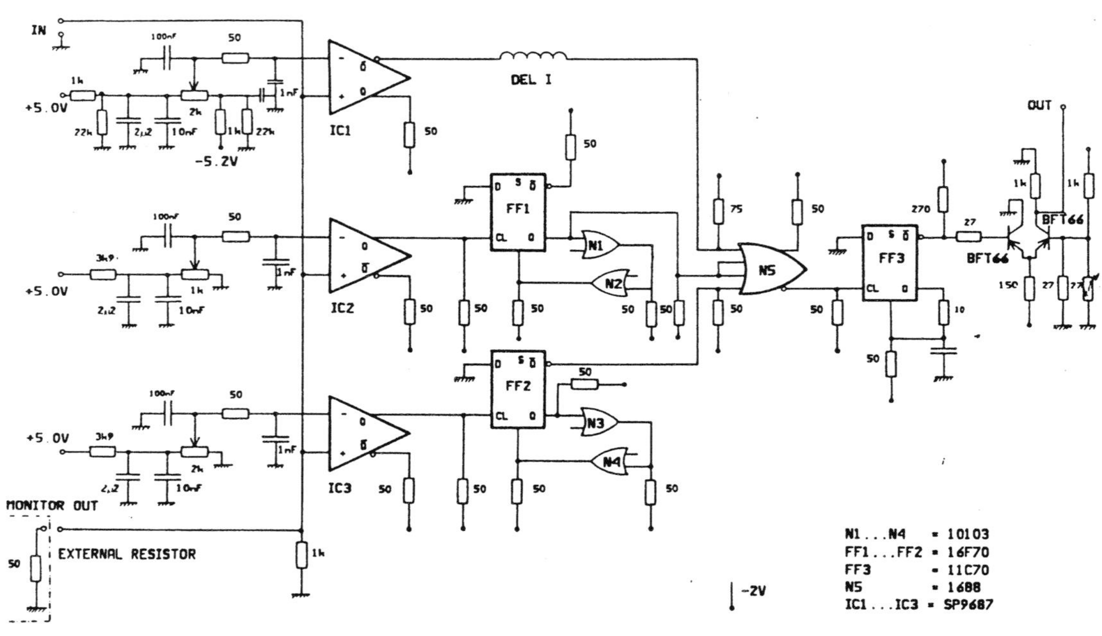

Figure 10.10 Diagram of the time discriminator. (From Bengtson, B. and Moszinski, M., *Nucl. Instrum. & Methods,* 204, 129, 1982. With permission.)

The PM R1635 is a miniature tube with a size of 9.7 mm $\phi \times 45$ mm long, and its rise (0.8 nsec) and fall times (7.8 nsec) are quite suitable for high counting rates. The original high-voltage divider circuit is shown in Figure 10.11(a) which is designed for a divider string current of 300 nA at a high voltage of 1000 V. This divider current is set only for the normal use; thus the responsibility for the high current is quite poor. The modified circuit diagrams are shown in Figures 10.11(b) (emitter-follower type) and 10.11(c) (booster type). The emitter-follower circuit is used in the last four stages with a combination of a high-frequency transistor 2SC507 and diode 1N914. The dividing resistors in the other stage are also replaced to produce higher current compared with the original ones. The test was achieved at a high voltage of 700 and 1000 V. In the case of the booster type, the voltages of the last three dynodes are supplied from a separated power supply. In the experiment, a high voltage of 1000 V was applied and 270, 180, and 90 V were supplied to the dynode number 6, 7, and 8.

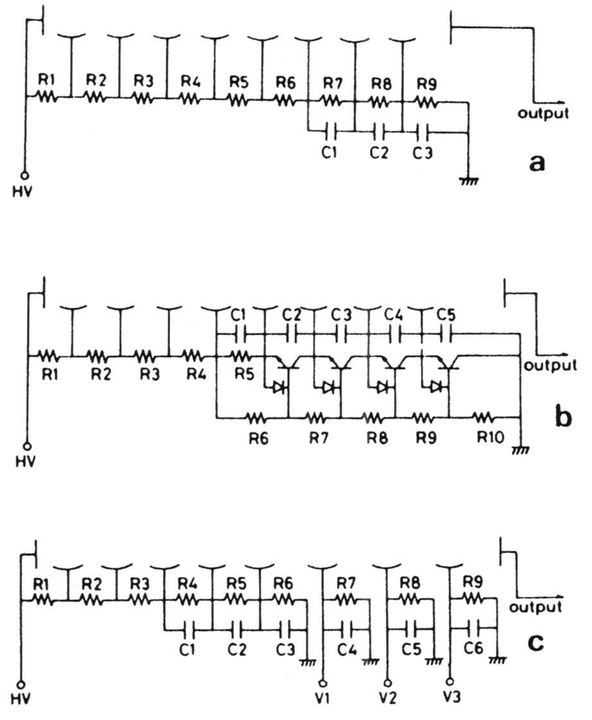

Figure 10.11 Three types of test dividers. (a) Is the original divider, (b) is the emitter-follower type, and (c) is the booster-type divider. The values of resistors and capacitors are the following. (a) R1, R2: 680 kΩ; R3–R9: 330 kΩ; C1–C3 = 0.002 μF. (b) Transistor: 2SC507, diode: 1N914, R1, R2, R6–R10: 180 kΩ; R3, R4: 82 kΩ; R5: 150 kΩ; C1–C5: 0.0047 μF. (c) R1, R2, R8: 680 kΩ; R3–R5, R9, 330 kΩ; R6: 1.3 MΩ; R7: 1.0 MΩ; C1–C6: 0.0047 μF. (From Ohmori, C. et al., *Nucl. Instrum. & Methods,* A256, 361, 1987. With permission.)

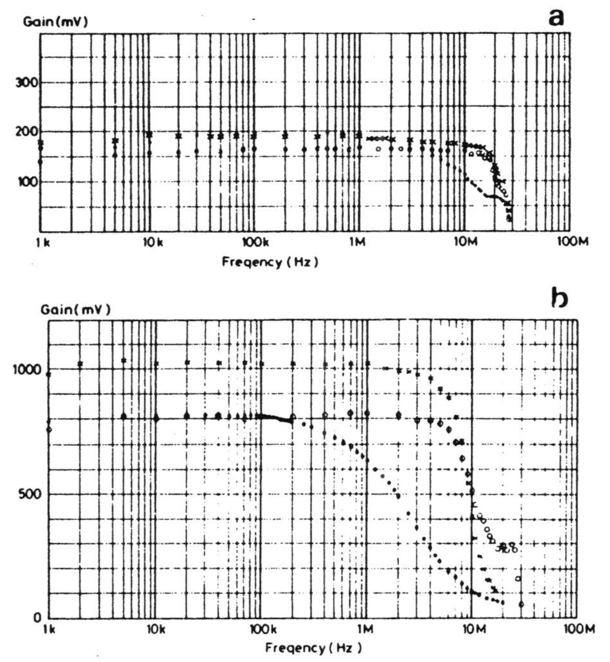

Figure 10.12 The gain of the PM in the case of three different dividers. (a) and (b) correspond to the PM high voltage of –700 and –1000 V. • shows the original divider, O and X correspond to the emitter-follower type and booster one. It shows that there are no essential differences between two modified dividers, but the differences between the original divider and the modified ones are remarkable. (From Ohmori, C. et al., *Nucl. Instrum. & Methods,* A256, 361, 1987. With permission.)

The gain variation of each counter system consisted of the PM R1635, and the high-voltage dividers were investigated. The test was done using a light-emitting diode (LED) HLP (30) (supplied by Hitachi Company, Ltd.) stimulated by a high-frequency pulser with a pulse width of 9 nsec. The LED emits a light with a wavelength of 780 nm and up to a maximum frequency of 70 MHz. The pulse height of the output signal vs. the input signal rate for the three different dividers is shown in Figure 10.12.

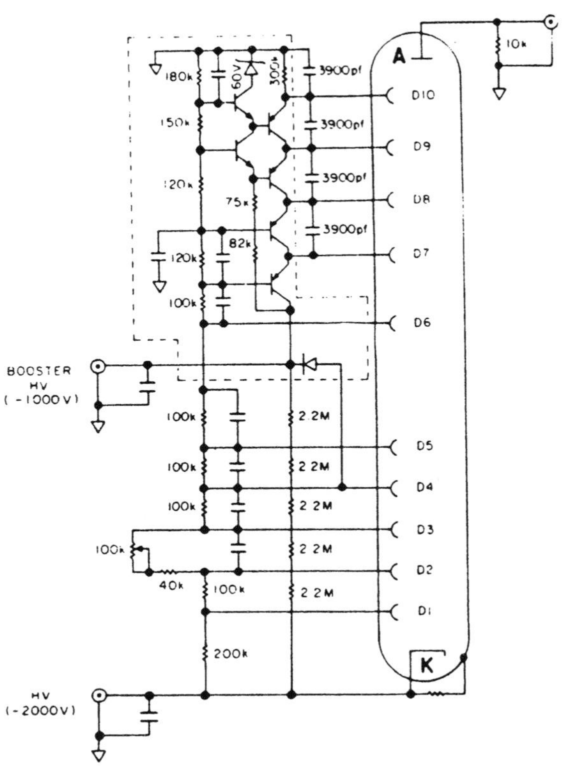

Figure 10.13 Design of the high-current voltage distribution circuit developed for use with the array of 49. All capacitors, unless otherwise indicated, are 0.01 μF. Dynodes 7 and 8 are buffered with MPS U60 transistors and dynodes 9 and 10 by MPS U60 and U10 combinations. The MPS U10s are heat sunk. The components within the dashed area are contained on a removable circuit card. Cathode voltage is applied to a conductive coating on the outside of the PMT envelope. (From Wilson, S. L. et al., *Nucl. Instrum. & Methods,* A264, 263, 1988. With permission.)

The gain of the original divider starts to decrease at 100 KHz, which has been expected. However, the modified dividers show a remarkable improvement, i.e., the gain is constant up to 15 MHz for HV = 700 V and 2 MHz for HV = 1000 V. Both figures show clearly that the signal frequency to the falloff of the constant gain depends on the high voltage, i.e., the divider current of the phototube.

The principal requirement of a photomultiplier tube (PMT) for use with an array of 49 modules (Wilson et al.[11]) is its ability to provide the highest possible line output current for gamma-ray incident in the range 1 MeV to several GeV. Secondary requirements are reasonable gain and moderate price. The output current requirements restricted the search to linear focused PMTs.

A high-current nonlinear voltage distribution circuit was developed in anticipation of operating the array of 49 at high count rates. The circuit is shown in Figure 10.13. Negative high voltage is used to permit direct coupling of the data acquisition electronics and operation at high count rates without baseline shift. All dynode voltages are determined by a resistive divider chain with the higher current dynodes buffered by emitter followers. The pnp transistor polarity for the buffers and npn/pnp transistors in cascade for the two highest current dynodes are selected so that the transistors increase in conduction for increasing PMT current. This technique permits minimum standby current in the PMT base. A booster power supply is used for the higher current dynodes. Bypass capacitors are appropriately placed to smooth peak currents and to maintain the average divider current essentially constant with varying pulse rates. The design criterion was that the interdynode voltage at any point on the divider chain changed by less than 1% for NaI(T1) pulses of peak anode current of 100 mA at a count rate of 100 kHz. When the Amperex 2202B is operated with this base and stimulated with LED pulses 200 nsec in width and 100 mA in peak amplitude, less than 1% reduction in output was observed for rates up to 25 kHz.

The faster Amperex type XP 2232B was used for the crystal box. This choice and that of polished module surfaces were made to enhance the rise time of the output pulse and thereby improve the time resolution of the detector. The circuit for the base used with the XP 2232B is shown in Figure 10.14. This base used only passive components to ensure high reliability, low-temperature dependence, and long-term stability. Two high-voltage supplies were used to provide the required stability at high counting rates. The Amperex recommended taper "B" was used to enhance linearity; and the divider chain was designed such that the current in the latter part of the chain, supplying voltage to the last seven dynodes, was equal to four times the current in the first part of the chain supplying voltages to the first five dynodes and to the

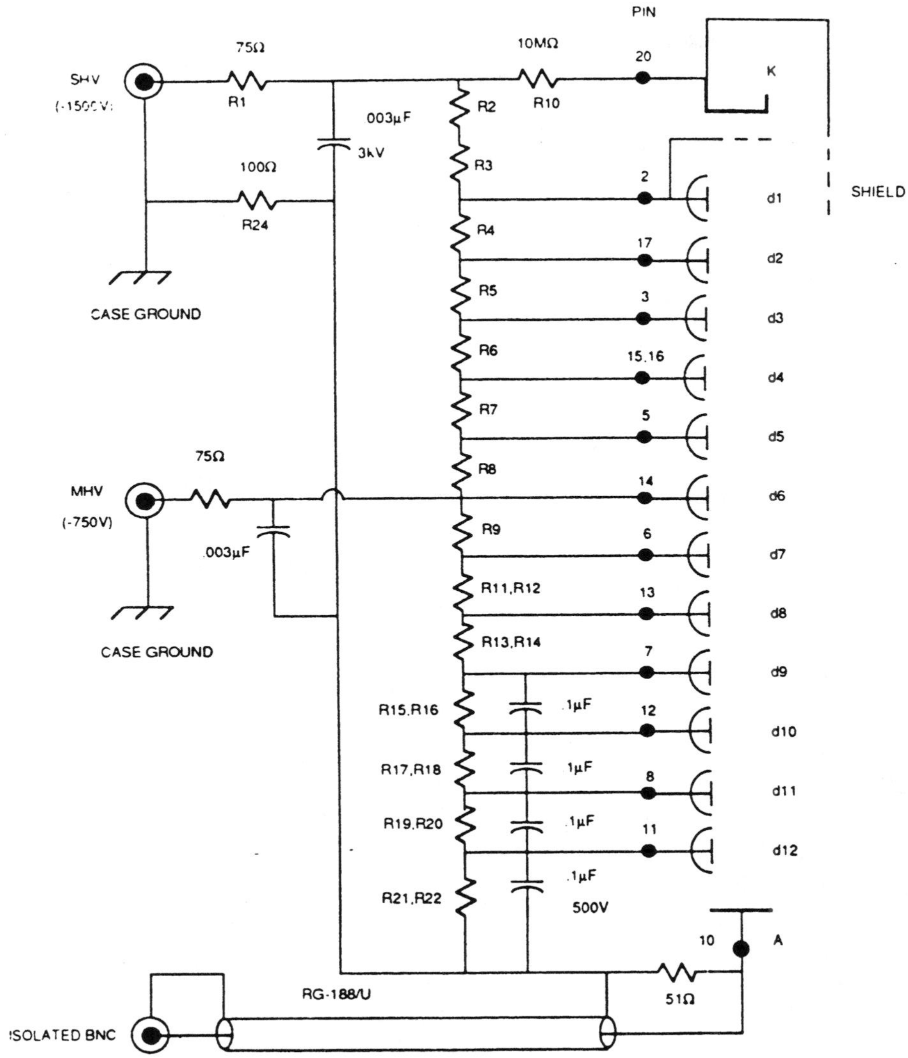

Figure 10.14 The high-voltage distribution circuit developed for use with the crystal box. Cathode voltage is applied to a conductive coating on the outside of the PMT envelope. (From Wilson, S. L. et al., *Nucl. Instrum. & Methods,* A264, 263, 1988. With permission.)

cathode. The output coaxial cable was direct coupled to the anode to avoid rate effects and back-terminated at the anode connection in 50 Ω. The signal ground was isolated from the case ground and the high-voltage power returns by a resistance of approximately 100 Ω to reduce the effects of ground loops.[11]

The pilot-U scintillator, with a decay constant of the light emission of 1.36 nsec, is known to be one of the fastest scintillators. The 1-mm thickness is unnecessarily larger than the range of the light ions to be measured, but needed to get a self-supporting and simple construction.

The photomultiplier, Hamamatsu H 2431, was chosen by Wattecamps et al.[12] for the good compromise of cost and the following characteristics: a gain of 10^6, a transit time of 7 nsec, a transit time spread of 0.22 nsec, and a 0.4-nsec anode pulse rise time (nominal values). The voltage divider is shown in Figure 10.15 together with the preamplifier. A linear signal is extracted from the sixth dynode, whereas a fast signal comes straight from the anode. The photomultiplier was operated at the maximum voltage of 3500 V. The observed rise time of the anode pulse is 3 nsec, and these pulses are fed into a constant-fraction timing discriminator TC 455 of Tennelec, with an external delay line of 1 nsec. The zero-crossing level discriminator is set at 10 mV, and the threshold at 40 mV.

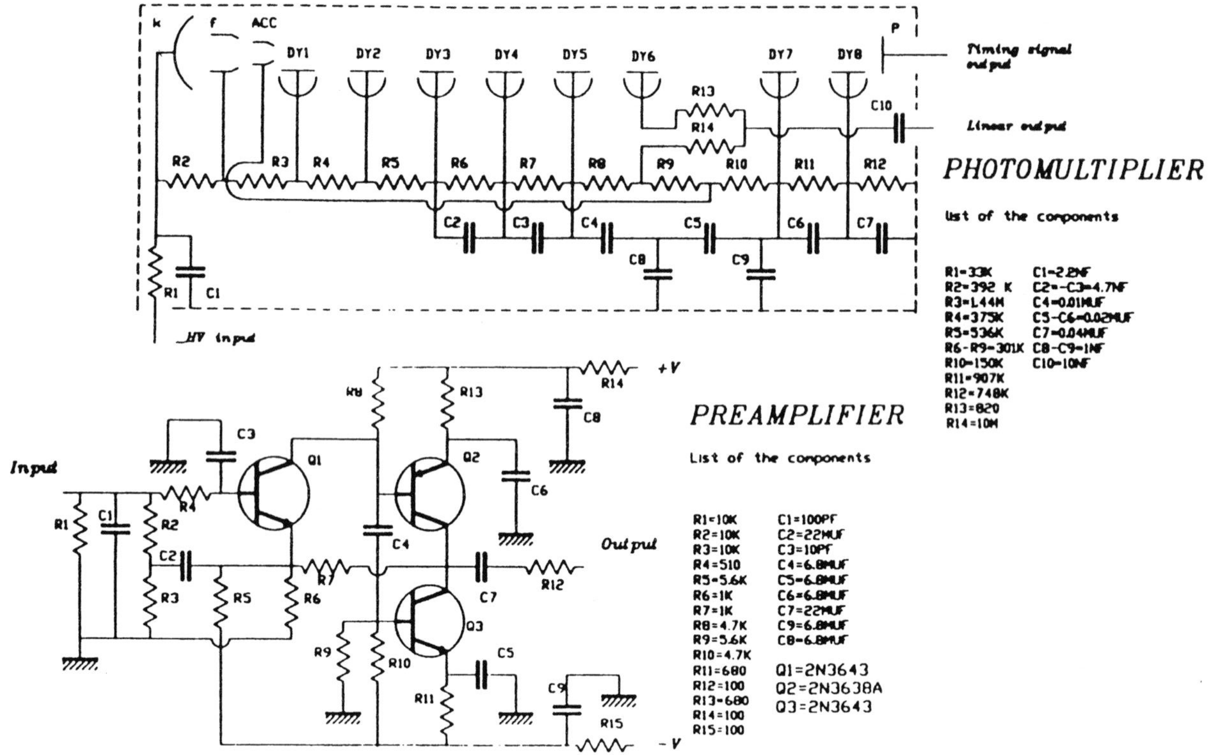

Figure 10.15 Circuit diagram of the photomultiplier and preamplifier. (From Wattecamps, E. et al., *Nucl. Instrum. & Methods,* A309, 209, 1991. With permission.)

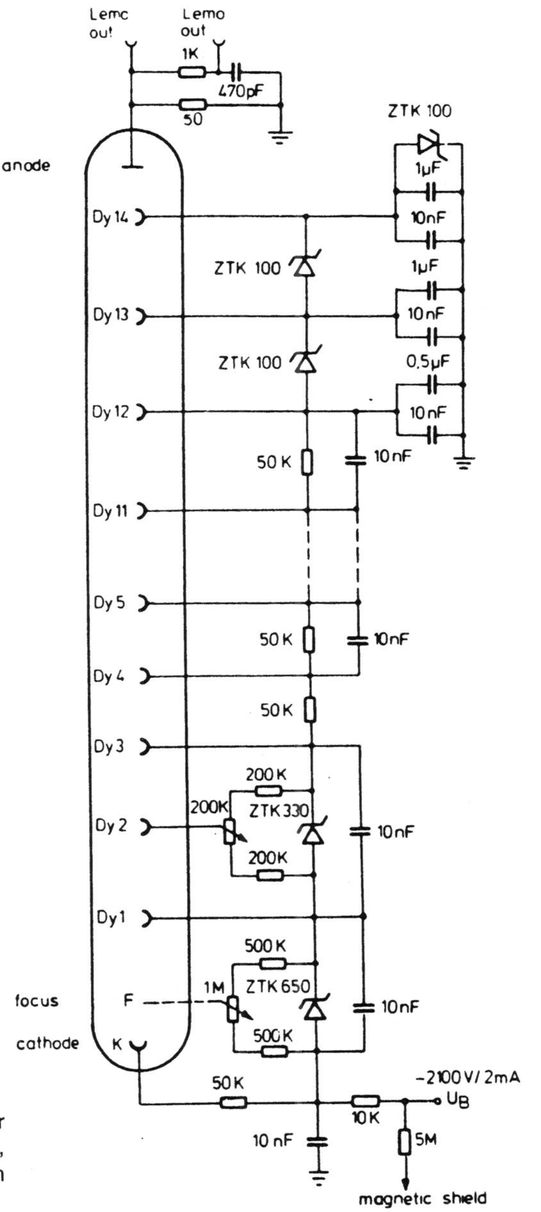

Figure 10.16 Schematic diagram of the voltage divider chain for the EMI 9823 QA photomultiplier. (From Wisshak, K. et al., *Nucl. Instrum. & Methods,* A251, 101, 1986. With permission.)

Wisshak et al.[13] found that much better timing is obtained with a PM of 5-in. diameter compared to a 3-in. tube in spite of the worse electron transport characteristics. In the first case 62% of the base surface is covered by the cathode while this fraction is reduced to only 18% for the 3-in. PM. Consequently, the time spread of the light reaching the cathode is significantly larger in the second case due to the many required reflections. The EMI tube 9823 QA offers the necessary combination of fast rise time, 5-in. cathode, and quartz window required for the ultraviolet scintillation light. A special voltage divider chain has been designed for use with BaF_2 crystals. The chain recommended by the manufacturer and normally used in NaI(T1) detectors yielded only poor energy resolution with BaF_2. A significant improvement was obtained by choosing larger steps in the potential at the first and last two dynodes compared to the others. At these important dynodes, the voltage is stabilized by means of zener diodes, and resistors with appropriate temperature dependence were selected in order to compensate for temperature-sensitive gain variations. At a typical bias voltage of –2000 V the current in our divider chain is 2 mA, sufficient for

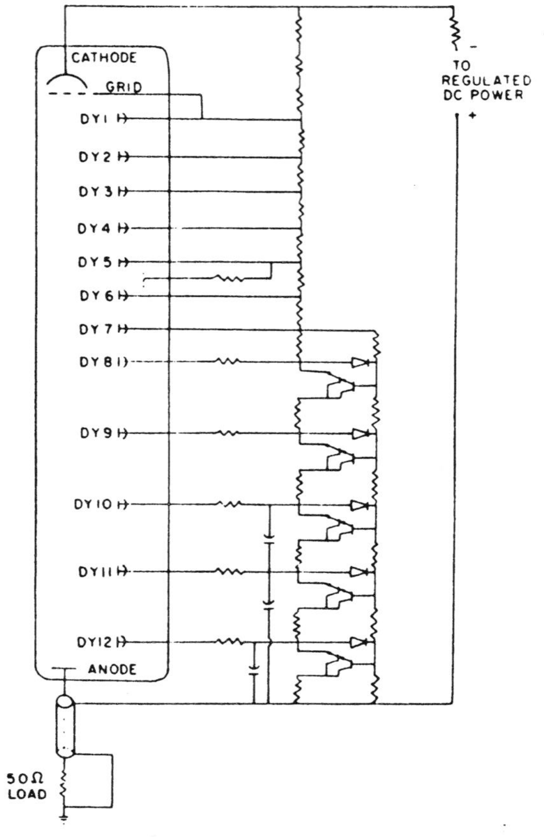

Figure 10.17 Design of the photomultiplier tube base on the TOF scintillation array. (From Engelage, T. et al., *Nucl. Instrum. & Methods,* A277, 431, 1989. With permission.)

stable operation in the planned experiments with expected count rates of less than 5 kHZ per BaF_2 crystal. A schematic diagram of the voltage divider chain is shown in Figure 10.16.

In an experiment done by Engelage et al.,[14] each scintillator was individually wrapped and was viewed at both ends by an Amperex XP 2230 photomultiplier tube. Each tube was connected to a base that was modified to operate over a large dynamic range while exhibiting only a small distortion in pulse shape. As shown in Figure 10.17, this modification involved the addition of a Darlington pair to each of the last five stages. This type of emitter-follower circuit increases the power level available to the dynode while maintaining the voltage gain as set by the resistor chain. The bases of the tubes were equipped with two anode outputs, one of which was connected to a LeCroy 2280 analog-to-digital converter (ADC) system. The second anode signal supplied the stop for the individual channels in a LeCroy 2228A TDC system.

An high-voltage switch has been installed by Banerjee et al.[15] which can ramp the high voltage within a fraction of a millisecond. The circuit diagram of this unit is shown in Figure 10.18. With this switch, the voltage is lowered by 600 V, and data taking is resumed 50 msec after bringing up the high voltage without degradation in time resolution.

Photocathodes are at negative high voltage to avoid capacitive elements on anode outputs. The photomultiplier base design is shown in Figure 10.19. It is based on a modified dynode Mark III design[6] (nontransistorized). Only resistive and capacitive elements are used in the divider chain. Output from the last dynode, used for pulse-height digitizing, is inverted with a ferrite toroid in the bases.

Dynode signals were fed to inputs of LeCroy 2285 ADCs to measure charge (least count 0.25 pC), and the anode signal was fed to LeCroy model 2228A TDCs (least count 50 psec and total range = 100 nsec) via a LeCroy 620L discriminator for time measurement.

An improved method for the stabilization of NaI photomultiplier gamma detectors against thermal and other drift was described by Zucker.[16] Alpha peaks have been used as part of servo systems to stabilize NaI photomultiplier gamma detectors against drift. However, alpha peaks shift with temperature change differently than do gamma peaks, thus spoiling what would otherwise be a workable scheme for stabilizing against probably the most serious source of NaI-PM detector drift, namely, thermal effects.

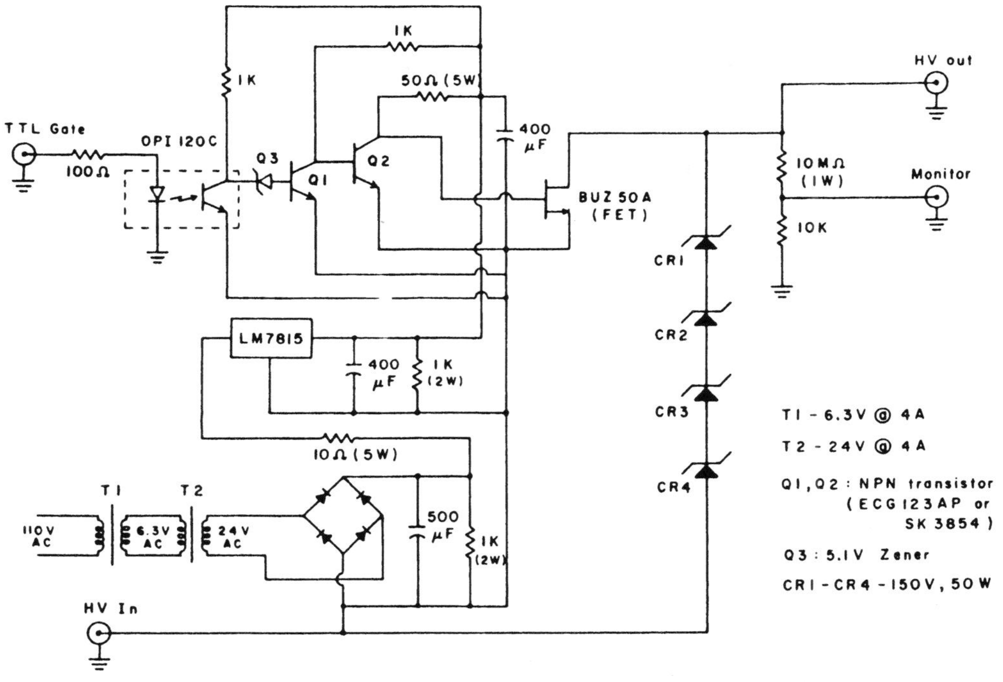

Figure 10.18 Circuit diagram of the fast high-voltage switch used in ramping the PMT HV supply. (From Banerjee, S. et al., *Nucl. Instrum. & Methods,* A269, 121, 1988. With permission.)

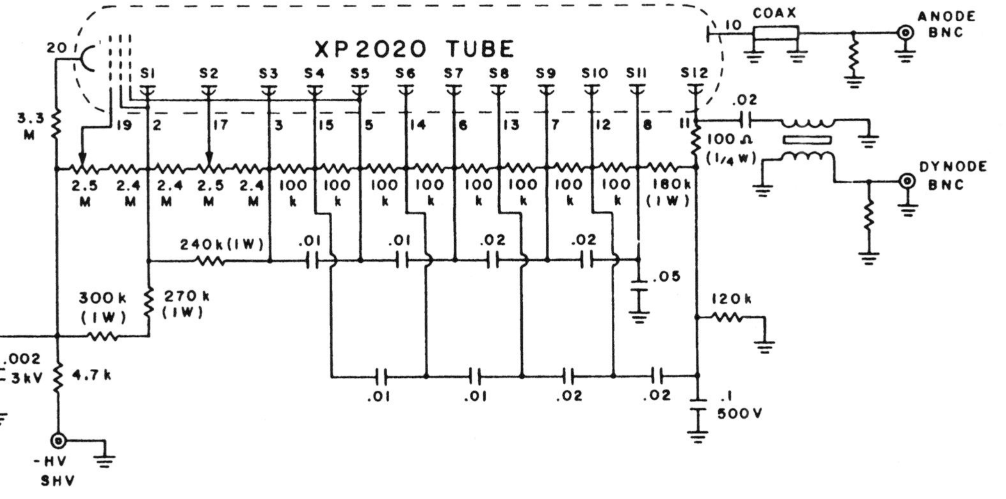

Figure 10.19 Circuit diagram of the phototube base. (From Banerjee, S. et al., *Nucl. Instrum. & Methods,* A269, 121, 1988. With permission.)

It has been found possible to accurately compensate for the difference in the shift with temperature vs. gamma peaks using the signal derived from thermistor in thermal contact with the NaI crystal to control the bias of a discriminator in the servo circuit. The servo circuit utilizing this principle has been used in commercial multichannel analyzers of the type intended for field use under adverse ambient conditions.

The gain stabilization circuit taking advantage of the alpha source and thermistor is shown in block diagram form in Figure 10.20. Alpha particles from the source produce pulses with amplitudes corresponding approximately to that of a 3-MeV gamma full-energy pulse. This alpha pulse is processed in

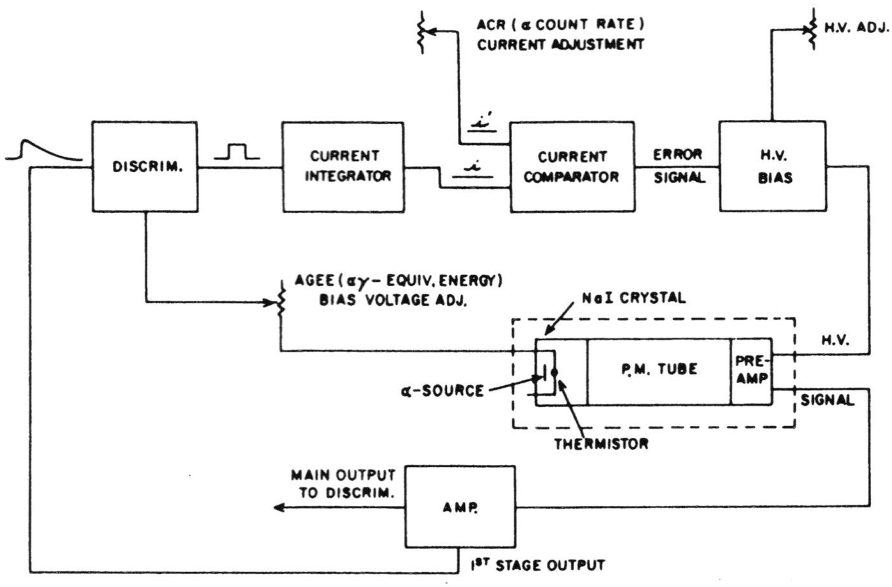

Figure 10.20 Block diagram of the NaI-PM tube gain stabilizer system. (From Zucker, M. S., *Nucl. Instrum. & Methods,* A242, 306, 1986. With permission.)

the same amplifier used for gamma pulse except that it is taken out at a point before it saturates. It is sent to a discriminator whose bias is set relative to the alpha peak position so that half the alpha pulses exceed it. This bias, chosen for convenience to be 1 V/MeV, depends on what the gamma equivalent energy is for the particular detector. Since the alpha-gamma equivalent energy (AGEE) may differ even among detectors of nominally the same design, an adjustment must be provided. The thermistor to compensate for peak shift with temperature is in good thermal contact with the NaI crystal. Standard pulses from the discriminator are converted into a current proportional to the rate at which they are generated by an integrator circuit. This current, proportional to the count rate of alphas, is sent to a comparator whose other input is a current adjusted to be that appropriate to the nominal alpha-count rate (ACR) of the source. The difference constitutes an error signal presumed to indicate the deviation of the system gain from its desired operating point. The error signal is used to accordingly adjust the amplitude of an oscillator energizing the primary of a step-up transformer whose rectified filtered output furnishes the high voltage (HV) for the detector bleeder string, therefore changing the PM tube gain.

Once the nominal settings for the alpha energy and count rate have been made for the particular detector in the laboratory or factory (AGEE and ACR controls), it is only required to set the HV adjustment of its nominal value. This is done with the aid of two LED indicators. These indicate that either the HV adjustment is too high or too low. Splitting the difference in the (10 turn helipot) setting gives the best nominal HV value.

The gain of PMs may change in time with a drift rate on the order of 20% per week.[17] For both calibration and stabilization, the PM gain can be adjusted by changing the PM high voltage. According to the block diagram of Figure 10.21 the cathodes of the PMs are connected to high voltage digital-to-analog converters (HV-DAC) for changing the gain of the PMs. The gains are adjustable by changing the window positions of discriminators in the stabilization circuit with other DACs.[17]

The stabilizing system is based on LED light pulses. At a frequency of 2.5 kHz blue ($\lambda = 430$ nm) light pulses are generated in an LED and are distributed via 16 light guides to every group of four PMs. The amplified anode signals of the light pulses have the same order of magnitude as the neutron signals. With the analog switches operating consecutively, the pulse of a particular PM is selected and its pulse height is analyzed in a dual-window discriminator. By comparing the pulse height with the window position that is set by a DAC, up and down counter pulses are generated which change the high-voltage DAC accordingly. It takes about 5 s per PM to obtain a match between the average pulse height and the window positions at the discriminators.

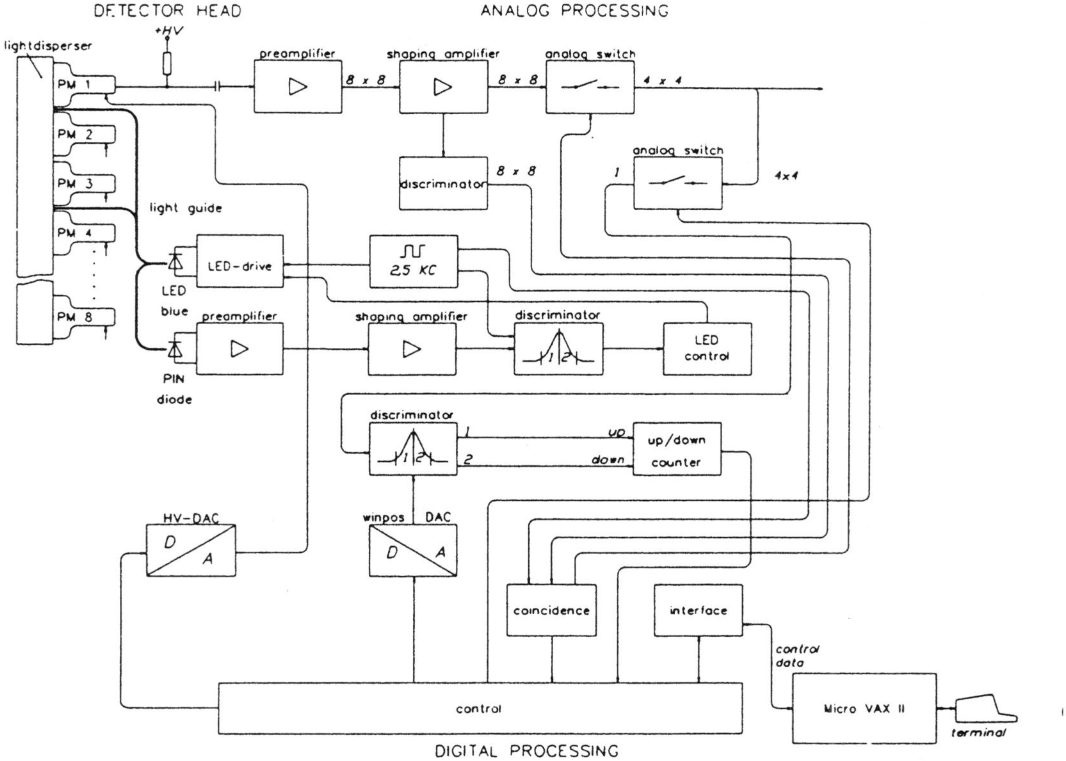

Figure 10.21 A block diagram of the gain-stabilizing circuitry. (From Heiderich, M. et al., *Nucl. Instrum. & Methods,* A305, 423, 1991. With permission.)

Furthermore, the LED light signals are controlled. For this purpose there is a light guide connection between the LED and a PIN diode. The amplified diode signal is analyzed by another discriminator. Its generated up and down counter pulses are used to change the LED current pulse accordingly.

REFERENCES

1. **Bengtson, B. and Moszynski, M.,** *Nucl. Instrum. & Methods,* 204, 129, 1982.
2. **Bonnet, L. et al.,** *Nucl. Instrum. & Methods,* A292, 343, 1990.
3. **Sené, M. R.,** *Nucl. Instrum. & Methods,* A278, 503, 1989.
4. **Randors-Pehrson, G. et al.,** *Nucl. Instrum. & Methods,* 215, 433, 1983.
5. **Finly, R. W. et al.,** *Nucl. Instrum. & Methods,* 198, 197, 1982.
6. Catalog P001/E79 (EMI), GENCOM Inc., 80 Express Street, Plainview, NY 11803, 1979.
7. **de Vries, J. and Kelling, F. E. T.,** *Nucl. Instrum. & Methods,* A262, 385, 1987.
8. **Bengston, B. and Moszinski, M.,** *Nucl. Instrum. & Methods,* 204, 129, 1982.
9. **de Vries, J. and Kelling, F. E. T.,** *IEEE Trans. Nucl. Sci.,* 35, 392, 1988.
10. **Ohmori, C. et al.,** *Nucl. Instrum. & Methods,* A256, 361, 1987.
11. **Wilson, S. L. et al.,** *Nucl. Instrum. & Methods,* A264, 263, 1988.
12. **Wattecamps, E. et al.,** *Nucl. Instrum. & Methods,* A309, 209, 1991.
13. **Wisshak, K. et al.,** *Nucl. Instrum. & Methods,* A251, 101, 1986.
14. **Engelage, T. et al.,** *Nucl. Instrum. & Methods,* A277, 431, 1989.
15. **Banerjee, S. et al.,** *Nucl. Instrum. & Methods,* A269, 121, 1988.
16. **Zucker, M. S.,** *Nucl. Instrum. & Methods,* A242, 306, 1986.
17. **Heiderich, M. et al.,** *Nucl. Instrum. & Methods,* A305, 423, 1991.

Chapter 11

Auxiliary Circuits

CONTENTS

11.1 Timing Circuits 149
11.1.1 Pulse Resolving 149
11.1.2 Time Spectroscopy 150
11.1.3 Pulse Shape Timing 150
11.1.4 Coincidence Circuits 151
11.2 Linear Gate 159
11.3 Sample-and-Hold Circuits 164
11.4 Pulse Stretcher 173
References 174

11.1 TIMING CIRCUITS

11.1.1 PULSE RESOLVING

An important parameter of the information obtained from the radiation detector is the pulse timing. One way of timing determination is by coincidence. When two pulses enter the system through different inputs, the coincidence system verifies their occurrence within a time interval, the resolving time. If the resolving time is tau and the count rates of both inputs are N_1 and N_2, then the number of coincidence events is

$$N_c = 2N_1N_2\tau \tag{11.1}$$

The resolution time τ can be obtained by measuring N_1, N_2, and N_c. It is possible that two simultaneously arriving pulses will be registered as arrived at two different moments due to different heights and thus different times of crossing the discrimination level as shown in Figure 11.1. Other delays can be due to different ionization location in the detector. Timing on the order of 1 μsec, is obtained with a monostable circuit. The coupling RC time constant sets the resolving time. Well-defined short pulses are produced by the delay line (as shown in Figure 11.2) Q_2 is conducting. When the circuit is triggered, a negative pulse of height $I_e/2R$ is produced in the delay coil (where I_e is the current in the emitter of Q_2). R is the impedance of the delay coil and the matching resistor. The pulse length is twice the delay time provided it is shorter than the monostable period. When the circuit returns to the initial condition, a positive pulse is generated.

A simple circuit for overlap determination is shown in Figure 11.3. It is an AND circuit, with both diodes conducting with a constant current. Every input pulse increases the voltage on the cathode; and the diode is cut off and the output voltage is raised, respectively. When two input pulses appear simultaneously both diodes are cut off and the rise in output voltage is doubled for the overlapping time. The circuit is called a mixer. To define the time of each input a monostable circuit can be connected with a delay line as described above. Such an AND gate can be used for anticoincidence if one diode is reversed.

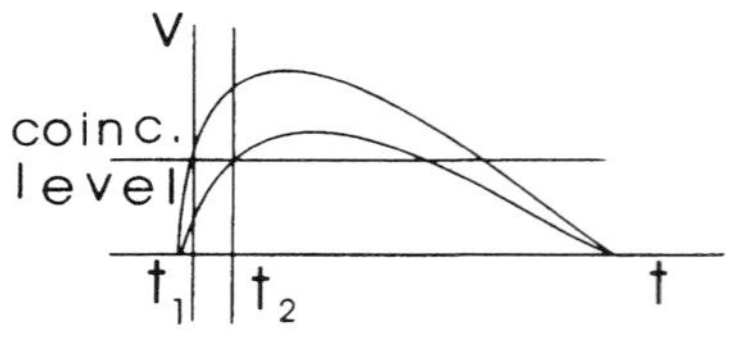

Figure 11.1 Coincidence walk.

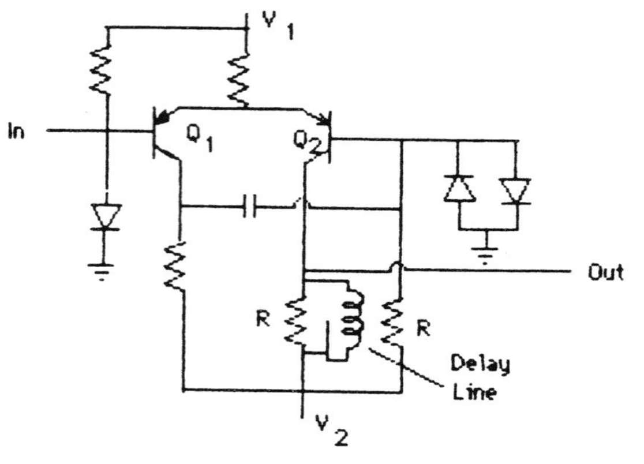

Figure 11.2 Timing circuit.

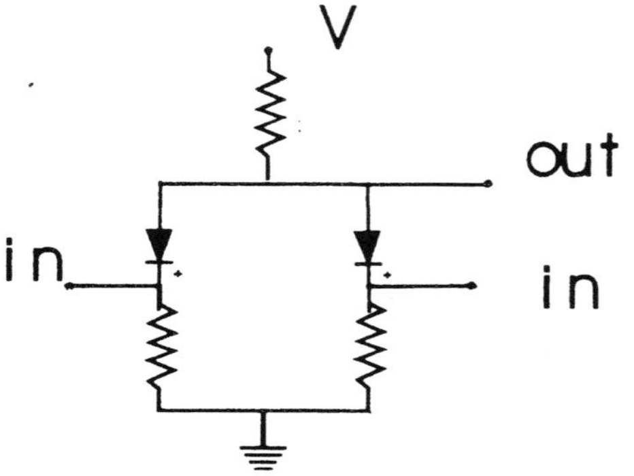

Figure 11.3 Pulses overlapping determination.

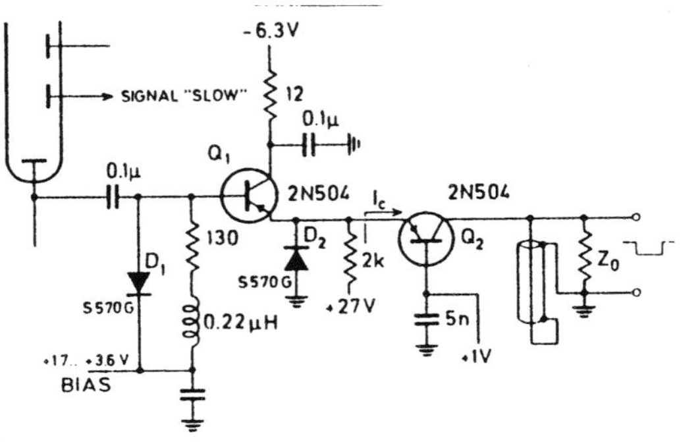

Figure 11.4 Limiter circuit. (From Barna, A. J. et al., *Nucl. Instrum. & Methods,* 7, 124, 1960. With permission.)

11.1.2 TIME SPECTROSCOPY

The second timing parameter is the time interval between pulses. Time of flight is a very common way of measuring this interval. Long time intervals are measured by counting oscillator cycles. Several techniques are used for short (<1 nsec) time intervals; most are based on capacitor charging. A capacitor is charged by a constant current, between two pulses. The capacitor voltage is relative to the time interval. Time pulse-height conversion is done with two switches, a start and a stop. The first is normally open and closes on receiving the start pulse. The second is normally closed and opens by the stop pulse. The capacitor voltage remains constant after the stop pulse at a level relative to the time interval between the two pulses.

In the overlap converter two pulses of length T are applied to a diode mixer providing an output pulse during the overlap only. The overlap pulse charges a capacitor. The shorter the time interval between pulses the longer is the overlap and vice versa. The advantage of this method is that there is no output if there are not two input pulses.

In the time expander method the discharge time of the capacitor is measured instead of its voltage. The measurement is done through a time constant 100 or more times longer than the charge time constant.

11.1.3 PULSE SHAPE TIMING

Pulse shaping for coincidence circuits requires input pulses of a standard shape, in particular, for a fast coincidence. Pulse timing is done by the leading edge of the pulse or by zero crossing. Using a photomultiplier tube, anode pulses are used for timing. A circuit which provides a fast square pulse and reduces the noise was designed by Barna et al.[1] (Figure 11.4). The current in Q_2 enters a clipping cable of impedance Z_0. Q_1 is cut off by low bias. A negative pulse switches Q_1 on and transfers the current in

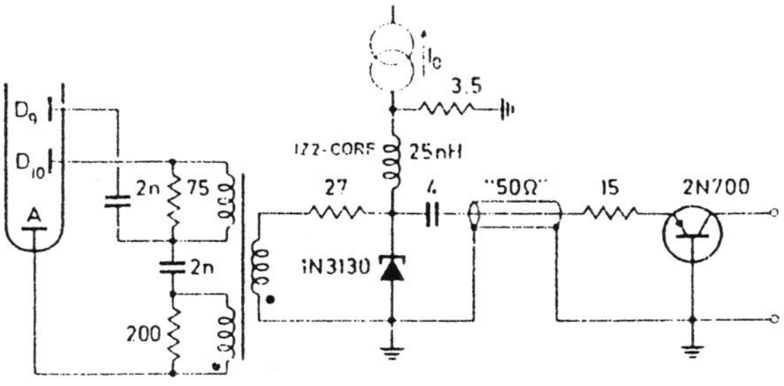

Figure 11.5 Zero-crossing pulse shaper. (From Barna, A. J. et al., *Nucl. Instrum. & Methods,* 7, 124, 1960. With permission.)

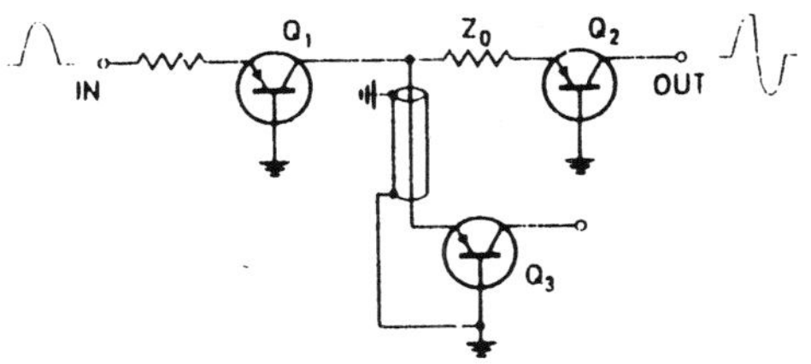

Figure 11.6 Fast bipolar pulse shaping by means of delay cable differentiation. (From Barna, A. J. et al., *Nucl. Instrum. & Methods,* 7, 124, 1960. With permission.)

Q_2 to Q_1; thus a standard pulse is obtained. The diode D_1 limits the positive voltage jump at the input. D_2 together with the base-emitter junction of Q_1 limits the negative voltage. The pulse rise time at the output is ~1 nsec. A common emitter circuit will be slower than a common base circuit.

Using a tunnel diode (TD), a trigger pulse is obtained instead of the rectangular pulse of the limiter. A tunnel diode enables fast switching and the output pulse is delayed relative to the input pulse according to the pulse height. A timing circuit using a tunnel diode by zero crossing is shown in Figure 11.5. A bipolar pulse is delivered through R_1 (=27 Ω) to the TD monostable which responds to the pulse undershoot. The monostable output is 3 nsec wide and the rise time is 150 psec. The pulse is differentiated by the capacitor C and is used to control the transistor. Another way of bipolar pulse shaping is obtained by cable clipping pulses from the photomultiplier (Figure 11.6). They are fed through a common base transistor and a shorted delay line of a proper length. The original pulse and the reflected one are picked up by a common base transistor. The junction emitter base of Q_3 serves as a termination of the clipping cable. The zero crossing is marked by the trailing edge.

The use of leading edge has a smaller error than the bipolar pulse zero crossing. The reason is that the leading edge will trigger when the pulse reaches a certain (low) level. Zero crossing is done by the bipolar pulse when it is doubled with an opposite sign by the delay line. It reaches the zero at half of its height (of the negative section) and thus is higher than the leading edge.

Bell[2] has calculated the resolution of these two methods for NaI(T1) and for a plastic scintillator and found that the resolution time using the zero-crossing method is 13.7 times larger than that with the leading edge.

11.1.4 COINCIDENCE CIRCUITS

The ideal coincidence circuit is a very fast AND gate receiving standard pulses of width δt, having a resolution time $2\tau = \delta t_1 + \delta t_2 = 2\delta$ (more than two inputs are possible). The circuit (Figure 11.7) is in a mix adder with a discriminator at the output. When the two input pulses overlap, the output pulse is double the input pulse height. The discriminator has an output pulse whenever there is a pulse of a certain height at its input. If the pulse is not square but has a certain rise time, the out pulse is lower. When the two pulses are almost separated, the resolving time depends on the discrimination level.

Many practical coincidence circuits can be found in the literature; a few examples are shown here. A medium fast coincidence circuit that was developed by Emmer[3] is shown in Figure 11.8. Zero-crossing discriminators (ZCDs) were used for pulse shaping. The output of the two ZCDs enters the circuit inputs. At first Q_2, Q_4, and Q_6 are conducting. A positive pulse at one of the inputs cuts off Q_2 or Q_4 so that the current passes through the clipping cable. The pulse length fits twice the cable length. The amplifier Q_5, Q_6 was used as a discriminator giving an output only when the pulse height was doubled. The load resistors of Q_2 and Q_4 were chosen to match the cable. The trailing edge of the input pulse creates a negative pulse in the cable which does not pass through the discriminator.

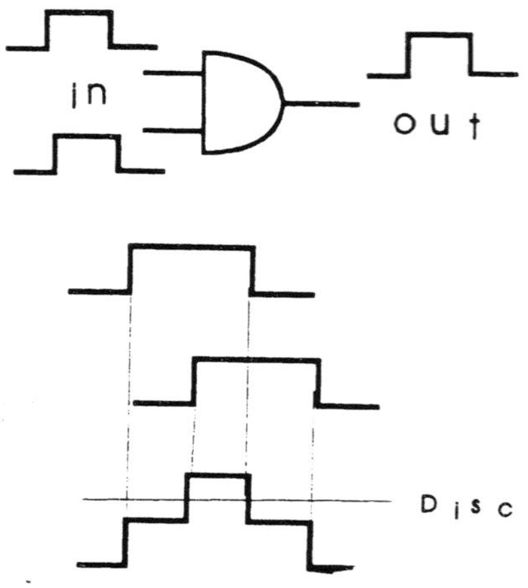

Figure 11.7 Ideal coincidence.

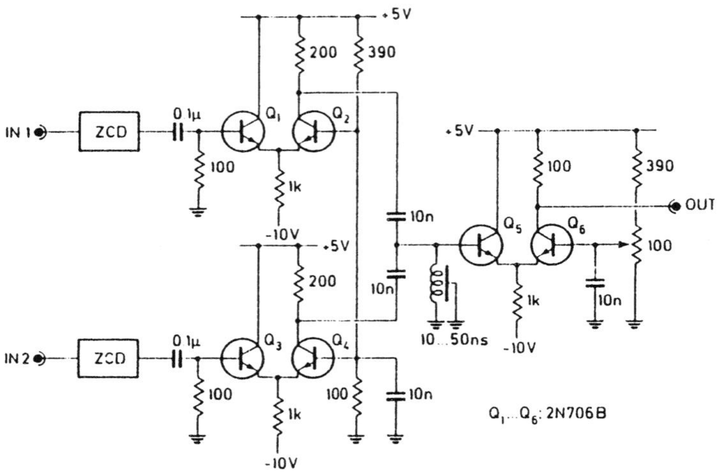

Figure 11.8 Medium fast coincidence stage for use with zero-crossing discriminators (ZCD).

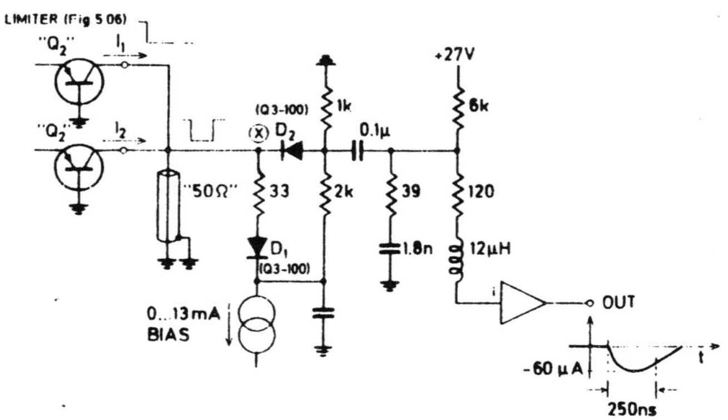

Figure 11.9 Fast coincidence stage.

A circuit based on a similar principle but with the discriminator composed of diodes was designed by Barna et al.[1] (Figure 11.9). Pulse shaping was done by a limiter circuit (common base transistor). The limiter pulses were differentiated and created a negative pulse at the diode junction.[4] The pulse height in coincidence was –0.6 V and it passed D_2 and amplified in a current amplifier at the output. The bias was supplied through D_1. It assured termination of the cable by one of the two diodes and the following resistor; the resolving time was 2τ = 2nsec.

When the limiter cannot be connected directly to the discriminator, an intermediate pulse-shaping circuit is connected through a cable and a common base transistor as shown in Figure 11.10. Such

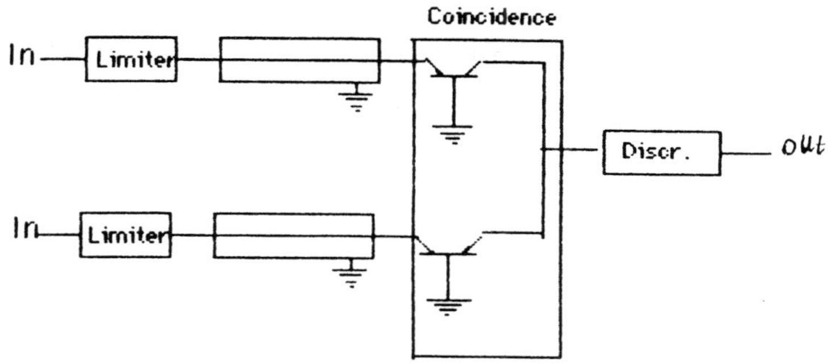

Figure 11.10 Coincidence circuit. (From Sugarman, R. M. et al., *BNL*, 711, 1962.)

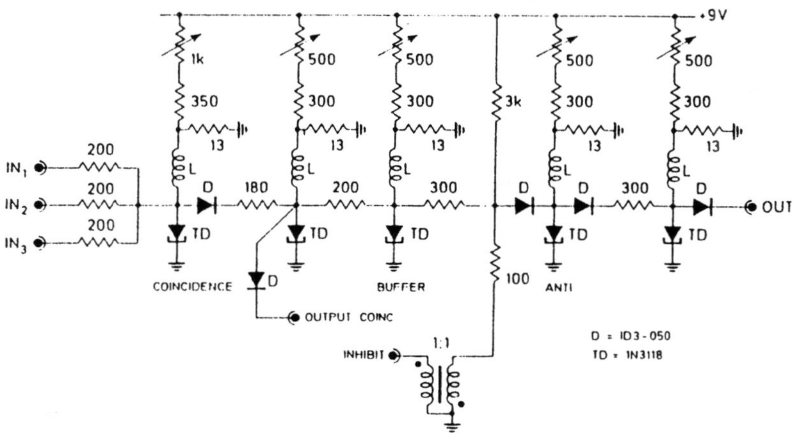

Figure 11.11 Tunnel diode coincidence circuit. (From Whetstone, A. and Kounosu, S., *Rev. Sci. Instrum.*, 33, 423, 1962. With permission.)

connection separates the limiter and the mixer. A tunnel diode was used by Whetstone and Kounosu[5] as shown in Figure 11.11. Standard voltage pulses after pulse shaping through 200-Ω resistors were fed into the first tunnel diodes. Adjustment was made so that a trigger was obtained by the sum of all three inputs or the sums of any two only. The second tunnel diode was used as a multivibrator and shaped the pulse. After a buffer stage an inhibit pulse could be introduced, i.e., anticoincidence. The transformer inverted the pulse polarity for the inhibition.

A series of nonlinear mixers is shown in Figure 11.12. This is a simple basic circuit. When all transistors are on by positive input pulses, a negative output pulse will appear. If three pnp transistors are used, negative pulses are needed at the input and a positive pulse will appear at the output. A similar circuit in parallel connection is shown in Figure 11.13. All transistors are in saturation. When they are all cut off, a positive pulse is obtained at the output. Here two pnp transistors can be used for positive polarity. The resolving time is determined by the time constants and switching times of the transistors.

Diodes can be used instead of transistors in the last two circuits. An improvement was obtained[6] by connecting a resistor and a diode in parallel to the load resistor. The current is reduced every time a transistor is cut off; and a small output pulse is obtained for one input pulse, a larger one for two input pulses, and still a larger one for three input pulses (all transistors cut off). Since the transistors are not in saturation, the switching is fast.

A three coincidence and one anticoincidence circuits are shown in Figure 11.14. The resolution time is set by the input pulse width. It is possible to disconnect any one of the channels. The tunnel diode is used as a mixer. When there are no input pulses, the currents I_1, I_2, and I_3 are 5 mA each and I_4 is 13 mA (reversed). The tunnel diode is conducting at 2-mA reverse current. Coincidence pulses remove the currents I_1, I_2, and I_3 so that the tunnel diode is switched to high-voltage state by the current I_4 and trigger to the output is attained circuit. When an anticoincidence pulse occurs, I_4 stops and the tunnel diode cannot switch to a high-voltage state. The switching off of a channel is the same as having a constant input pulse in it. When the anticoincidence is switched off, I_4 is maintained.

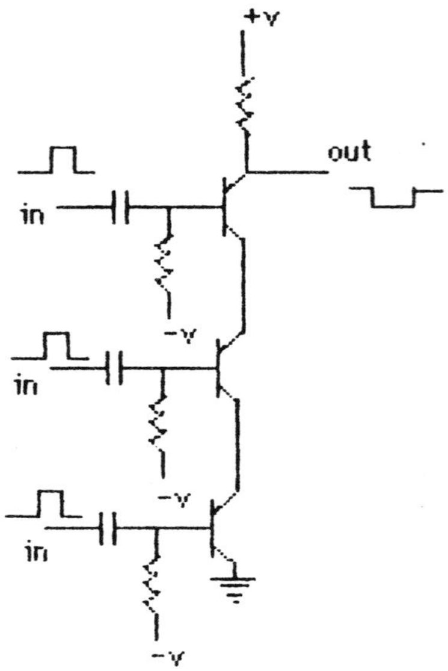

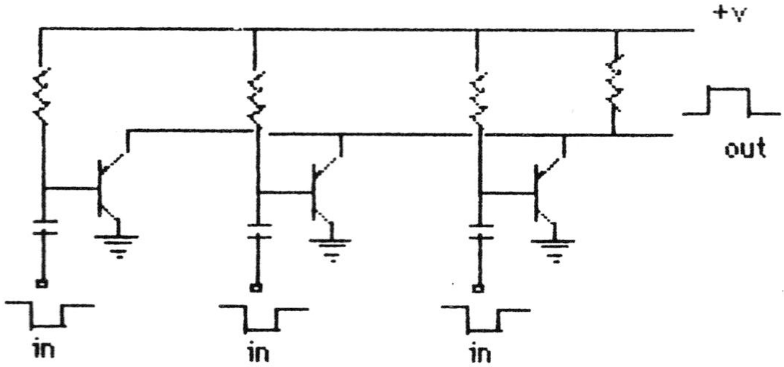

Figure 11.13 Parallel coincidence.

Figure 11.12 Series coincidence.

A coincidence of two pulses with a certain delay is obtained using a number of delay lines. Two versions are shown in Figure 11.15. If two delay lines are used, the output is obtained through an AND gate. If one cable is used, the output is obtained through a discriminator. The delay in each stage is Δt. A multichannel coincidence with a width of $2\Delta t$ is obtained. Since the resolving time is twice the pulse width, i.e., shorter than $2\Delta t$, one pulse can span over two channels.

Time coincidence measurements on the order of nanoseconds were discussed by Kirsten.[7] A Schmitt trigger is often used to generate a logic level equal in length to the overlap of input pulses. In other cases a one-shot is used to generate a logic pulse of standardized length each time a coincidence is discovered.

Figure 11.16 shows an example in which the switches that control the constant currents are in the form of diodes. If input E1 is at 0 V, the constant current of 5 mA flows through diode D_1, and diode D_2 is cut off. If the input voltage goes to –0.7 V, diode D_1 cuts off; and the 5 mA flows through D_2 into the current summing point, the emitter of transistor Q_1. The emitter input impedance of transistor Q_1 is approximately $r_e + r_b/\beta$ and is typically of the order of a few ohms at frequencies much larger than f_t. The transistor therefore effectively isolates the 100-Ω resistor from the current summing point. The advantage of this isolation is that the capacity associated with the 100-Ω resistor can thereby be reduced, particularly as the number of coincidence inputs is increased.

Figure 11.17[8] is another example of coincidence logic. In this case diodes are used as current switches. A logic 1 input causes the current to switch into the left-hand diode and out of the right-hand diode. A tunnel diode is used at the current-summing point. The tunnel diode has a peak current of 5 mA. It receives sufficient current to switch into its high-voltage state only when both coincidence inputs are at the logic 1 state.

Figure 11.18 shows an example of transistors used as current-switching pairs. It is possible to detect the presence of n or more simultaneous input signals where n is less than the number of input connects, N. This is shown in Figure 11.19. The threshold of discriminator C (for example) is set so that it trips any time three or more inputs are in the logic 1 state.

The use of high-speed integrated circuits (ICs) (such as MECL-III) can result in coincidence circuits which have performances as good or better than most discrete component designs. Figure 11.20 shows three examples of input circuits.[7] Figure 11.20(a) is derived from Figure 11.17. In this circuit, diodes D_1 and D_2 are normally conducting. The input impedance is a complicated, nonlinear function of the input voltage not only at this input, but at other inputs of the coincidence circuit. The value of the 75-Ω resistor has been chosen to minimize reflections, which according to the designer, are less than 20% (capacitive). In Figure 11.20(b) the transistor provides some isolation. The 56-Ω resistor and the inductor, L, are chosen so that they, in parallel with the capacitive and resistive base input impedance of the transistor, provide a fairly good 50-Ω termination.

The circuit of Figure 11.20(c) apparently uses faster transistors than that of Figure 11.20(b). The diodes in Figure 11.20(c) are used for input transistor protection.

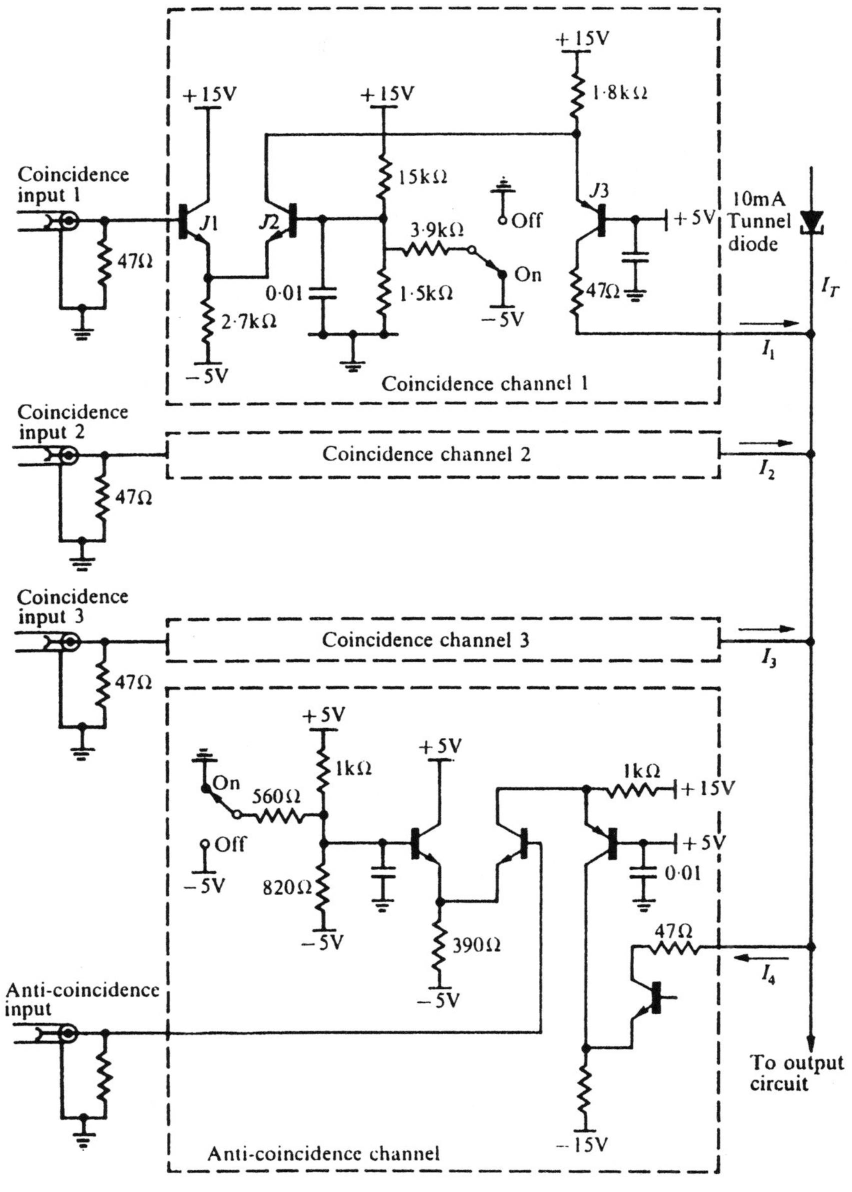

Figure 11.14 Fast coincidence circuit.

The input circuit of Figure 11.21 includes series diode protection for 100-V, 20-nsec transients as well as shifting from nuclear instrument module (NIM) to MECL voltage levels. Reflections are quoted as being 10% for 1-nsec rise time and 25% for 100-psec rise-time inputs. In this circuit impedance does not change much as the diode protection takes over.

Figure 11.22[9] shows an output circuit typical of fast logic modules. In response to input stimuli not shown, the transistors can switch 34 mA into or out of the two outputs, each of which are intended to have 50-Ω loads. If one output has a 50-Ω resistor load, the other is back-terminated.

An improved shift-register (SR) coincidence-counting logic circuit, was developed by Swansen et al.[10] for use with thermal neutron well counters. A simplified block diagram of the shift-register coincidence counter logic is shown in Figure 11.23. The pre-delay and gate are shift registers operating with a 2-MHz clock; e.g., a 32-μsec gate uses 64 shift-register stages with 0.5 μsec per stage. The pre-delay SR has a

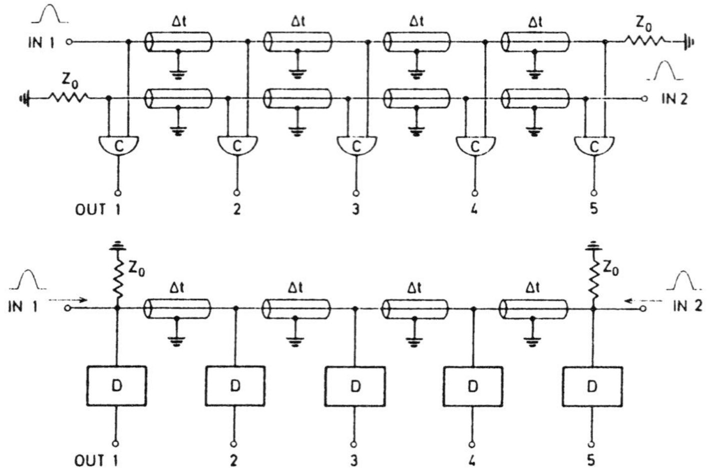

Figure 11.15 Two versions of a multichannel coincidence array.

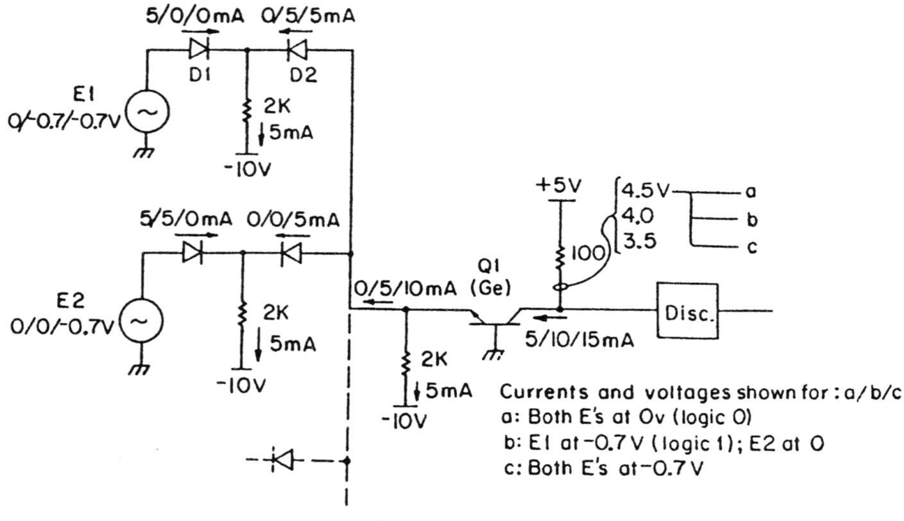

Figure 11.16 An example of a coincidence circuit using diodes as switches, and a common base-connected transistor as current summer. (From Kirsten, F. A., *IEEE Trans. Nucl. Sci.,* NS20(5), 22, 1973. With permission.)

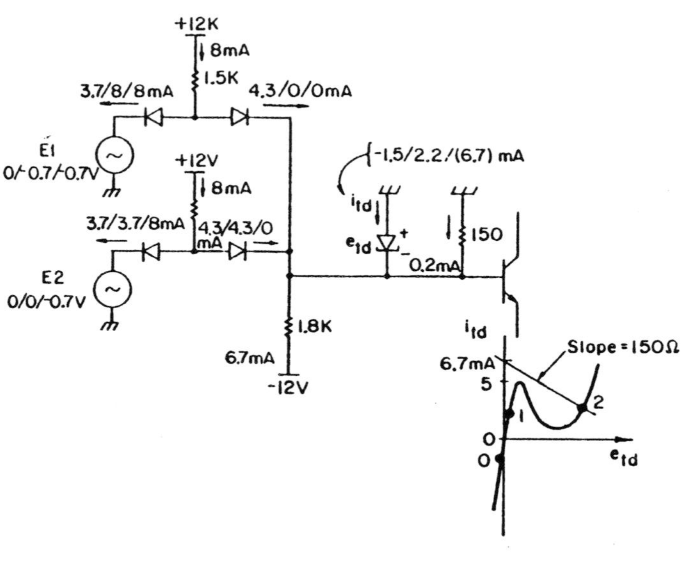

Figure 11.17 Another example of diode switching. This circuit has a tunnel diode used as a current summer and Schmitt trigger. (From Nanni, F. and Verweij, H., Report CERN 69-31, 1969. With permission.)

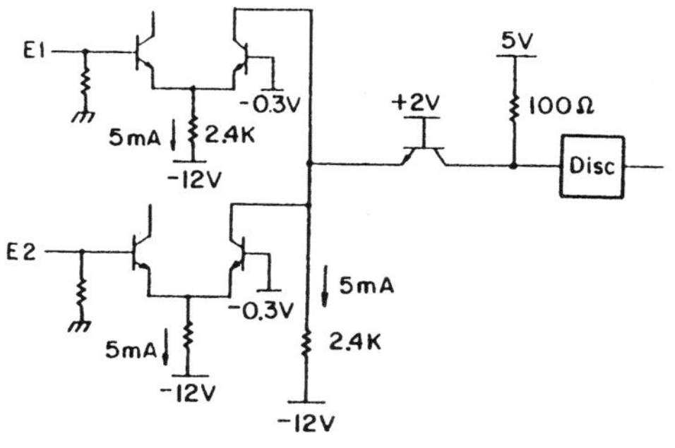

Figure 11.18 An example of transistors used as current switches. (From Kirsten, F. A., *IEEE Trans. Nucl. Sci.*, NS20(5), 22, 1973. With permission.)

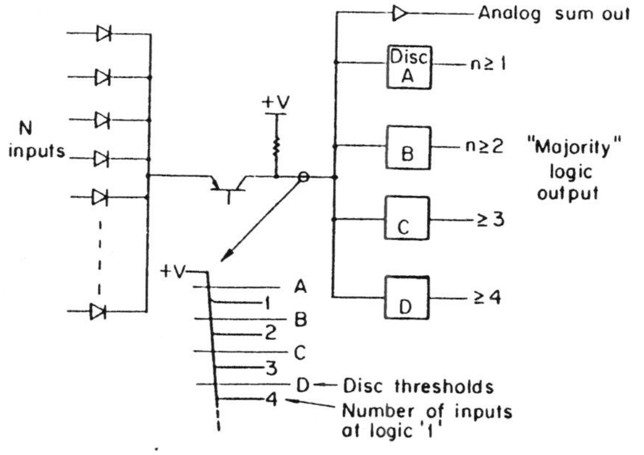

Figure 11.19 The current summing technique can be used to perform majority-logic decisions. Four discriminators detect four levels of "n out of N" input signals present. (From Kirsten, F. A., *IEEE Trans. Nucl. Sci.*, NS20(5), 22, 1973. With permission.)

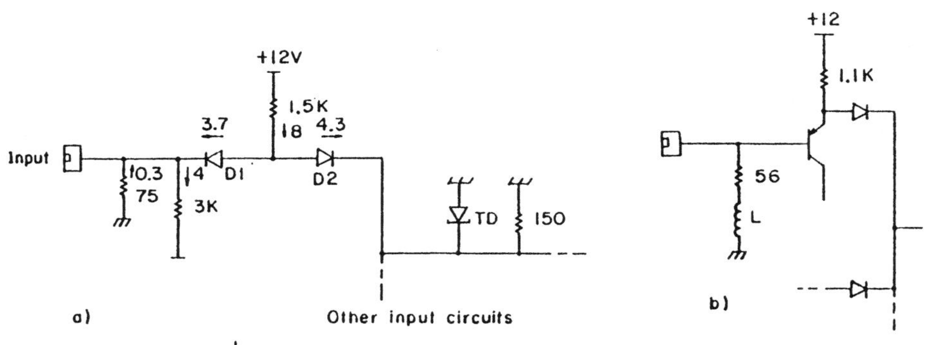

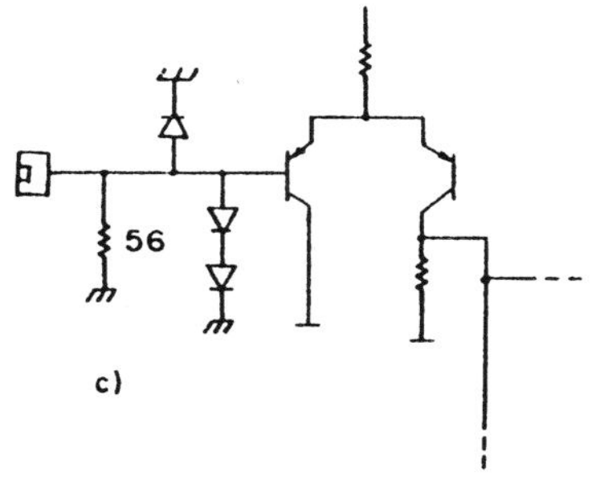

Figure 11.20 (a–c) Examples of input circuits for coincidence circuit modules, showing some techniques used in matching the 50-Ω transmission lines. (From Kirsten, F. A., *IEEE Trans. Nucl. Sci.*, NS20(5), 22, 1973. With permission.)

selectable length from 1.0 to 32.5 μsec in 0.5-μsec steps. The gate SR has a selectable length of 8, 16, 32, 64 or 128 μsec.

The standardized logic input pulses first synchronized with the 2-MHz SR clock and then sent through the pre-delay SR and gate SR. The contents of the up-down counter are added to the R + A and A accumulators for each input pulse as indicated by the strobe arrows in Figure 11.23; the strobe to the R + A accumulator occurs immediately as the pulse leaves the synchronizer, whereas the strobe to the A accumulator is delayed by 1024 μsec.

The principal advantage of the shift-register coincidence circuitry is that each input pulse generates its own R + A and A gates; it is not necessary for one gate to close before the next can open. This feature greatly reduces the system dead time and allows useful operation of the circuit to counting rates of several hundred kilohertz.

The multibit nuclear instruments modules (MBNIM) have been developed by Beer et al.[11] for use in trigger logic for high energy physics experiments. The miniature coaxial cable was replaced by a 40-wire flat cable bus carrying 16 counter signals, thus significantly reducing the cost and the wiring effort.

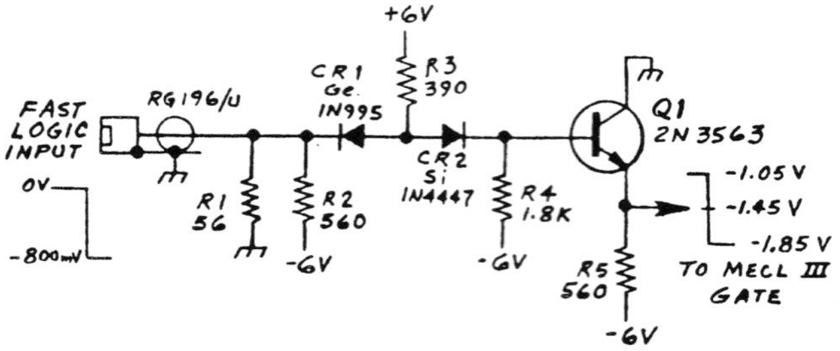

Figure 11.21 Another example of an input circuit. (From Althous, R. F. and Nagel, L. W., *IEEE Trans. Nucl. Sci.*, NS19(1), 520, 1971. With permission.)

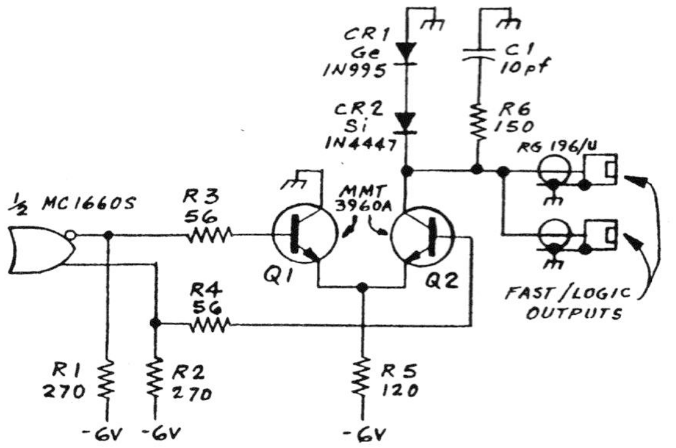

Figure 11.22 An example of an output circuit capable of driving two 50-Ω transmission lines. (From Althous, R. F. and Nagel, L. W., *IEEE Trans. Nucl. Sci.*, NS19(1), 520, 1971. With permission.)

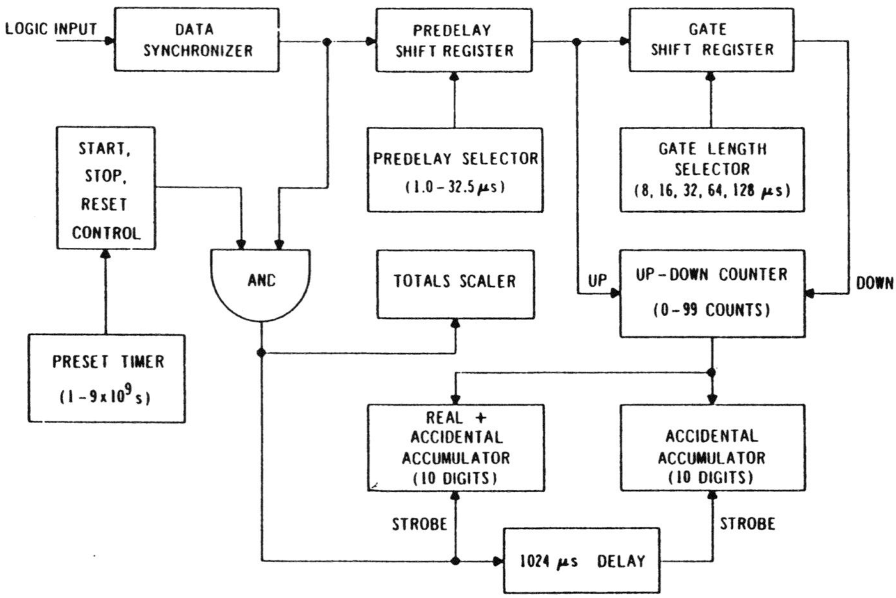

Figure 11.23 Simplified block diagram of the shift-register coincidence logic. (From Swansen, J. E. et al., *Nucl. Instrum. & Methods,* 176, 555, 1980. With permission.)

The MBNIM are characterized by the fact that they drive or read signals from a standard 40-wire flat cable bus carrying 16 data lines, 2 synchronization lines, and 1 control line. In order to have clean and fast signals the bus is (1) driven by means of ECL 10 K gates of the type that is off in the low state and (2) is terminated at both ends. Each signal line is separated from the neighboring one by a ground line.

The connectors are 40 pin stamp-on socket connectors, SCOTCHFLEX type or the equivalent. The cost of a 1-m cable equipped with connectors at both ends is five times lower than the solution involving

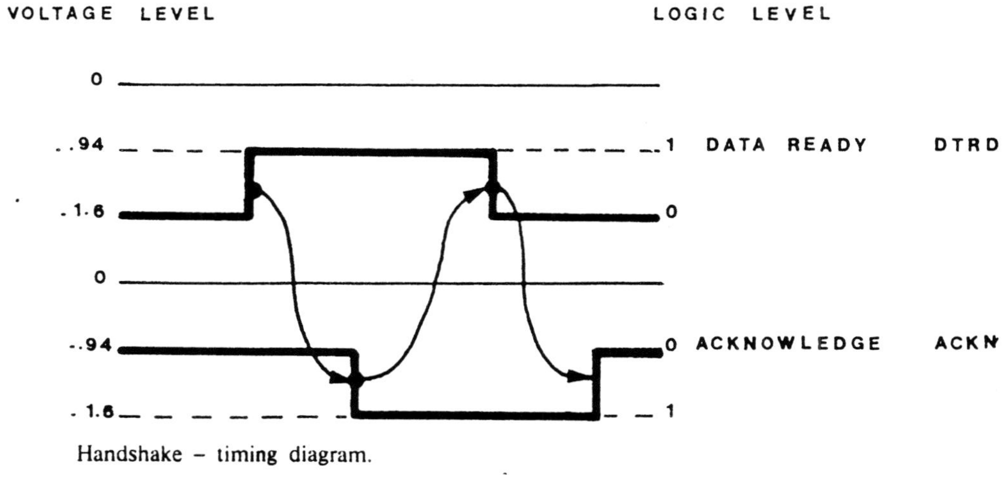

Figure 11.24 Handshake-timing diagram. (From Beer, A. et al., *Nucl. Instrum. & Methods,* 160, 217, 1979. With permission.)

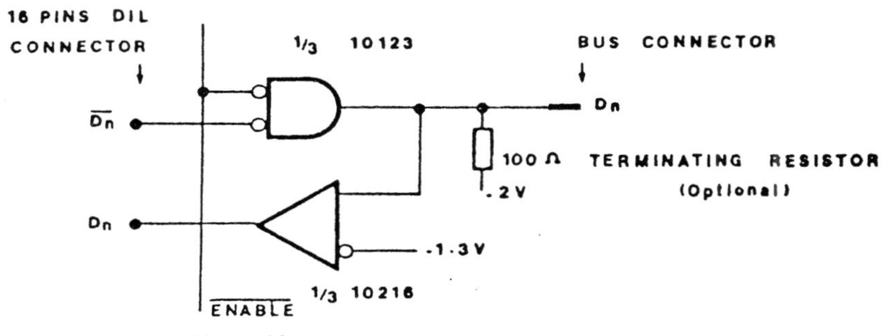

Figure 11.25 Bus driver/receiver circuit. (From Beer, A. et al., *Nucl. Instrum. & Methods,* 160, 217, 1979. With permission.)

16 miniature coaxial cables. The two synchronization lines are optionally used by the modules for full handshake purposes. The first one is the data ready line (DTRD); the second one is the acknowledge line (ACKN). The handshake timing diagram is shown in Figure 11.24. It can be seen that if more than two ports are connected to the same bus segment the DTRDs are wire-ORed while the ACKNs are wire-ANDed. Standard interface boards have been designed: the single-port interface which may also receive 8 miniature coaxial connectors giving access to 4 NIM to ECL 10 K level converters and 4 ECL 10 K to NIM level converters, and the dual-port interface.

These interface boards have been designed so that 10 ports can be connected onto the same bus. At the reef of each interface port, two rows of three dual in-line 16-pin connectors were used to transmit the bus signals to the function board in the user module. The basic bus driver/receiver circuit is shown in Figure 11.25.

11.2 LINEAR GATE

Both linear gate (LG) and coincidence circuits are designed to provide a signal when two events (or more) take place simultaneously. However, the linear gate is designed to deliver the input pulse exactly as it appeared (shape and amplitude) at the circuit input, where the coincidence circuit creates an output signal whenever two (or more) input signals appear within its resolving time.

A linear gate is a system designed to pass pulses or stop them according to instruction received. The transfer of the pulses must be accurate without any distortion. The instruction is generally a gate pulse. The system is open as long as the gate pulse is above a certain level. If the system is not linear, the limit between opening and closing the gate is not well defined, and small signals will be distorted. Linearization is obtained by feedback.

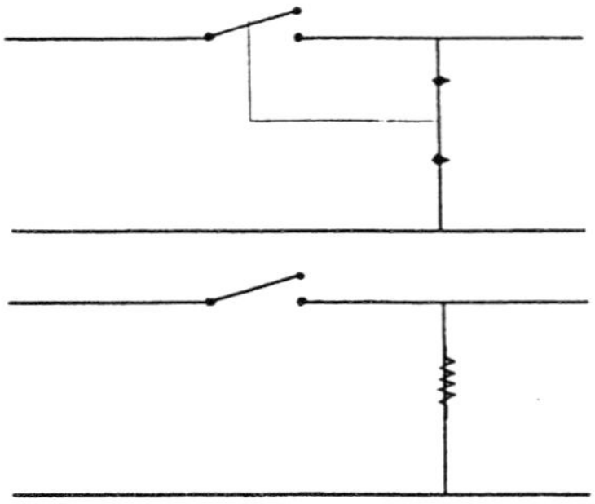

Figure 11.26 Linear gate.

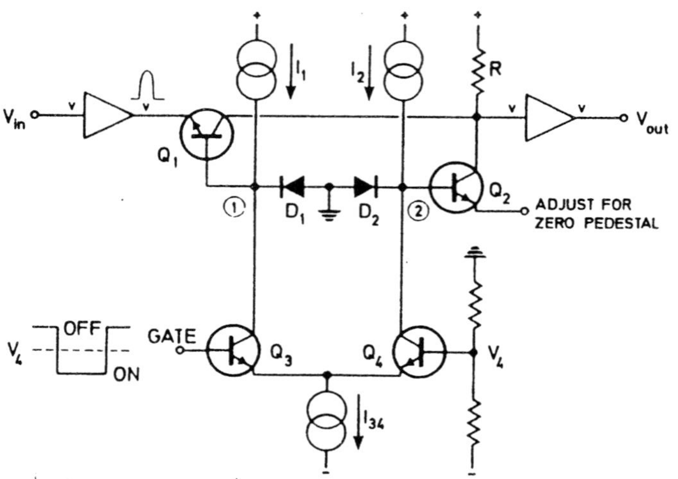

Figure 11.27 Operating principle of a series-parallel gate with transistor switches.

An ideal linear gate should give an output which is linearly related to the input when the gate is open and 0 V (or other constant reference voltage) when the gate is closed. Generally linear gates are realized by opening and closing an ideal gate or by routing the input signal to ground. A bipolar transistor, a field effect transistor, or a diode are used as gates. All these means are not ideal, because of the series resistance, stray capacitance, and nonlinearities in these devices.

The simple forms of linear gates are switches. Two possible switches are shown in Figure 11.26. The switches are electronic circuits. In the first switch the resistance is the output resistance of the switching circuit. In the second case the output voltage is the voltage drop on the resistor as a part of a voltage divider made of this resistor and the output resistance of the switching circuit. This output resistance is very small when the gate is closed and very large when the gate is open. An example of a gate of the first kind is shown in Figure 11.27. When the gate is closed, Q_3 is conducting, Q_4 is cut off, and the current I_3 flows through Q_3. A part of the current comes through the diode; thus junction 1 is a little negative and Q_1 is cut off. Q_2 is in saturation due to I_2 and positive input pulses do not reach the output. When the gate opens (ON), Q_3 is cut off and I_1 causes saturation in Q_1. Junction 2 is in the previous state of junction 1, Q_2 is cut off, and the gate opens.

The opposite switch state is obtained by the circuit shown in Figure 11.28.[12] In the OFF state Q_1 is in saturation and Q_2 is cut off. In the ON state Q_2 is conducting, Q_1 is cut off, and an output pulse is available. Two symmetrical input pulses are needed to operate the gate. When the threshold is determined by current, the system is more stable. A system designed by Kandiah[13] is shown in Figure 11.29. The gate current flows through Q_3, Q_2, and the diode. Up to a certain gate current level Q_1 is cut off. Above threshold it turns ON and the input current passes to the amplifier.

A linear gate for a bipolar pulse is shown in Figure 11.30. To avoid baseline shift a bipolar pulse is used. In the OFF state all the diodes are cut off; in the ON state all diodes are conducting. The current in all diodes should be the same.

Battista[14] suggested the use of an operational amplifier gating at the level at which the spikes generated during the opening, and closing transitions do not appear at the output. The idea was to use an operational

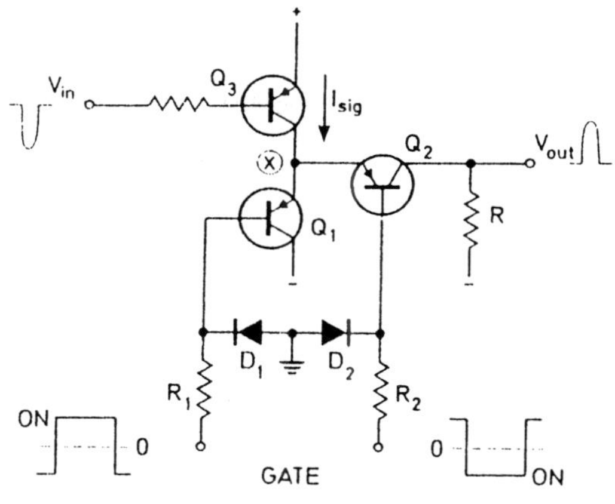

Figure 11.28 Series-parallel gate for current input.

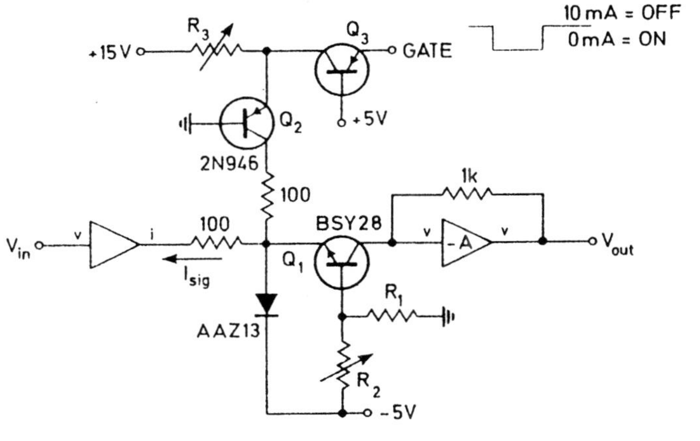

Figure 11.29 Linear gate. (From Kandiah, K., NAS-NRC Pub. 1184, 119, 1964. With permission.)

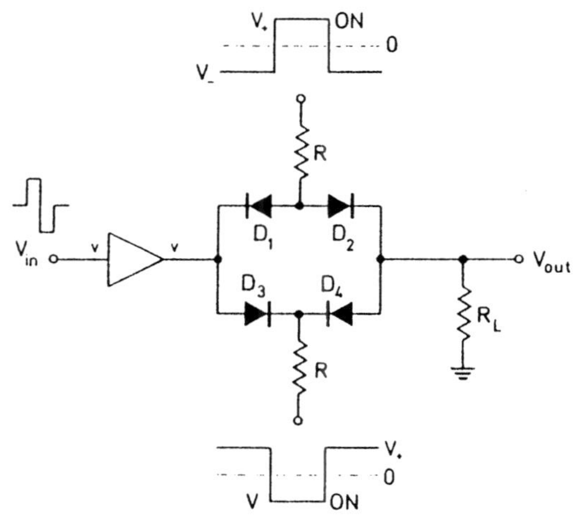

Figure 11.30 Diode-bridge linear gate.

amplifier with two input stages; one was connected to the input signal source and the other to the gate-closed reference (0 V). The switching from one input stage to the other was performed by a diode switch arrangement (diodes and switches in parallel on the signal path). This arrangement caused large spikes at the output. To avoid the spikes the switching was done out of the signal path by connecting the constant current source, Q_5, (Figure 11.31) alternatively to the signal input stage onto the standby input stage. The output stage represented by A was common to both inputs; and its low impedance output was connected to two feedback networks, one for each input stage. The circuit is shown in Figure 11.32.

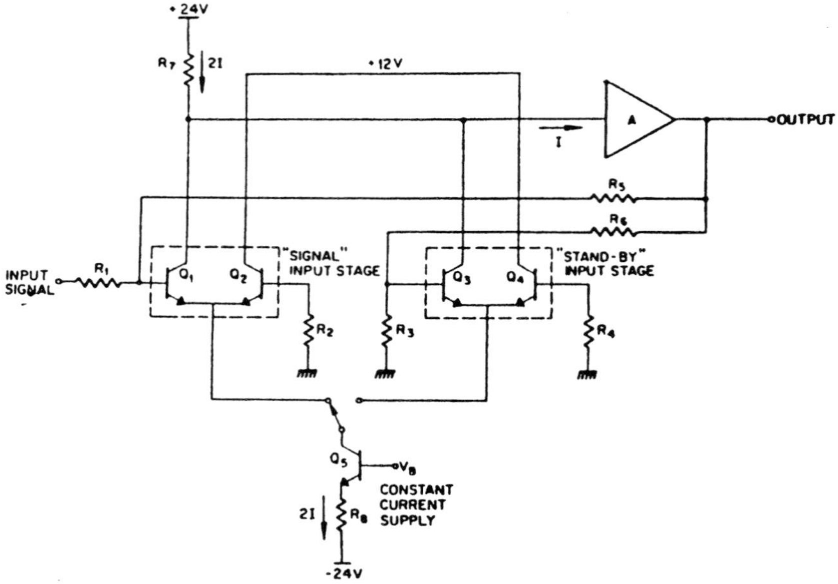

Figure 11.31 Linear gate with gate-switching arrangement out of the signal transmission path. (From Battista, A., *Nucl. Instrum. & Methods,* 80, 172, 1970. With permission.)

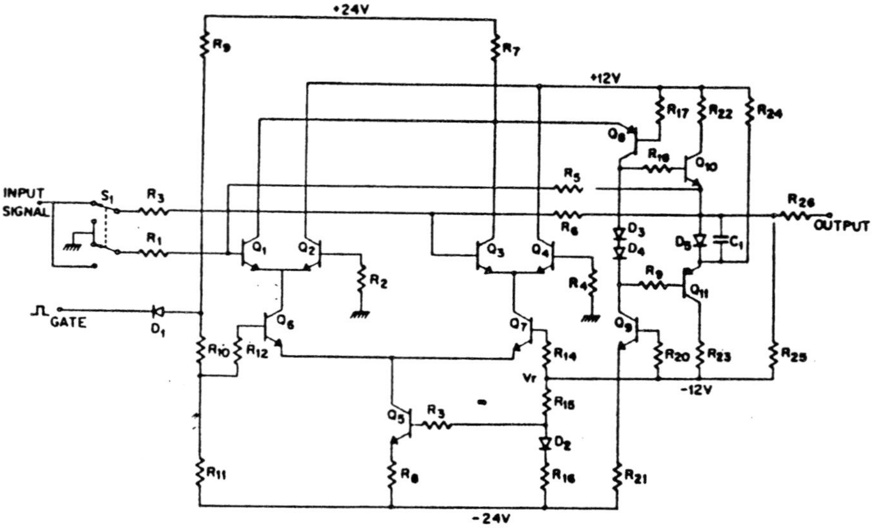

Figure 11.32 Linear gate schematic. (From Battista, A., *Nucl. Instrum. & Methods,* 80, 172, 1970. With permission.)

Avrahami et al.[15] designed an active linear gate which had better merits over the passive linear gate in linearity, gain-stability pedestal level, transients, feedthrough, DC stability, and switching time. The idea of the active linear gate using a feedback amplifier is shown in Figure 11.33.

When the linear gate is open, its dynamic impedance is low, the loop is closed, and A_d becomes a voltage follower improving the linearity and gain stability. The output signal is analyzed with the equivalent circuit (Figure 11.34):

$$V_o = \frac{v_p R_L}{r_c + R_L + A_d R_L} + \frac{v_i A_d R_L}{r_c + R_L + A_d R_L} \tag{11.2}$$

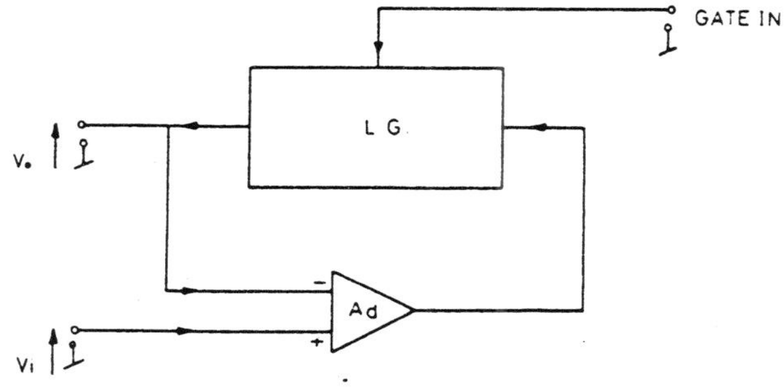

Figure 11.33 Block diagram of basic configuration. LG, passive linear gate; A_d, operational amplifier. (From Avrahami, Z. et al., *Nucl. Instrum. & Methods,* 95, 61, 1971. With permission.)

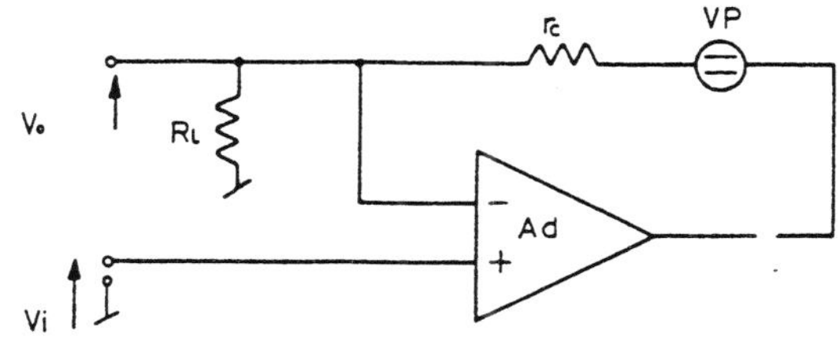

Figure 11.34 Equivalent circuit of basic configuration. r_c, dynamic resistance of LG + output resistance of A_d; V_p, DC shift due to switching. (From Avrahami, Z. et al., *Nucl. Instrum. & Methods,* 95, 61, 1971. With permission.)

If A_d is large, $A_dR_L \gg r_c + R_L$ and $v_o = v_p/A_d + v_i$; r_c is the dynamic resistance of the linear gate plus output resistance of A_d and v_p is the DC shift due to switching.

The transient analysis is done as follows: v_p is replaced by v_{tr}-transient voltage and r_c is replaced by R which is $R = r_c + r_{nc}(1 - e^{-t/\tau})$ for ON → OFF switching and $R = r_c + r_{nc}e^{-t/\tau}$ for OFF → ON switching, r_{nc} is the dynamic resistance in the nonconducting state (OFF).

For the ON → OFF switching:[15]

$$v_o = \frac{v_{tr} \cdot R_L}{r_c + r_{nc}\left(1 - e^{-t/\tau}\right) + R_L + A_dR_L} + \frac{v_iA_dR_L}{r_c + r_{nc}\left(1 - e^{-t/\tau}\right) + R_L + A_dR_L} \qquad (11.3)$$

For the OFF → ON switching:

$$v_o = \frac{v_{tr}R_L}{r_c + r_{nc}e^{-t/\tau} + R_L + A_dR_L} + \frac{v_iA_dR_L}{r_c + r_{nc}e^{-t/\tau} + R_L + A_dR_L} \qquad (11.4)$$

The configuration described thus far does not improve the feedthrough since even at the OFF stage the amplifier is still active as a unity gain follower and its gain increases to maintain v_o constant. The result is that the amplifier goes to saturation for small input signals. To overcome this drawback two passive gates were used as shown in Figures 11.35A and B. An inverted output terminal was added to the amplifier and drove the input of LG 2. The control of the two gates is complementary so that when one is conducting the other is not. When LG 1 is closed and LG 2 is open, thus closing the loop, then:

$$\frac{(v_L - v_o)A_dR_L}{R_L + 1/sC} = v_o \qquad (11.5)$$

$$v_2 = \frac{(v_o - v_2)A_dRi\|R}{r_c + Ri\|R} + \frac{v_iR_ir_c}{R_i\|r_c + R} \qquad (11.6)$$

or

$$v_o = \frac{v_ia_3a_1A_d}{1 + A_d(a_1 + a_2)} \qquad (11.7)$$

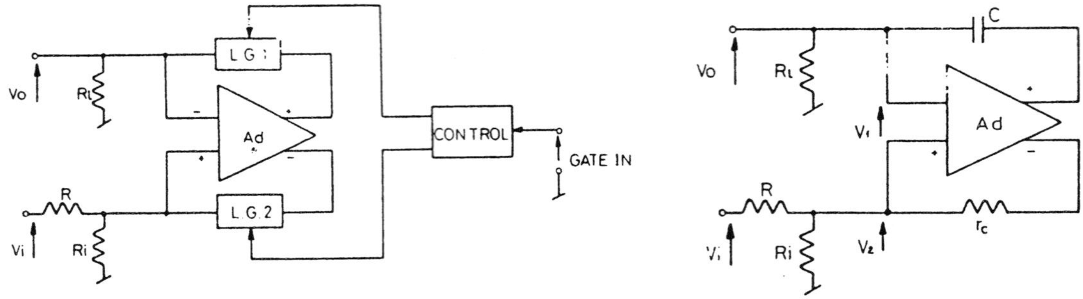

Figure 11.35 (A) Improved version of basic configuration. (B) Equivalent circuit of improved version of LG 1 OFF and LG 2 ON. C, effective capacitance of LG 1. (From Avrahami, Z. et al., *Nucl. Instrum. & Methods,* 95, 61, 1971. With permission.)

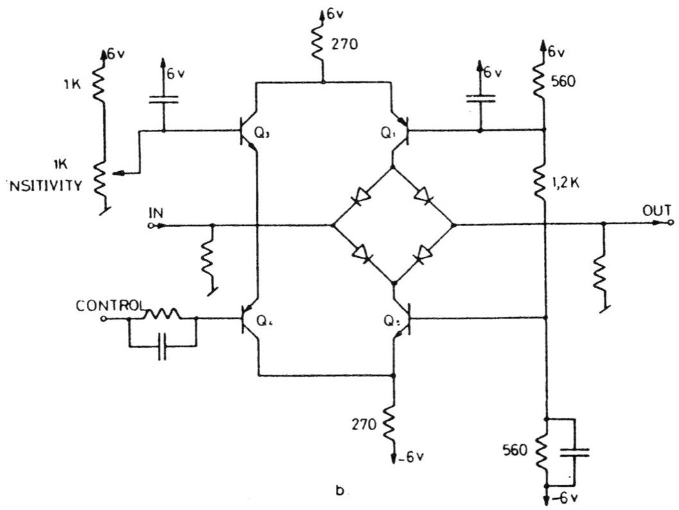

Figure 11.36 Passive linear gate circuit. (From Avrahami, Z. et al., *Nucl. Instrum. & Methods,* 95, 61, 1971. With permission.)

where

$$a_1 = \frac{R_L}{R_L + 1/sC}, \quad a_2 = \frac{R_i \| R}{r_c + R_i \| R}, \quad a_3 = \frac{R_i \| r_c}{R_i \| r_c + R}$$

The passive linear gate circuit is shown in Figure 11.36.

11.3 SAMPLE-AND-HOLD CIRCUITS

Sample-and-hold amplifiers "sample" an analog input signal and then "hold" the instantaneous input value upon the command of a logic control signal. Basically the sample-and-hold is an analog memory where a capacitor serves as the storage element. Applications in which a time-varying input cannot be tolerated require sample-and-hold circuits. A fast successive approximation analog-to-digital converter (ADC) is one application.

Precision Monolithic Incorporated (PMI)[16] sample-and-hold amplifiers are functionally identical to track-and-hold circuits. They continuously track input signals during the sample mode. When an AC-controlled sample-and-hold is commanded to sample, it will take a fast sample and immediately return to the hold mode. It cannot continuously track an input signal.

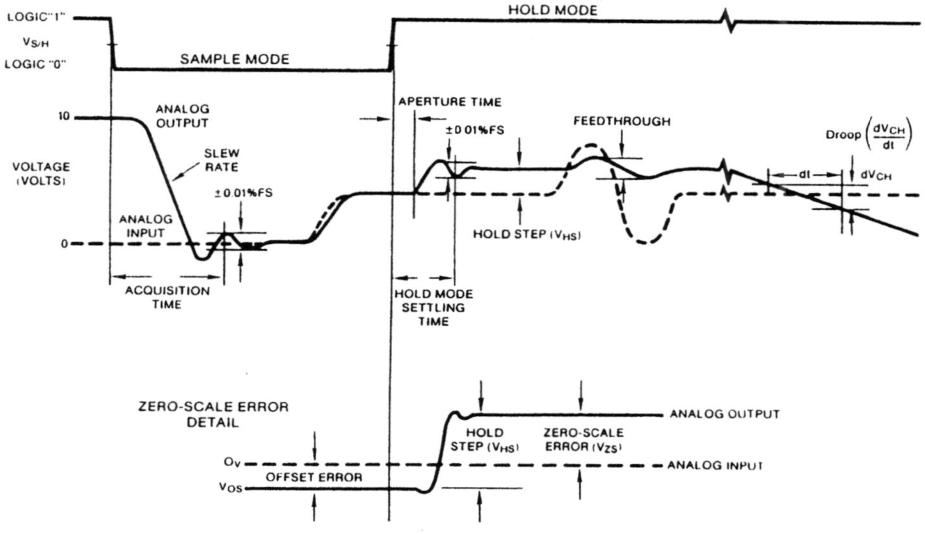

Figure 11.37 Timing diagram of sampling and hold circuit. (From *1988 PMI Databook.* Reprinted with permission of Analog Devices, Inc.)

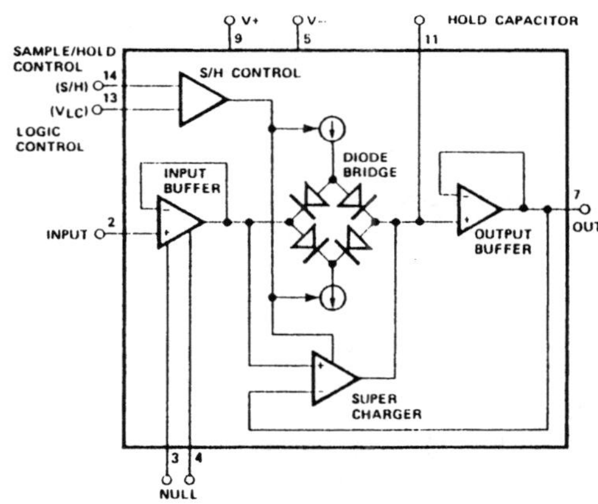

Figure 11.38 Functional diagram of SMP-10/11. (From *1988 PMI Databook.* Reprinted with permission of Analog Devices, Inc.)

S/H	MODE
0	Sample
1	Hold

V_{LC} = 0V for TTL input compatibility.

A sample-and-hold circuit consists of an amplifier, switch, and capacitor. Many specifications are similar to those of switches and operational amplifiers — bias currents, voltage gain, and charge injection are examples.

Slew rate is the maximum possible rate of change of the output voltage when supplying the rated output. For a sample-and-hold circuit, slew rate must be defined with a specified value of holding capacitor C_H. Slew rate can either be measured by operating the circuit in the sample mode and applying a step function to the input, or by applying an input voltage which differs from the output voltage, with the circuit in the hold mode, and then switching to the sample mode and observing the rate of change of the output voltage (Figure 11.37)

The SMP-10/11 are precision sample-and-hold amplifiers that provide the high accuracy, the low-droop rate, and the fast acquisition time required in data acquisition and signal processing systems. Both devices are essentially noninverting unity gain circuits consisting of two very high-input impedance buffer amplifiers connected together by a diode bridge switch.

The high-input impedance and the low-droop rates of the SMP-10 and the SMP-11 (Figure 11.38) are achieved by using bipolar Darlington circuits and an ion implant process that creates superbeta transistors.

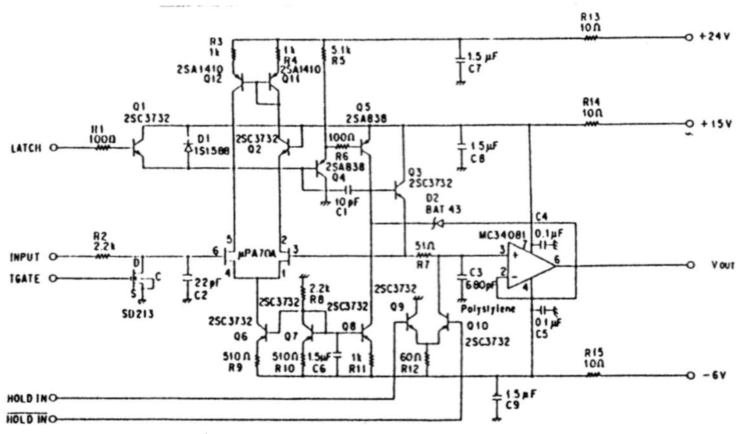

Figure 11.39 Circuit diagram of the peak-hold circuit. (From Taniguch, T. et al., *IEEE Trans. Nucl. Sci.*, NS36, 657, 1989. With permission.)

The output buffer input stage converts to a superbeta Darlington configuration during the hold mode, which results in a very low-droop rate with no penalty in acquisition time. The use of bipolar transistors achieves a low change in droop rate over the operating temperature range.

A unique supercharger provides up to 50 mA of charging current to the hold capacitor, which results in smooth, fast charging with minimum noise. As the hold capacitor voltage nears its final value, the low-current diode bridge controls the final setting time. This unique combination of linear function in a monolithic circuit enables the system designer to achieve superior performance.

The SMP-81[16] precision sample-and-hold amplifier provides the high accuracy, low-droop rate, and fast acquisition ideally required for PCM encoders. The SMP-81 is a non-inverting unity gain circuit consisting of two buffer amplifiers of very high-input impedance connected by a diode bridge switch (Figure 11.38).

The high-input impedance and low-droop rate of the SMP-81 are achieved by the ion implant superbeta process. The high-input impedance permits high-source impedance applications without degrading accuracy, and low-droop rate.

A unique supercharger or transconductance amplifier provides up to 50-mA charging current to the hold capacitor. As a result, smooth charging of the hold capacitor is achieved with minimum noise. The supercharger, in conjunction with the high-slewing rate input and output buffer amplifiers, permits fast acquisition operation. The adjustable logic input threshold makes the SMP-81 compatible to all logic families.

A peak-hold stage circuit diagram is presented in Figure 11.39[17] The input stage is a linear gate followed by a differential amplifier with a matched pair JFET (μPA70A:NEC). The second stage is a voltage-gain stage with an emitter follower and a common-emitter amplifier to drive a transistor which charged up the hold capacitor. A polystyrene capacitor was used. The overall feedback reduces the nonlinearity to a minimum. The bases of Q_4 and Q_5 were clamped with diodes for minimum recovery time.

The hold voltage was buffered by a field effect transistor (FET)-input operational amplifier. The output signal was fed into an ADC through analog multiplexers. The peak-hold section must be inactive during the A-to-D conversion. This is done by bringing the LATCH input to higher than 9 V.

The discharge of the hold capacitor is done through a differential switch. The maximum time to discharge was set to about 200 nsec which was short enough for practical applications.

A multichannel sampled hold calorimeter data unit (CDU) was presented by Haller and Freytag.[18] All of the sample values are retained in MOS capacitors on the chip, and can be read out serially to an ADC. Special care has been taken in the IC design to minimize error from the sampling process and following stages, and to provide a maximum signal-to-noise capability.

Figure 11.40 shows the diagram of a single storage cell for an analog sample. The main inputs to the storage cell are the analog signal input and the write clock ϕ_w. While the write clock is high, the signal input is connected through Q_1 to the storage capacitor C1, and the reference input through Q_8 to the storage capacitor C2. The nodal time constant at the capacitors is approximately 5 nsec. With the falling edge of ϕ_w, Q_1 and Q_8 are turned off and the most recent values of the signals are held on the capacitors. Since both the signal and reference inputs are sampled at the same time, offsets due to gate overlap

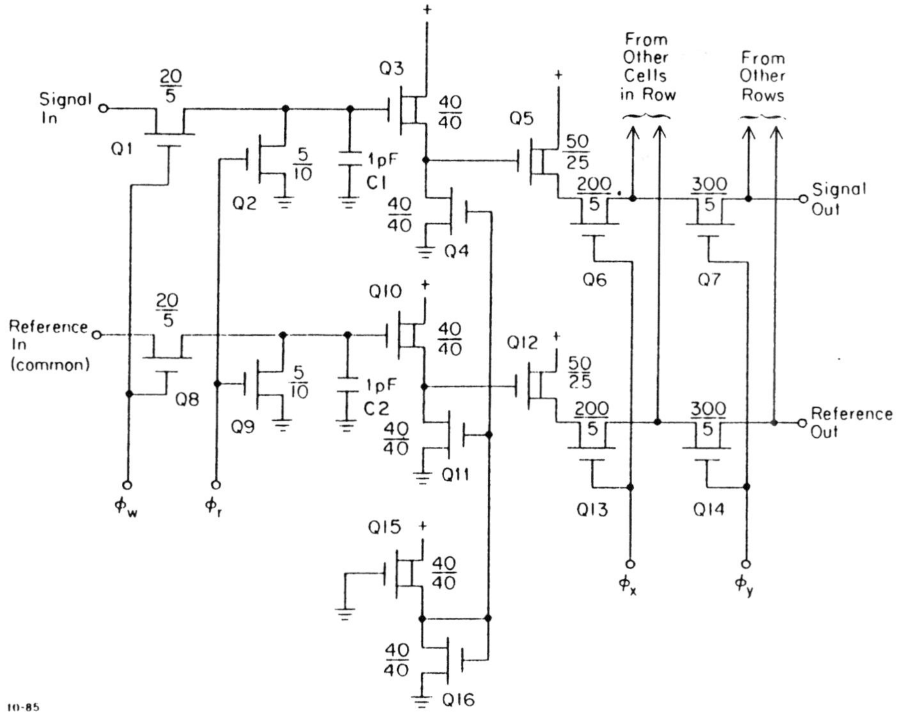

Figure 11.40 One storage cell. Transistor size W/L (microns); depletion transistors are marked with a bar. (From Haller, G. H. and Freytag, D. R., *IEEE Trans. Nucl. Sci.*, 33, 221, 1986. With permission.)

capacitance, charge pumping, and input common mode noise are canceled. The signal and reference samples receive identical amplification and gating in the remainder of the device.

The signal voltage stored on C1 is buffered by voltage follower Q_3 to isolate it from charge-sharing effects, which the following signal multiplexer stages could cause. A constant sink current for biasing Q_3 is provided by Q_4 to keep Q_3 always at the same operation point. This bias voltage is generated by Q_{15} and Q_{16}, which have a geometry and layout identical to Q_3 and Q_4. Changes in the fabrication process which affect the transistor conduction characteristics will have a compensating effect on the bias generator, so that the proper bias current for the source followers is automatically generated. The output of Q_3 is connected to a second source follower Q_5, which provides the actual signal current readout of the chip when this cell is addressed. Transmission gate Q_6 connects the cell to the row bus, which is routed to the output by Q_7.

A high-performance peak-detect and hold circuit for pulse-height analysis was described by Buckens and Veatch.[19] A transconductance amplifier configuration was adopted to minimize proliferation of poles in the transfer function. A fully symmetrical push-pull output stage was adopted in order to double the slew rate for equal output stage quiescent current. Minimizing the quiescent current is particularly important in minimizing the pedestal, and, to a lesser degree, the droop rate.

A monolithic input transistor pair Q_{15} (see Figure 11.41) was used to minimize the offset drift. The quiescent current through each input transistor is 5 mA, while the quiescent current in the output stage is 9 mA. The remainder of the transconductance amplifier design is classical and will not be elaborated further except to mention that, in the tracking mode, the quiescent current path follows CR9 and Q_{13}.

The signal on the storage capacitor is fed back to the input via Q_{16} and U47. Q_{16} has a very low gate leakage current (to minimize droop), while U47 is necessary to buffer the source of A16 from the nonlinear characteristic of the input load during the transition from peak-detect to hold modes.

The novel architecture resides in the current steering circuit composed of Q_6, Q_7, Q_{13}, CR6-11, CR13-16, and U46. CR9 and Q_{13} are tracking the signal during the tracking mode; and so do CR8 and Q_7 (the node at the cathode of CR8 is at high impedance in this mode).

When the mode is switched to peak-detect and hold, Q_6 sources current, which turns on Q_7, thereby diverting the path for the output stage quiescent current to CR8 and Q_7, and also turning off CR9 and Q_{13}.

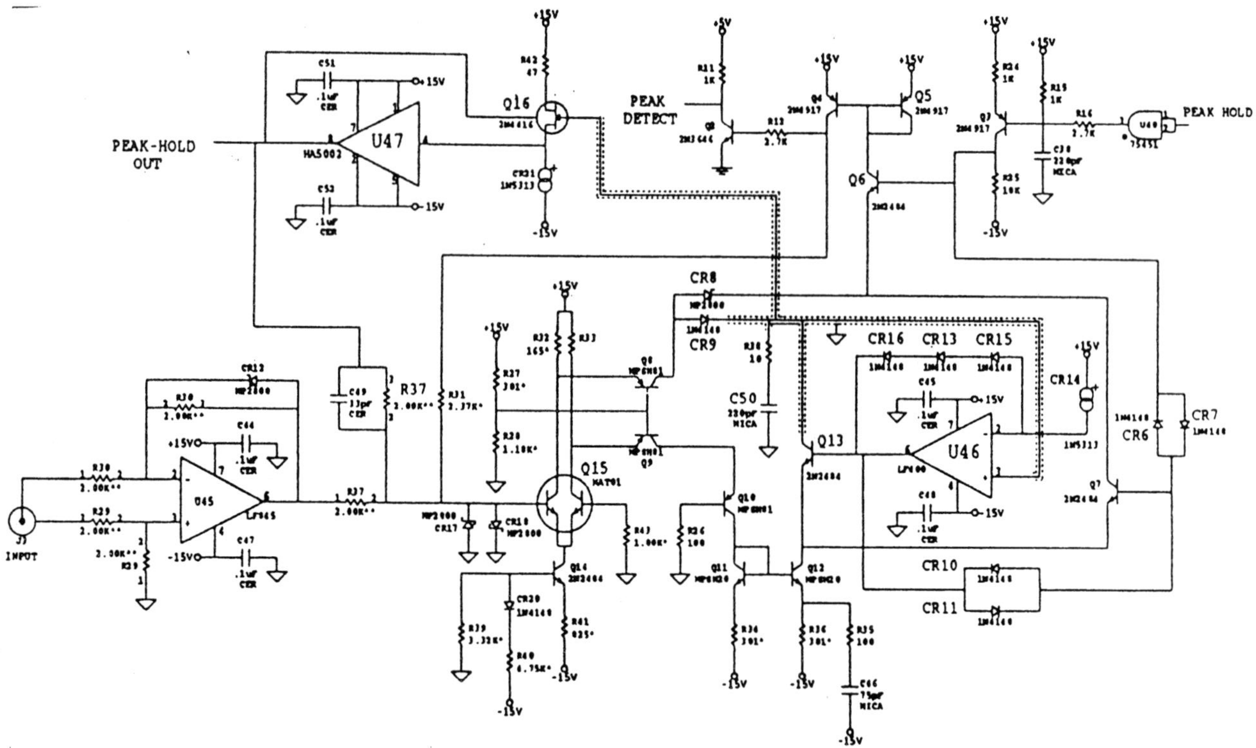

Figure 11.41 Schematic of novel peak-detect and hold circuit. (From Buckens, P. F. and Veatch, M. S., *IEEE Trans. Nucl. Sci.*, 39, 753, 1992. With permission.)

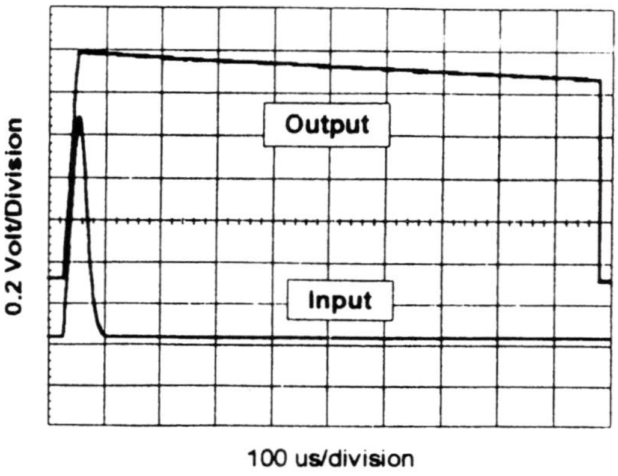

Figure 11.42 Peak-detect and hold droop. (From Buckens, P. F. and Veatch, M. S., *IEEE Trans. Nucl. Sci.*, 39, 753, 1992. With permission.)

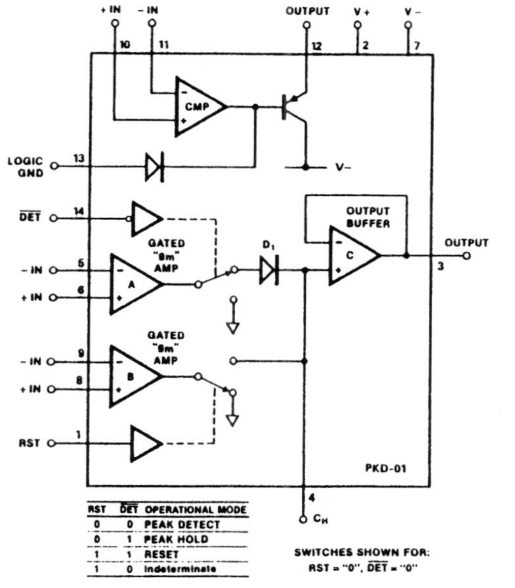

Figure 11.43 Functional diagram. (From *1988 PMI Databook*. Reprinted with permission of Analog Devices, Inc.)

Note that during this transition, the switching components are all tracking the voltage on the storage capacitor. No change in quiescent current or output voltage is required to turn CR9 off. This is the key in achieving a very small pedestal. U46 is a low leakage FET input, 40-MHz op-amp; it is required for providing the bias to Q_{13} and for handling the current needed to bias Q_7, while allowing these to track the output voltage without affecting the flow of currents around the hold capacitor or in the amplifiers output.

In the peak-detect mode, an input signal higher than the output will result in an excess of sourcing current at the output of the amplifier, which will turn on CR9, thereby charging C50. Similarly, an input signal smaller than the output will produce an excess of sinking current; this current will flow through Q_7, thereby turning on Q_5 and thus creating a PEAK DETECT signal. This signal is used to flag the following ADC to quantize the held voltage. PEAK DETECT is also used to inject a current at the input of the system through R31 to prevent a rise at the input, subsequent to a peak, to affect the stored value. Consequently, the latter also creates a nonlinear load at the input of the circuit, therefore the need for U47 buffering. With such an architecture, the only contribution to a pedestal is the charge injection on the hold capacitor by the switching of CR9. The resulting pedestal is $0.5 \text{ V} \cdot 2 \text{ pF}/220 \text{ pF} = 4.5 \text{ mV}$. This is confirmed in the actual circuit.[19]

Figure 11.42 illustrates the peak-detect and hold mode. Here a peak was held for 900 µsec to display the droop rate (125 µV/µsec). This is typical of one of the highest droop rates found among 50 production units.

The PKD-01[16] tracks an analog input signal until a maximum amplitude is reached. The maximum value is then retained as a peak voltage on a hold capacitor. Being a monolithic circuit, the PDK-01 (PMI) offers significant performance and package density advantages over hybrid modules and discrete designs without sacrificing system versatility. The matching characteristics attained in a monolithic circuit provide inherent advantages when charge injection and droop rate error reduction are primary goals (Figure 11.43).

Innovative design techniques maximize the advantages of monolithic technology. Transconductance (g_m) amplifiers were chosen over conventional voltage amplifier circuit building blocks. The g_m amplifiers simplify internal frequency compensation, minimize acquisition time, and maximize circuit accuracy. Their outputs are easily switched by low glitch current steering circuits. The steered outputs are clamped to reduce charge injection errors on entering the hold mode or exiting the reset mode. The inherently low zero-scale error is reduced further by active "zener-zap" trimming to optimize overall accuracy.

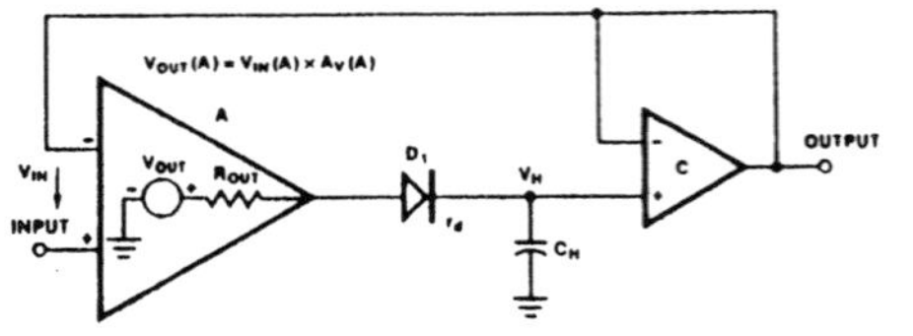

Figure 11.44 Conventional voltage amplifier peak detector. (From *1988 PMI Databook.* Reprinted with permission of Analog Devices, Inc.)

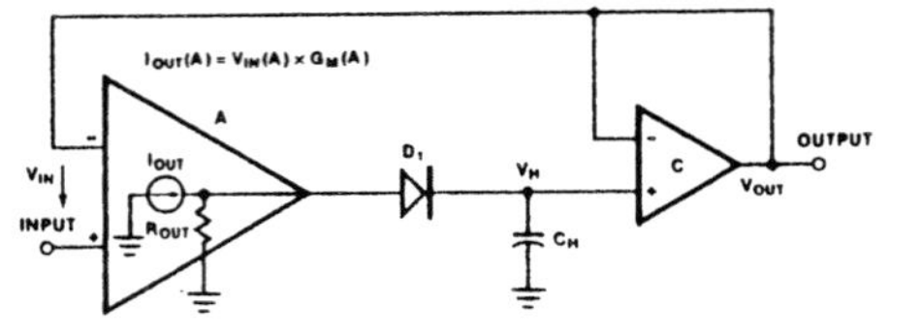

Figure 11.45 Transconductance amplifier peak detector. (From *1988 PMI Databook.* Reprinted with permission of Analog Devices, Inc.)

The output buffer amplifier features an FET input stage to reduce droop rate error during lengthy peak hold periods. A bias current cancellation circuit minimizes droop error at high ambient temperatures.

The typical peak detector uses voltage amplifiers and a diode or an emitter follower to charge the hold capacitor, C_H, unidirectionally (Figure 11.44) (PMI). The output impedance of A plus D_1 dynamic impedance, r_d, make up the resistance which determines the feedback loop pole. The dynamic impedance is $r_d = kT/qI_d$. where I_d is the capacitor-charging current.

The pole moves toward the origin of the S plane as I_d goes to zero. The pole movement in itself will not significantly lengthen the acquisition time since the pole is enclosed in the system feedback loop.

When the moving pole is considered with the typical frequency compensation of voltage amplifiers, there is, however, a loop stability problem. The necessary compensation can increase the required acquisition time. The PMI approach replaces the input voltage amplifier with a transconductance amplifier (Figure 11.45).

The PKD-01 transfer function can be reduced to:

$$\frac{V_{OUT}}{V_{IN}} = \frac{1}{1+\dfrac{sC_H}{g_m} - \dfrac{1}{g_m R_{OUT}}} \approx \frac{1}{1+\dfrac{sC_H}{g_m}} \tag{11.8}$$

where $g_m \approx 1\ \mu A/mV$ and $R_{out} \approx 20\ M\Omega$.

The diode in series with the A output, (Figure 11.45) has no effect because it is a resistance in series with a current source. In addition to simplifying the system compensation, the input transconductance amplifier output current is switched by current steering. The steered output is clamped to reduce and match any charge injection.

Figure 11.46 shows a simplified schematic of the rest g_m amplifier, B. In the track mode, Q_1 and Q_4 are ON and Q_2 and Q_3 are OFF. A current of 2I passes through D_1; I is summed at B and passes through Q_1, and is summed with $g_m V_{IN}$. The current sink can absorb only 3I; thus, the current passing through D_2 can only be: $2K\text{-}g_m V_{IN}$. The net current into the hold capacitor node then is $g_m V_{IN}$, $CH = 2I - (2I - g_m V_{in})$. The hold mode, Q_2 and Q_3 are ON while Q_1 and Q_4 are OFF. The net current into the top of D_1 is $-I$ until D_2 turns ON. With Q_1 OFF, the bottom of D_2 is pulled up with a current until D_4 turns ON; thus D_1 and D_2 are reverse biased by ~0.6 V, and charge injection is independent of input level.

The monolithic layout results in points A and B having equal nodal capacitance. In addition, matched diodes D_1 and D_2 have equal diffusion capacitance. When the transductance amplifier outputs are switched open, points A and B are ramped equally but in an opposite phase. Diode clamps D_3 and D_4 cause the swings to have equal amplitudes. The net charge injection voltage change at node C is therefore zero.

The peak transconductance amplifier, A, is shown in Figure 11.47. Unidirectional hold capacitor charging requires diode D_1 to be connected in series with the output. On entering the peak hold mode D_1 is reverse biased. The voltage clamp limits charge injection to approximately 1 pC and the hold step to 0.6 mV.

Minimizing acquisition time dictated a small C_H capacitance. A 1000-pF value was selected. Droop rate was also minimized by providing the output buffer with an FET input stage. A current cancellation circuit further reduces droop current and minimizes the gate current tendency to double for every 10°C temperature change.[16]

The hold capacitor (C_H) serves as the peak memory element and compensating capacitor. Stable operation requires a minimum value of 1000 pF. Larger capacitors may be used to lower droop rate errors,

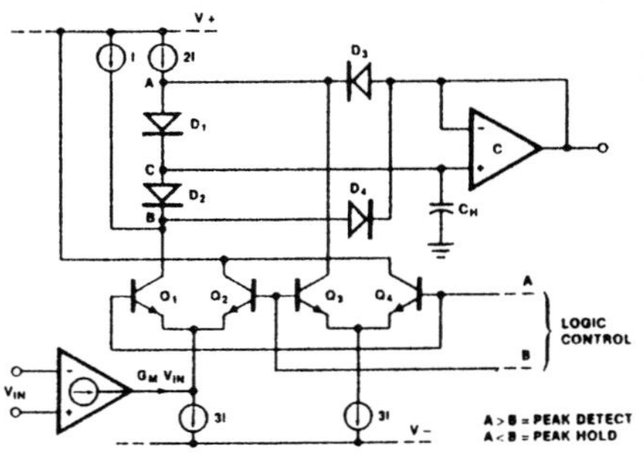

Figure 11.46 Transconductance amplifier with low glitch current switch. (From *1988 PMI Databook.* Reprinted with permission of Analog Devices, Inc.)

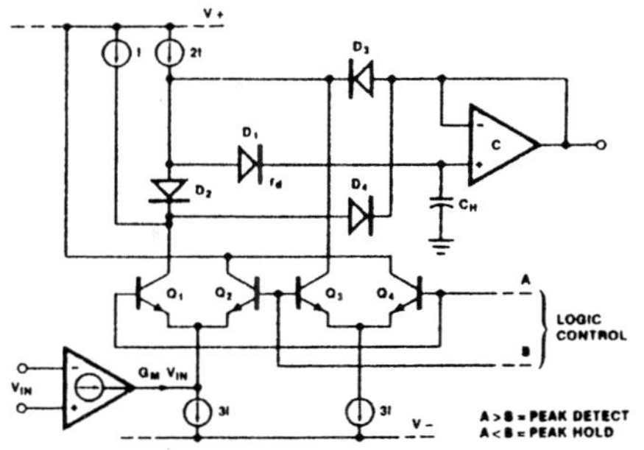

Figure 11.47 Peak-detecting transconductance amplifier with switched output. (From *1988 PMI Databook.* Reprinted with permission of Analog Devices, Inc.)

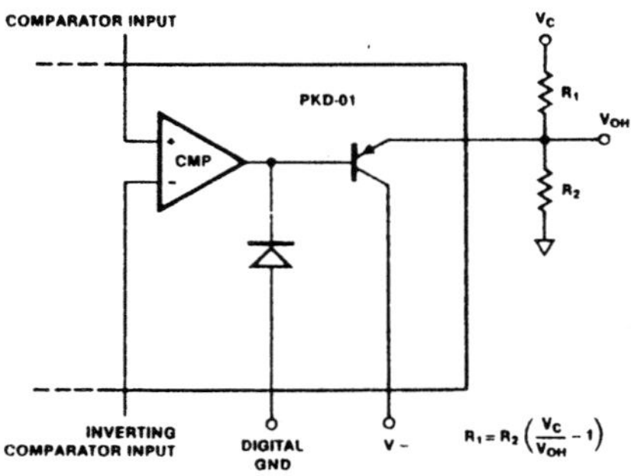

Figure 11.48 Comparator output with external level setting resistors. (From *1988 PMI Databook.* Reprinted with permission of Analog Devices, Inc.)

Table 11.1 Comparator Output

V_C	V_{OH}	R_1	R_2
5	3.5	2.7K	6.2K
5	5.0	2.7K	∞
15	3.5	4.7K	1.5K
15	5.0	4.7K	2.4K
15	7.5	7.5K	7.5K
15	10.0	7.5K	15K

but acquisition time will increase. Zero-scale error is internally trimmed for C_H = 1000 pF. Other C_H values will cause a zero-scale shift which can be approximated with the following equation:

$$\Delta V_{zs}(\text{mV}) = \frac{1\times 10^3\ (\text{pC})}{C_H\ (\text{nF})} - 0.6\ \text{mV} \tag{11.9}$$

The peak hold capacitor should have very high-insulation resistance and low-dielectric absorption. For temperatures below 85°C, a polystyrene capacitor is recommended, while a Teflon capacitor is recommended for high-temperature environments.

The comparator output high level (V_{OH}) is set by external resistors. It is possible to optimize noise immunity while interfacing to all standard logic families — transistor-to-transistor logic (TTL), DTL, and complementary metal oxide semiconductor (CMOS). Figure 11.48 shows the comparator output with external level setting resistors. Table 11.1 gives typical R_1 and R_2 values for common circuit conditions.

The maximum comparator high-output voltage (V_{OH}) should be limited to V_{OH} (maximum) $<V^+ - 2.0$ V. With the comparator in the low state (V_{OL}), the output stage will be required to sink a current approximately equal to V_C/R_1. The transconductance amplifier outputs are controlled by the logical logic signals RST and $\overline{\text{DET}}$. The PKD-01 operational mode is selected by steering the current (I_1) through Q_1 and Q_2, thus providing high-speed switching and a predictable logic threshold. The logic threshold voltage is 1.4 V when digital ground is at 0 V.

Other threshold voltages (V_{TH}) may be selected by applying the formula:

$$V_{TH} \approx 1.4\text{V} + \text{digital ground potential}$$

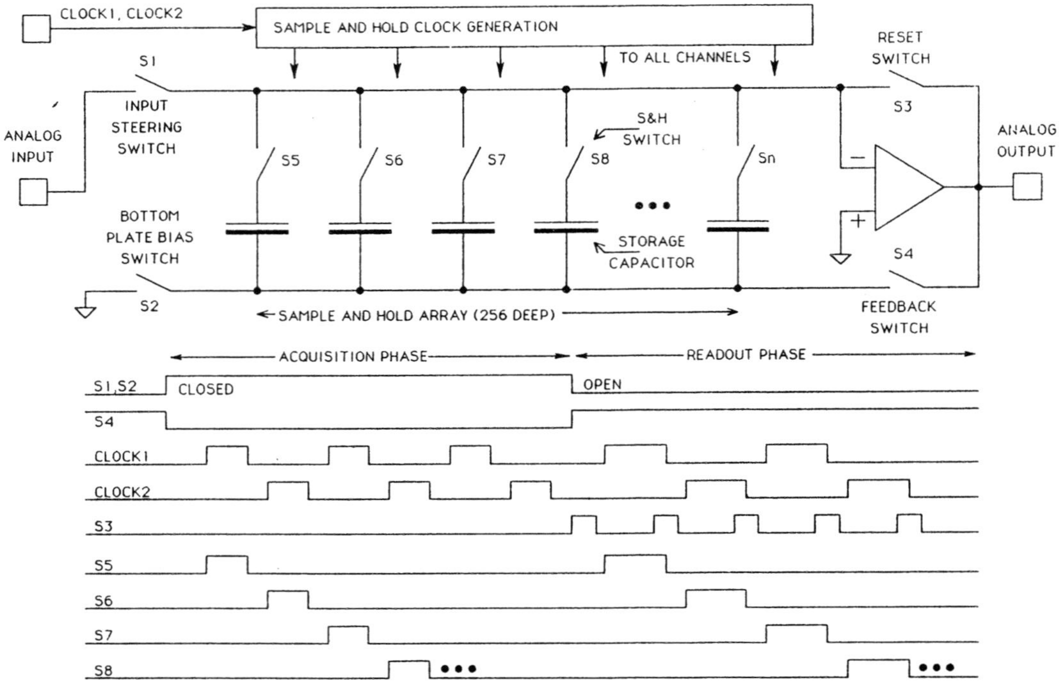

Figure 11.49 Simplified circuit and timing diagrams. (From Kleinfelder, S. A., *IEEE Trans. Nucl. Sci.*, 37, 1230, 1990. With permission.)

For proper operation, digital ground must always bet at least 3.5 V below the positive supply and 2.5 V above the negative supply. The RST or $\overline{\text{DET}}$ signal must always be at least 2.8 V above the negative supply.

Operating the digital ground at other than 0 V does influence the comparator output low voltage. The V_{OL} level is referenced to digital ground and will follow any changes in digital ground potential:

$$V_{OL} \approx 0.2\,\text{V} + \text{digital ground potential}^{16}$$

A full custom switched capacitor transient analog waveform storage and reconstruction integrated circuit containing 4096 sample and hold cells has been designed, fabricated, and tested by Kleinfelder.[20] The switched capacitor array (SCA) is organized as 16 parallel channels of 256 serially addressed samples per channel with multiplexed and buffered analog output. Signal sampling frequencies of up to 100 MHz have been accomplished, with a dynamic range (measured at 10 MHz) of at least 8000 to 1.

The switched capacitor approach may offer lower cost, lower power, and higher density while providing fast sampling speeds, wider dynamic range, and a high degree of flexibility. High-channel count per chip permits low cost per channel and high-density installations. The high-performance, low-cost CMOS process used, available from prototype foundries such as MOSIS, encourages experimentation and specialization.

The SCA integrated circuit uses a 2-μm double metal CMOS process with two layers of polysilicon forming the high-density capacitors. The device was prototyped through the MOSIS service. It has 16 parallel analog signal inputs, 16 parallel analog outputs, and 1 multiplexed and buffered analog output. Two digital clock inputs, four steering switch and rest control inputs partially redundant, and four channel address inputs are used to operate the circuit. Figure 11.49 shows a simplified circuit diagram and timing information.

The circuit contains 4096 sample-and-hold cells subdivided into 16 parallel channels of 256 cells per channel. Each of the 16 channels has a dedicated analog input which is steered onto a bus distributing the input signal to 256 sample-and-hold cells. Steering is accomplished by transmitting the input signal through a low on-resistance complementary switch (S1). Each sample-and-hold cell consists of a

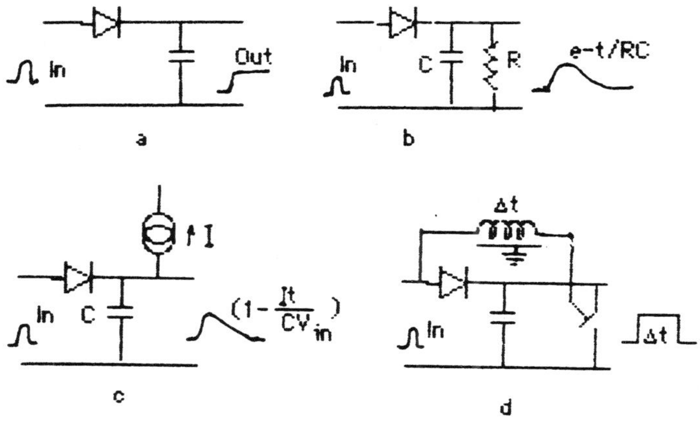

Figure 11.50 (a–d) Pulse stretchers.

complementary CMOS transmission gate (complementary switch) and a 1.5-pF double polysilicon capacitor. The capacitors use a high-quality silicon oxide dielectric of 700-Å thickness. Complementary switches consist of an n-channel and a p-channel transistor whose sources and drains are paralleled. When complementary, rail-to-rail clocks are applied to the gates of both transistors, transmission of analog signals from rail to rail can be accomplished. An externally supplied reference voltage is applied to the bottom plate of all capacitors through another set of steering switches (S2). The voltage sorted on each sample-and-hold capacitor then corresponds to the difference between the input signal from a given channel at the time its sample-and-hold switch is opened and the applied reference voltage. The reference voltage can be adjusted to shift the baseline of the acquired signal for better level matching.[20]

A shift-register-based clocking mechanism sequentially turns ON and OFF sample and hold switches using a controllable break before make action. Break before make ensures that no charge sharing between subsequently engaged capacitors can occur. Break before make can be controlled by varying the amount of time between nonoverlapping clock phases. Normally, only a few nanoseconds are required. The 256 pairs of sample-and-hold clock signals are generated once and shared by all channels for lower power consumption and higher density. The sample-and-hold clock generation is designed to capture one sample per clock phase; thus two samples are captured per clock cycle, reducing external clock generation demands (100-mHz sample rates are attained with a 50-MHz clock speed). All clock and control input receivers are buffered internally for minimal external loading.

Before readout, the analog input and reference potential steering switches (S1 and S2) are turned off, disconnecting the signal bus and bottom capacitor plate from the input side of the circuit. The reconstruction operational amplifiers, one per channel, can then be connected to the sample-and-hold arrays (via S4). The readout sequence then proceeds in much the same way as the acquisition phase; under control of the shift registers, each sample capacitor is placed sequentially in the feedback path of the reconstruction op-amp. Once the amplifier outputs have settled, multiplexing of all 16-channel amplifier outputs through a single buffered analog output can be accomplished, selected via the four digital binary encoded channel multiplex inputs. This allows all data from one or many SCA chips to efficiently share a single high-resolution analog-to-digital converter. After reconstruction and readout of each sample, stray charge on the sample capacitors and signal distribution buses are cleared by shorting each amplifier's inverting input nodes and output nodes (via S3). This prevents charge remaining on parasitic nodes from the previous samples from influencing the next readout sample. Reset of the sample capacitors also prevents any memory effect from influencing the next acquisition; this is only strictly necessary at the highest acquisition speeds.[20]

11.4 PULSE STRETCHER

When the pulse height has to be maintained for a short time, a stretcher is used. It is done by charging a capacitor and stopping it from discharging. The basic circuit is shown in Figure 11.50: (a) consists of a capacitor and a diode; the addition of a large resistor in parallel to the capacitor (b) causes an exponential decay of the output pulse. Linear decay is obtained with a constant current withdrawal (c). A square pulse is obtained by delaying the original pulse by the time t and shutting the output gate (d).

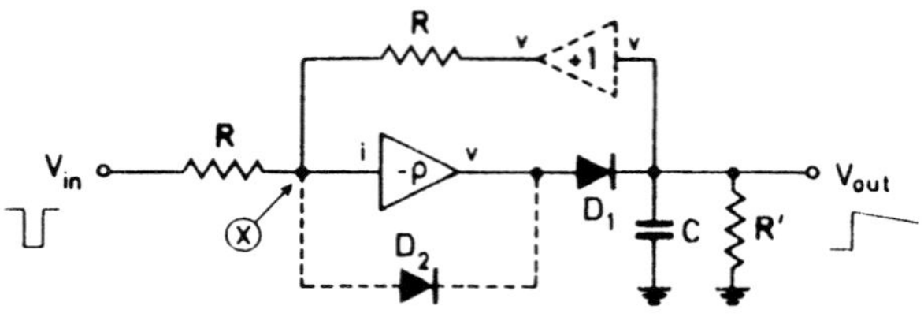

Figure 11.51 Pulse stretcher with feedback.

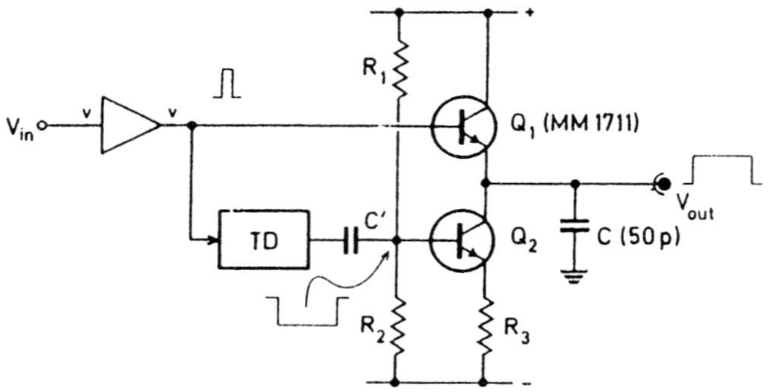

Figure 11.52 Practical pulse stretcher circuit. (From Weddigen, C. and Haase, E. L., *Nucl. Instrum. & Methods,* 33, 157, 1965. With permission.)

Instead of a diode, the base-emitter junction of a transistor can be used. The smallest input pulse width is limited by the diode resistance (conducting) plus the output resistance of the pulse source times the capacitor capacitance.

To reduce the input resistance a feedback is added, as shown in Figure 11.51. The linearity of the stretcher is also improved. A practical pulse stretcher is shown in Figure 11.52.[21] It has a delay to obtain a long square output pulse. When the two transistors are conducting, the capacitor potential is determined by the base of Q_1 and the current through Q_2 as set by R_1, R_2, R_3. The input pulse starts a tunnel diode multivibrator which supplies a 1-μsec negative pulse, cutting off Q_2. The capacitor is charged through Q_1. C is charged until Q_2 opens again and C discharges.

REFERENCES

1. **Barna, A. J. et al.,** *Nucl. Instrum. & Methods,* 7, 124, 1960.
2. **Bell, R. E.,** *Nucl. Instrum. & Methods,* 42, 211, 1966.
3. **Emmer, T. L.,** NAS-NRC Pub. 1184, 112, 1964.
4. **Sugarman, R. M. et al.,** *BNL,* 711, 1962.
5. **Whetstone, A. and Kounosu, S.,** *Rev. Sci. Instrum.,* 33, 423, 1962.
6. **Garwin, R.,** *Rev. Sci. Instrum.,* 24, 618, 1953.
7. **Kirsten, F. A.,** *IEEE Trans. Nucl. Sci.,* NS20(5), 22, 1973.
8. **Nanni, F. and Verweij, H.,** Report CERN 69-31, 1969.
9. **Althous, R. F. and Nagel, L. W.,** *IEEE Trans. Nucl. Sci.,* NS19(1), 520, 1971.
10. **Swansen, J. E. et al.,** *Nucl. Instrum. & Methods,* 176, 555, 1980.
11. **Beer, A. et al.,** *Nucl. Instrum. & Methods,* 160, 217, 1979.
12. **Barna, A. and Marshal, J. H.,** *Rev. Sci. Instrum.,* 35, 881, 1964.
13. **Kandiah, K.,** NAS-NRC Pub. 1184, 119, 1964.
14. **Battista, A.,** *Nucl. Instrum. & Methods,* 80, 172, 1970.
15. **Avrahami, Z. et al.,** *Nucl. Instrum. & Methods,* 95, 61, 1971.
16. Precision Monolithic Inc. (PMI) Catalog, 1990.
17. **Taniguch, T. et al.,** *IEEE Trans. Nucl. Sci.,* NS36, 657, 1989.
18. **Haller, G. H. and Freytag, D. R.,** *IEEE Trans. Nucl. Sci.,* 33, 221, 1986.
19. **Buckens, P. F. and Veatch, M. S.,** *IEEE Trans. Nucl. Sci.,* 39, 753, 1992.
20. **Kleinfelder, S. A.,** *IEEE Trans. Nucl. Sci.,* 37, 1230, 1990.
21. **Weddigen, C. and Haase, E. L.,** *Nucl. Instrum. & Methods,* 33, 157, 1965.

Chapter 12

Data Acquisition

CONTENTS

12.1 CAMAC 175
12.1.1 Basic Concepts 175
12.1.2 CAMAC Systems 177
12.2 FASTBUS 201
12.2.1 Introduction to FASTBUS 201
12.2.2 FASTBUS Applications 203
12.3 Other Data Acquisition Systems 214
References 232

12.1 CAMAC

12.1.1 BASIC CONCEPTS

CAMAC is a modular data handling system used at almost every nuclear physics research laboratory and many industrial sites all over the world. It represents the joint specifications of the U.S. Nuclear Instruments Modules (NIM) and the European ESONE Committees.

The primary application is data acquisition but CAMAC may also be used for remotely programmable trigger and logic applications. The CAMAC standard covers electrical and physical specifications for the modules, instrument housings or crates, and crate backplane.

Individual crates are controlled by slave or intelligent controllers. The controllers are tied together with a parallel branch highway that ends in a branch driver. The branch driver is interfaced directly to a data acquisition computer. Alternatively, tree or parallel data acquisition architectures may be created by connecting secondary CAMAC branches via CAMAC branch driver modules.

Timing and protocol specifications permit up to 1 megaword/s transfers of 16 or 24-b words for both the dataway and CAMAC branch. GPIB timing is usually limited by the host computer and typically runs at 500 kilobytes/s.

CAMAC is an international standard of modularized electronics as defined by the ESONE Committee of the JRC, Ispra, Italy. Its function is to provide a scheme to allow a wide range of modular instruments to be interfaced to a standardized multireceptacle which, in turn, may be interfaced to a computer. In this way, additions to a data transfer and control system may be made by plugging in additional modules and making suitable software changes. Thus, CAMAC allows information to be transferred into and out of the instrument modules.

CAMAC modules may be plugged into a CAMAC crate which has 25 stations, numbered 1–25. Station 25, the rightmost station, is reserved for a crate controller, whereas Stations 1–24 are normal stations used for CAMAC modules (see Figure 12.1). Usually, Station 23 is also used by the controller in that most controllers are double width (no. 2 CAMAC). The purpose of the controller is to issue CAMAC commands to the modules and transfer information between a computer (or other digital device) and other CAMAC modules.

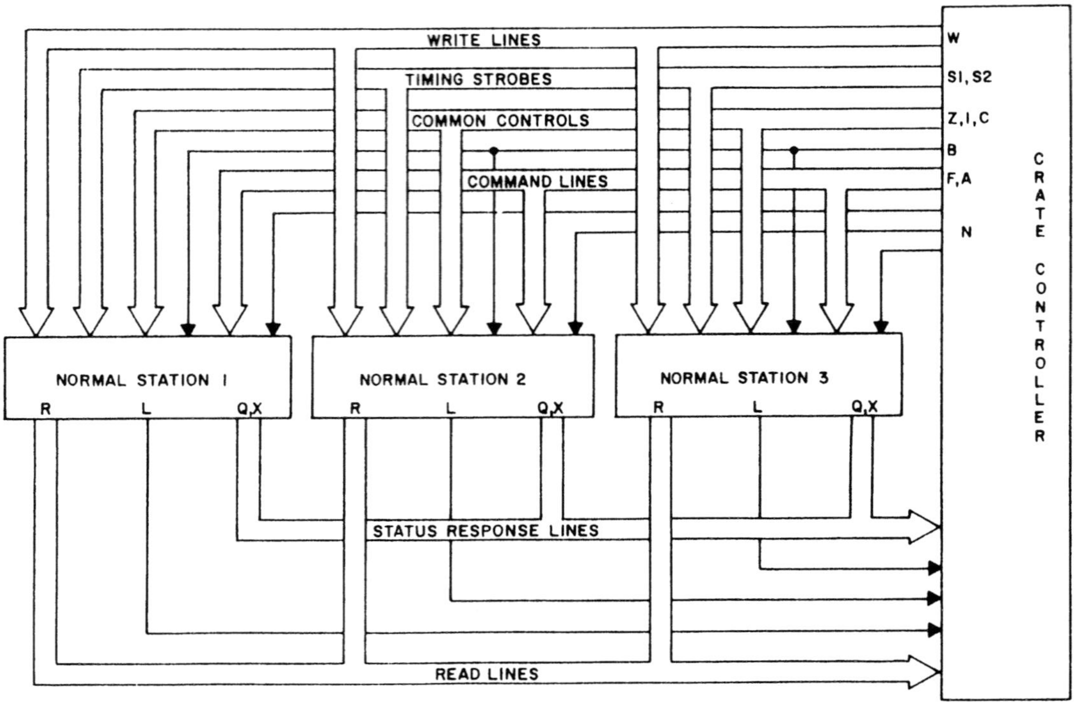

Figure 12.1 Block diagram and signal paths for three CAMAC modules and a crate controller. (From LeCroy Corporation Research Instrumentation Catalog, 1992. With permission.)

Data transfer, control functions, and module powering are affected via the dataway. This is a series of bus and individual lines across the back of the crate. The dataway lines include digital data transfer lines, strobe signal lines, and addressing and control lines.

In a typical dataway operation, the crate controller issues a CAMAC.command which includes specifying a station number (N), a subaddress (A), and function codes (F). In response, the subaddress of the module will generate an X response (valid command accepted) and act on the command. If this command requires data transfer, the READ (R) or WRITE (W) line will be used. Note that the terms READ and WRITE apply to the controller, not the module. For example, under a READ command, the controller reads data contained within a module.[1]

Communication with plug-in units takes place through the dataway. This passive multiwire highway is incorporated in the crate and links the 86-pin sockets to all stations. The bus lines link corresponding pins at all normal stations and, in some cases, the control station. Individual lines link one pin at a normal station to one pin at the control station. The patch pins have no specified dataway wiring but can be connected to individual points to which patch leads may be attached.

During a dataway operation the controller generates a command consisting of signals on individual station number lines to specify one or more modules, signals on the subaddress bus lines to specify a subsection of the module or modules, and signals on the function bus lines to specify the operation to be performed. The command signals are accompanied by a signal on the busy bus line, which is available at all stations to indicate that a dataway operation is in progress.

When a module recognizes a READ command calling for a data transfer to the controller, it establishes data signals on the read bus lines. When a controller recognizes a WRITE command calling for a data transfer to a module, it establishes data signals on the write bus lines. In addition, regardless of whether there is transfer on the R or W lines, the module may transmit 1 b of status information on the response bus line.

Two timing signals, strobes S1 and S2, are then generated in sequence on separate bus lines. The strobes are used to transfer data from the dataway into modules (on WRITE commands) and into the controller (on READ commands). They may also initiate other actions within the controller and modules. Whenever there is no dataway operation in progress (indicated by the absence of the busy signal) any module may generate a signal on its individual look-at-me line to indicate that it requires attention. Three common control signals are available at all stations, without requiring addressing by a command, in order to initialize all units (typically after switch-on), to clear data registers, and to inhibit features such as data taking.

A command consists of signals on the dataway lines which specify at least one module (by individual station number lines), a subsection of the module or modules (by the four subaddress bus lines), and the function to be performed (by the five function bus lines). The command signals are maintained for the full duration of the operation on the dataway. They are accompanied by a signal on the busy bus line which indicates to all units that a dataway operation is in progress.

Each normal station is addressed by a signal on an individual station number line (N) which comes from a separate pin at the control station. The stations are numbered in decimal code from the left-hand end as viewed from the front, beginning with Station 1.

Two strobe signals S1 and S2 are generated in sequence on separate bus lines. These signals are used to transfer information between plug-in units via the dataway or to initiate operations within units. In either case the specific action is determined by the command present on the dataway. Both strobes are generated during each dataway command operation, and all plug-in units which accept information from the dataway do so in response to these strobes. The first strobe S1 is used for actions which do not change the state of signals on the dataway lines. All units which accept data from the dataway in a READ operation, or in a WRITE operation do so in response to S1. The second strobe S2 is used to initiate any actions which may change the state of dataway signals, for example, clearing a register whose output is connected to the dataway.

Look-at-me (L), like the N line, is an individual connection from each station to a separate pin at the control station. When there is no dataway operation in progress (no B present) any plug-in unit may generate a signal on its L line to indicate that it requires attention. When B is present, each L signal is gated off the dataway line by the unit which generates it.

A look-at-me (LAM) request can be reset by clear look-at-me, by the initialize signal, or by the performance of the specific action which generated the request.

The busy signal (B) is used to interlock various aspects of a system which can compete for the use of the dataway.

The Q bus line is used during a dataway operation to transmit a signal indicating the status of a selected feature of the module.

The initialize signal (Z) has absolute priority over all other signals or controls. It sets all units to a basic state by resetting all registers, whether data or control, to a defined state; and by resetting all L signals and disabling them where possible. Units which generate Z must also cause S2 and B to be generated. Modules which accept Z gate it with S2 as a protection against spurious signals on the Z line.

The presence of an inhibit (I) signal inhibits any activity (for example, data taking). It must either not change when B is present or have rise and fall times not less than 200 nsec.

The sequence of events during a single dataway operation is shown in the timing diagram (Figure 12.2). Points t_0, t_3, t_6 represent the initiation of the negative going of the Command, Strobe 1, and Strobe 2 signals, respectively. They are the times at which the signals would be received from an ideal dataway with no capacitive loading.

Points t_9, t_5, and t_8 represent similarly the initiation of the positive-going edges of the same signals.

Points t_2 and t_{11} are the latest time at which the data source is permitted to initiate the negative-going and positive-going edges of the data signals.

Points t_1, t_3, t_4, and t_7 represent the latest times at which the received signals are permitted to reach a maintained "1" state; and therefore refer to the last negative-going transition across the +0.8-V threshold.

Points t_6, t_9, t_{10}, and t_{12} represent the latest times at which the received signals are permitted to reach a maintained "0" state; and therefore refer to the last positive-going transition across the +2.0-V threshold. The next dataway operation must not start before t_9.[1]

12.1.2 CAMAC SYSTEMS

A real-time data acquisition system for a very high-energy gamma-ray astronomy experiment was described by Bhat et al.[2] The detection of an extensive air shower initiated by TeV gamma-rays or more abundant cosmic-ray protons in the earth's atmosphere is through the Cherenkov light from the secondaries (e+ and e–). It is identified by multiplicity logic of the signals from an array of parabolic mirrors (or a group of mirrors called a bank) at the foci of which a fast (RCA 8575) phototube is placed. Whenever

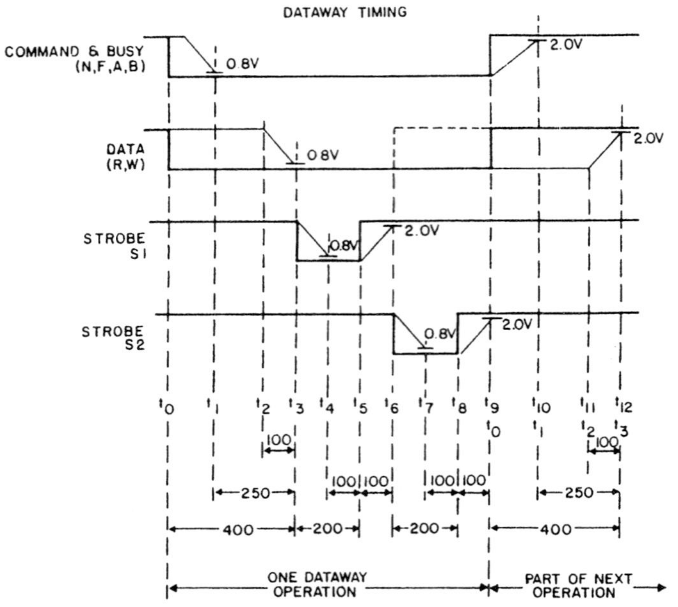

Figure 12.2 Dataway timing. (From LeCroy Corporation Research Instrumentation Catalog, 1992. With permission.)

such an event occurs, they record: the time of occurrence correct to a microsecond, the scalers which record various counting rates, the latch information signifying which mirrors or banks have triggered during the event, the pulse height in each phototube to be read from a CAMAC analog-to-digital converter (ADC) module, and the relative time delays between various mirrors or banks from a CAMAC TDC.

The hardware of the data acquisition system (DAS) was based on a 16-b processor PDP 11/23. The processor has the advantage of a bus structure that provides position-independent priority as the peripheral device interfaces are connected to the I/O bus and fast interrupt response without device polling. The simple LSI-11 I/O bus is fast and easy to use as an interface between LSI-11 microcomputer and memory and peripheral interface modules. The system was configured with three basic interface module options each of which has unique device and vector addresses: (1) serial line unit interfaces which connect to the processor with EIA-compatible devices (modems) like the video display unit (VDU), dot matrix printer, etc.; (2) DRV-11 general-purpose parallel line unit interface for data I/O operations; and (3) DRV-11B DMA interface incorporated in a streamer tape coupler for block data transfers.

Figure 12.3 explains the overall block diagram of the DAS.

The modular composition of the system manager is shown in Figure 12.4. It consists of

- General-purpose parallel interface (DRV11): The parallel interface between the system manager and the master control unit (MCU) comprises memory-mapped 16-b O/P port (used for generating addresses and control signals) and 16-b I/P port (used for data). It has two vector interrupts, namely, Request A (EVENT INTPT) and Request B (MON. INTPT).
- PDP 11/23 CPU: It is a 16-b data, 20-b address on-line processor. It supports memory-mapped I/O port and vectored interrupt structure.
- Video display unit (VDU): It is an interface between the user and the system.
- Printer: This is self-explanatory.
- Magnetic tape: It is used to store the data.

The master control unit (MCU) plays a vital role in the data acquisition system. Figure 12.5 shows a functional block diagram of the MCU. It consists of

- Bus director: This routes the I/O data from the DRV11 to the standard structured HERGROI bus.
- Event interrupt generator: Trigger pulses T1–T4 are generated by the analog electronics system.
- Control unit: It consists of four subunits: (1) mode generator which gives out eight modes, (2) mode "0" called the EVENT MODE, (3) INHIBIT generator, and (4) the CLEAR generator.

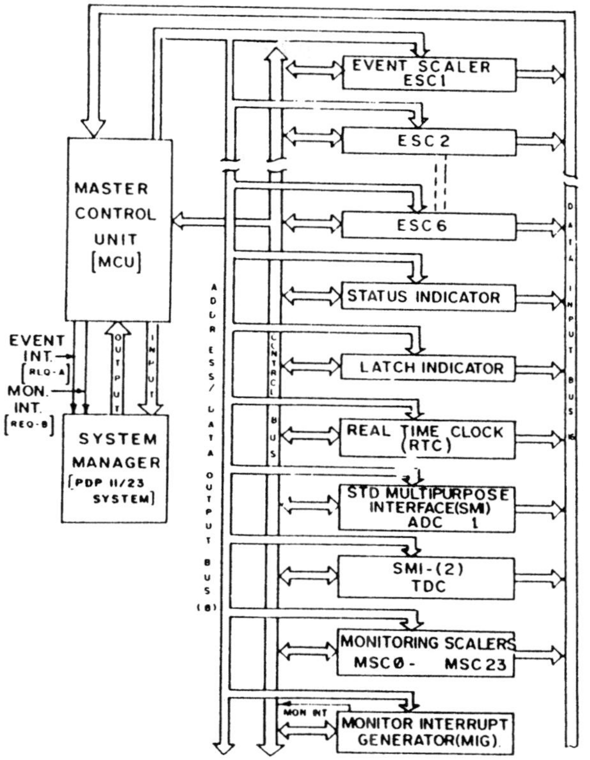

Figure 12.3 An overall functional block diagram of the data acquisition system (DAS). (From Bhat, P. N. et al., *Nucl. Instrum. & Methods,* A292, 494, 1990. With permission.)

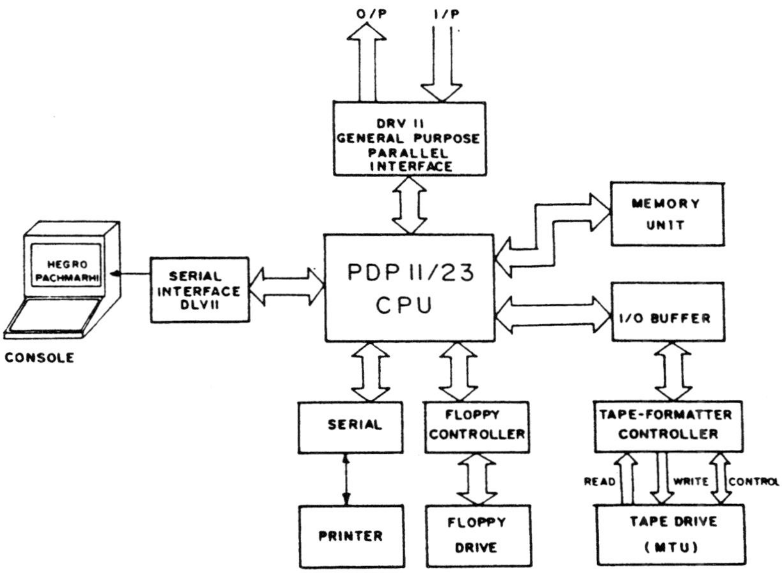

Figure 12.4 Block diagram showing the modular composition of the system manager. (From Bhat, P. N. et al., *Nucl. Instrum. & Methods,* A292, 494, 1990. With permission.)

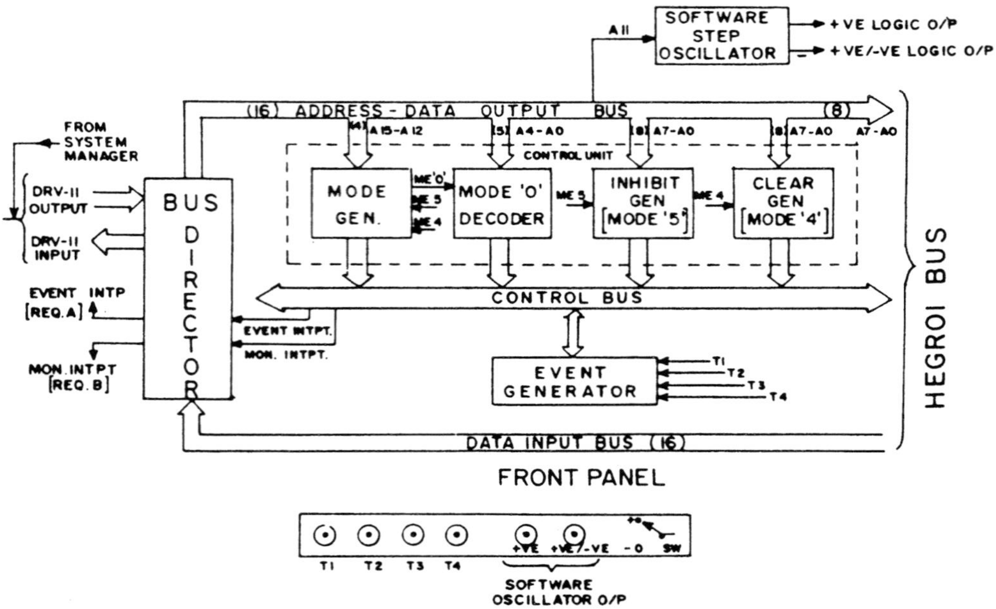

Figure 12.5 A functional block diagram of the master control unit (MCU). (From Bhat, P. N. et al., *Nucl. Instrum. & Methods,* A292, 494, 1990. With permission.)

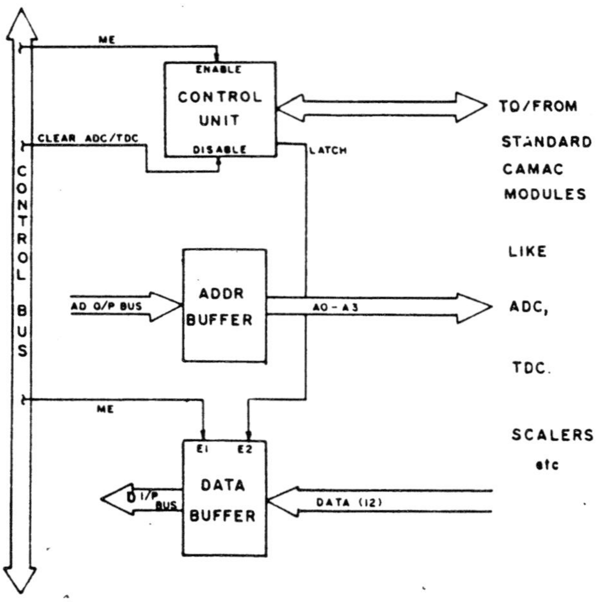

Figure 12.6 A block diagram of the standard multipurpose interface (SMI). (From Bhat, P. N. et al., *Nucl. Instrum. & Methods,* A292, 494, 1990. With permission.)

Each of the six scalers is a seven-digit decade counter with visual display. The latch indicator gives out a quick visual display of the shower recorded using an event latch displaying the mirrors of the group of mirrors which gave rise to the event trigger.

Figure 12.6 shows the block diagram of the standard multipurpose interface (SMI). This is used for interfacing the HEGROI bus to standard fast, multichannel CAMAC modules such as 8-channel time-to-digital converter (TDC) and 12-channel fast (24-b) scalers.

The block diagram of the real-time clock (RTC) module is shown in Figure 12.7. The event latch from the MCU latches the RTC up to microsecond accuracy. The five 16-b latched words are put on the data input bus (DIB) using mode "0" and enables E16–E20. Enables E21–E31 are reserved for future use.

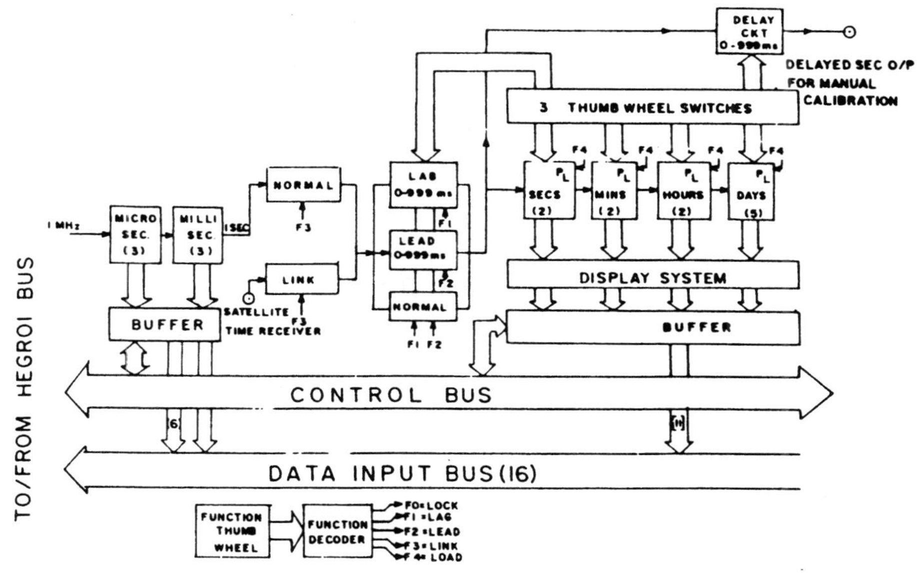

Figure 12.7 A functional block diagram of the real-time clock (RTC). (From Bhat, P. N. et al., *Nucl. Instrum. & Methods,* A 292, 494, 1990. With permission.)

HEGROSOFT is a special-purpose real-time operating system designed to accomplish the following functions: real-time acquisition of the gamma-ray event data with a specified response time, continuous monitoring of individual detectors in sequence and output information, attending to run time service requests through commands from the console terminal, and regularly performing a series of jobs in a low-priority job queue making use of the idle time of the processor. These functions are performed at four different levels of task priorities.[2]

A CAMAC module designed to accumulate data from multiple ADCs was described by Nagashima et al.[3] The module has a coincidence logic for multidimensional data taking up to eight dimensions and can be operated in either of two coincidence modes: complete coincidence mode (AND mode) or time-limiting mode (OR mode). Each module is designed to interface each one of the ADCs. Up to eight modules can be combined with each other for taking multiparametric data.

The schematic diagram of the interface module is shown in Figure 12.8. The terminals are classified into three groups shown by different symbols. The triangles show the CAMAC standard signals and are connected to the CAMAC data highway. The terminals connected to the standard NIM electronics such as an ADC are shown by the open squares. The solid squares show the terminals for connecting the interface modules to each other. The terminals marked by the squares are located on the front panel of the module. There are four points, from J1 to J4, present with jumping wires. The wires are connected to the line with the same number as that of the corresponding ADC.

Five addressed and two nonaddressed instructions are prepared and summarized in Table 12.1. The F(16) command is used to set the information in the table and the latches. The F(17) command controls the interface and ADC in accordance with the information given in Table 12.1. The data of the ADC are read in the memory of the computer by the F(0) command. F(8) is the test command for the LAM circuit. The command F(10) reset a LAM flip-flop (FF). There are two ports named EVENT and ADC READY for accepting the different trigger signals.

There are two coindence modes named AND mode and OR mode. In the AND mode, only data satisfying the following condition are accepted: either the ADC READY signals or the EVENT signals of all ADC, joining together in a multiparametric mode, appear within the present time. On the other hand, in OR mode, the data are accepted only if at least one of the ADC READY signals or the EVENT signals of any ADCs in the multiparametric mode appears. The flip-flop named MODE FF selects two modes under the program control.

A timer circuit is located in the middle part of Figure 12.8. The timer circuit, consisting of the clock, the scaler, the time latch, and the time over detector, is used for judging whether a multiparametric phenomenon occurred in coincidence. The oscillator named clock supplies the clock pulse to the scaler.

Table 12.1 List of CAMAC Instructions

CAMAC Instructions	Comments
N(i)F(0)	Read data of ADC through a DATA RECEIVER
N(i)F(16)	Write data in COINCIDENCE and TIME LATCHES W1: SetADC 0 in coincidence ... ... W8: SetADC 7 in coincidence W9: time duration in ...coincidence W12:
N(i)F(17)	Control ADC and interface W1: LAM enable/disable W2: ADC start/stop W3: Initialize all W4: Coincidence mode AND/OR
N(i)F(8)	Test LAM
N(i)F(10)	Clear LAM FF
I	Inhibit all ADCs
Z	Initialize all ADCs and IF

From Nagoshima, Y. et al., *Nucl. Instrum. & Methods,* 206, 147, 1983. With permission.

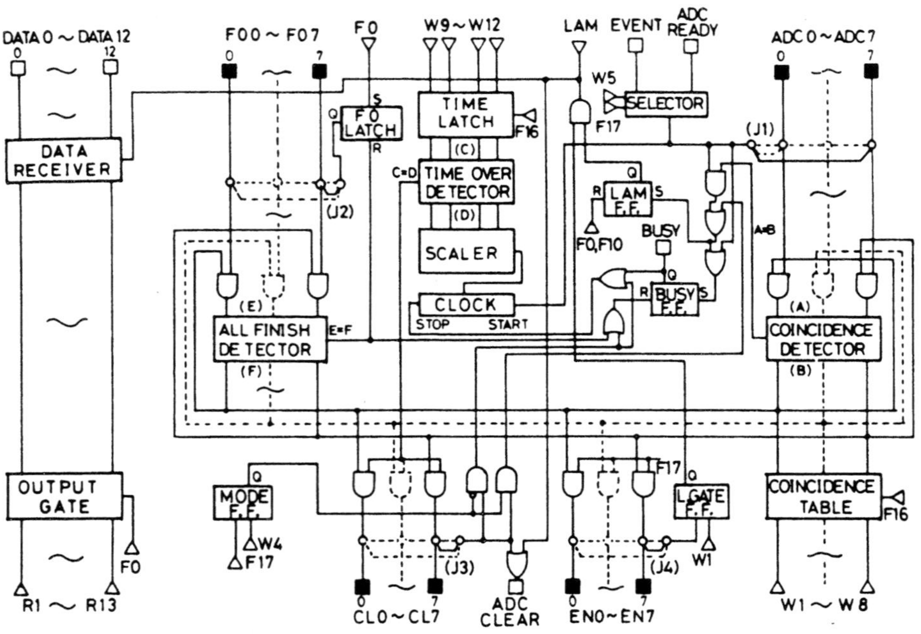

Figure 12.8 Schematic circuit diagram of the CAMAC interface. (From Nagoshima, Y. et al., *Nucl. Instrum. & Methods,* 206, 147, 1983. With permission.)

It generates two pulse trains having different intervals. If the only ADC READY signal is used as a trigger, it generates a pulse train, the interval of which is 10 μsec. On the other hand, it generates a pulse train whose interval is 100 nsec if the EVENT signal is used as a trigger. A preset time is stored in the register named time latch. The range of the preset time is from 100 to 1600 nsec in 100 steps or from 10 to 160 μsec in 10-μsec steps.[3]

The data handling circuit is shown on the left-hand side of Figure 12.8. Once the look-at-me interrupts issue from the interfaces to the computer through the LAM grader and the branch driver, the computer

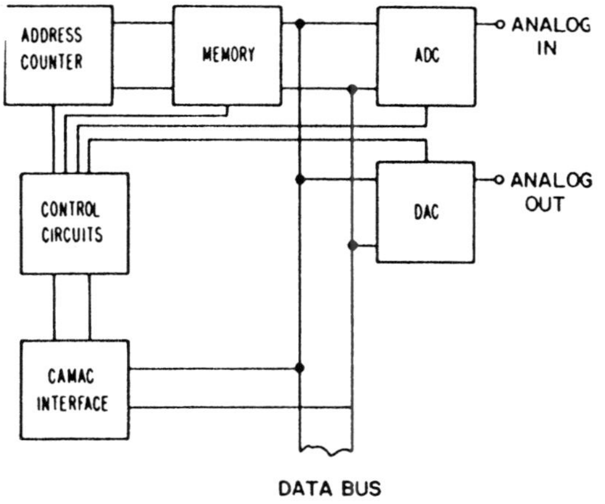

Figure 12.9 Block diagram of the digitizer. (From Iqbal, M. Z. and Gimlett, L. J., *Nucl. Instrum. & Methods,* A248, 434, 1986. With permission.)

starts reading the data of the ADCs by F(0) commands. The F(0) command opens the output gate and sets the F0 latch. The state of the F0 latch shows whether the data in the data receiver have been transferred to the computer, and is sent to other ADCs through the line named F00-F07. When the computer finishes reading the data in all related interfaces, it issues a pulse to set the interface to an initial state so as to be able to accept the next trigger signal.

A low-cost pulse shape digitizer for a time projection chamber was designed by Iqbal and Gimlett.[4] It was designed and constructed using a 6-b flash ADC (TRW 1014), 16K static RAMs, and standard transistor-to-transistor (TTL) logic.

The block diagram for a single digitizer is shown in Figure 12.9. The digitizer is comprised of three separate functional units, the CAMAC interface unit, the control unit, and the digitizer unit. The CAMAC interface allows for interfacing with a computer via the CAMAC dataway. An internal bus consisting of control lines and databus enables communication between the CAMAC interface and the controller. The controller contains the timing and mode selection circuitry for controlling several digitizer units in parallel. The digitizer unit consists of a 6-b flash ADC and a memory bank of four 2K × 8 random access memories (RAMs). Data storage is accomplished by sequentially accessing each RAM, thereby reducing the access rate by a factor of 4. For diagnostic purposes a DAC is connected to the databus for monitoring the analog sample at the clock speed. The digitizer is triggered by a START signal and stops after a preselected time. A triggered response is sent if it digitizes a signal with an amplitude exceeding a predetermined discriminator level. The digitizer is changed into READ mode, where the memory contents can be read. Once the memory is read the digitizer can then be set to READY mode for the next event.

A schematic diagram of the CAMAC interface is shown in Figure 12.10(a). The F lines are decoded with IC1. Data to be written to the controller is latched onto the internal bus with S1 and strobed with S2. All control functions are ANDed with S2. For CAMAC function a commercial transient recorder (LeCroy TR88377F) was used.

READ, WRITE, and DISPLAY are set with CAMAC commands F(17), F(9), and F(18), respectively.

The digitizer has three different modes — READ (R), WRITE (W), and DISPLAY (D) — selectable by CAMAC commands which the CAMAC interface reinterprets by producing a low pulse in normally high lines W-EN (write enable), R-EN (read enable), or D-EN (display enable). The control signals are used by two RS latches (IC7) to produce high state in W (WRITE), R (READ), or D (DISPLAY) when selected, as shown in Figure 12.10(c). The direction of the arrow indicates whether the signals are coming from or going to the CAMAC interface.

The flash ADC is strobed by the clock ϕ_0, which is the highest frequency in the entire circuit. To accomplish this timing, the clock frequency is divided by four and shifted by zero or three clock cycles to produce clocks of four different phases, ϕ_1, ϕ_2, $\overline{\phi}_1$ and $\overline{\phi}_2$, respectively.[4]

The address counter (Figure 12.11(a)) is designed to increment the address of the four RAM ICs sequentially and with the appropriate phase relation to the data latch. The address of the first (top) RAM is incremented with ϕ_1. The same counter is used to set the address of the other ICs sequentially, at clock frequencies ϕ_2, $\overline{\phi}_1$, and $\overline{\phi}_2$, by using three latches strobed with these signals. Normally the counter is cleared by a W-EN signal or the overflow of the same counter.

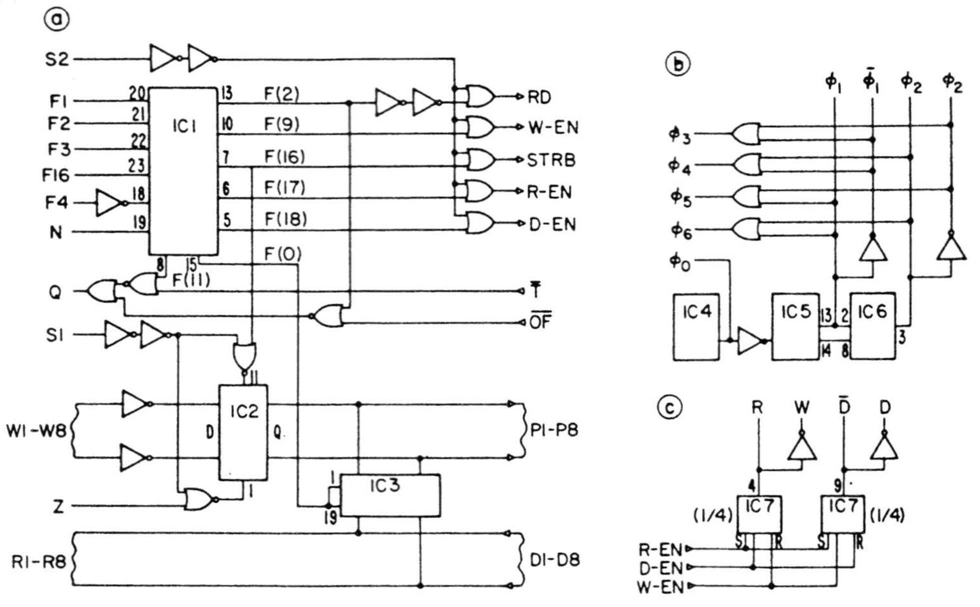

Figure 12.10 Schematic diagram of (a) the CAMAC interface, (b) the clock, and (c) the mode selector. (From Iqbal, M. Z. and Gimlett, J. L., *Nucl. Instrum. & Methods,* A248, 434, 1986. With permission.)

In display mode (D is high) the RS latch controlling the counter is inhibited to change state so that a continuous spectra can be produced by cycling the memory over and over. During the READ mode (R is high) the address counter, instead of incrementing according to ϕ_1, is incremented by one address for every RD command. Once the memory is read the digitizer is reset to the WRITE mode with a W-EN command, enabling it for the next event.

The ADC memory arrangement is shown in Figure 12.11(b). The analog circuity for the TRW (1014) ADC is adapted as suggested by TRW. The ADC is strobed by the primary clock ϕ_0 which latches the data at its leading (positive) edge. The digitized output is fed to four latches which are strobed sequentially by ϕ_1, ϕ_2, $\bar{\phi}_1$, and $\bar{\phi}_2$, respectively. If the ADC digitizes a signal larger than 25% of its full range (1 V), one of the two most significant bits of the data latches is high, resetting a latch. This latch output is used as a triggered signal T, and supplied to the CAMAC interface for generation of a Q response indicating that the digitizer has been triggered. This latch is reset at the beginning of a digitizing cycle by the W-EN. During the READ mode the four latches are kept at high Z state for isolating the databus from the ADC. The memory bank consists of four 2K × 8 static RAMs. During the WRITE mode, when both the address and data lines have stabilized, the RAMs are switched to the WRITE state and held long enough for data access. This is done by using ϕ_1, ϕ_2, $\bar{\phi}_1$, and $\bar{\phi}_2$ to control the WRITE state.[4]

A data acquisition system with high-amplitude resolution (12 b), high-speed conversion (clock frequency 50 MHz), used in an experiment with a large number of counters (4200) was described by Alde et al.[5] GAMS-4000 is a large photon (electron) detector. Each high-energy (≥0.2 GeV) gamma-ray (or electron) incident on the detector develops an electromagnetic shower inside the lead glass. The electrons and positrons from a shower emit Cherenkov light on their way through the glass, the amount of light being proportional to the energy of the incident gamma-ray.

By a suitable summation of these signals it is possible to obtain the energy of the incident photon. The anode signal from each photomultiplier (PM) is transmitted through a 50-m long coaxial cable to its associated charge-to-time converter (QTC) circuit.

The largest pulse delivered at a QTC input has an amplitude (max) of –0.8 V, a rise time of 16 nsec, a width of 30 nsec (FWHM), and a full width of 60 nsec. This corresponds, in a 50-Ω load, to a maximum instantaneous current of 16 mA and to an integrated charge of ~500 pC.

Each QTC is followed by a 12-b scaler (49704 + SN74LS303 circuits). The output signal width from the QTC is measured by counting the number of pulses delivered by a 50-MHz quartz oscillator.

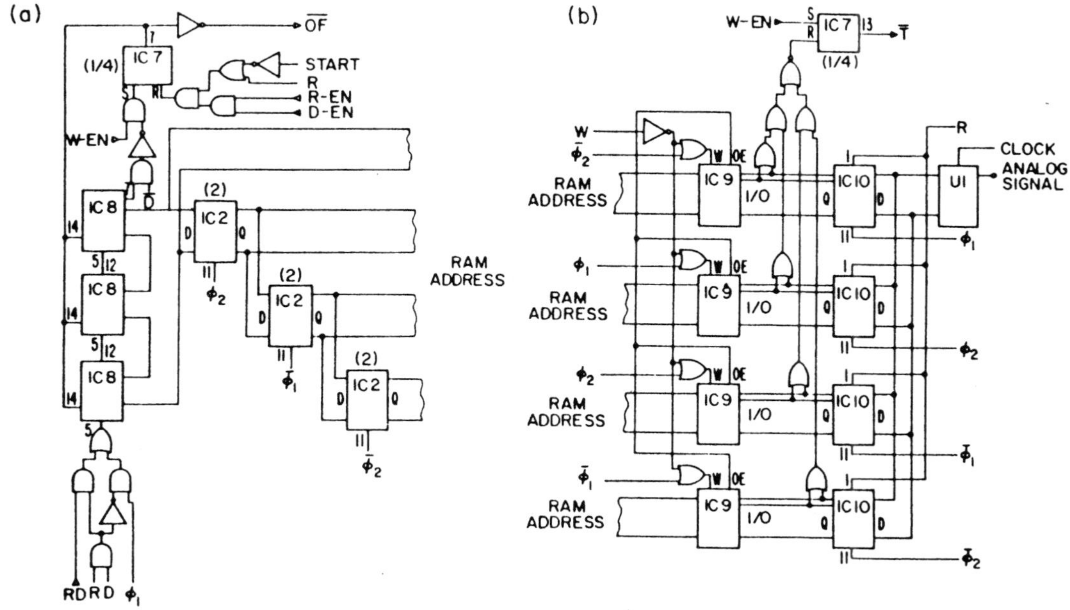

Figure 12.11 Schematic diagram of (a) the address counter and (b) the ADC-memory arrangement. (From Iqbal, M. Z. and Gimlett, J. L., *Nucl. Instrum. & Methods,* A248, 434, 1986. With permission.)

The whole system consists of 11 Euronorm crates. Each crate has 27 slots, with 24 slots being used for the charge-to-time converters and their associated scalers. The connection between QTC and its corresponding scaler is made by an optocoupler.

Each crate controller gives access to the 384 scalers of the crate in a CAMAC addressing mode (Figure 12.12). It also contains two sets of RAM memories: (1) a 384- $\times$ 9-b RAM memory for the pedestals; and (2) a 64- $\times$ 9-b RAM memory for indirect addressing of a sample of cells, used for checking purposes. Several functions are performed by the crate controller: (1) loading and reading of the pedestal memories; (2) loading of the address of each cell or autoincrementation of this address at the next access; (3) loading or reading of Δ, where Δ is a 4-b constant value which takes into account the pedestal instabilities; and (4) reading of the content of the scalers, step by step or automatically in sequential mode, this last mode being commonly used during data taking. A comparator allows the information from channel i to be transferred only if the following condition is fulfilled: amplitude (i) — pedestal (i) $- \Delta > 0$.

The CAMAC module (MIPCAM) is designed to accept or deliver data in the form of 16-b words. The crate controller can also be controlled in parallel by the MIPROC 16S processor.[5]

The data are sent to a first-in-first-out (FIFO) buffer memory, through a fast bus of 24 b (12 addressing b, 12 data b, plus two control lines, namely, a data valid line to the FIFO and a ready line from the FIFO). A block diagram of the complete system is given in Figure 12.13.

A technique for generating linear pulses which can be random in both amplitude and time is described by Abdel-Aal.[6] Values for both pulse amplitude and spacing are set for the individual pulses by the software on a pulse-by-pulse basis.

The VAX sends an integer corresponding to the amplitude of the next pulse to be generated to a CAMAC digital-to-analog converter (DAC) module which is configured to produce a linear pulse of a fixed width at the desired amplitude. A delay corresponding to the required pulse spacing is inserted by the VAX between the CAMAC write operations.

The pulse generator module uses a 12-b DAC IC which produces an output voltage corresponding to the desired pulse amplitude. To generate a pulse output, rather than a DC value, the data to the DAC is cleared to zero after a preset short duration equal to the pulse width. Figure 12.14 is a simplified block diagram of the pulser CAMAC module and Figure 12.15 shows the main timing waveforms relevant to pulser operation. A 12-b integer representing the desired pulse amplitude is written into the CAMAC module using function F(16). The CAMAC timing signal S1 loads the data into a set of data latches; and

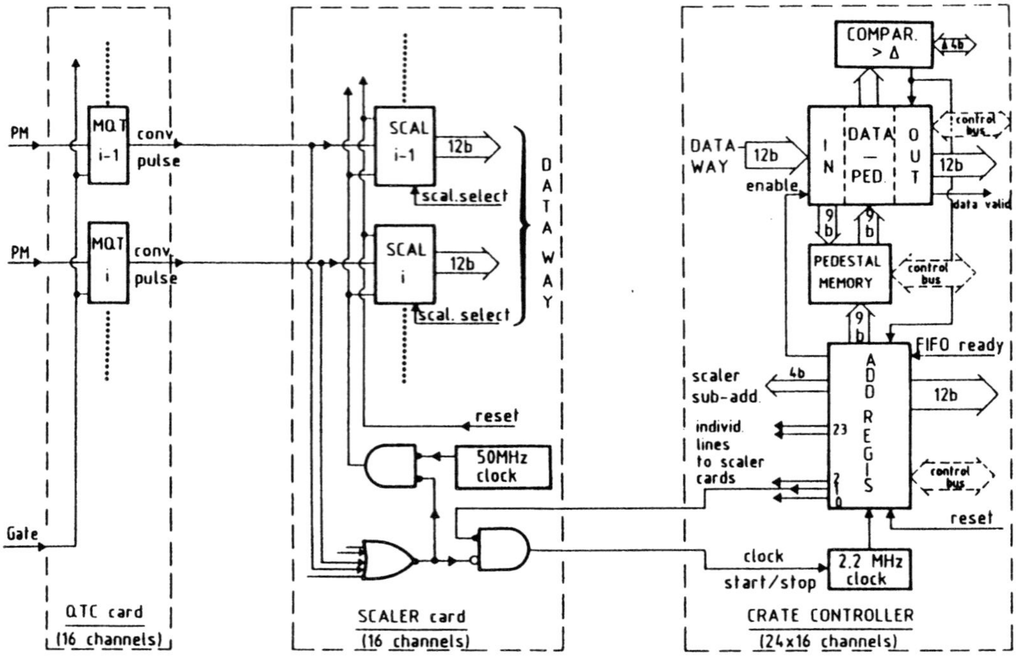

Figure 12.12 Crate flow diagram. (From Alde, D. et al., *Nucl. Instrum. & Methods,* A240, 343, 1985. With permission.)

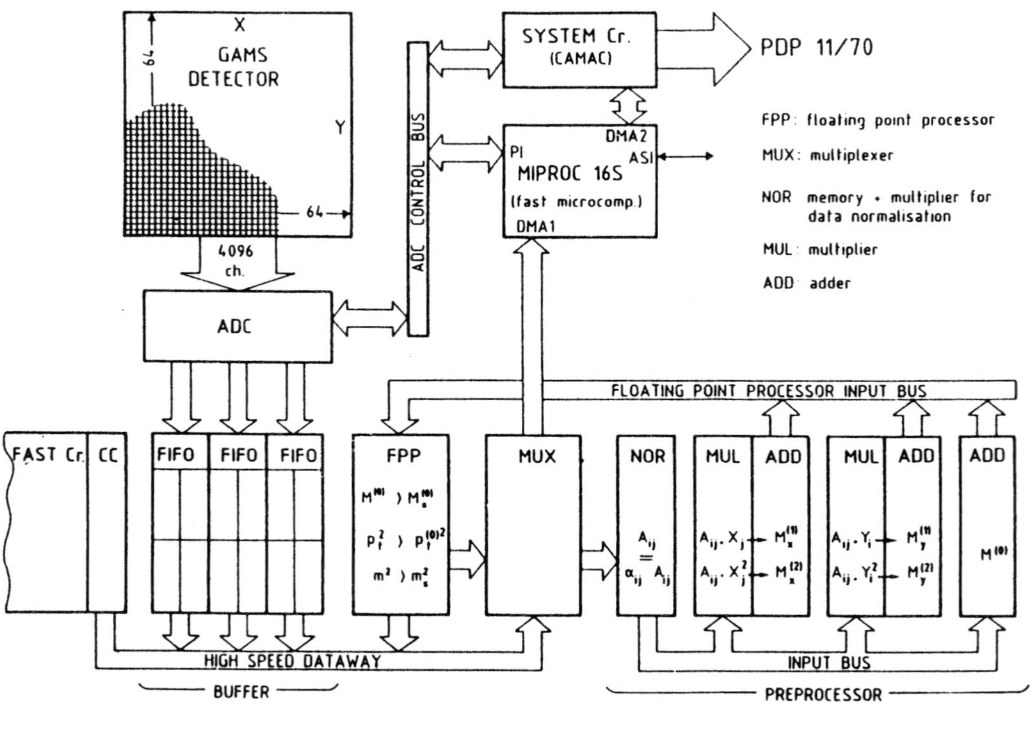

Figure 12.13 GAMS-4000 data acquisition system. (From Alde, D. et al., *Nucl. Instrum. & Methods,* A240, 343, 1985. With permission.)

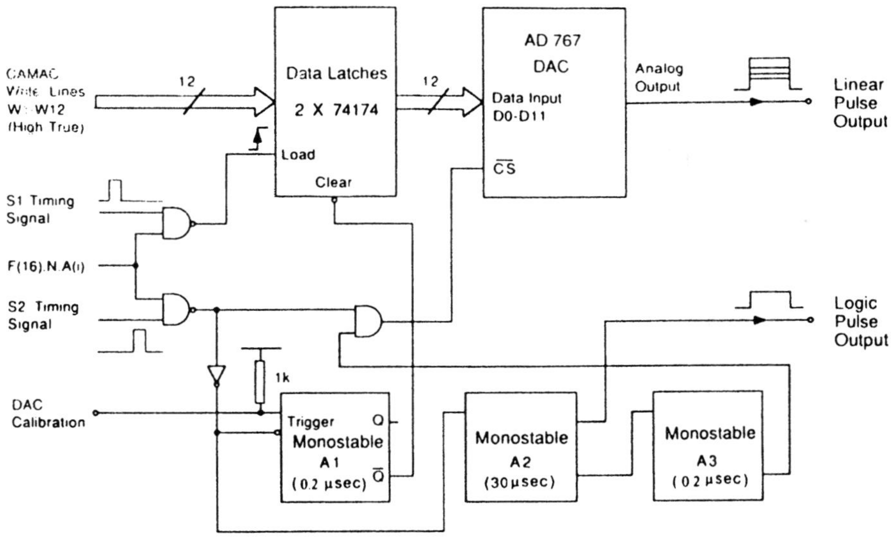

Figure 12.14 Block diagram for the programmable random pulser CAMAC module. (From Abdel-Aal, R. E., *Nucl. Instrum. & Methods,* A309, 331, 1991. With permission.)

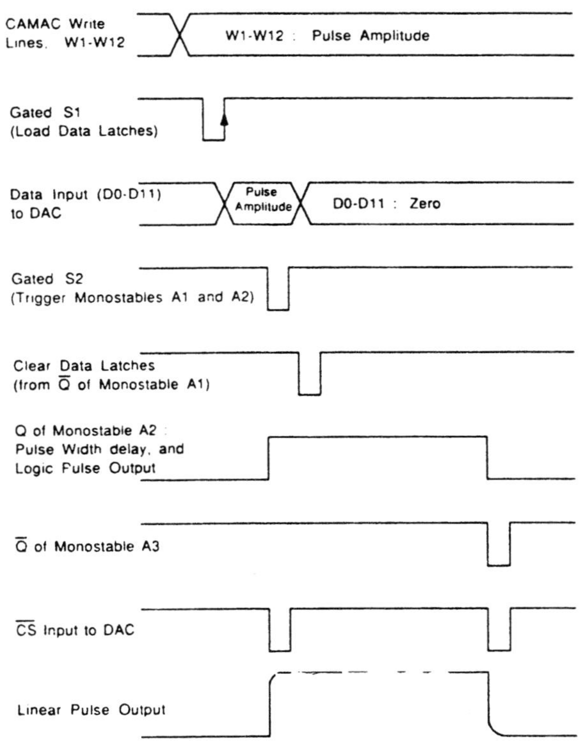

Figure 12.15 Timing waveforms for the operation of the CAMAC pulser module. (From Abdel-Aal, R. E., *Nucl. Instrum. & Methods,* A309, 331, 1991. With permission.)

the subsequent timing signal S2 enables the DAC IC to convert, and therefore the DAC output rises to the desired pulse amplitude. The trailing edge of S2 triggers a short pulse from monostable A1 which clears the data latches, thus setting all the data input lines to the DAC to zero. The total propagation delay through this monostable and the data latches should be large enough to meet the hold time requirements on the DAC data input with respect to the positive-going edge of the $\overline{CS}$ pulse. The DAC does not convert this zero data until it receives another enable conversion pulse at the $\overline{CS}$ input. This pulse is derived from monostable A3 following a delay interval equal to the desired pulse width which is generated by monostable A2. At the end of this delay interval, the DAC output reverts to zero until the next CAMAC WRITE operation for the next pulse. Therefore, this output is a linear pulse of the desired amplitude with a width equal to the width of the pulse generated by monostable A2. For this application this width is fixed at approximately 30 μsec which is about 10 times the maximum full-scale settling time for the DAC device. In addition to the linear output pulse from the DAC, the output of monostable A2 is used as a coincident logic pulse.[6]

A smart CAMAC crate controller has been designed by Conetti and Kuchela.[7] Its distinctive feature is a hard-wired list processor performing block transfers by executing lists of prestored CAMAC instruction. Data collected from the crate are sent to an external port: controllers organized in serial or parallel channels transmit their data to buffer memories. Processing of interactive CAMAC operations, downloading of CAMAC instruction lists, and supervision are provided by a Motorola 6809 microprocessor linked to the controller via the STD bus and to a host computer via an RS-232 serial interface.

The smart controller is driven by a hard-wired processor executing, on reception of an even trigger, lists of prestored CAMAC instructions at the rate of 1 MHz, the maximum allowed by CAMAC. The lists are downloaded from a host computer through the serial interface of a 6809-based single-board microcomputer (7911/SP9 Matrix Corporation) linked to the controller via the STD bus.

In addition to the fast data acquisition mode the crate controllers can be operated in interactive CAMAC mode to emulate a standard A-1 controller. Single CAMAC instruction can be executed by the host, communicating to the controller via the serial port of the 7911/SP9 microcomputer. Host software has been written to run under PDP-11/RSX-11M and VAX/VMS systems. A block diagram of the crate controller indicating major data paths and CAMAC dataway signals is shown in Figure 12.16. The controller has two basic modes of operation: list and interactive. When in list mode, the control circuit, shown in Figure 12.17B regulates the execution of CAMAC instructions stored in the list in the following way: in addition to the FNA information (14 b), a 2-b mode (M) field is present. A CAMAC cycle generator circuit is shown in Figure 12.17A.

A waveform sampler module (WSM) for the measurement of signal shapes coming from the multihit drift chambers of the SLAC SLD detector was described by Freytag et al.[8] The module uses a high-speed, high-resolution analog storage device (AMU) University. The AMU devices together with high-speed TTL clocking circuity are packaged in a hybrid which is also suitable for mounting on the detector. The module is in CAMAC format and provides eight signal channels, each recording signal amplitude vs. time in 512 cells at a sampling rate of up to 360 MHz. Data are digitized by a 12-b ADC with a 1-μsec conversion time and stored in an on-board memory accessible through CAMAC.

The hybrid (HAMU II) used in this instrument contains 16 AMU chips with a total of 4096 analog storage cells, four fast TTL shift registers, one OR chip, and a dual JK flip-flop. The block diagram of Figure 12.18 shows the subdivision of the hybrid into two groups of AMUs (A0,..., A7, B0,..., B7) with the input to each AMU externally accessible. The hybrid can thus be operated in a 16×256 or 8×512 cell mode. Each group has 16 fast write clock signals (F0A,..., F15A, F0B,..., F15B), which are generated by four fast shift registers 54F164. The shift registers are controlled by four external clocks (CP1,..., CP4), and two JK flip-flops (54F109). The slow write clock signals for each group of AMUs (ϕ_{1SA}, ϕ_{2SA}, and ϕ_{1SB}, ϕ_{2SB}) are derived from the appropriate fast clocks via the four gates of a 74LS32 chip. For the read cycle, each group of eight memory chips is controlled by a serial data-in signal, one set of noninterleaving read clocks, and the enable read signal ENA (ENB), respectively.

The maximum frequency at which data can be written into the eight signal channels of the HAMU II is 360 MHz. This limit is determined by the four 54F164 shift registers (rated at 90 MHz), which are driven by four interleaving clocks.

The waveform sampler module processes right analog signals under the control to three signals provided by a central WSM fast clock generator (Figure 12.19). The three signals are (1) master reset, (2) a train of clock pulses (140 MHz for the drift chamber tests), which causes the acquisition of the data samples; and (3) a convert signal, which initiates digitization of the analog data.

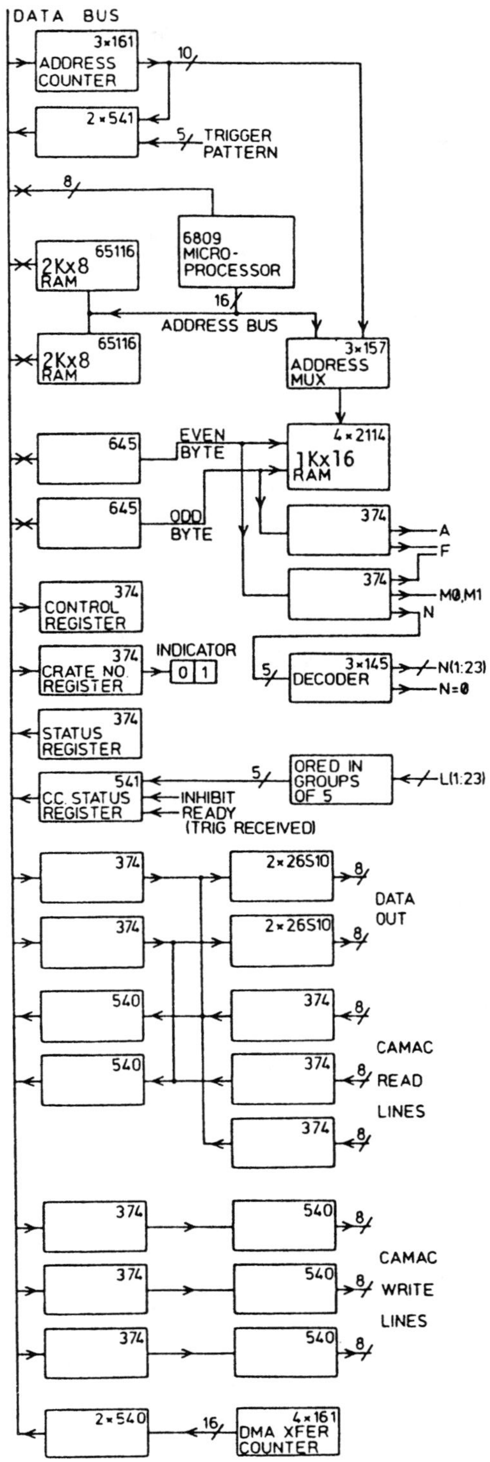

Figure 12.16 Block diagram of smart crate controller. (From Connetti, S. and Kuchela, K., *IEEE Trans. Nucl. Sci.*, NS32, 1326, 1985. With permission.)

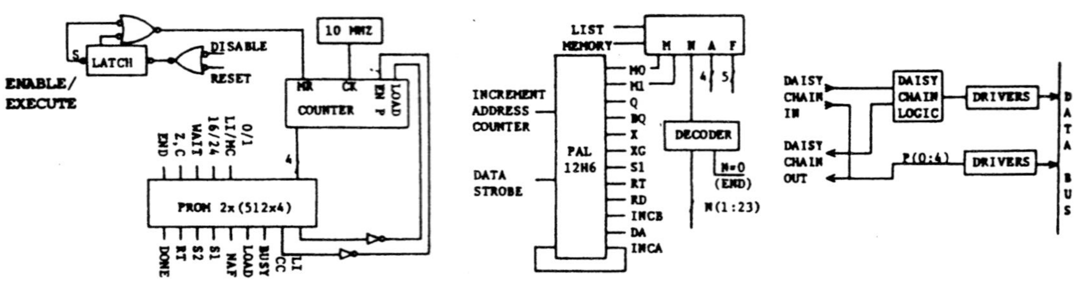

Figure 12.17 (A) Dataway cycle generator. (B) List processor. (From Conetti, S. and Kuchela, K., *IEEE Trans. Nucl. Sci.,* NS32, 1326, 1985. With permission.)

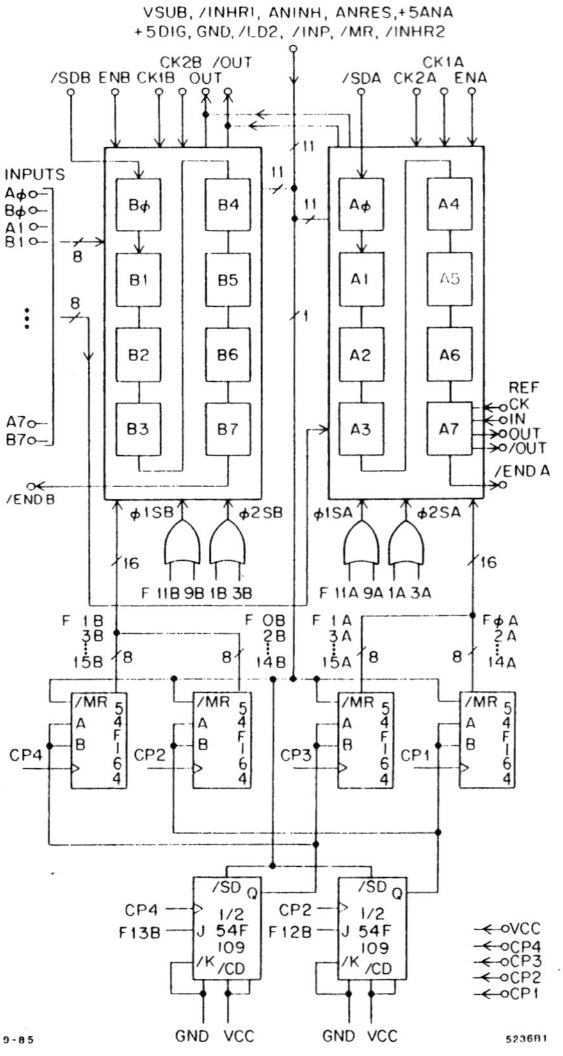

Figure 12.18 Block diagram of the hybrid HAMU II. AO,..., B7 are AMUs. The remaining devices are the 8-b shift-register 54F164, the J-K flip-flop 54F109, and the quad OR gate 74LS32. (From Freytag, D. R. et al., *IEEE Trans. Nucl. Sci.,* 33, 81, 1986. With permission.)

An acquisition cycle for analog data is started by the master reset (MR) signal which powers up VCC and +5ANA and initializes the logic (Figure 12.20). After a time interval (<1 μsec) the WSM receives a train of clock pulses, which are divided into four sets of interleaving clock pulses (CP_1, CP_2, CP_3, and CP_4) driving the hybrid. Writing into the hybrid proceeds in interleaved mode into the two AMUs assigned to each signal channel. After completion of the analog write cycle, power to the hybrid is dropped. The time constant for retention of charge on the sampling capacitors is on the order of 1 s.[8]

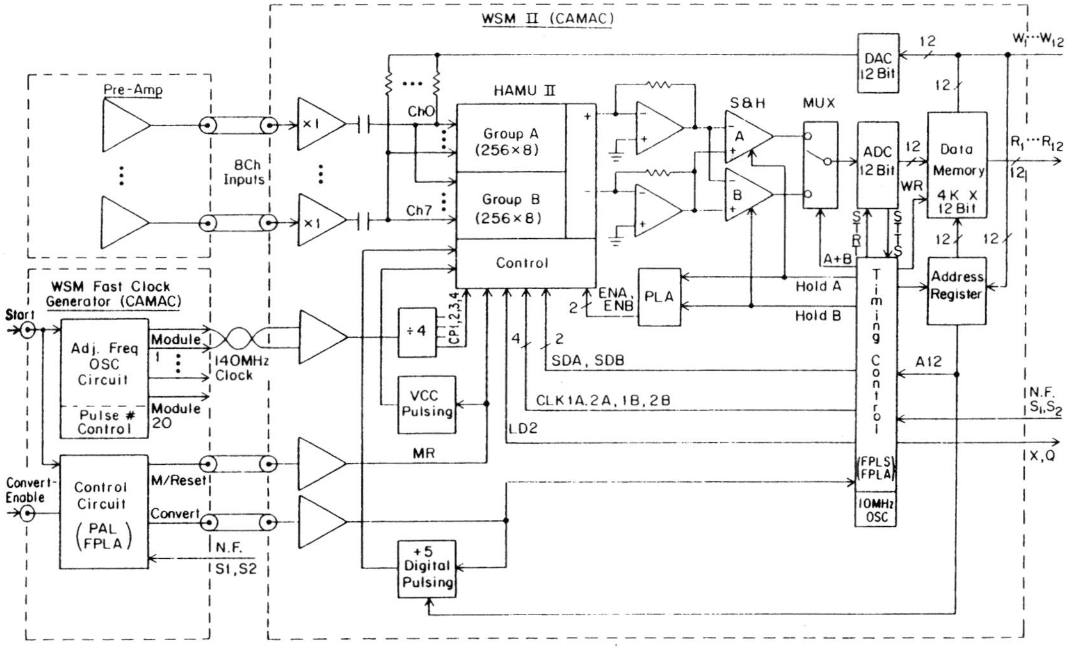

Figure 12.19 Block diagram of the CAMAC module WSM II. (From Freytag, D. R. et al., *IEEE Trans. Nucl. Sci.*, 33, 81, 1986. With permission.)

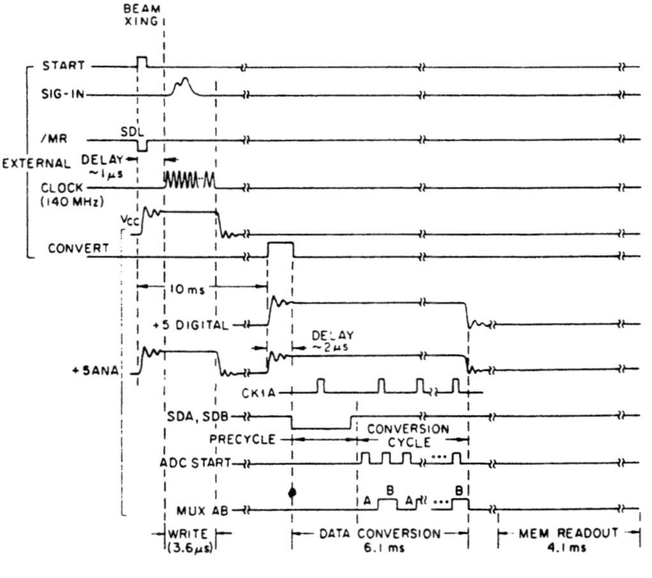

Figure 12.20 Timing diagram of external and internal signals in WSM II. (From Freytag, D. R. et al., *IEEE Trans. Nucl. Sci.*, 33, 81, 1986. With permission.)

A data acquisition system for recording multiparameter digital data into a large memory array at over 2.5 MHz is described by Hendricks et al.[9] The system consists of a MOSTEK MK8600 2048K × 24-b memory system, I/O ports to various external devices including the CAMAC dataways, a memory incrementor/adder, and a daisy chain of experiment-specific modules which calculate the memory address to be incremented. The design of the daisy chain permits multiple modules and provides for easy modification as experimental needs change.

A block diagram of implementation of the CAMAC interface is shown is in Figure 12.21 and the function codes to which it responds are given in Table 12.2. There are three on-board registers: the memory address register (AR), the memory address increment register (AIR), and the transfer count register (TC). There is also a data buffer which is filled in response to memory read requests. The data

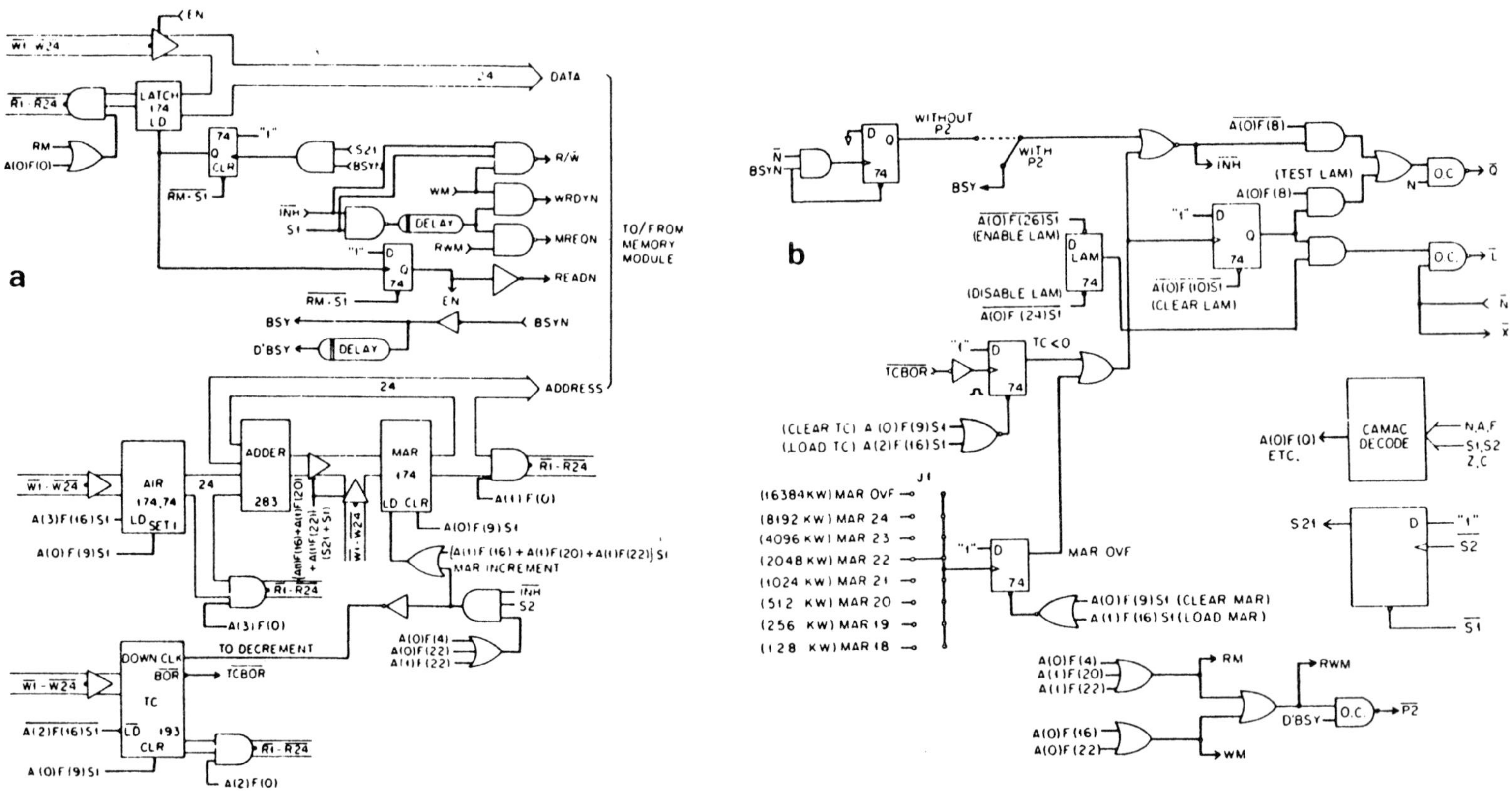

Figure 12.21 Block diagram of CAMAC-MOSTEK I/O port interface. (a) Memory module interface; (b) CAMAC strobe lines. (From Hendricks, R. W. et al., *Nucl. Instrum. & Methods,* 201, 261, 1982. With permission.)

Table 12.2 CAMAC Function Codes for MOSTEK I/O Port Interface

Function	Operation
A(0)F(0)	Read data buffer
A(0)F(4)	Read data buffer; AR = AR + AIR; TC = TC − 1; initiate memory read request; if TC < 0,AR overflow, or BSY then Q = 0; otherwise, Q = 1; LAM generated if enabled
A(0)F(16)	Write data buffer; initiate memory write request; if $\overline{\text{BSY}}$, Q = 1
A(0)F(22)	Write data buffer; initiate memory write request; AR = AR + AIR; TC = TC − 1; if TC < 0, AR overflow, or BSY, then, Q = 0; otherwise, Q = 1; LAM generated if enabled
A(1)F(0)	Read address register (AR)
A(1)F(16)	Load AR
A(1)F(20)	Load AR; initiate memory read request; if $\overline{\text{BSY}}$, Q = 1
A(1)F(22)	Load AR; initiate memory read request; AR= AR + AIR; TC = TC − 1; if $\overline{\text{BSY}}$, Q = 1
A(2)F(0)	Read transfer count register (TC)
A(2)F(16)	Load TC
A(3)F(0)	Read address increment register (AIR)
A(3)F(16)	Load AIR
A(0)F(8)	Test LAM; if present, Q = 1 (LAM generated on TC < 0 or AR overflow)
A(0)F(9)	Initialize; AR = 0, AIR = 1, TC = 0
A(0)F(10)	Clear LAM
A(0)F(24)	Disable LAM
A(0)F(26)	Enable LAM

From Hendricks, R. W. et al., *Nucl. Instrum. & Methods,* 201, 261, 1982. With permission.

buffer for memory write requests resides in the I/O port interface. Two types of CAMAC operations are decoded in the interface: those which actually initiate on I/O request to the memory I/O port, and those which operate on information contained exclusively in the CAMAC module. As can be seen in the timing diagram of Figure 12.22, when an I/O request is initiated, the I/O port immediately (within 30 nsec) asserts BSY. If there is an attempt to execute a CAMAC command which initiates an I/O request while the I/O port is busy, the CAMAC system may take one of two actions depending on the design of the crate controller. The module described here, on decoding A(0)[F(4) + F(16) + F(22)] + A(1)[F(20) + F(22)], immediately sets the crate controller line P2 low if BSY is asserted by I/O port (see Figure 12.22). If the crate controller has the P2 "stretch cycle" feature implemented and then if P2 goes low before the S1 strobe, the dataway cycle is held and does not resume until P2 returns high. Thus, dataway functions which initiate I/O port activity are forced to wait for completion of a prior request. The module generates a Q = 1 response to indicate successful completion of the requested function. If the crate controller does not respond to P2, the dataway cycle continues without pause. In this case, module (1) makes no data transfer, (2) does not initiate any I/O port request, and (3) generates Q = 0 for that dataway operation. Because the timing the logic is slightly different in each of these cases, a jumper-selectable logic was implemented for both. Clearly, it is the programmer's responsibility always to check for a Q = 1 response to be sure the I/O request was actually performed.[9]

In a data acquisition system described by Iqbal et al.,[10] the TTL outputs of the discriminators are stored in a memory bank housed in a single-width CAMAC module. Each module has 48 input lines to match the output of a single discriminator module. It can run by an internal or external clock at a maximum speed of 8 MHz and can store information in a programmable memory size up to 2048. These memory banks have sufficient intelligence in their control logic to select a valid event. Each CAMAC module starts storing data once it receives the start trigger generated from the anode wire signal. The modules reject an event if it receives a veto signal during the event or within the drift time before the event has started. The data are saved and a look-at-me (LAM) signal is produced only if a second trigger is received during

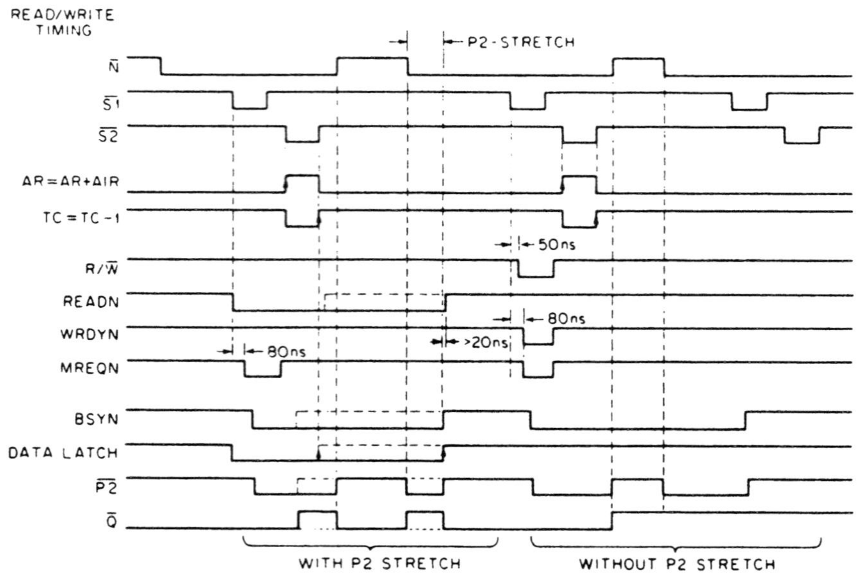

Figure 12.22 Timing diagram for CAMAC-MOSTEK I/O port bus. (From Hendricks, R. W. et al., *Nucl. Instrum. & Methods,* 201, 261, 1982. With permission.)

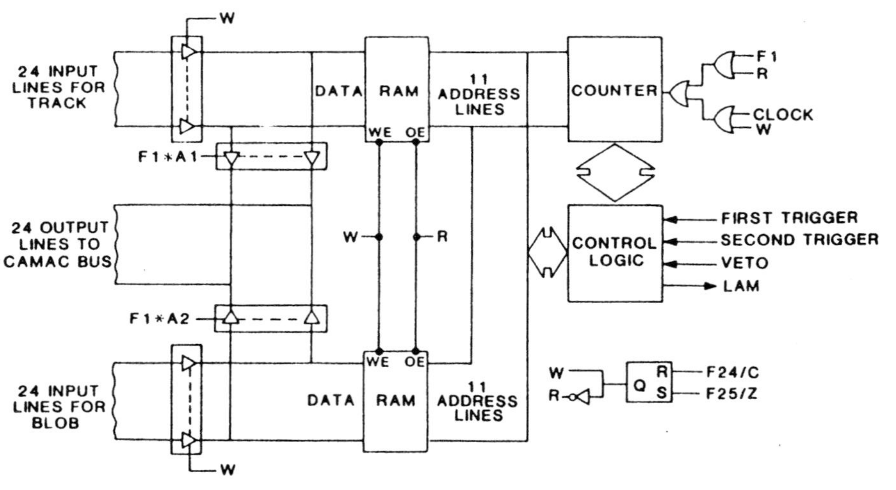

Figure 12.23 The schematic diagram of the CAMAC interface module showing the data handling process. (From Iqbal, M. A. et al., *Nucl. Instrum. & Methods,* A263, 387, 1987. With permission.)

the event. The CAMAC modules are designed so that the output of the discriminators for the low- and high-threshold levels are stored in two different memory blocks and can be read independently.

The schematic diagram in Figure 12.23 shows the basic operation of the CAMAC interface. The module can either be in the WRITE state (W line is low) or the READ state (R line is low). The WRITE state can be selected using the CAMAC command F24 (or C for the whole crate) which enables the two input buffers and the RAM ICs. Once the control logic unit receives the proper trigger signals, it allows the clock pulses to increment the counter which in turn advances the RAM address for a predetermined time. The trajectory information is stored in memory location as a function of time in the increasing address of the RAM IC. At the end of an event the control logic produces a LAM. At this point the status is changed into READ (R is low by the CAMAC command F25 (or Z for the whole crate) which disables

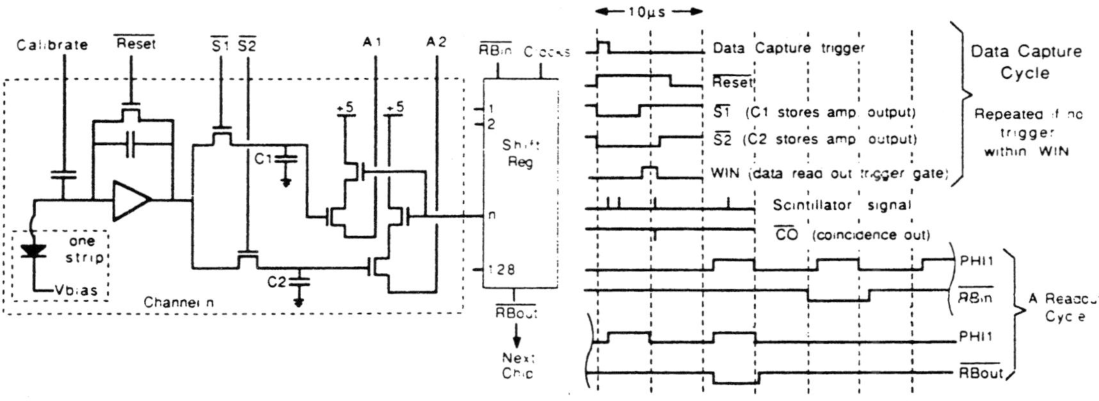

Figure 12.24 Schematic (a) and timing (b) diagrams. (From Joyce, D. et al., *Nucl. Instrum. & Methods,* A279, 189, 1989. With permission.)

the input buffer and enables the output of the RAM ICs. Now the modules are checked to see whether they contain information in any block by the CAMAC command F8 × A1 for the track and F8 × A2 for the blob. Those blocks with data are then read by F1 × A1 and F1 × A2, respectively, when the counter is incremented by F1 pulses.

The RAL Microplex readout chip was used by Joyce et al.[11] to test a silicon microstrip detector. The readout chip provided amplification with double-correlated sampling for each microstrip channel and multiplexing of 128 signals.

The RAL MX2 Microplex chip is a very large system integration (VLSI) integrated circuit chip using complementary metal oxide semiconductor (CMOS) technology with a 3-μm feature size. It is a newer version of the Microplex chip using NMOS technology. CMOS has a low-power consumption, enabling its use in a >10,000-channel microvertex detector for collider experiments.[12]

The RAL chip, 6.0 × 6.4 mm^2, has 128 input bonding pads on one end and 25 bonding pads on the other end for control lines, and ground and power supplies.

The readout chip provides a double-correlated sample of the charge collected on each of the 128 strips. The output of a current-integrating amplifier is recorded on two storage capacitors; one holds the value seen just before a possible signal arrives and the other, just after. During readout, signals from these capacitors are treated as a differential pair and subtracted, thus canceling out noise characterized by a frequency lower than 1 over the sampling period. In particular, it reduced the effects of leakage current (a white noise source).

A simplified circuit diagram of one channel and a basic timing diagram appear in Figure 12.24. A data capture trigger into the driving box generated a data capture cycle. The $\overline{\text{RESET}}$ control line low zeroes the output of the amplifier; and the $\overline{\text{S1}}$ and $\overline{\text{S2}}$ signals low places the output of the amplifier on C1 and C2, respectively. After 4 μsec C1 was disconnected by the rise of S1 and retained the voltage value it had at that time. C2 was allowed to follow the output value for an additional 2 μsec and was then disconnected. In order to test operation of the chip at the SPS bunch crossing rate, the entire data capture cycle has been run as short as 3.8 μsec.

Multiplexing was performed by a 128-stage shift register integrated onto the chip. The two analog levels stored on C1 and C2 were connected to the gates of high input-impedance field effect transistors (FETs), whose power supplies were switched on in sequence as the shift moved the "read bit" through each amplifier stage. The two signals were converted from current to voltage at the preamplifier stage and then subtracted at the analog stage inside the SIROCCO digitizing module.[11]

The SIROCCO is a CAMAC module specially designed to digitize and record the difference between the two analog signals. It is designed to digitize and store data for up to 2048 detector channels.

A 16-channel CAMAC module has been designed and constructed by Crittenden et al.[13] The unit is standard double with a CAMAC module. The block diagram is shown in Figure 12.25. The module has the following features (Table 12.2): (1) the input channels have a common voltage threshold; (2) the multiplicity calculation and comparison is done with emitter-coupled logic (ECL) digital logic; (3) the voltage and multiplicity thresholds are set manually; and (4) NIM gate and reset inputs are required.

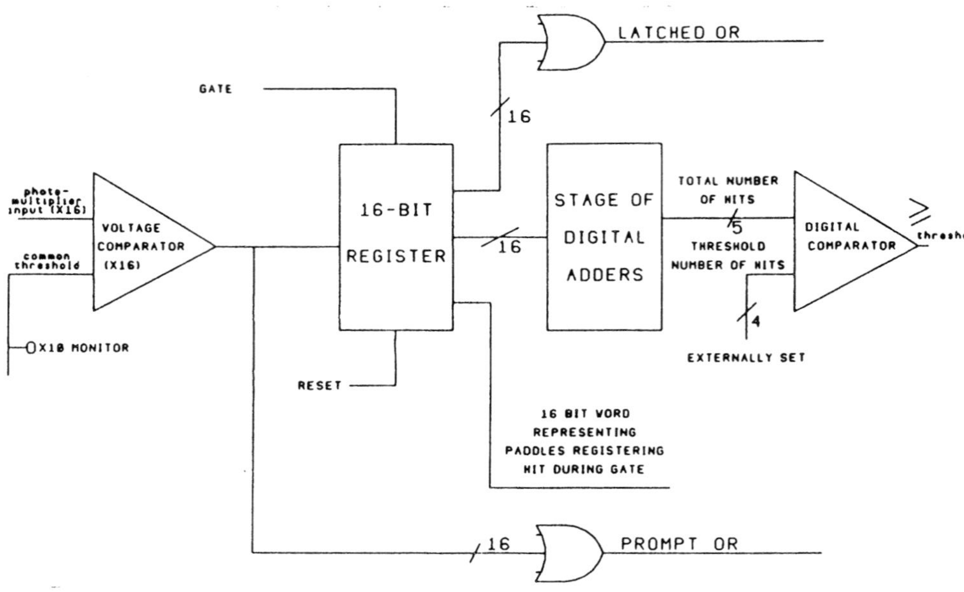

Figure 12.25 A 16-channel CAMAC module. (From Crittenden, R. R. et al., *Nucl. Instrum. & Methods,* A270, 99, 1988. With permission.)

Advanced Micro Devices AM 687 voltage comparators discriminate signals from noise. The common threshold voltage for the discriminators, which may be set between 30 and 500 mV, is adjustable by a potentiometer screw. The register is made up of 10135 ECL JK flip-flops.

A versatile and flexible data acquisition system called the Kakuken Online Data Acquisition system (KODAQ) has been developed by Omata et al.[14] The system runs with CAMAC and a personal computer, PC9801 (NEC), similar to the IBM PC/AT. The system is designed to easily set up a data acquisition system for various kinds of nuclear physics experiments.

Figure 12.26 shows the program flow. The configuration of the KODAQ system is normally defined in two files: one is a text file INITKDQ.KDQ consisting of KODAQ commands and the other is PROCAMAC.C written in the C language.

The INITKDQ.KDQ is called and executed at the system start and defines all environmental conditions such as: how many and what kind of display windows are to be opened at which part of the CRT display, how many areas with what memory size are to be reserved for data storage, and which data are to be displayed in these windows.

The other file written in the C language contains CAMAC data-read and monitoring routines which must be compiled and linked to the main program as a part of the interrupt handler in advance.

A flow diagram of the interrupt-handling procedure is shown in Figure 12.26(a). In data-acquisition-enabled state, the interrupt-handling routine is activated as a foreground task each time an event-trigger signal is sent to the CPU.

The hardware setup of the KODAQ is shown in Figure 12.27. The interrupt module and output module are indispensable. File devices, fixed disk, magnetic optical (MO) disk, and RAM disk are optional for raw-data dump. CAMAC crates are extensible up to four crates, using the crate controller CC/7000 in each crate. All the CPU models of the personal computer PC9801, 8086/V30 to 80386, with clock frequency up to 33 MHz, have been found to run the KODAQ system without any problem. The interrupt module at Station No. 1 is used to realize the multievent triggering. Only the input signals to the interrupt module generate the LAM signal and interrupt the system. The LAM signals from all other modules are inhibited. The output module at Station No. 2 is used to send out a computer busy signal. These two modules are the minimum required for the standard KODAQ system.

A fast data acquisition system was designed by Conetti et al.[15] composed of several CAMAC smart controllers sending their data in a combined serial/parallel organization to a VME-based data collection and forwarding system. It was designed to handle data rate in excess of 10^5 16-b CAMAC words per

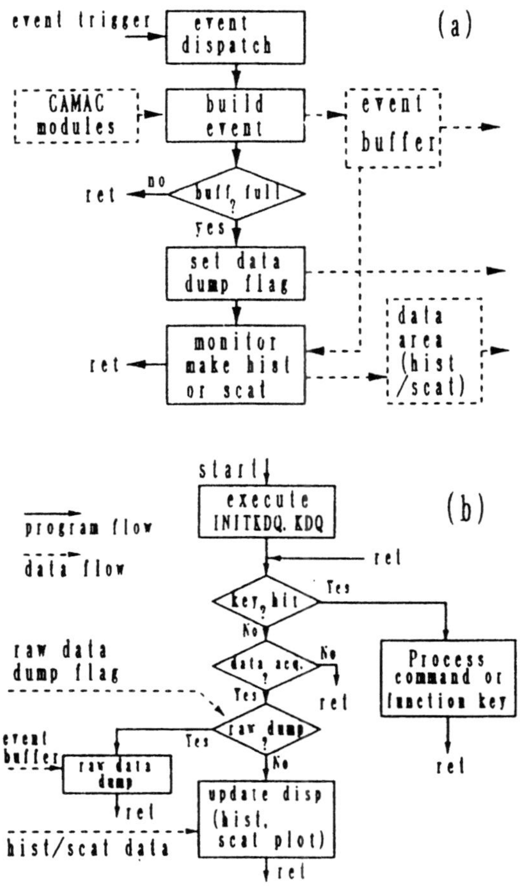

Figure 12.26 (a) Foreground and (b) background program flow of the KODAQ system. (From Omata, K. et al., *IEEE Trans. Nucl. Sci.*, 39, 143, 1992. With permission.)

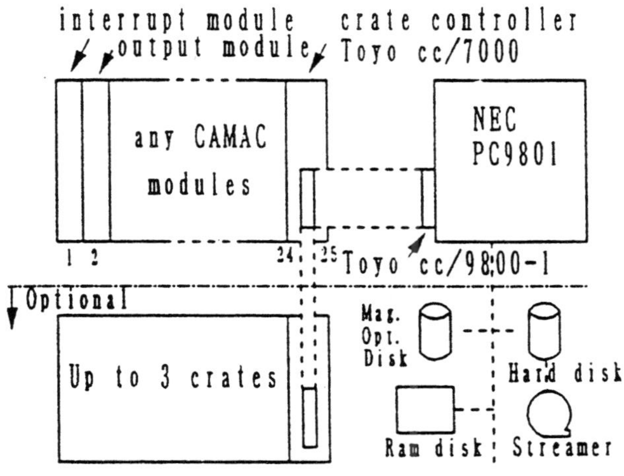

Figure 12.27 Minimum hardware requirement of the KODAQ. (From Omata, K. et al., *IEEE Trans. Nucl. Sci.*, 39, 143, 1992. With permission.)

second with peak rates of up to 2×10^6 words per second. The dead time due to data acquisition in the experiment is kept to less than 20%.

The system consists of four to six parallel channels of CAMAC crates, which are read concurrently into a set of channel buffer memories contained in a VME-based data collection system (Figure 12.28). The crates are read out at maximum CAMAC bus speed (1 MHz) by smart CAMAC crate controllers. By arranging the crates in parallel, a typical E705 event of ~2000 words can be digitized and unloaded

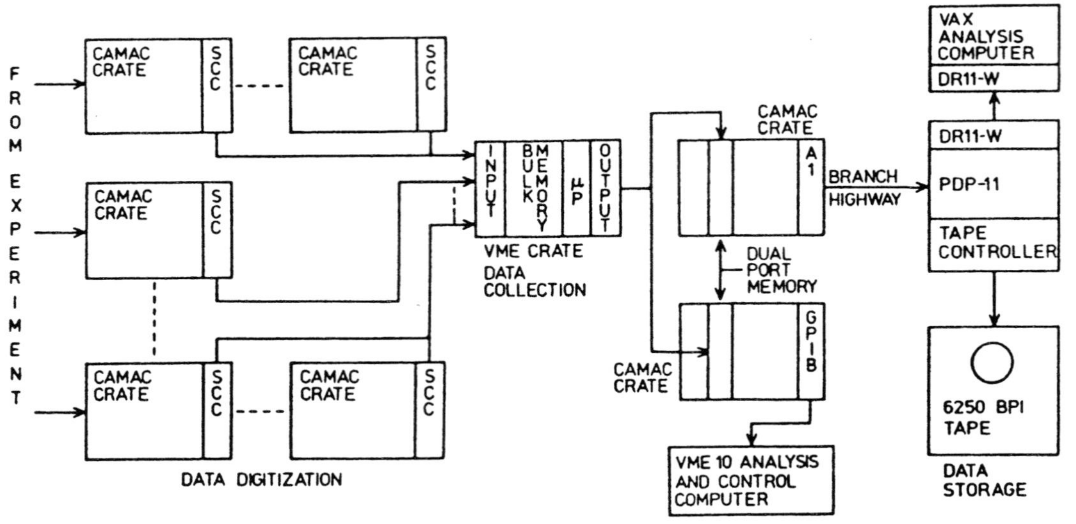

Figure 12.28 Data flow in E705. (From Conetti, S. et al., *IEEE Trans. Nucl. Sci.*, NS32, 1318, 1985. With permission.)

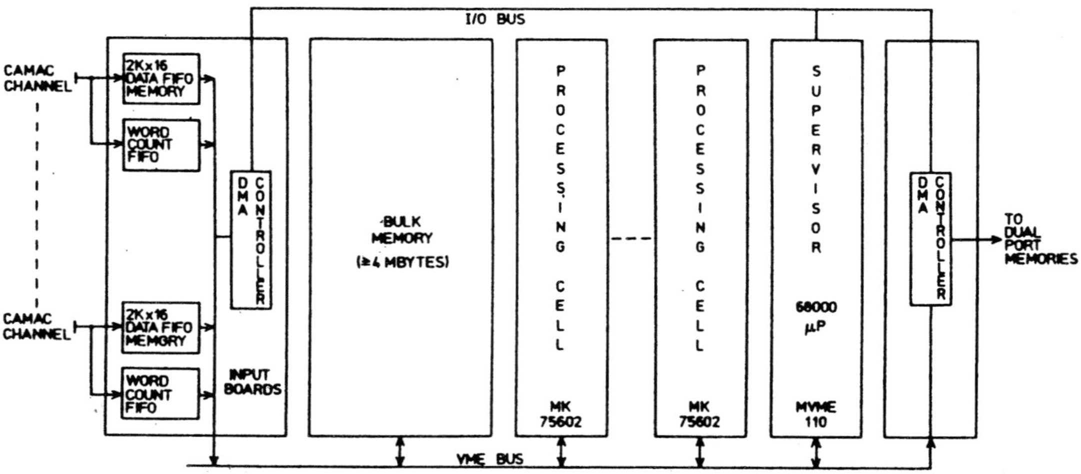

Figure 12.29 Data flow of VME crate. (From Conetti, S. et al., *IEEE Trans. Nucl. Sci.*, NS32, 1318, 1985. With permission.)

from CAMAC in less than 1 msec. The VME system collects the parallel stream of data from the buffer memories, arranges it into separate events, and adds a data descriptor to each event.

The VME system is composed of a single VME crate containing a Motorola MVME 110 single-board computer which acts as system supervisor, input and output boards, a large bulk memory, and (optionally) several MOSTEK MK75602 single-board computers which can be used for additional data processing (Figure 12.29).

The input boards house a 2 K × 16 buffer memory for each CAMAC channel. This memory is treated as a FIFO buffer by using a Signetics 8 × 60 FIFO controller (SCC) which can thus write sequentially into the memories simply by strobing the data. After all the data has been transferred from a crate, the SCC transmits a word count, which is loaded into a separate 32 × 18 word count FIFO for that channel. The only control the VME system has over the CAMC data transfer is a WAIT line to the SCC, which is activated when the memories are full or whenever the supervisor wants to suspend the data flow.

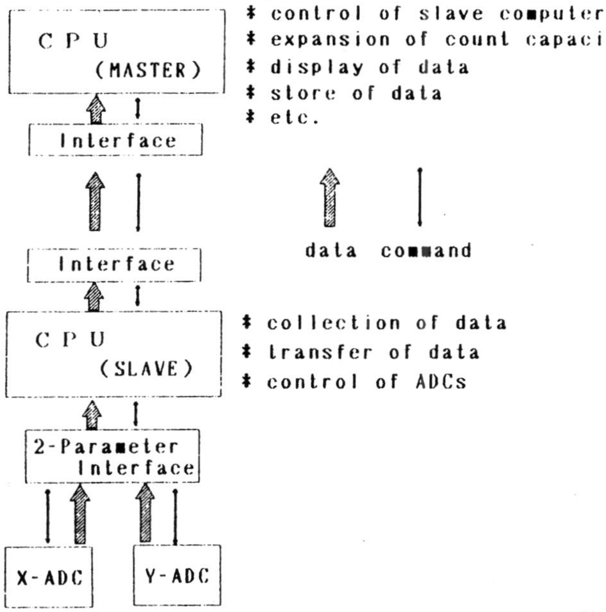

Figure 12.30 Block diagram of system. (From Akimoto, T. et al., *IEEE Trans. Nucl. Sci.*, 37, 1594, 1990. With permission.)

A dual-parameter multichannel pulse-height analyzer (MCA), which uses two central processing units (CPUs), was developed by Akimoto et al.[16] One computer (master computer) controls the other computer (slave computer), and also displays, records, and analyzes the data. The slave computer controls two analog-to-digital converters (ADCs), collects the dual-parameter data, and then transfers the collected data to the master computer. Three methods of data transfer between the two CPUs are examined and discussed. The count capacity at the data collection by the slave computer was 2^{16}-1/ch (2-byte-long data), and this was expanded to 2^{32}-2/ch (4-byte-long data) at the master computer.

Large data array handling was simplified by flexible application of the segment register of the 820286 CPU. The processing speed was improved by transferring data from the ADC to CPU in 16-b words and by a simultaneous reading of the status flag and data. The graphic display processes were speeded up by directly writing the data corresponding to locations into the graphic video random access memory (VRAM).

The system is composed of two ADCs (Toshiba NAIG E-551, 50-MHz Wilkinson type); three parallel peripheral interface boards (Interface Company Ltd. 98PPI(I/M)AZI-120, or ELM DATA Company Ltd. EP-98DIOA); and two personal computers (NEC Corporation PC9801VX21 and PC9801UV2). The master computer (PC9801UV2, CPU: μPD70116-10, equivalent to an 8086) controls the slave computer (PC9801VX21, CPU:80286) and displays, records, and analyzes the data. The slave computer controls the two ADCs, collects the dual-parameter data, and transfers the data to the master computer under the control of the master computer. The slave computer is operated by the master computer except for booting of the operating system (OS) and loading of the program. A program written in Basic is mainly employed, but a program written in machine language and called by the Basic program is also used when high-speed processing such as collection of dual-parameter data, graphic display of the data, and data transfer, etc. are required. A block diagram of the system is shown in Figure 12.30.

Data reading and processing involve the slave computer controlling the two ADCs and collecting the data through a parallel peripheral interface board. The speed of data processing is increased by transferring data from ADC to CPU in 16-b words and by a simultaneous reading of the status flag (ADC-BUSY signal and DATA-READY signal) and data. This dual-parameter MCA system can handle large data arrays of 128 kbytes. The dual-parameter information is graphically displayed during measurement with a 16-color contour display or an isometric display at the master computer by transferring the data as necessary to the master computer.

Data transfer between two CPUs is the process of data collection in the slave computer being interrupted when the data from the slave computer are transferred to the master computer. Therefore, the time required to transfer data strongly affects the performance of the system.

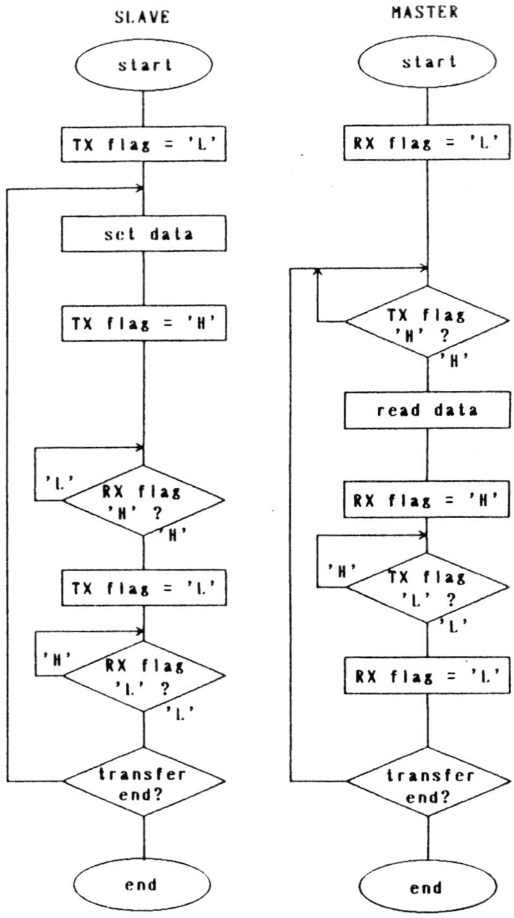

Figure 12.31 Flow chart of data transfer routine by software handshaking. (From Akimoto, T. et al., *IEEE Trans. Nucl. Sci.*, 37, 1594, 1990. With permission.)

The software handshaking method is highly flexible for data transfer because the logic of signal and timing of data transfer can be decided freely by the software. It is particularly effective when the processing speed of CPU and peripheral equipment are different. However, this method loses time when changing the state of flags by the software.

In the case of the method with software handshaking, commercially available parallel peripheral interface boards (Interface Company Ltd. 98PPI(I/M)AZI-120) are employed as the interface between the master and slave computer. Mode 0 of the two programmable peripheral interfaces (PPIs) installed on the board are adopted. This PPI is a μPD71055 (equivalent to 8255) with three parallel 8-b ports. Mode 0 is the basic operation mode that all ports can use as input or output lines. The flow chart and the timing diagram of the data transfer routine are shown in Figure 12.31 and Figure 12.32, respectively. Here the TX flag indicates transmission of data from the slave computer, and the RX flag indicates reception at the master computer.

First, both flags TX and RX are set to L. The slave computer prepares the data and informs the master computer of the situation (TX flag = H). The master computer confirms the TX flag, reads the data, and informs the slave computer of data acceptance (RX flag = H). The slave computer confirms the data acceptance by the master computer and informs the master computer about the state of preparation for data setting (TX flag = L). The master computer confirms preparation of data transmission from the slave computer (RX flag L). The slave computer sets the new data after confirming the state of the master computer. Reliable data transfer can be attained without duplication and loss of data by this handshaking.

Automatic handshaking is a method in which automatic handshaking is carried out by the hardware of the interface board. Data are transferred by automatic changes of flag states corresponding to the data input and output. The same board, 98PPI(I/M)AZI-120, as employed in software handshaking, was also used as the interface board with Mode 1. Mode 1 is the operation mode that each port can use as input or output lines and that has control and status lines corresponding to the input and output. In automatic

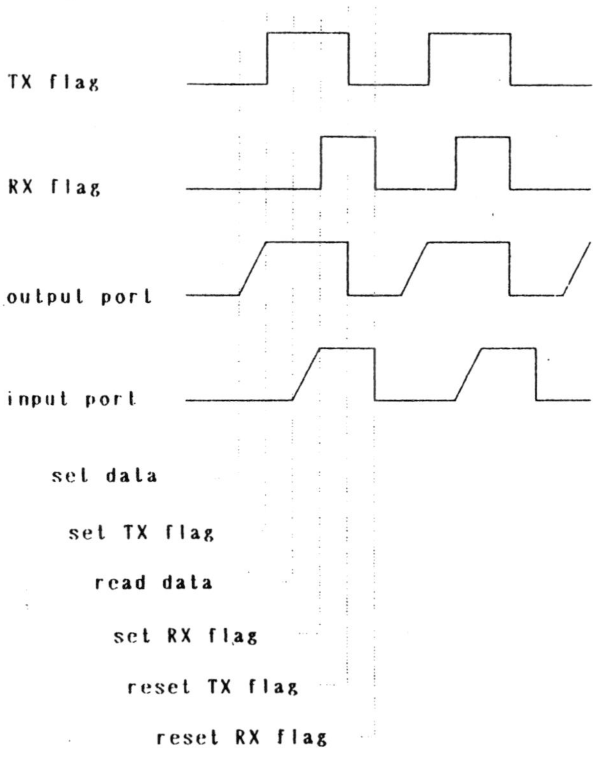

Figure 12.32 Timing diagram of data transfer routine by software handshaking. (From Akimoto, T. et al., *IEEE Trans. Nucl. Sci.*, 37, 1594, 1990. With permission.)

handshaking, only a check for permission to input and output data and transfer of the data are performed by the software. The slave computer confirms that the WINT state (Write Interrupt, C3 port of the PPI on the board) becomes H and sets the 16-b data to the A ports. The master computer confirms that RINT (Read Interrupt, C3 port of the PPI on the board) is H and reads the 16-b data from the A ports. The C6 port of the slave computer and C4 port of the master computer must be set to H. The flow chart is shown in Figure 12.33.[16]

12.2 FASTBUS

12.2.1 INTRODUCTION TO FASTBUS

FASTBUS is a sophisticated data acquisition system standard developed by the U.S. NIM Committee in collaboration with the European ESONE Committee (ANSI/IEEE STD 960-1986). FASTBUS was designed to keep features of older important standards while extending the capabilities of data acquisition systems. FASTBUS provides for a more densely packed system, reducing dramatically the per-input cost. This and other design goals have been achieved. FASTBUS meets the requirements of the next generation of data acquisition systems by incorporating several powerful features, including:

- Modularity, 32-b address and data fields
- High-speed, asynchronous ECL backplane with 32-b wide dataway
- Multiple, parallel processor bus architecture with multiple bus segments that operate independently but link together for passing data
- Asynchronous handshaking for compatibility with modules having different data transfer speeds operating on the same bus
- Synchronous nonhandshake data transfer for maximum speed
- Broadcast operations for initializing, clearing, etc. several modules in one operation
- A polling structure for fast scanning of sparse data from a large number of modules
- Easy links to computers for host intervention or data transfer

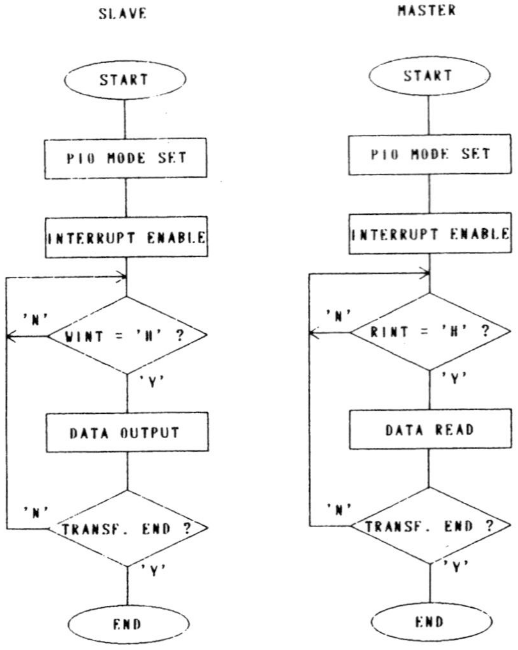

Figure 12.33 Flow chart of data transfer routine by automatic handshaking. (From Akimoto, T. et al., *IEEE Trans. Nucl. Sci.*, 37, 1594, 1990. With permission.)

Essentially, the FASTBUS backplane is intended to be an extension of a computer backplane. By its nature, therefore, it is an expandable system. The backplane is called a segment. Connections between segments are made by an interface, called a segment interconnect (SI). An SI in one segment is connected to a SI in another segment by a cable segment. The segment interconnect can be either a master or slave, depending on the operation it is performing. Moreover, it can be a master on the FASTBUS segment and simultaneously a slave on the cable segment. The designations master and slave here are dynamic, in that the designation is defined by the role played by the SI at some given time.

FASTBUS permits multiple masters to have access to the segment. The standard provides a protocol for arbitration when more than one master requests control of the segment. Once mastership is granted, the master may then proceed to establish a communications "lock" with any other device on the segment which will act as a slave. Alternatively, the master may broadcast a message to all slaves which respond to the broadcast (e.g., clear, polling for data ready, etc.).

The integrity of communication between master and slave modules is ensured via a handshaking scheme. This frees the standard from a speed requirement. Communication occurs at whatever speed the master/slave system will support. The versatility and expandability are sufficient to meet the most demanding requirements of data acquisition systems.

The FASTBUS power standard allows for +5, –5.2, –2, ±15, and +28 V supplies. The backplane of the FASTBUS crate consists of two parts, the segment and the auxiliary backplane. The FASTBUS segment is the high-speed digital data bus of the system. It provides for the control and flow of data. On the rear of the crate are two cards that are called ancillary logic. The geographic address control card (GAC) is used to terminate the FASTBUS signal lines with 100-Ω resistors. It should be located in either slot 24 or 25 on the rear side of the FASTBUS backplane. The arbitration timing control card (ATC) supplies the required timing signals for bus arbitration and broadcast addressing. It should be located in either slot 0 or slot 1 on the rear side of the FASTBUS backplane, and also terminates the bus in 100 Ω.

Because a segment may contain more than one master, FASTBUS provides for arbitration between them to determine the order in which they will gain control over the bus. The master asynchronously requests use of the segment, specifying the priority of the request. This initiates an arbitration cycle. On obtaining mastership, it performs its operations.

The FASTBUS segment includes a 32-b wide data bus, used both for data and for addressing. Any slot within a FASTBUS crate can contain a master. For this reason sophisticated addressing schemes have been implemented in FASTBUS. A master first must establish a communication lock with one or more modules during a primary address cycle. Three methods of primary addressing are implemented in FASTBUS: geographic addressing, logical addressing, and broadcast addressing. In each case the primary address cycle causes one or more slaves to become attached to the master and to participate in subsequent data cycles.

Primary addressing includes geographic addressing of modules referring to the establishment of communication between a master and a slave in a particular slot. Logical addressing provides the user with the flexibility to relate a module address to a logical name. Broadcast addressing is the method used for communication with more than one slave. Secondary addressing involves specifying an address to an attached slave during a data cycle, and is indentified by a mode select (MS) code.

Two address spaces are defined in FASTBUS: control space and data space. The space being accessed by a given FASTBUS operation is defined by the status of the MS lines at the time of the primary address cycle. Once an address lock is established with a slave, all data cycles access control space or data space, but not both.

To reduce system dead time and increase data throughput, the AUTOBLOCK transfer protocol is used. The AUTOBLOCK mode uses the FASTBUS protocol to transfer data at high rates even in the presence of a relatively slow FASTBUS master. Normally when a module has completed data acquisition and is ready to transfer data to the event builder, it must wait until a FASTBUS master decides all modules are ready and then takes the data from them, usually in a fixed sequence. With the AUTOBLOCK feature, the module can be programmed with a destination address and enabled. As soon as the data are ready, the module becomes a master on the FASTBUS and initiates a block data transfer to its preprogrammed destination. Thus the transfer of data to the event builder does not need to wait for the slowest module, usually the one with the most data, but may begin with the fastest.

A system using the AUTOBLOCK protocol will execute the following steps:

- A FASTBUS master initializes all data acquisition modules which use the AUTOBLOCK protocol. This includes setting the destination address, master status codes, and arbitration priority for each module.
- A FASTBUS master receives a command from the external trigger logic indicating that an event is ready to be read out. This master then arbitrates for the bus (at a low priority, possibly waiting for a previous readout to complete) and then broadcasts a command (setting a CSRO bit in all slaves using the AUTOBLOCK protocol) to initiate an AUTOBLOCK transfer.
- All AUTOBLOCK data modules which have data for the current event (or a header indicating no data) arbitrate for the FASTBUS (temporarily becoming the FASTBUS master) and transfer one block of data to the programmed destination address.
- The FASTBUS master notes the completion of the entire event readout by its ability to win an arbitration for the FASTBUS (at a priority lower than any of the data modules). At this time any required end protocol is completed.[1]

12.2.2 FASTBUS APPLICATIONS

An application of FASTBUS was described by Verweij.[17] The detector was composed of two large planar drift chambers, a liquid argon calorimeter, a hadron calorimeter, and some other elements. The first level (real-time) trigger rate was $\approx 5 \times 10^4$/s, which should be reduced to ≈ 300/s by multilevel processing. FASTBUS was found ideal because of its flexible data transfer possibilities at a high speed.

A symbolic diagram of the system is shown in Figure 12.34. It is a two-crate arrangement, with approximately 30 modules of 12 different types.

The motivation for choosing FASTBUS for front-end electronics (ADC, TDC, etc.) is primarily because the valid data is sparsely distributed over the detector channels, and FASTBUS offers the possibility of scanning these channels at high speed. Filtered and formatted data from the front-end electronics are then buffered and made easily accessible for second, third, and possibly higher level trigger processors and emulators. In this part of the data acquisition system it is essential that data can be efficiently transferred between the various buffers and processors. Here, FASTBUS offers the required large addressing space, high speed, multimaster capability with flexible arbitration, a segmented structure allowing undisturbed parallel processing and setting up of data routes, etc.

The starter kit is composed of the following elements: crate with power supply and cooling, ancillary logic boards, CAMAC to FASTBUS interface FIORI (European version of IORFI11), CAMAC I/O register (active extender board), bus monitor, FASTBUS display module with test box extensional, kluge boards with or without bus interface, and software (FASTBUS extensions to ESONE CAMAC Fortran). A symbolic diagram of a starter kit implementation is shown in Figure 12.35.

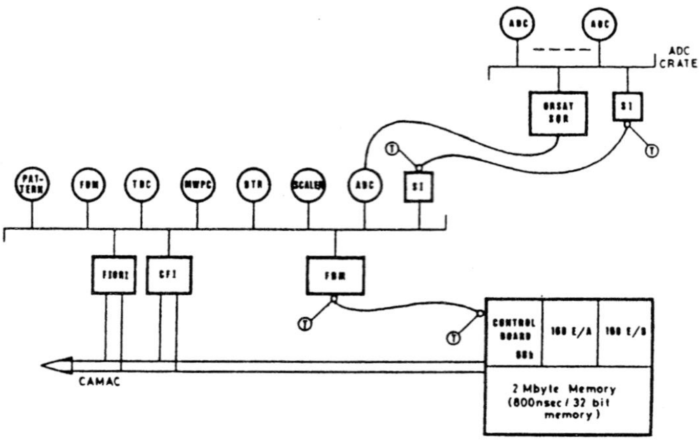

Figure 12.34 Block diagram of the FASTBUS system for the experiment NA-31. FDM, display module; MWPC, multiwire proportional chamber read-out; DTR, drift time recorder; FBM, data block mover; Orsay sqr, sequencer; other items go without explanation. (From Verweij, H., *IEEE Trans. Nucl. Sci.,* NS31, 211, 1984. With permission.)

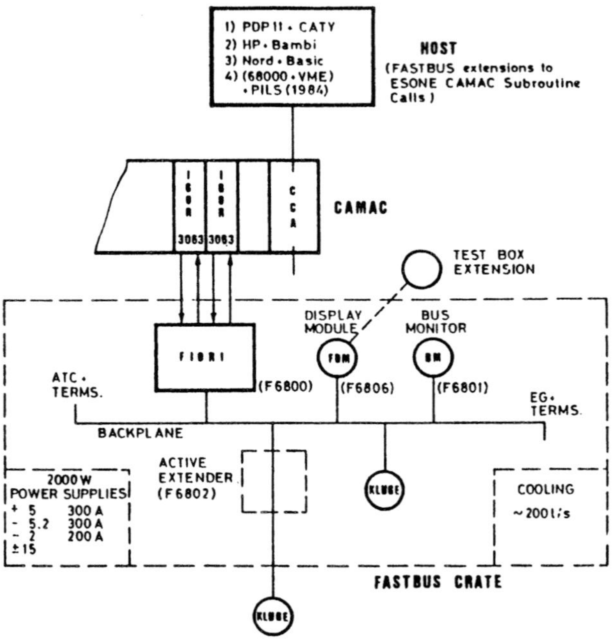

Figure 12.35 Block diagram of FASTBUS starter kit and its connection via CAMAC to a host computer. (From Verweij, H., *IEEE Trans. Nucl. Sci.,* NS31, 211, 1983. With permission.)

The design of a general-purpose CAMAC to FASTBUS interface (CFI) is based on the M6800 for port-to-port control and the AMD 2910 for handling the external bus protocols. It contains a buffer memory to decouple the CAMAC and FASTBUS cycles, and offers therefore a higher instantaneous data throughput than the FIORI.

The FASTBUS readout module for the SLD limited streamer tubes was described by Beconcini et al.[18] To be able to keep under control 10^5 channels of digital electronics, a readout module was designed with the following aims:

1. Minimizing the number of modules and connections needed for the whole device
2. Achieving a sufficiently high speed to be able to read the strips twice per beam crossing, the first time searching for e^+e^- annihilation data and the second time looking for cosmic rays
3. Maximizing the amount of preliminary data elaboration on the module itself, thus reducing the load on the other links in the data acquisition chain.

The above requirements led to the design of an intelligent FASTBUS module (WICDRM) based on the Motorola MC68020 microprocessor, to read the previously described front-end electronics for streamer tube cathode strips.

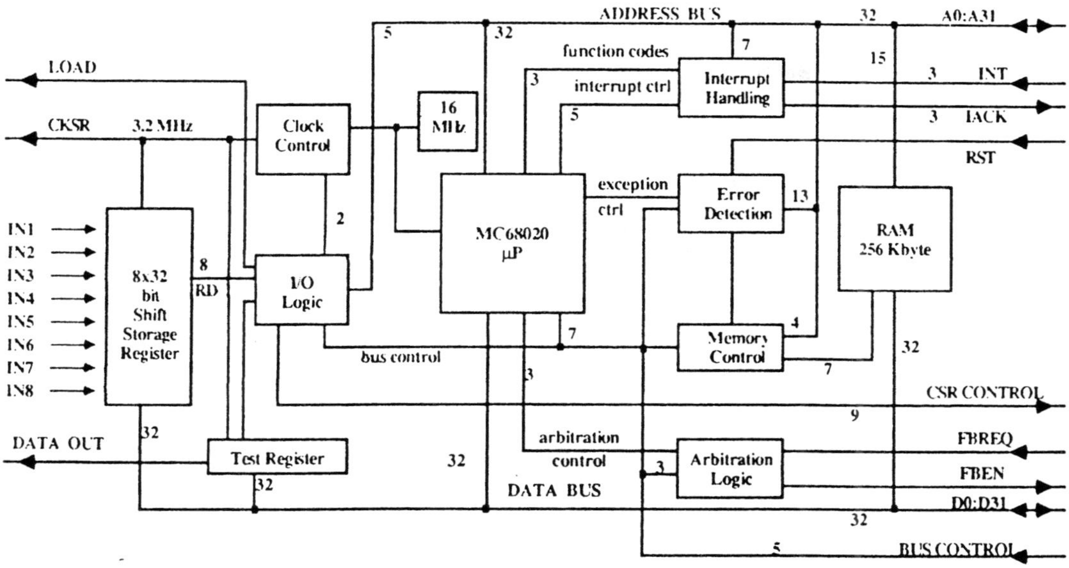

Figure 12.36 Block diagram of the WICDRM front-end processor. (From Beconcini, F. et al., *Nucl. Instrum. & Methods,* A277, 222, 1989. With permission.)

The strip information, digitally stored into the front-end shift registers, is daisy chained and clocked with a 3.2-MHz (and possibly faster) rate into eight parallel input buffers inside the module, under the full control of the processor.

The module is a FASTBUS slave and can be addressed primarily in two modes, geographic and broadcast. It contains 128 kbytes of program memory and 128 kbytes of data memory. Both memories are mapped in data space and can be accessed from FASTBUS in both WRITE-and READ-cycle modes. The microprocessor operates at its maximum clock speed of 16 MHz, and the memories are static RAMs (32K × 8.85-nsec access time) which introduce no wait cycles.

A block diagram of the front-end processor is shown in Figure 12.36. The signals drawn on the left of the figure are directed to the front-end SGS electronics; those on the right are the internal connections with the FASTBUS interface.

On reception of an interrupt, the processor can send a LOAD signal to the front-end electronics, latching the digital information. Subsequently the shift-register clock (CKSR) is started and serial data from all eight input channels (IN1–IN8) are shifted into eight 32-b wide latches. When the eight input buffers are full, the processor transfers the latched data into memory while optionally performing a preliminary zero suppression. The readout procedure continues until all the daisy-chained data have been transferred and processed. If the trigger is accepted, the zero-suppressed data are further elaborated by the MC68020. The microprocessor evaluates the size of each cluster of ajacent strips fired, and its location inside the apparatus. This information, properly coded, is read by the master and eventually transferred to mass storage.[18]

The module also has a 32-b read-write register, provided for the test of the front-end shift-register chain. It is possible to load any 32-b pattern and serially inject it at the beginning of each chain. The memory is divided into two banks, one of which is dedicated to programs and the other to data storage.

The arbitration logic handles the FASTBUS request (FBREQ) to access the memory by generating a bus request (BR) to the MC68020. The processor answers by asserting the bus grant (BG) to indicate that the bus will be available at the end of the current bus cycle, and enters an idle state. The BG generates a FASTBUS enable (FBEN) signal that opens the control, data, and address bus buffers, giving FASTBUS access to the memory.

A FASTBUS master module (68K20FPI) based on a 68020 microprocessor was developed by Arai and Uehara.[19] By introducing the 68K20FPI interface in the DA crate at the electronics hut, all of the data are read out by the 68K20FPI module. Although the maximum data transfer speed of the module is almost the same as that of the VAX-FPI, due to the short length of cable segment between the buffer module and the 68K20FPI and very little overhead, the effective speed has been increased. All of the data can

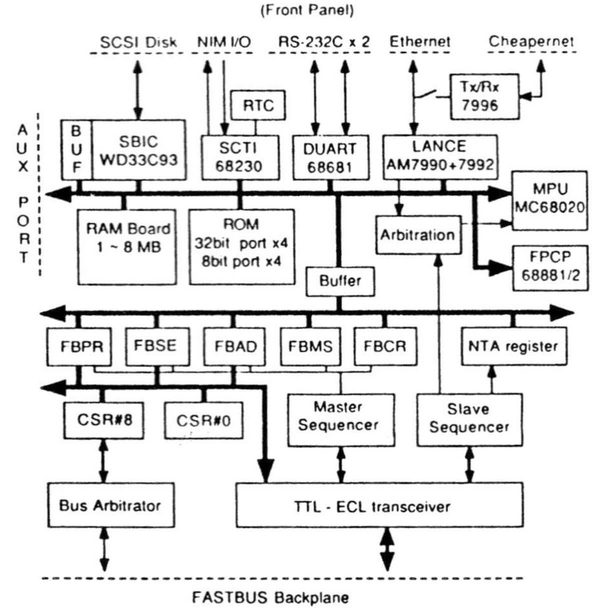

Figure 12.37 Block diagram of the 68K20FPI module. (From Arai, Y. and Uehera, S., *Nucl. Instrum. & Methods*, A301, 497, 1991. With permission.)

be read into the 68K20FPI within about 6 msec. These data are sent to the VAX after releasing a trigger busy signal.

By utilizing the data processing time (analog-to-digital conversion time, etc.) at the front-end electronics, it was possible to carry out second-level trigger processing in the 68K20FPI. After receiving a trigger signal, trigger information is read out soon from the trigger circuits. The 68K20FPI module then checks the trigger information in a more elaborate way than the hard-wired logic does. If the event fails the trigger criteria, the data are abandoned.

A block diagram of the 68K20FPI is shown in Figure 12.37 and the main features of the module are summarized in Table 12.3.

Battaiotto et al.[20] chose the DSP6001 as a basis for the design of a FASTBUS module, intended for trigger applications in the DELPHI forward electromagnetic calorimeter (FEMC). The interfacing to FASTBUS and to the detector front-end electronics was made more compact and reliable by an extensive use of XILINX XC3020 programmable gate arrays. All the circuits could fit in less than one half FASTBUS board, at a cost roughly an order of magnitude lower than that of an equivalent processor of the previous generation.

The design of the module was kept as simple as possible. A block diagram is shown in Figure 12.38. The DSP microprocessor performs fast trigger computations, executing programs normally loaded from EPROMs. The input trigger data are provided through an INPUT interface which in that specific application reads calorimeter trigger data digitizers (fast data digitizers and processors (FDDP)). The timing signals for the synchronization with the experiment are obtained from the local trigger supervisor (LTS), through the LTS interface. A special ECL bus interface, following the ECL line protocol, makes the DSP results available to other trigger processors.

The data acquisition system (DAS) required FASTBUS access to the module. This is obtained via a standard FASTBUS slave coupler plus control and status registers (CSR) and a dual-port memory acting both as a front-end buffer (FEB) and as a mailbox for communications between DSP and the DAS. This structure allows downloading of DSP programs crossassembled or compiled in an host computer. The dashed boxes in Figure 12.38 indicate the circuits that are integrated in programmable gate arrays.

The DSP56001 is a reduced instruction set (RISC) microprocessor with a Harvard-type pipeline architecture, as sketched in Figure 12.39.

Data and address manipulations are performed by three units:

Table 12.3 Main Features of the 68K20FPI

MC68020 32-b microprocessor (16 MHz)
MC68881/2 floating point co-processor (12.5/16 MHz)
1 ~ 8 Mbyte RAM board
Max. 512-kByte ROM with 8 sockets (long-word/byte ports)
Ethernet controller + Cheapernet transceiver
SCSI interface, dual RS-232C, NIM level input and output
A user-programmable timer and a battery backup clock
FASTBUS master: all addressing mode
FASTBUS slave: geographic addressing
Block transfer speed of 6(20) MHz as master (slave)
CSR#0, CSR#8, and 22-b NTA registers
Single-width FASTBUS module
Local bus extension via auxiliary connector

From Arai, Y. and Uehera, S., *Nucl. Instrum. & Methods,* A301, 497, 1991. With permission.

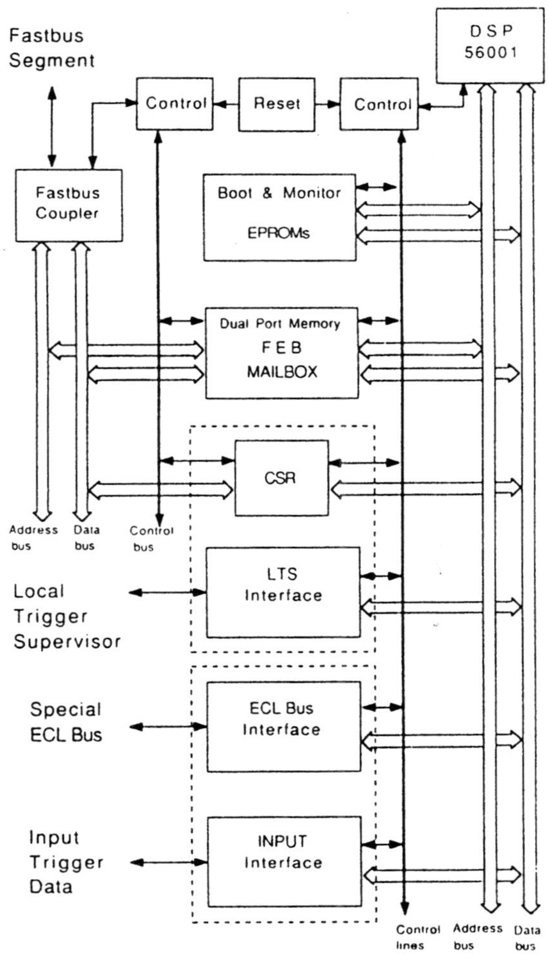

Figure 12.38 The architecture of the FASTBUS module. (From Battaiotto, P. et al., *Nucl. Instrum. & Methods,* A301, 265, 1991. With permission.)

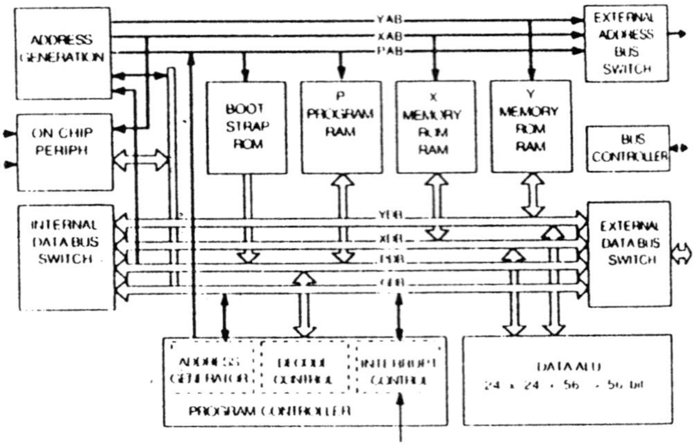

Figure 12.39 A simplified block diagram of the DSP56001. (From Battaiotto, P. et al., *Nucl. Instrum. & Methods,* A301, 265, 1991. With permission.)

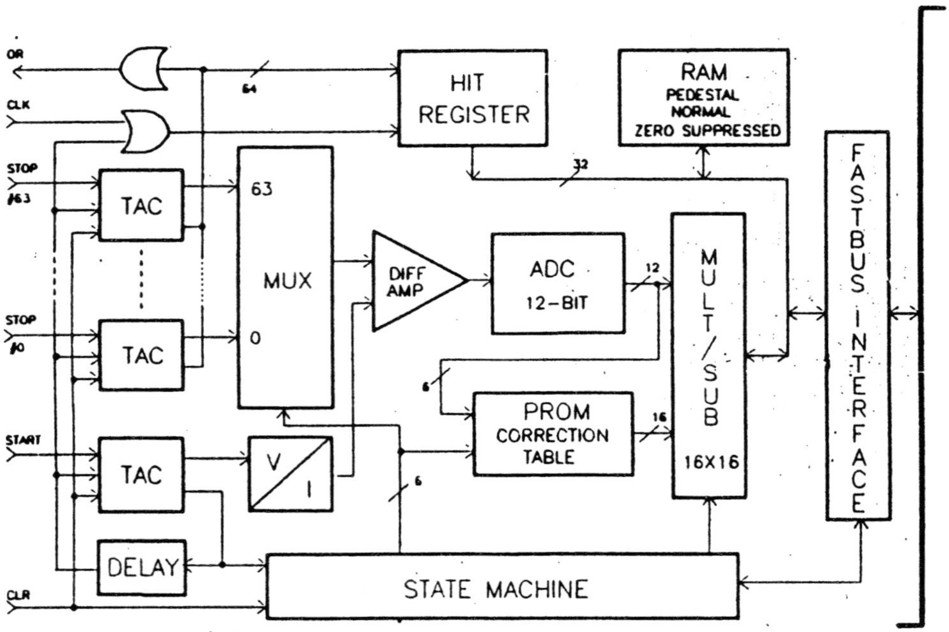

Figure 12.40 Block diagram. (From Sobczynski, C. W. et al., *IEEE Trans. Nucl. Sci.,* 36, 1989. With permission.)

1. A data ALU, with 24-b input registers and 56-b accumulator registers, performing logic and arithmetic operations, including single-cycle multiplication, between 56-b operands
2. An address generation unit, offering several addressing modes for program and data memories
3. A program controller, prefetching and decoding instructions, executing loops, and controlling interrupts

Three different memory address spaces are defined: two for the data memory, called X and Y; and one for the program memory, called P. X and Y data memories include internal RAMs (two, 256×24 b each) and read-only memories (ROMs) (two, 256×24 b each), and can be expanded externally to $64K \times 24$ b each. The 31 registers of the interfaces are mapped to the external X space and the dual-port memory (1024×16 b) of the FASTBUS interface to the external Y space. The P program memory includes a 512- $\times$ 24-b internal RAM and a smaller "bootstrap" ROM. An external expansion up to $64K \times 24$ b is possible. The separate internal buses for X, Y, P data and addresses are multiplexed to one external bus for addresses and one for data.[20]

A 25-psec resolution, 12-b, 64-channel FASTBUS time-to-digital converter (F432) was described by Sobczynski et al.[21] Information flow through the F432 TDC is shown in Figure 12.40. The time delay between the common start input and each of 64 stop inputs is measured differentially. Each of these 65 signals is routed to a separate time-to-amplitude converter (TAC) circuit which uses the capacitor charge/discharge technique (Figure 12.41). The stop channel TAC voltages are multiplexed to the noninverting

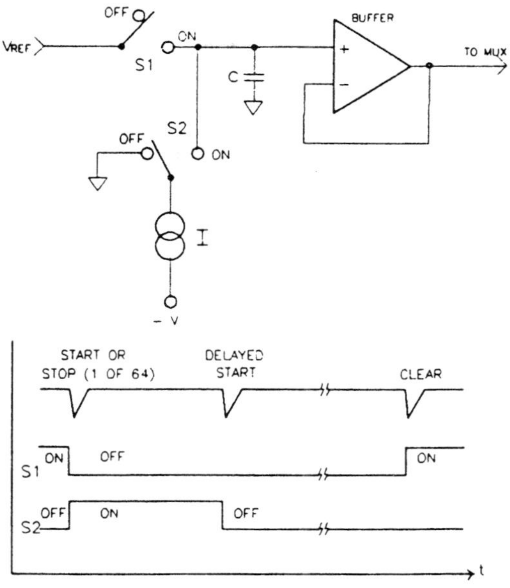

Figure 12.41 Simplified schematic of TAC circuitry and timing. (From Sobczynski, C. W. et al., *IEEE Trans. Nucl. Sci.*, 36, 1989. With permission.)

input of a precision differential amplifier, while the start TAC voltage is converted to a current and directed to the inverting input of this amplifier. A separate output from each stop channel TAC is routed to a hit register which is clocked internally or externally as described below. Meanwhile, the result of the start/stop analog subtraction is converted by a 12-b ADC and routed to a digital multiplier/subtracter. Another bus carries a channel-dependent pedestal, previously stored in the pedestal area of RAM, into a register within the multiplier/subtracter. The upper bits from each ADC conversion, along with the multiplexer channel information, simultaneously address a programmable read-only memory (PROM) containing correction factors. These factors are multiplied with the full 12-b ADC data to produce a result from which the currently loaded pedestal data are subtracted.

The TDC design employs a differential technique, using the start TAC as a reference. This provides three performance advantages. First of all, time interval measurements are possible over the entire full-scale range of the device, beginning at 0 psec, without any deterioration of linearity at the range limits. Second, the subtraction of start and stop TAC voltages cancels, to a large degree, the voltage droop caused by the leakage of charge from the hold capacitor of each TAC. Consequently, the resultant difference voltage is more nearly static, especially important in successive approximation A/D systems. The third desirable advantage of this technique over single-ended approaches is the improved compensation of thermal effects and power supply instabilities.[21]

The TAC circuity needed to achieve the desired performance requires a combination of 10 KH ECL logic, fast transistors, and precise passive components.

One minor but, nevertheless, essential design addition to the start/stop subtraction circuity is the translation of the output voltage signal of the start TAC into a current signal. This eliminates the unbalancing effects that the nonnegligible ON resistance of the multiplexer would have on this subtraction circuity.

Kurashige et al.[22] have developed a data acquisitions system using a TKO box as a front-end system and FASTBUS as an on-line data transfer bus. Good ground isolation between the front-end system and FASTBUS was the key for noise reduction in a large detector system.

FASTBUS is suitable for an on-line data bus because of its flexibility and fast data transfer speed. However, its complex bus protocol and high-power requirement are not welcome features for front-end electronics.

Ground isolation between detector sections is essential to reduce pickup noise due to ground loops.

Pedestal subtractions and zero suppression in the early stages of data transfer are essential to reduce data size of an event to a manageable level. Parallel processing is important for a large system to reduce time spent for data handling.

The block diagram of the system is shown in Figure 12.42. The memory partner (MP) is a FASTBUS memory module and links to the host computer through FASTBUS. The controller head (CH) is a TKO box controller located in each KTO box, and all front-end modules are in TKO boxes. A CH links to an

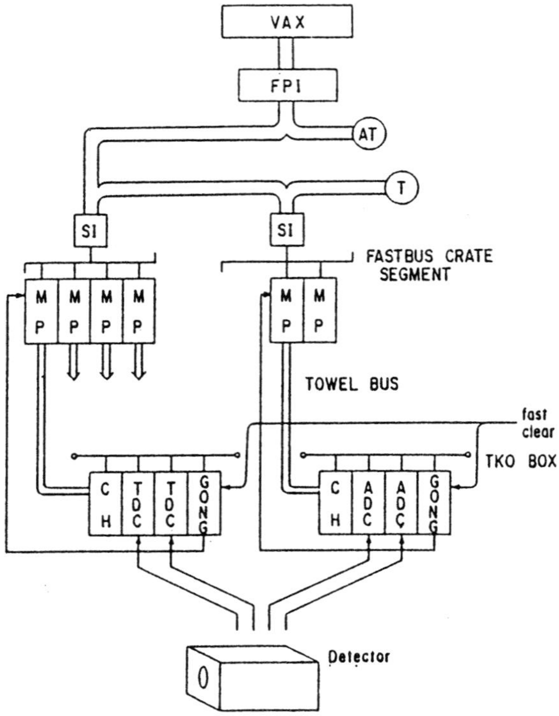

Figure 12.42 Schematic of the system for VENUS in TRISTAN. (From Kurashige, H. et al., *IEEE Trans. Nucl. Sci.*, 35, 257, 1988. With permission.)

MP module through a TKO on-line word encoding (towel) bus. Thus, each CH-MP pair handles a box full of front-end modules. Optical isolator modules decouple the grounds between MP and CH.

Data processing of this system is separated into two stages to execute parallel processing and data reduction in the TKO box. First, data acquisition in TKO is initiated without any action of FASTBUS. After a sparse data scan and data compaction, the CH transfers only the meaningful data to the MP memory. Data transfer between each CH and MP pair is performed in parallel. In the next step, the MP is accessed through FASTBUS operation.

The go-or-no-go (GONG) module provides accurate timing signals for the KTO A-bus. The GONG issues a trigger signal to MP for the readout operation. A CH is the master of the KTO box which in turn is controlled by an MP. An MP is a memory module on FASTBUS and also a controller of the towel bus. The towel bus is a specified cable bus to link a CH to an MP.

A FASTBUS digital readout module (WICDRM) has been designed by Bilei and Castaldi[23] to read the front-end electronics of the streamer tube cathode strips. The data, digitally stored into the front-end shift registers, are daisy chained and clocked into eight input buffers inside the module, under the full control of the MC68020 microprocessor. During the buffer readout, the processor writes into RAM memory only those words which contain at least one cluster (defined as a set on one or more adjacent hit strips), together with coordinates of these words within the daisy chain.

Immediately after each beam crossing, all the digital data streams are read and, if the trigger has in the meantime been accepted, the MC68020 performs a second stage of elaboration, compacting the data and preparing it for further analysis by the FASTBUS master processor.

The module is a FASTBUS slave and can be primary addressed in two modes: geographic and broadcast. It contains 128 kbytes of program memory and 128 kbytes of data memory. Both memories are mapped in data space and can be accessed from FASTBUS in both WRITE and READ cycle modes.

The FASTBUS interface has been designed using the TTL ADI chips. A processor interrupt by FASTBUS is implemented, and three different sources of service requests are used to flag the status of the current process to the FASTBUS master.

A block diagram of the front-end processor is shown in Figure 12.43. The signals drawn on the left of the figure are directed to the front-end SGS electronics; those on the right are the internal connections with the FASTBUS interface. On reception of an interrupt, the processor can send a load signal to the

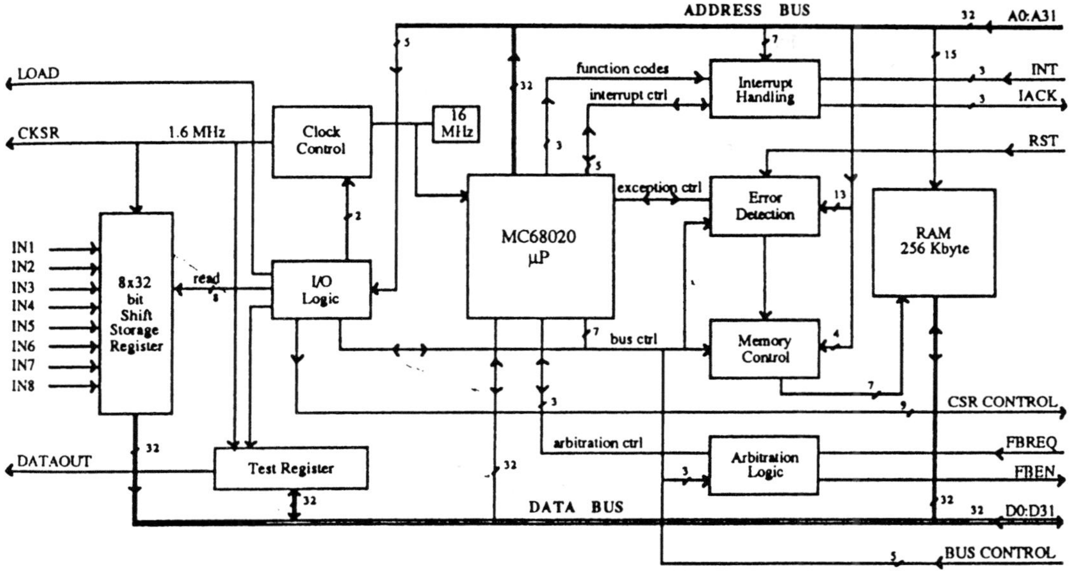

Figure 12.43 Block diagram of the front-end processor. (From Bilei, G. M. and Castaldi, R., *IEEE Trans. Nucl. Sci.*, 35, 282, 1988. With permission.)

front-end electronics, latching the digital information. Subsequently the shift-register clock (CKSR) is enabled and serial data from all eight input channels (IN1–IN8) is shifted into 8- × 32-b wide latches.

The WICDRM is a single-width slave module that performs most of the standard operations specified by the FASTBUS protocol, such as broadcast operations (either general, Class N, or TP); geographic addressing in data and control spaced secondary addressing (READ-WRITE); random data operations (READ-WRITE); handshake block transfers (READ-WRITE), and slave status responses (implemented SS responses are 0, 1, 2, 3, 6, and 7). Parity is neither checked nor generated by the device.

The block diagram of the FASTBUS interface is shown in Figure 12.44. The bus-driving hardware is a mixture of 10- and 100-K ECL logic. The 100-K ECL-TTL bus transceivers are used only for the AD lines; 10-K ECL logic is used on all the other information and timing lines.

The interface has been implemented using two ADI (Fujitsu MB114F307) chips. The ADI low (ADIL) handles the 16 low-order lines (AD0 through AD15), while the ADI high (ADIH) handles the high-order lines (AD16 through AD31)

The control and timing circuitry generates all the necessary signals for the control of the ADI chips (for example to load, read the increment the NTA or to read and write CSR No. 7, to open and close data buffers etc.). In response to a FASTBUS cycle this circuitry also generates the handshake signals AK and DK according the FASTBUS protocol. In the WICDRM five control and status registers are implemented: CSRs numbers 0, 1, 7, 10, and 12. All the CSRs can be read or written by FASTBUS.[23]

A time-digitizing system was developed by Ikeno et al.[24] The system is based on 64-channel single-hit TAC modules with analog multiplexers and a scan ADC (SADC). Up to 1536 TAC channels in a FASTBUS crate were scanned and an AD converter by a SADC was in the same crate. Hardware data processing such as pedestal and slope corrections (Ax + B), address conversions, sparse data scanning, and test hit-pattern loading are available.

SADC is a full-fledged FASTBUS master and slave module supporting all three FASTBUS transfer modes, random access, handshake, and pipeline transfers. Figure 12.45 shows its block diagram. The memory space for data are from 0 to 2047. Each word consists of an address part (bit 31–16) and a pulse-height part (15–0). The pulse height is an ordinary 2-byte integer, 2's complement. It can be negative when the Ax + B circuit is active.

AD conversion procedure starts immediately after the SADC gets a common-stop (or an end of S/H gate) signal from the front panel input. When the SADC encounters a hit channel of a TAC, it saves the pulse height on the analog bus in a sample-and-hold circuit (Harris HA5320). The timing of sample and hold is adjusted to be about 2 µsec after the analog multiplexing to allow the pulse ringing to be settled.

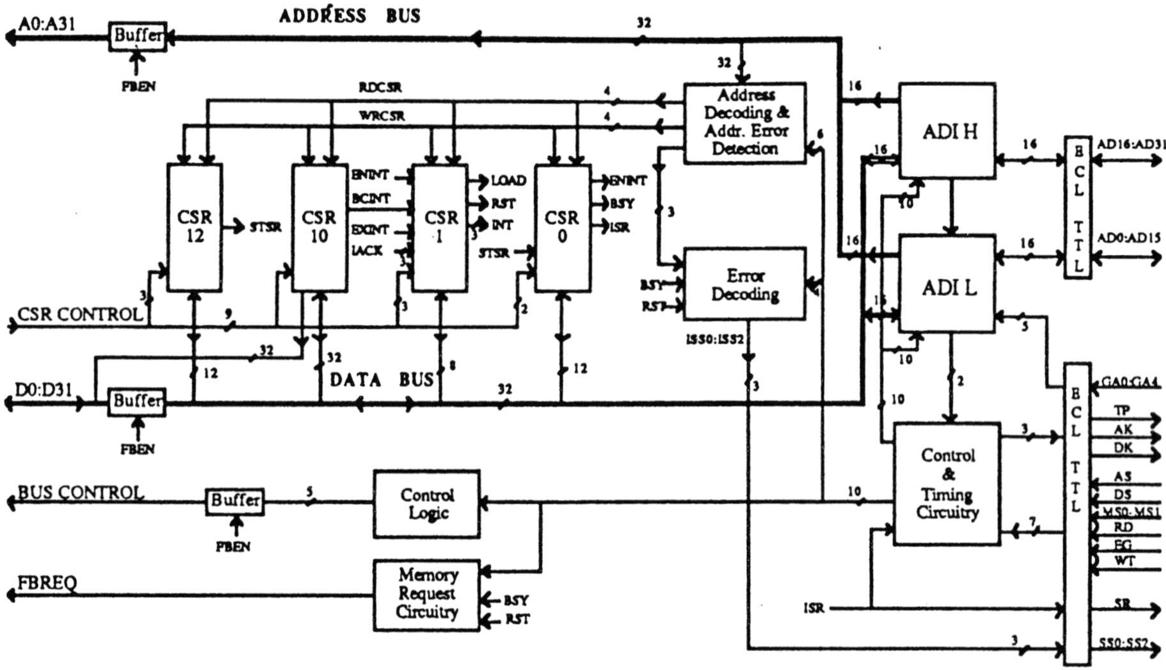

Figure 12.44 Block diagram of the FASTBUS interface. (From Belei, G. M. and Castaldi, R., *IEEE Trans. Nucl. Sci.*, 35, 282, 1988. With permission.)

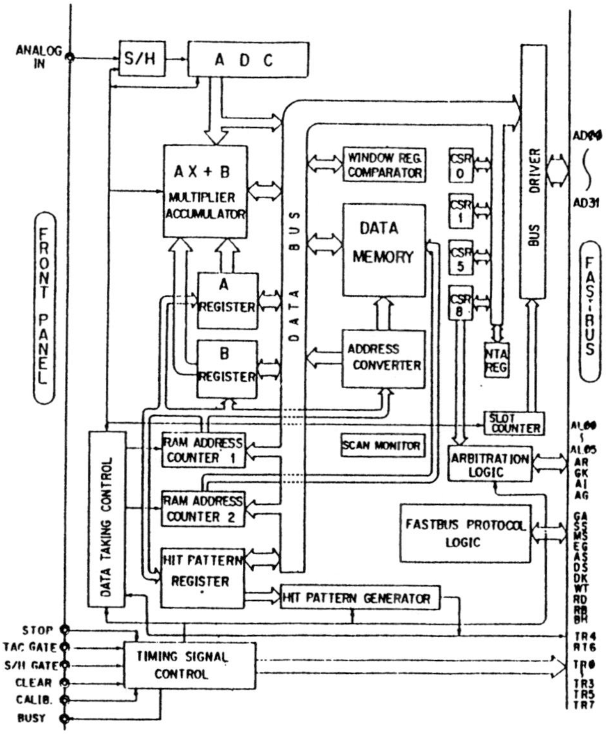

Figure 12.45 SADC block diagram. (From Adam, W. et al., *IEEE Trans. Nucl. Sci.*, 39, 166, 1992. With permission.)

After the completion of sample and hold, the SADC restores its search to the next hit channel and simultaneously A-to-D conversion starts in a 12-b ADC (Micro-Network MN5245, conversion time 900 nsec). Completion of the ADC conversion initiates a digital data conversion $A_i x_i + B_i$ using a 16-b × 16-b monolithic multiplier-accumulator (NPC SM5810-D). Two coefficients A_i and B_i are provided for CSR memories for individual channels. This Ax + B operation takes 65 nsec. A bypass mode is available, allowing one to see raw data directly. The scanning and conversion sequence could be canceled at any stage and the whole system could be initialized by a front panel clear signal.[24] The upper 16-b of data memories are used for channel addresses of corresponding data.

It takes 3 µsec for each hit channel, while less than 0.5 µsec for nonhit channels. The maximum conversion time, therefore, will be 4.5 msec when all 1536 channels get hits.

Design and performance of the DELPHI data acquisition system were described by Adam et al.[25] The data acquisition system of the DELPHI experiment at LEP allows recording data from the front-end electronics of 15 different detectors housed in about 100 FASTBUS crates. Each detector has the possibility of monitoring acquisition as well as of running in a stand-alone mode.

The DELPHI data acquisition system (DAS) has a tree-like structure. Each level of the DAS forms an elementary cell. All cells have the same basic structure described in Figure 12.46. It consists of two sets of buffers (source and destination), a transfer process, and a control process. The role of each cell is to transfer data from the source buffer to the destination buffer. It is triggered by the upstream cell which, in turn, triggers the downstream cell.

The front end receive data from the detector itself into a digitization module. The transfer process is here purely hardware. The destination buffer is called front-end buffer (FEB). Each FEB has a capacity of four events. The control process is called front-end freeing controller (FEFC) and is implemented in hardware in the HPC and in software for all other detectors. The filling of the FEBs is the only process in the chain which is synchronous with the actual event.

For those detectors which have a large number of electronic channels, one has added in the front-end electronics itself a second level of buffering to the FEBs after a data reduction stage (zero suppression). The output of this stage is called board event buffer (BEB) and has a detector dependent depth (one or four events). The control of the board processing is made by a mechanism attached to the FEFC such that the control protocol to the crate level is standard.

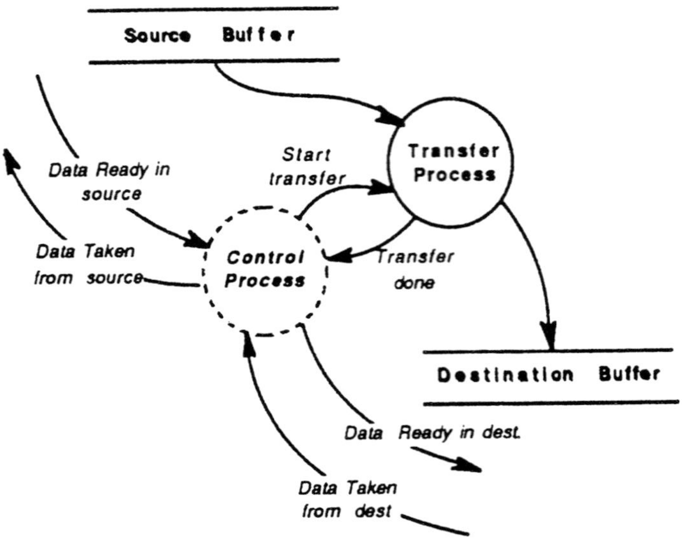

Figure 12.46 Basic readout cell. (From Adam, W. et al., *IEEE Trans. Nucl. Sci.*, 39, 166, 1992. With permission.)

In each FASTBUS crate, data are transferred from the FEB (or the BEB) into a crate event buffer (CEB) under control of a crate processor (CP). This control and transfer level is implemented using a standard piece of software running for all detectors in the same FASTBUS module called FASTBUS intersegment processor (FIP). Figure 12.47 shows all levels down to the crate.[25]

Each partition is either a full bus detector or part of it. Data are transferred from the CEBs into a multievent buffer (MEB) and a spy event buffer (SEB), under control of a standard piece of software called local event supervisor (LES).

Data from the MEBs are transferred into a global event buffer (GEB) under control of the global event supervisor (GES). The transfer itself is performed by a fast FASTBUS master called block mover (BM) in order to decouple the FASTBUS transactions for control and data transfer. Figure 12.48 shows the central partition data flow diagram. The full DELPHI event is transferred from the GEB into a set of emulators where it is first analyzed mainly in order to tag and classify interesting events.

The data from the emulators or directly from the GEB are transferred into a data acquisition host computer (VXDEAQ) for monitoring and logging onto mass storage. The DELPHI readout is built around a FASTBUS-embedded processor called FASTBUS intersegment processor (FIP).

The FIP processor is based around a Motorola 68020 processor running at 16 MHz. It owns 1.5 Mbytes of dynamic RAM, 256 kbytes of static RAM dual ported between the processor, and a FASTBUS cable slave port (note that the dynamic RAM is also accessible as the whole MC68020 address field through the SCR space of the slave). A FASTBUS crate master port allows all FASTBUS transactions (except single-line handling).[25]

A single-chip implementation of the general-purpose FASTBUS slave interface (FSI) has been developed by Nelson et al.[26] in ECL gate-array technology. The FSI occupies only 1.6% of the available circuit board space while providing a complete 32-b interface to the FASTBUS. All mandatory slave-interface requirements of IEEE 960 are supported, in addition to several nonmandatory requirements and the optional, extended MS code features. Geographic, logical, and broadcast addressing are implemented using on-chip registers.

The user interface provided by the FSI allows control to slave status-response and connection timing for both address and data cycles. The BIT1 ECL array technology used for the FSI allows direct connections to the FASTBUS, eliminating the need for external driver/receiver buffers. Nelson et al. partitioned the FSI into several major functional blocks, to make the design process more tractable.

12.3 OTHER DATA ACQUISITION SYSTEMS

An IBM PC AT interface for data acquisition from diode arrays was described by Hovander and Bohm.[27] The system operates as a multiscaler, allowing two processes to be recorded at the same time. The interface was designed using a programmable gate array (an 1800-gate-equivalent logic cell array to

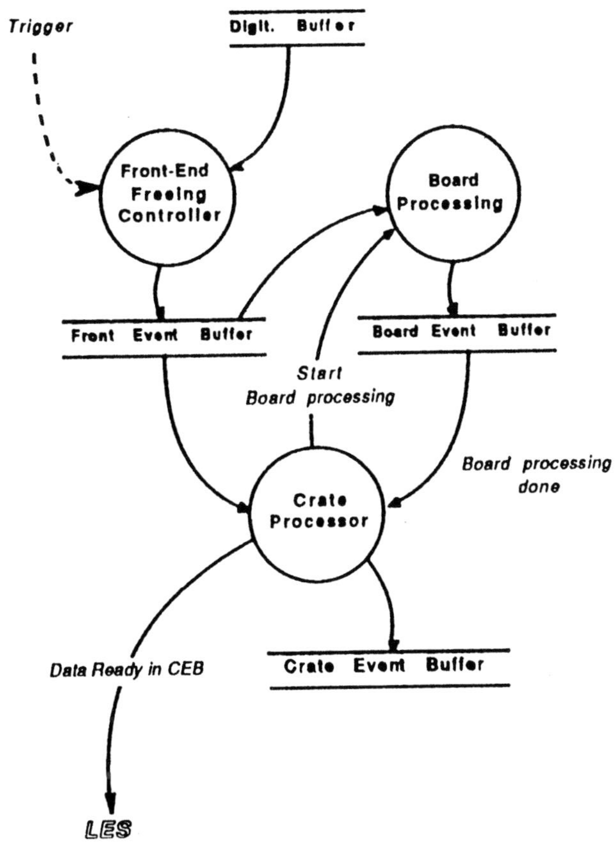

Figure 12.47 DFD of the front-end and crate levels. (From Adam, W. et al., *IEEE Trans. Nucl. Sci.*, 39, 166, 1992. With permission.)

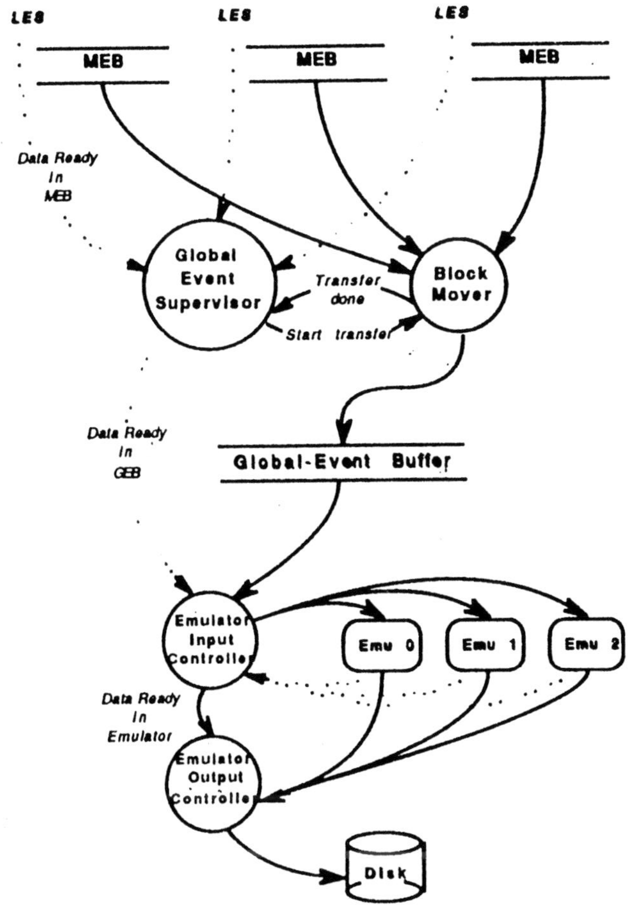

Figure 12.48 DFD of the central partition. (From Adam, W. et al., *IEEE Trans. Nucl. Sci.*, 39, 166, 1992. With permission.)

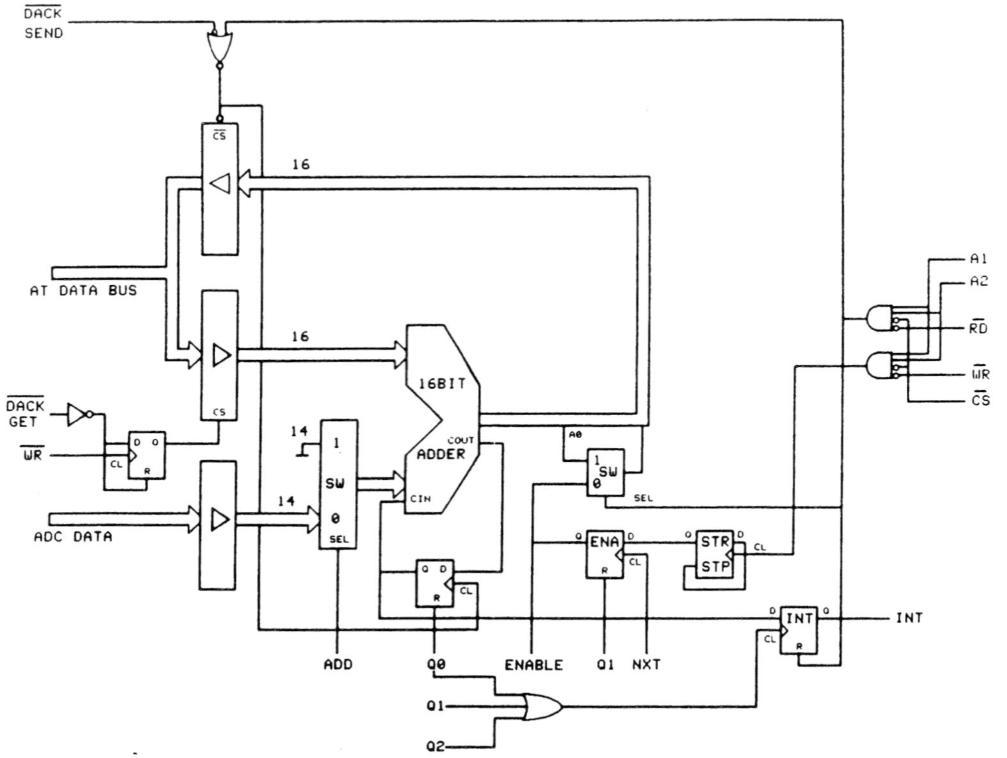

Figure 12.49 LCA scheme. (From Hovander, B. and Bohm, C., *IEEE Trans. Nucl. Sci.,* 37, 2191, 1990. With permission.)

implement most of the digital circuitry and a 12-b ADC to sample the analog inputs). A 12-b DAC was used for analog outputs and for diagnostic testing purposes.

To provide the required maximum transfer rate, 50 KHz or faster, a direct memory address (DMA) or coprocessor-based interface must be used.

To reduce hardware complexity and shorten the development time, the interface was built around a programmable gate array, an XILINX 2018 logic cell array (LCA) (Figure 12.49). The LCA contains programmable logic blocks with programmable interconnections. The configuration information is stored in an on-chip RAM, which means that the LCA must be reprogrammed at each power-up. The LCA is programmed when fed a valid stream of configuration data after reset. The data can be supplied by a regular programmable read-only memory (PROM), or, if it is desired to save LCA pins and component space, by a specially designed serial PROM. In this design, however, the programming is accomplished by downloading the configuration data from the controlling computer.

The programmable gate array used in this interface design, the XILINX 2018 LCA, consists of 100 logical blocks:[28] 56 inputs and outputs where every input can be configured with or without a register. Every logic block contains a flip-flop and four 5-input 2-output PROMs. Inputs, outputs, logical function blocks, and the interconnection among them are programmed by the configuration data to achieve the desired function. In all, the 2018 LCA corresponds to about 1800 logic gates.

The interface (see Figure 12.50) is equipped with an address decoder and bus drivers as well as some bus contention logic, all designed with traditional logic, which is a necessary condition to be able to configure the LCA from the system controller.

The interface uses DMA transfer to achieve sufficient high-speed data acquisition capability. Two of the PC AT 16-b DMA channels are used for this purpose. One channel is used to sequentially and repetitively write data from the memory array to the interface. The other channel performs the reverse task. By synchronizing the two DMA activities in a sequential process, it is possible to add a digitized value to the corresponding array value just being read from the memory and to write the result back into

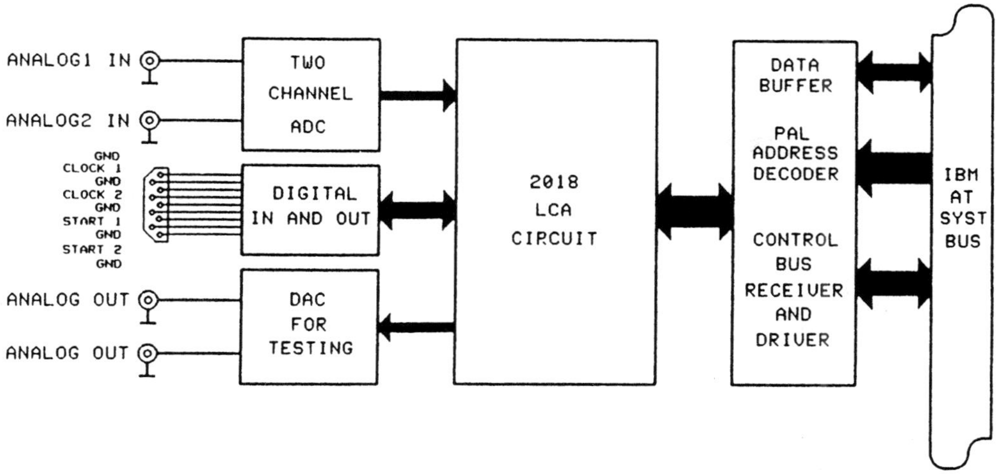

Figure 12.50 Block diagram of interface. (Hovander, B. and Bohm, C., *IEEE Trans. Nucl. Sci.*, 37, 2191, 1990. With permission.)

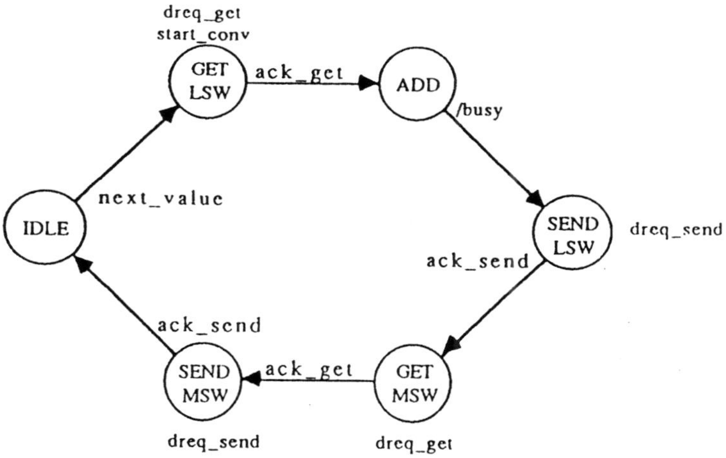

Figure 12.51 State diagram for DMA cycle. (From Hovander, B. and Bohm, C., *IEEE Trans. Nucl. Sci.*, 37, 2191, 1990. With permission.)

the same position. With double precision representation of data, six different states are needed in this process (see Figure 12.51).

The analog I/O circuitry (Figure 12.52) contains some relatively fast operational amplifiers, analog switches, and a 5-μsec and 12-b CMOS analog-to-digital converter (ADC from analog devices). The analog output is generated by a 1-μsec and 12-b CMOS digital-to-analog converter (DAC). The switching and handshaking logic is contained in the 70-MHz CMOS XILINX 2018 LCA circuit.

Belusevic et al.[29] have described hardware and software aspects of a very fast and versatile, yet conceptually simple, data transfer and processing system for use with future accelerators. It consists of a transputer-based crate controller (CC), which includes an Intel i860 microcomputer; and of a set of readout cards (RC), each containing a digital signal processor (DSP) for fast data parameterization and compaction. The reduced data are written into a dual-port memory (DPM), where it can be accessed concurrently by the transputer and transferred to a common DPM on the CC card. A crate full of data thus assembled at one place can further be processed by the powerful i860 microcomputer. Address generators (simple binary counters) are included on the crate controller and each readout card to enable direct memory access (DMA) operations, resulting in a considerable increase in data transfer speed (maximum 80 Mbytes/s). The use of a transputer as the sole controlling processor, in conjunction with DPMs, renders bus arbitration unnecessary, leading to very simple interfacing logic and operating software. The four

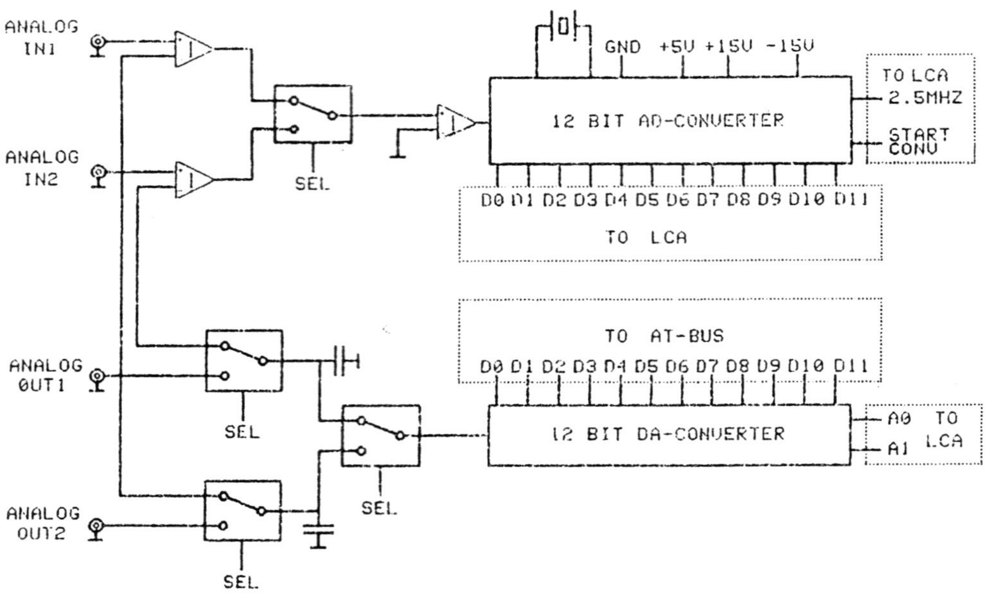

Figure 12.52 Measurement I/O port of interface. (From Hovander, B. and Bohm, C., *IEEE Trans. Nucl. Sci.*, 37, 2191, 1990. With permission.)

high-speed serial links of the transputer greatly facilitate downloading of programs and intercrate communications. An Intel i960CA processor, situated on the CC card, is used for fast data transfer between crates by means of its 32-b wide DMA channel. The operating software is written in the Occam language, which was specially developed for programming concurrent systems based on transputers.

The block diagram showing its main components and data flow within a crate, indicating also intercrate communication links, is in Figure 12.53. The most important characteristics of the design are

- Inherent concurrency with respect to data transfer and processing, achieved by interfacing the embedded microcomputers to the data bus via dual-port memories, thus avoiding bus arbitration (note only one controlling processor)
- Easy access to data for one of the most advanced rise microcomputers (the i860), resulting in an efficient use of processing power
- Simplicity of hardware and operating software
- Very high-data transfer rates even for the smallest block sizes, due to extremely low real-time overheads

To provide wideband communication links between crates, the crate controller incorporates an Intel i960CA microprocessor, whose 32-b wide DMA channel can achieve a maximum data throughput of 106 Mbytes.

With an execution rate of 66 million instructions per second (mips), the i8960CA is a powerful processor in its own right and may in some circumstances share the trigger processing task with the i860, in addition to performing data transfers.

The circuit diagram of the crate controller is shown in Figures 12.54 and 12.55. The Inmos T414 transputer, which is used both as the controlling processor and for DMA initiation, is a 20-MHz, 32-b CMOS microcomputer with 2 kbytes of on-chip RAM, a multiplexed data and address bus, and four serial links for interprocessor communications. The standard link speed is 10 Mb/s, but it can also be used at 5 or 20 M/s. The 32-b side configurable external memory interface of the transputer can access an address space of 4 Gbytes.

The circuit diagram of the readout card is shown in Figure 12.56. Front-end data (from 8-b, 100-MHz FADCs, for example) are placed, using the 16-b wide external data bus of a DSP, into a 32-b readout DPM. The memory is divided into two 16-b parts by means of the least-significant address bit A0, and is connected via its second port to the backplane bus. Data transfers (and downloading of DSP programs) are carried out via this DPM, thereby avoiding bus contentions.

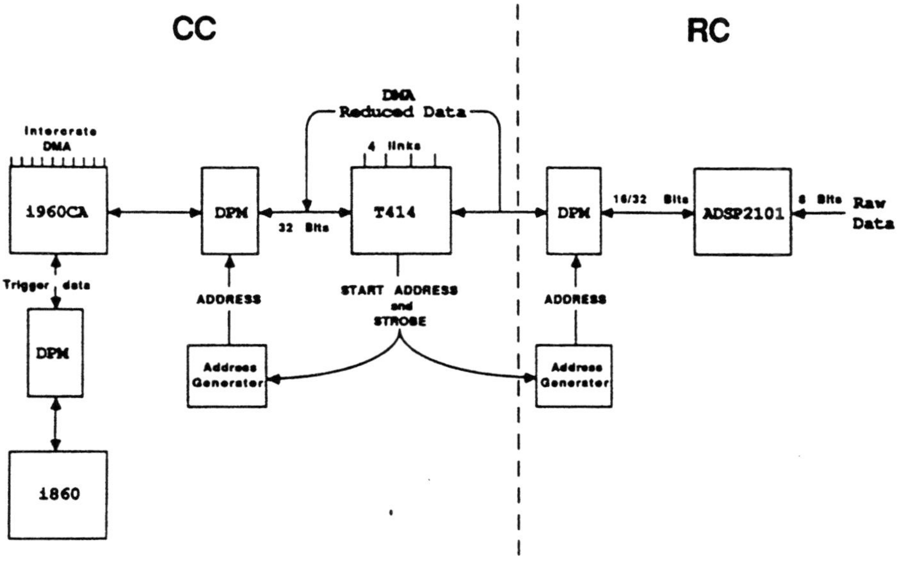

Figure 12.53 Block diagram showing system main components and data flow within a crate. (From Belusevic, R. et al., *Nucl. Instrum. & Methods,* A295, 391, 1990. With permission.)

The DSP used in this application is the Analog Devices ADSP2101, a 12.5-MHz microprocessor with separate internal data and instruction buses, 6 kbytes of on-chip program memory, and two serial ports for interprocessor communications. The main function of the DSP is to perform data compaction.

Readout control for high-luminosity accelerators was designed by Belusevic and Nixon.[30]

The most important requirements a readout system must satisfy include data transfer speed of at least 100 Mbytes/s, including real-time overheads; processing power to match the above; concurrency with respect to data transfer and processing independence of detector and crate type, and data compaction and trigger processing before readout by the crate controller.

To achieve the above goals Belusevic and Nixon suggested the following hardware and software solutions:

- Reduce interfacing logic and eliminate bus arbitration by using:
  - Single controlling microprocessor with an advanced instruction set and simple operating software which supports multitasking
  - Dual-ported memories throughout the system
  - LSI circuits, e.g., versatile microsequencers and PALs
- Reduce real-time overheads by means of specially designed hardware
- Provide easy access to data for the most powerful (vector) microprocessors
- Form a pipeline of reduced data (using digital signal processors or superscalar microcomputers for data parameterization and compaction) on the readout cards, which results in a better utilization of board space and avoids the transfer of unwanted data
- Used fast (1 Gb/s intercrate serial links

The crate controller (CC) contains two address generators: one for the local dual-port memory (DPM), and one for the DPMs on the readout cards, thereby reducing the amount of hardware per crate. The address generators enable direct memory access (DMA) operations, running at the maximum data transfer speed of 80 Mbytes. Fast intercrate data transfers can be accomplished by connecting the DPMs directly to a 1 Gb/s serial link (or transceivers, whose outputs may drive a set of twisted pairs), as shown in Figure 12.57. Each readout card (RC) contains a DSP for data compaction and trigger preprocessing. The reduced data, pipelined inside dual-ported memories, can be accessed by the transputer and transferred to a common DPM on the CC card for further processing by a powerful i860 microcomputer. The use of a transputer as the sole controlling processor, together with DPMs, renders bus arbitration unnecessary and leads to very simple interfacing logic and operating software. The four high-speed serial links of the transputer greatly facilitate downloading of programs and intercrate communications.

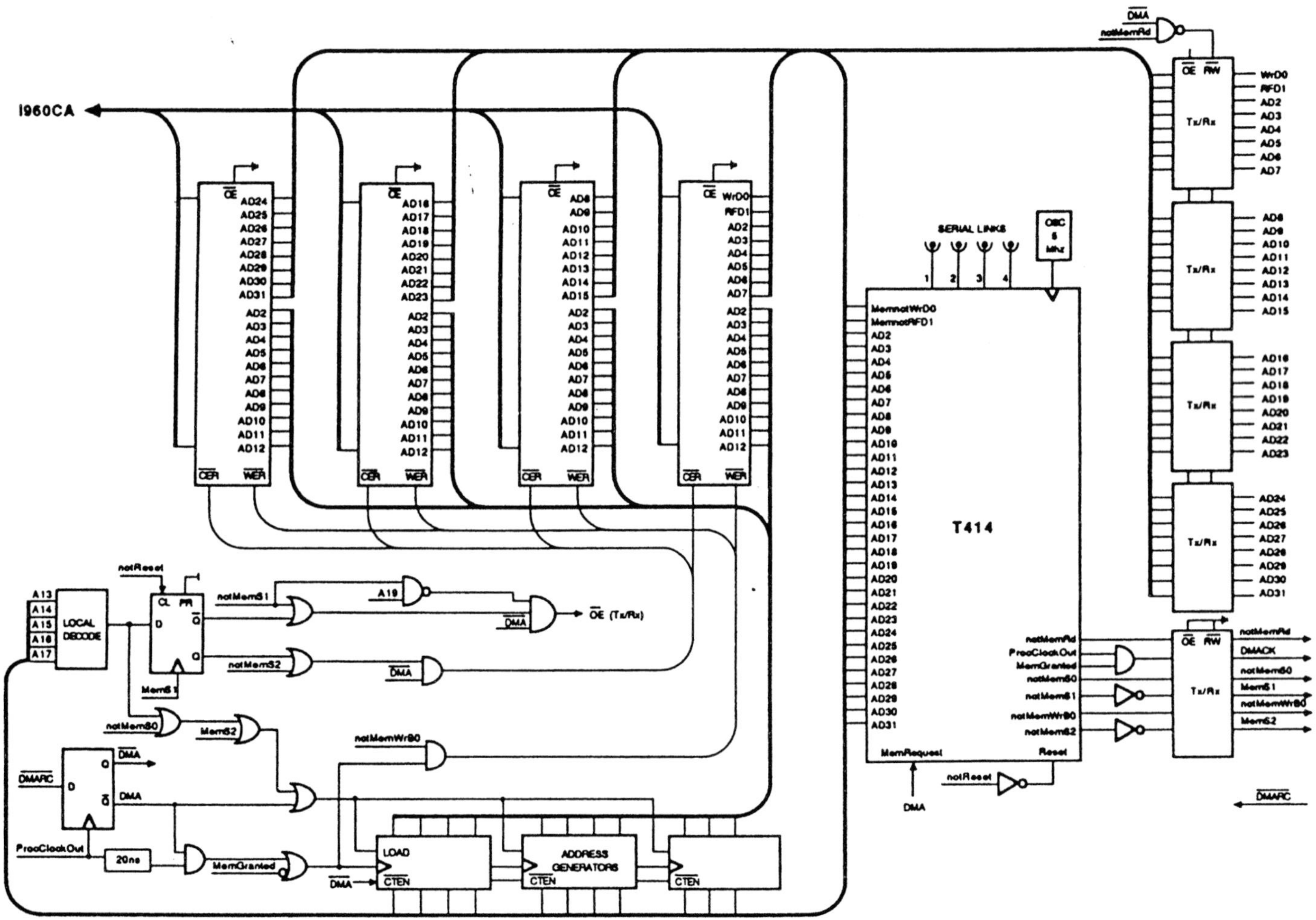

Figure 12.54 Crate controller: circuit diagram of the data transfer hardware. (From Belusevic, R. et al., *Nucl. Instrum. & Methods,* A295, 391, 1990. With permission.)

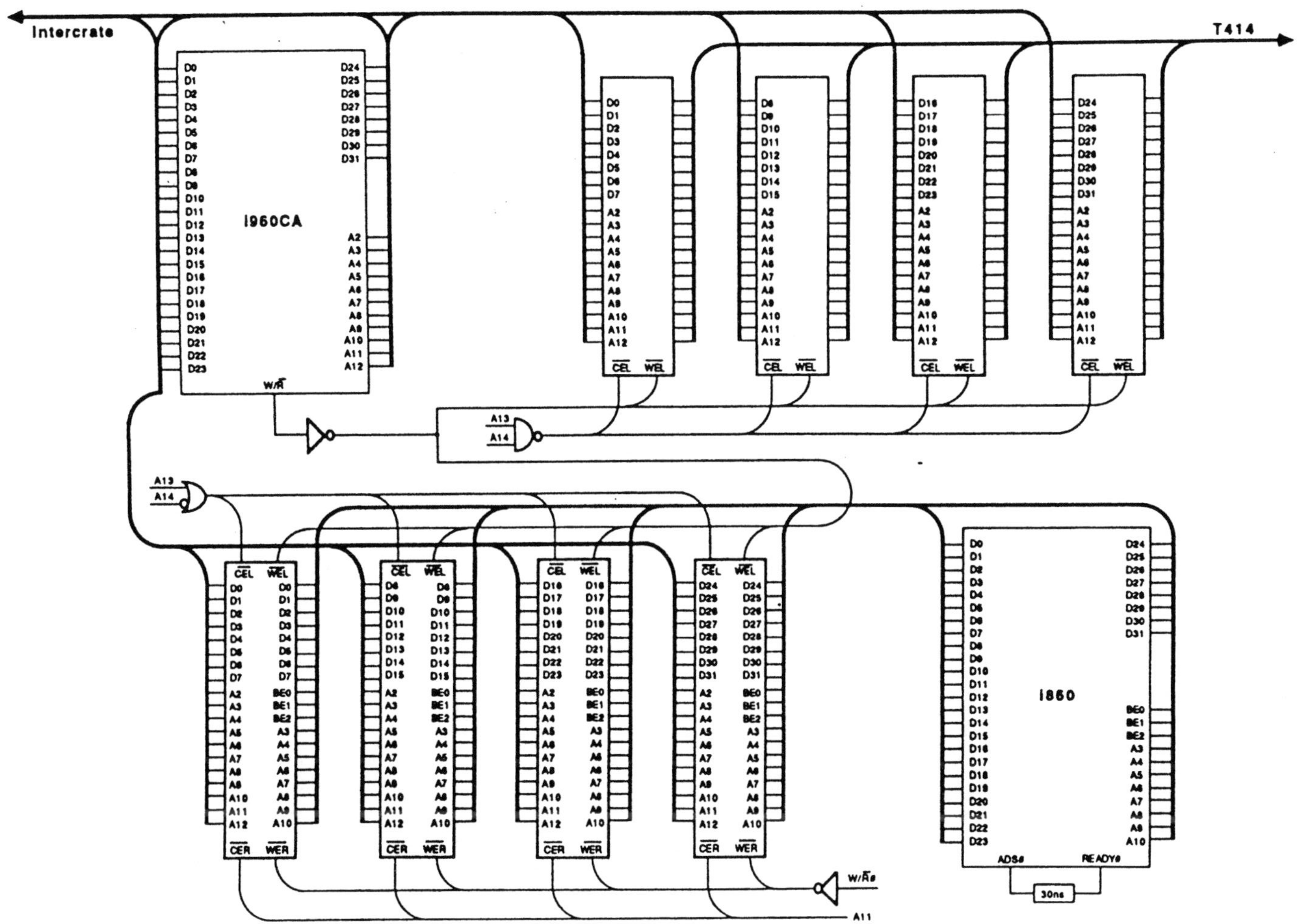

Figure 12.55 Crate controller: circuit diagram showing interfaces between processors. (From Belusevic, R. et al., *Nucl. Instrum. & Methods,* A295, 391, 1990. With permission.)

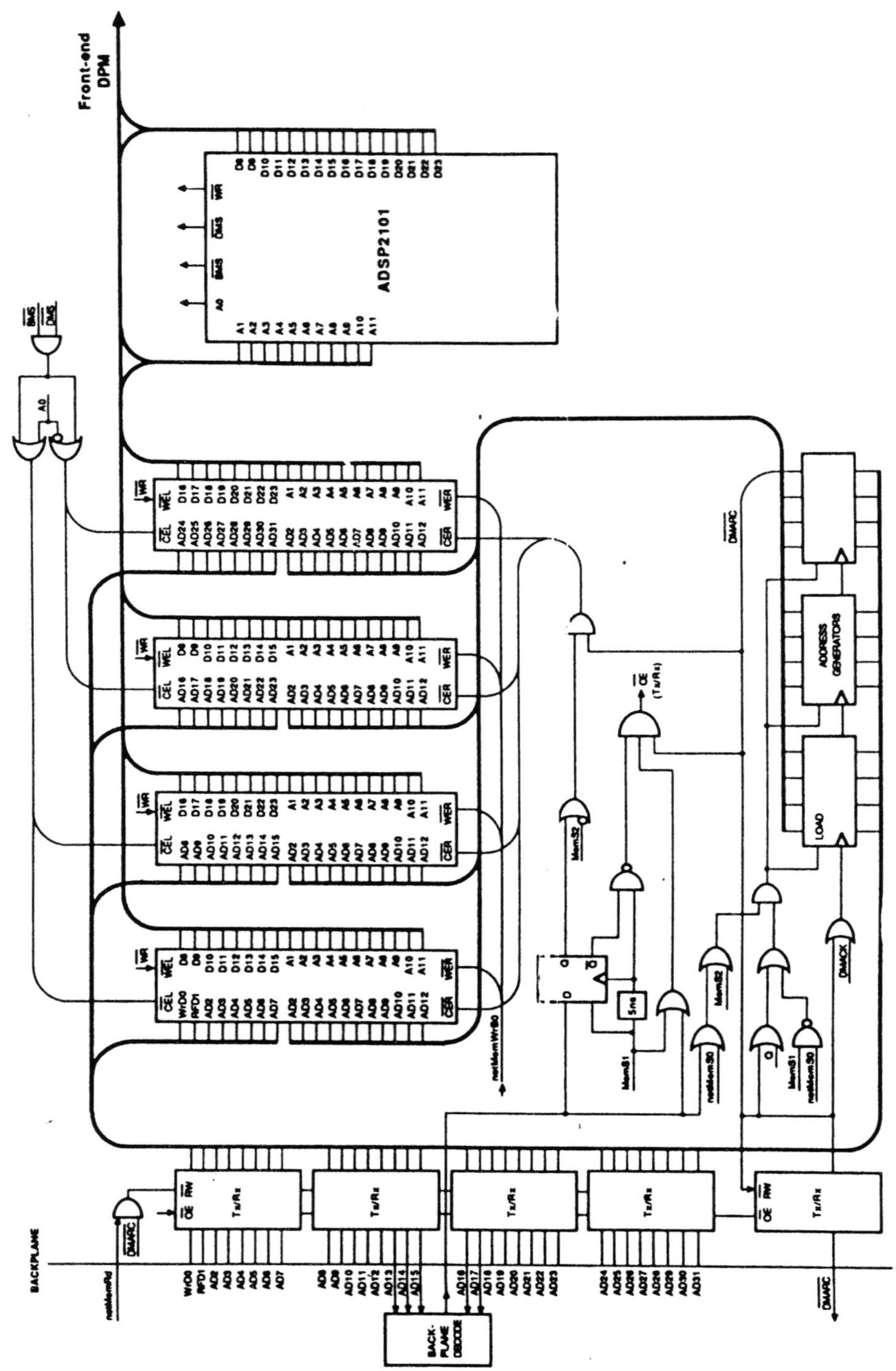

Figure 12.56 Circuit diagram of the readout card. (From Belusevic, R. et al., *Nucl. Instrum. & Methods*, A295, 391, 1990. With permission.)

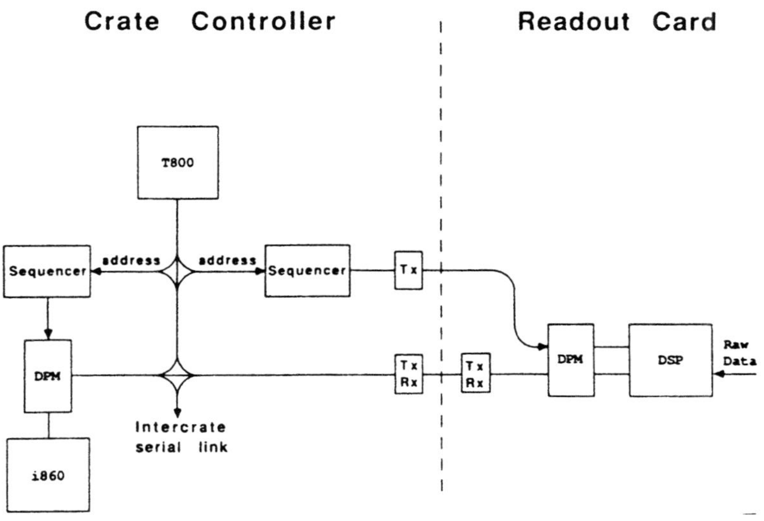

Figure 12.57 Block diagram showing system main components and data flow within a crate. (From Belusevic, R. and Nixon, G., *Nucl. Instrum. & Methods,* A307, 110, 1991. With permission.)

The circuit diagram of the readout controller is shown in Figure 12.58. The controlling processor is an Inmos T800 transputer, a 20-MHz, 32-b microcomputer with 4 kbytes of on-chip RAM; a multiplexed data and address bus; and four serial links (maximum link speed 20 Mb/s) for interprocessor communications. The 32-b wide memory interface of the transputer, which can access an address space of 4 Gbytes, can be configured both internally and externally. The internal configuration achieved by connecting the MemConfig pin of the transputer to the address/data pin AD10. Since this configuration generates dynamic RAM refresh cycles, the ordinary external accesses are qualified by the notMemRf signal (see Figure 12.58).

Two microsequencers, Texas Instruments type 74AS890, on the crate controller are used to latch transputer-generated addresses, and also as address generators for high-speed data transfers. One microsequencer serves the memories on the readout cards, and the other the local memory on the crate controller. The memory used throughout the system is a 32-b, 8-kbyte static DPM (4 times VT7132A; access time 30 nsec).

The microsequencer features two configurable registers, of which we use only one. A 3-b input (RCO-2) is decoded internally to provide load, hold, and decrement functions for the register. The register of one microsequencer (AG1) is connected via the backplane to the address lines of the readout cards, and that of the other (AG2) to the local address lines. Since the register input and output share the same internal bus, a unidirectional buffer serves to disconnect the latter from the external data bus when appropriate. Any type of TTL-driven backplane (VME-JI in this case) properly terminated can be used, and the line drivers are fast, high current (160 mA) Signetics 74F30245 transceivers.

The transputer-generated addresses consists of three parts. The least significant 11 bits (8 b in the case of the readout memories) —AD2–AD12 — from the internal DPM address. The second part is the board address (bits AD13–AD17), including when appropriate the readout memory sector address (bits AD18–AD20). The third part determines whether the current access will be a random access cycle or a DMA setup cycle (bits AD21–AD27); see Figure 12.58).[30]

The circuit diagram of the relevant part of the readout card is in Figure 12.59. Shown are the dual-port memory and the means of access to it from the backplane. A set of four transceivers drives the data lines from the readout card to the crate controller. The board address is decoded on the backplane from the address bits A13–A17 and latched to form a board select signal using a single chip (74AS885). This signal is used to enable both the readout memory and the transceivers. The memory address lines A1–A11 are connected directly to the backplane. The other side of the dual-port memory is connected to a DSP and to the outputs of readout pipelines. The data transfer hardware on the readout cards (of which there can be up to 20 per crate) has been significantly reduced by placing the RC address generator on the crate controller.

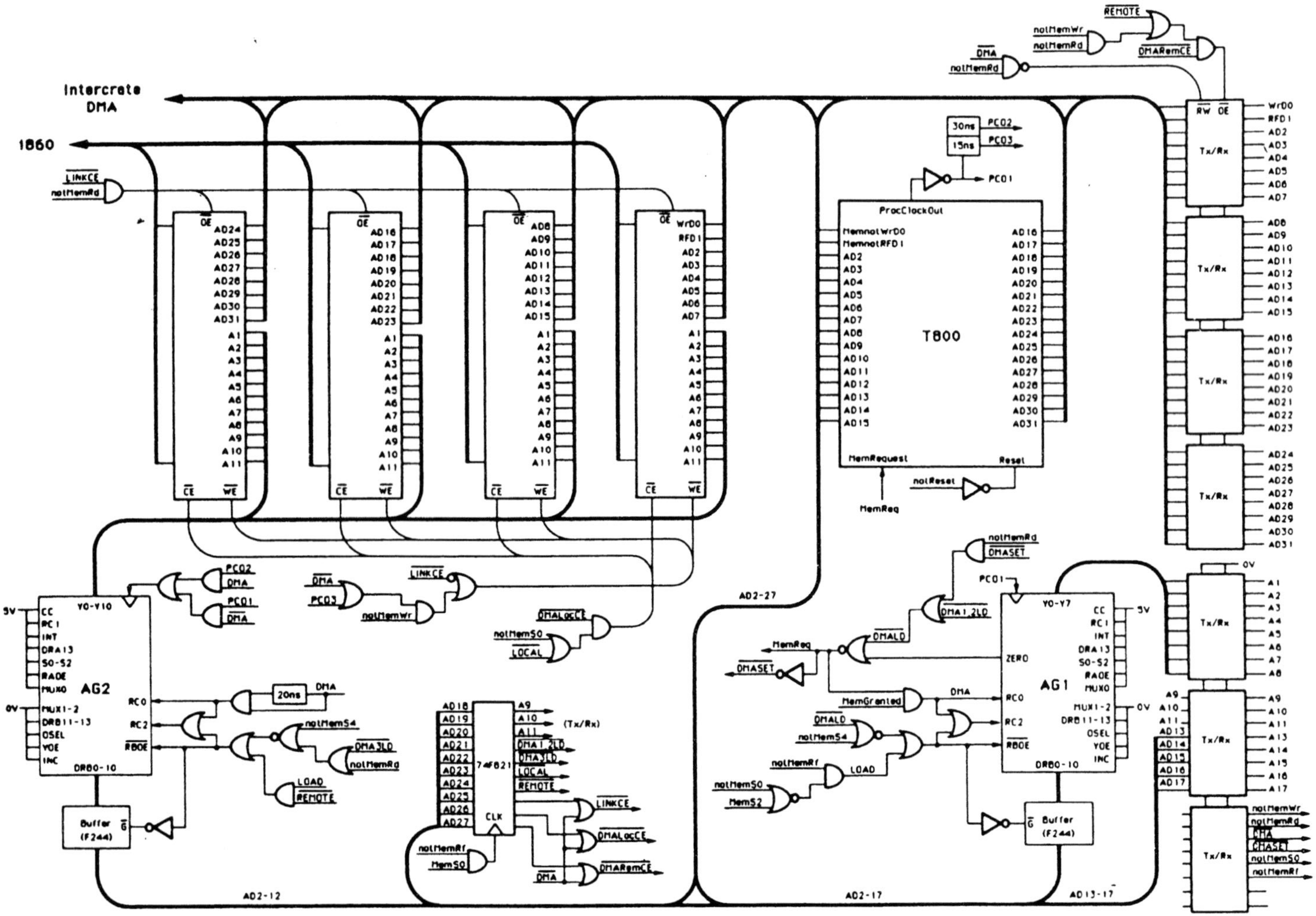

Figure 12.58 Crate controller: circuit diagram of the data transfer hardware. (From Belusevic, R. and Nixon, G., *Nucl. Instrum. & Methods,* A307, 110, 1991. With permission.)

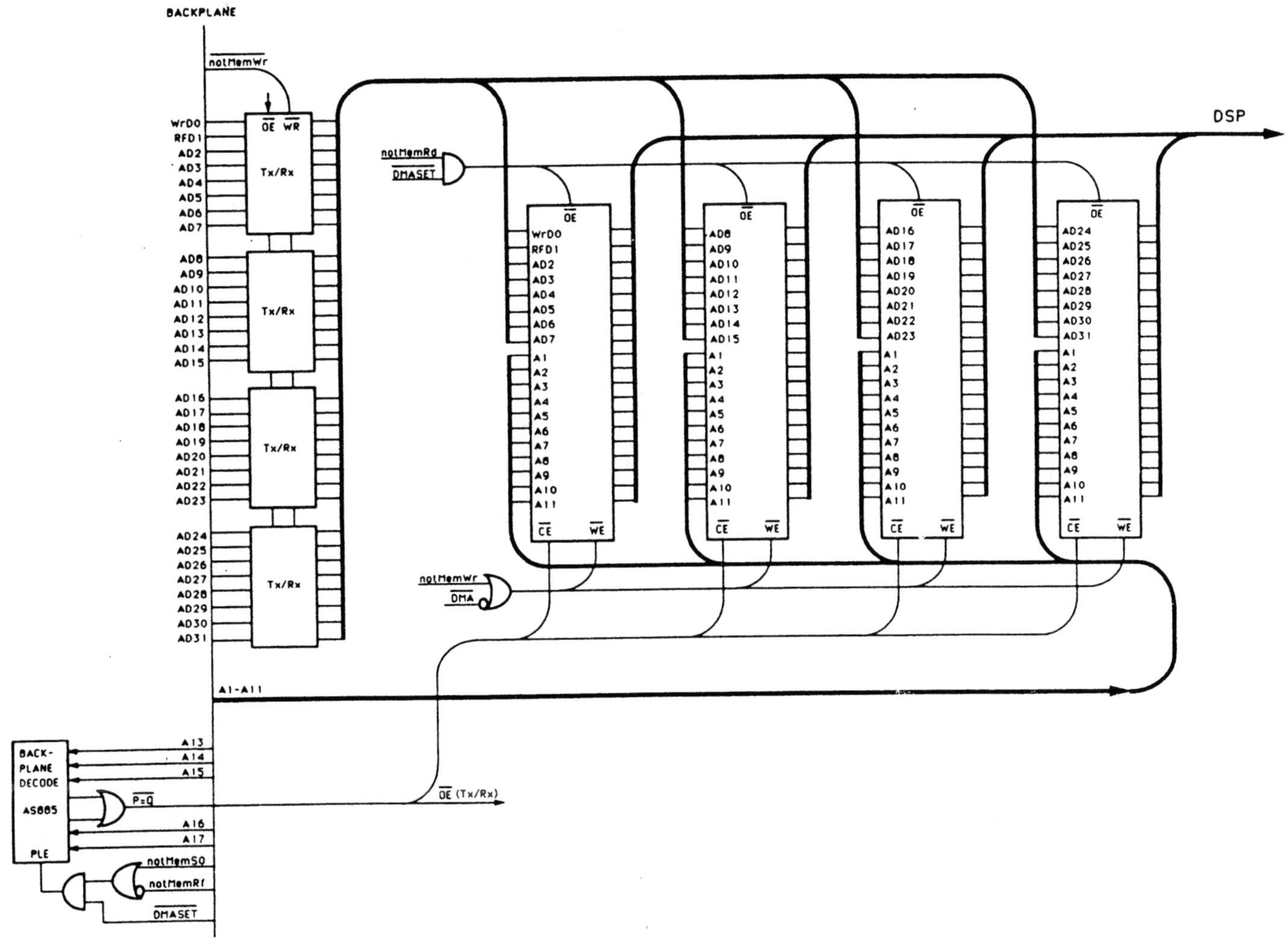

Figure 12.59 Readout card: circuit diagram of the data transfer hardware. (From Belusevic, R. and Nixon, G., *Nucl. Instrum. & Methods,* A307, 110, 1991. With permission.)

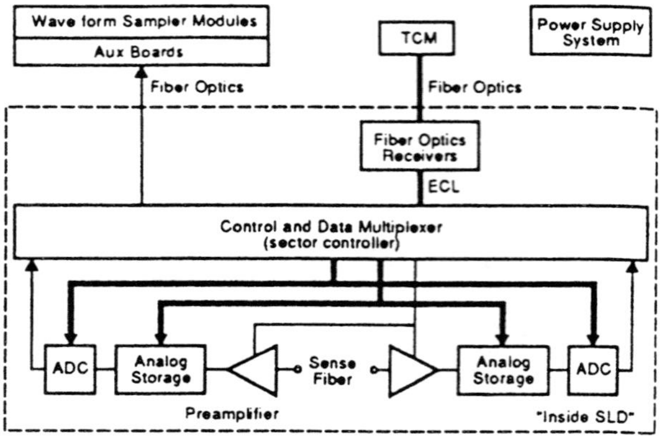

Figure 12.60 Block diagram of the data acquisition system for one CRID sense wire. (From Hoelflich, J. et al., *IEEE Trans. Nucl. Sci.*, 38, 348, 1991. With permission.)

Hoeflich et al.[31] described the front-end electronics for the Cherenkov ring imaging detector (CRID) of the SLD at the Stanford Linear Accelerator Center. The system receives signals from a highly efficient single-photo electron detector. These signals are shaped and amplified before being stored in an analog memory and processed by a digitizing system. The data from several ADCs are multiplexed and transmitted via fiber optics to the SLD FASTBUS system.

Each CRID detector has 93 sense wires on a 0.125-in. pitch. To accommodate charge division, the front-end electronics have 186 waveform digitizer channels, one for each wire end. Each channel consists of a preamplifier, analog storage, digitizer, and calibration circuitry for channel-to-channel gain and pedestal variations.

A FASTBUS plant controls the systems and processes the data from the CRID detectors. Figure 12.60 shows a block diagram of the overall acquisition system for one CRID sense wire. The timing and control module (TCM) delivers control and timing information to the detectors electronics via digital fiber optics. The communication is based on a three-wire protocol with lines named command, clock, and data. A fourth fiber is redundancy; there are two sets of such fibers for each end of the barrel and each end cap. When requested, the digitized data from the detectors are also sent via fiber optics to waveform sample modules (WSM). The WSM auxiliary board is used to demultiplex the incoming data.

The CRID amplifier schematically shown in Figure 12.61 was designed as a single-channel hybrid. The preamplifier functions are implemented using a low-noise JFET, followed by an analog semicustom integrated circuit that includes the remaining bipolar circuitry. The integrated circuit consists of a gain selectable amplifier, load driver, and power-down circuit. This, with a precision calibration circuit for the amplifier and HAMUs, is packaged on an 11-pin SIP hybrid with surface-mount components.

A versatile and automated monitoring environment has been developed to complement the data acquisition and trigger system employed in a nucleon decay experiment designed by Adarkar et al.[32] The 3834 counters of the detector are individually instrumented to process pulses from the central anode wires and thereby to measure the time of arrival and the ionization of charged particles.

The analog pulses from individual counters are shaped to have a uniform fall-time (decay constant) and fed to linear amplifiers of gain 80. The amplified pulses are converted by the discriminators to TTL pulses with widths logarithmically proportional to the input pulse heights. The amplifier and discriminator output pulses provide the basic inputs to the pulse-height and pulse-width measurement systems.

The log pulses from a layer are driven to the digital data storage module (DDSM), where the front-end circuits for event data recording, trigger generation, and monitoring are situated. A Z-80 based front-end processor (FEP) controls and provides the infrastructural facilities to eight such DDSMs; a total on nine FEPs are used in the system. An NEC PC 8801 computer, which is configured as the host, controls these FEPs through a dedicated network and provides the central data storage as well as user interface facilities. On the other hand, multiplexed linear pulses from all the SAD units of layer are fed to the analog data storage module (ADSM), where they are further multiplexed to obtain the linear pulses of a layer. Eight ADSMs are controlled by the analog control module (ACM), which is interfaced to the FEP through an extended bus to provide the control and A-D conversion functions for the ADSMs.

The trigger controller, employing template-matching algorithms, generates the trigger on identifying any one of the preprogrammed trigger patterns in the detector. The FEPs, on being interrupted by the trigger, collect event data and send them to the host. These data, along with other header information, are

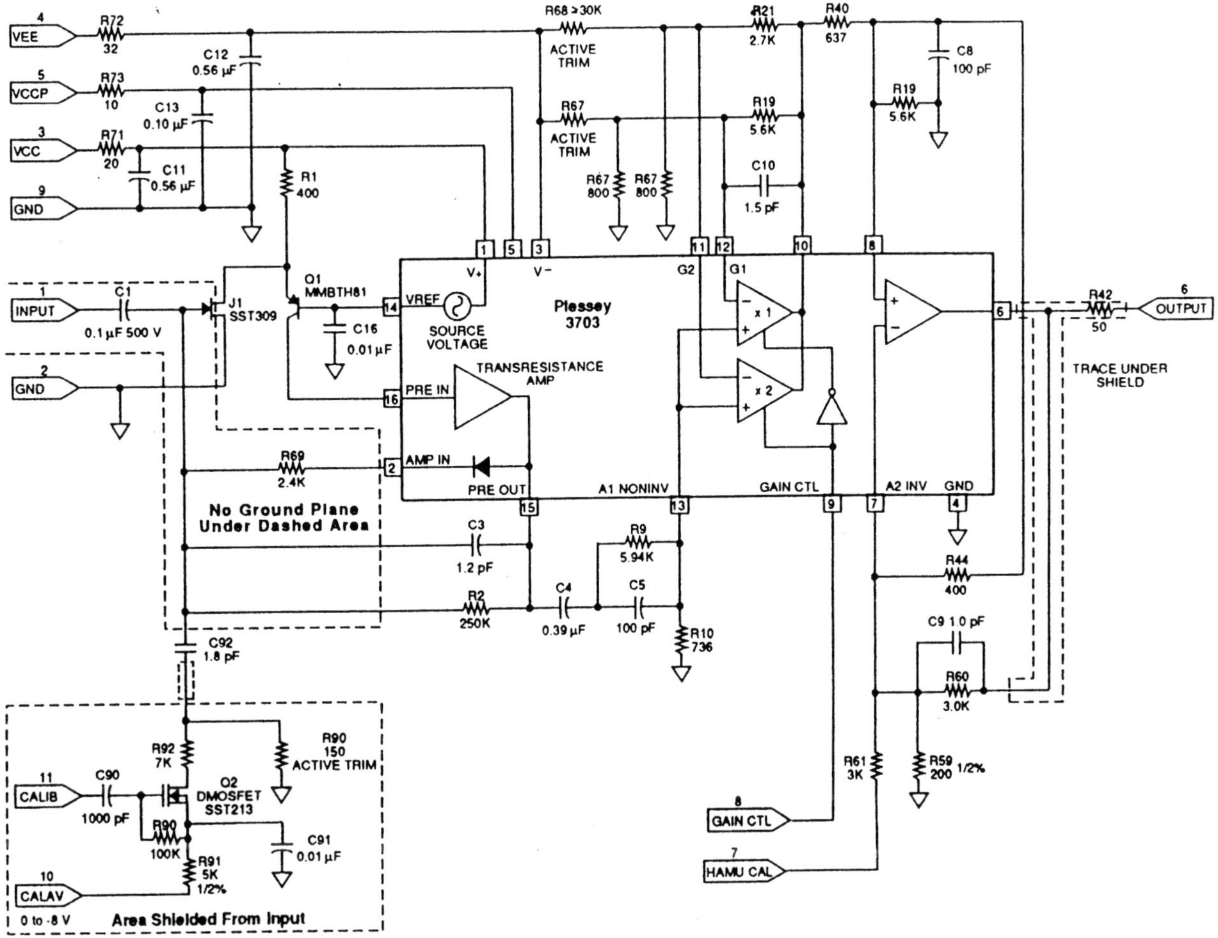

Figure 12.61 Schematic of the CRID hybrid amplifier. (From Hoelflich, J. et al., *IEEE Trans. Nucl. Sci.*, 38, 348, 1991. With permission.)

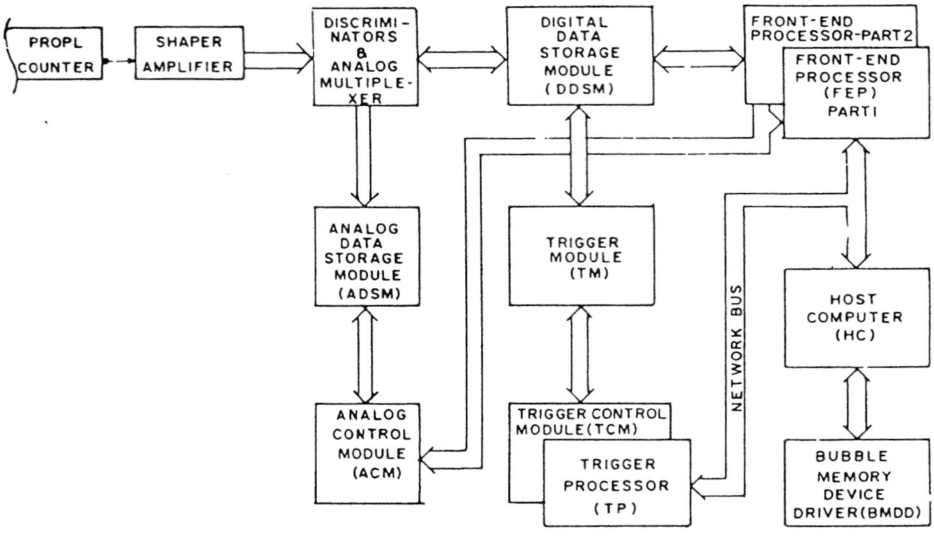

Figure 12.62 Schematic of the data acquisition system (DAQ). (From Adarkar, H. et al., *Nucl. Instrum. & Methods,* A308, 574, 1991. With permission.)

stored in a magnetic bubble cassette. The trigger controller also generates and transmits a universal clock (5 MHz), which is used by the FEPs for timing and pulse width measurements. Figure 12.62 shows the block diagram of the readout architecture.

The design considerations behind the DAQ are (1) simplicity due to the handling of digital data, (2) wider dynamic range of measurements by the use of discriminations, (3) timing information with 200-nsec accuracy, and (4) cost-effectiveness of digital ICs. Though the measurements are made mainly using the log pulses, ionization information obtained from the linear pulses of only one channel per layer for reasons of economy serves the purpose of cross-checking the log pulse data.[32]

A mixer unit for data acquisition (MUDA) was designed by Guazzoni et al.[33] This analog unit is designed to be used with multiparametric data acquisition systems, and with usual MCAs, to reduce the number of ADCs needed.

Each module consists of eight independent inputs, each one having the analog input section of an ADC (threshold, linear gate, stretcher, holder, etc.); an analog multiplexer; and a control unit for handshaking with an external ADC. It can accept and also process coincident pulses.

The working mode of MUDA is explained by the timing diagram (Figure 12.63) and general schematic (Figure 12.64). Each input is DC coupled and has its own independent lower threshold, coupled with a passive DC restorer. With threshold, a monostable rise-time protection (RTP) starts, together with the coincidence circuit with a trigger pulse that is the OR signal of the eight thresholds. A holder flip-flop starts at the end of RTP. It calls for the conversion, by means of an encoder, used for the handshaking with the ADC. Following input signals, they pass a linear gate, always open, with the stretcher connected in discharge.

Before the data processing will begin, it is necessary that all holders are set to avoid priority mistakes. To speed up MUDA, the signals are transmitted to ADC on the basis of the priority of collection, and not following an *a priori* chosen transmission code.

All holders interested are enabled at the end of RTP. The starting of holder switches off the discharge of the stretcher to allow the level of input signals to stand high. Holder closes also the input linear gate, to insulate the stretcher after RTP. In coincidence with Holder, data processing enables (DPE) starts. It enables the priority encoder for choosing the parameter of highest priority. Consequently MOS is enabled in order to put out the signal to the ADC. Just after, i.e., the time required for analog outputs to be flat (~1.5 μsec), its SAMP, used as an analog strobe for the ADC, starts. A new output to the ADC can be sent only when the ADCs rise-time protection is ended. At this moment, in fact, a new pulse starts and is held until its own SAMP arrives, in order to enable the conversion. At the end of the connected ADC RTP, another pulse is put out, waiting for SAMP and so on, up to the end of all parameters, for each event. An ADC RTP starts in connection with a leading edge of SAMP. The second one will start with the

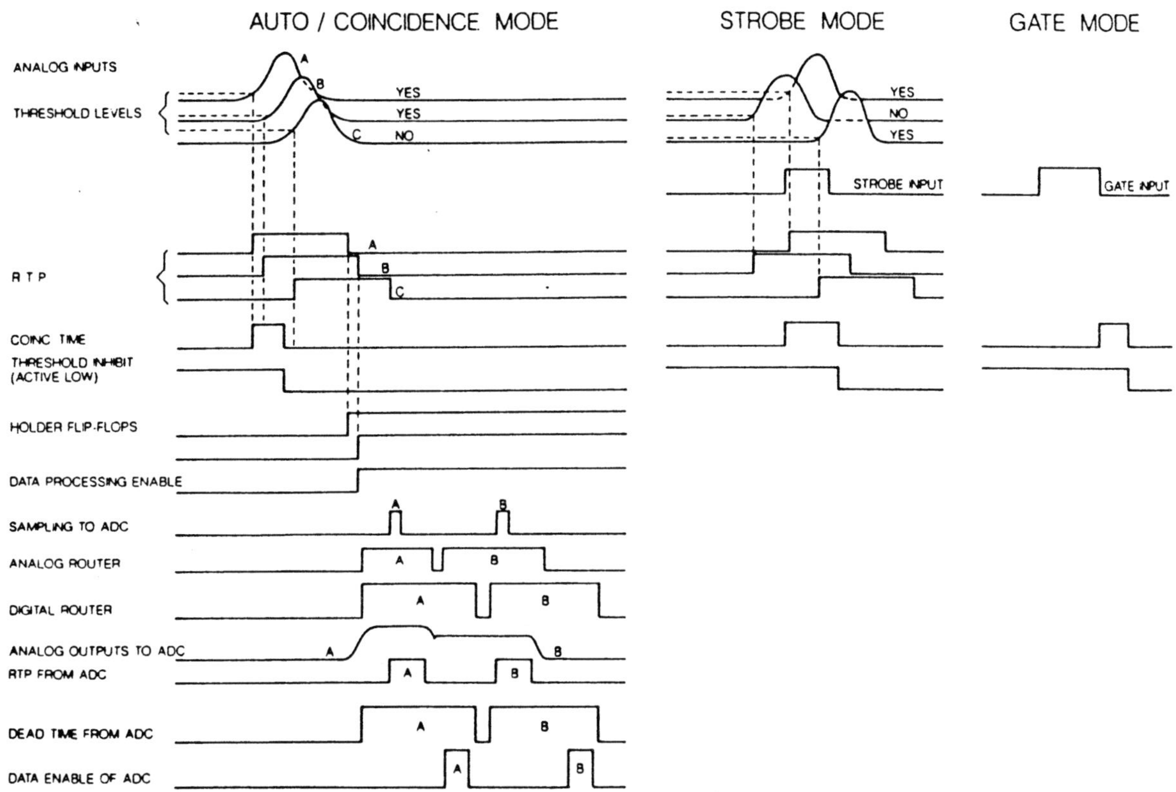

Figure 12.63 Timing diagram. (From Guazzoni, P. et al., *Nucl. Instrum. & Methods,* A305, 442, 1991. With permission.)

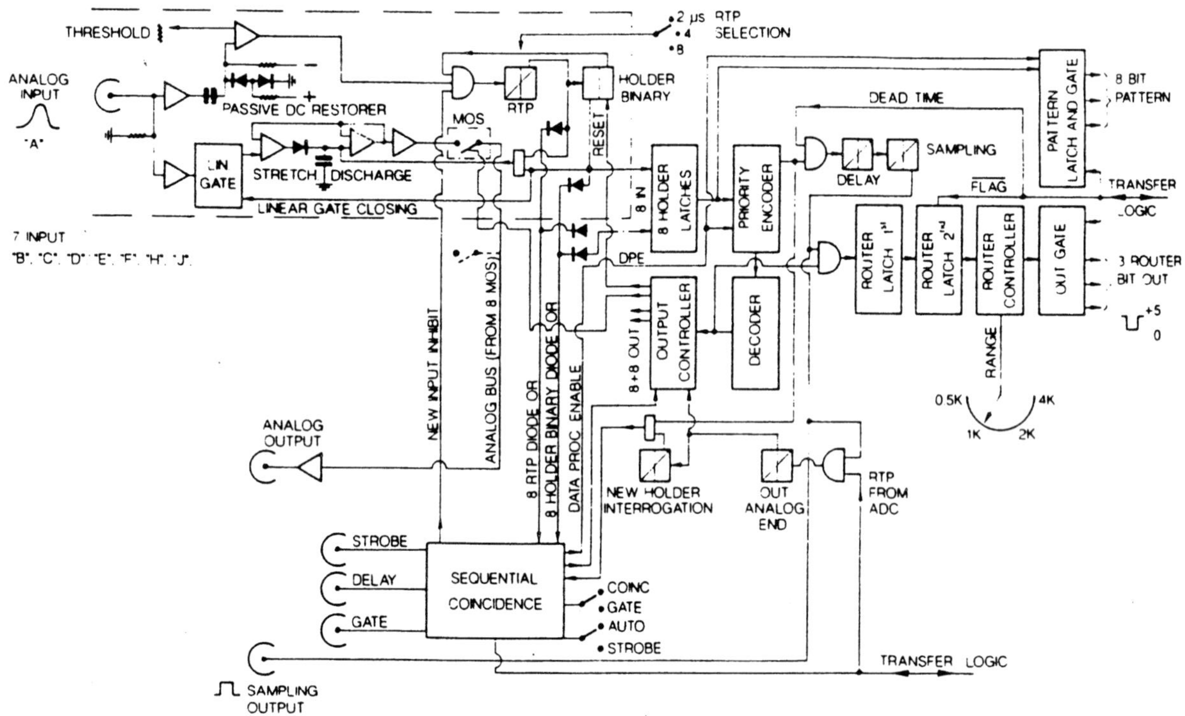

Figure 12.64 General schematic. (From Guazzoni, P. et al., *Nucl. Instrum. & Methods,* A305, 442, 1991. With permission.)

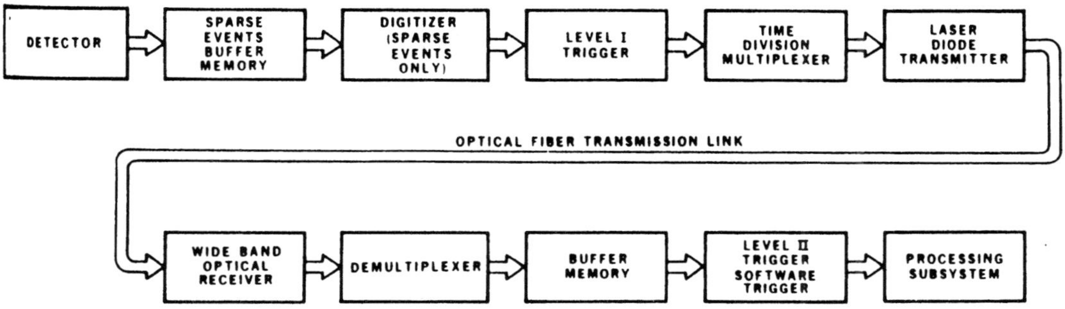

Figure 12.65 Simplified block diagram for optical fiber transmission link in SSC detector system where signals are multiplexed after the Level I trigger. (From Leskovar, B., *IEEE Trans. Nucl. Sci.*, 37, 271, 1990. With permission.)

second SAMP, etc. At the trailing edge of the first ADC RTP, the ADC will convert the A pulse, while the B one will be converted at the end of the second SAMP, and so on.[33]

The concept of guided lightwave communication along optical fibers has stimulated a major new technology over the past two decades. This technology profoundly impacts communication and instrumentation systems as well as computer interconnections and system architecture. Fiber-optic links provide several major advantages over conventional electronic systems. These include immunity to electromagnetic interference, and low transmission losses for very high-data rates. It also makes possible thinner and lighter cables and has a strong potential for long data transmission link capabilities extending to the gigahertz region.

The emergence of optical communication using fibers was made possible by the parallel development of low-loss fibers, heterojunction lasers, light-emitting diodes (which emit in spectral regions of low-fiber loss), and sensitive photon detectors.

Systems for local communication, such as computer interconnections, rf distributions in phased array radars, and instrumentation for basic and applied research, required a high-data rate capability.

Optical data transmission at the superconducting super collider detector system was discussed by Leskovar.[34] It will contain calorimetry, particle tracking, electron and muon identification subsystems, each involving typically 100,000 to 200,000 channels of readout electronics. The overall detector system should be capable of operating with luminosity corresponding to an interaction rate on the order of 10^8 events per second.

Typically, the dynamic range will be on the order of 8 b for the particle tracking chamber and 14 b for the calorimeter subsystems. The nonlinearity should be less than 1% over the operating range for some subsystems.

An example of the use of fiber-optic transmission links is given in Figure 12.65 where signals are multiplexed onto the fiber-optic waveguide after the Level I trigger. Only sparse data are digitized from all signals emanating from the detector and stored in the buffer memory. After the Level I trigger has selected the potential events of interest, the data are passed on to the time division multiplexer. The time division multiplexer converts the data from parallel bits to serial bit form for transmission over a fiber-optic waveguide. After reception of the optical signals, the demultiplexer converts the serial signals back to parallel bits for storage in another buffer memory. The data are then passed through the Level II trigger, the software trigger, and on to the processing subsystem.[34]

A time division multiplexer scheme using parallel to serial data conversion is shown in Figure 12.66. This multiplexer scheme demonstrates the principle by which data can be prepared for transmission over optical fiber cables. Appropriately controlling the parallel in/serial out lines for the two shift registers (SRA and SRB), SRA is parallel loaded from the digital buffer register A and then serially read out. In the meantime while SRA is being read out, SRB is parallel loaded with data from digital buffer register B and awaits its turn to be read out. Then while SRB is being read out, SRA is again parallel loaded with new data. SRA now awaits its turn to be read out serially. This procedure is repeated until all of the data from the buffer registers are transmitted. Control signals A and B steer the serial data through their respective gates, and the data are combined in the OR gate for presentation to the optical transmitter.

In this manner relatively slow digital operations handle the data to and from the digital buffer registers while extremely high serial data rates are handled by the optical fiber transmission system.

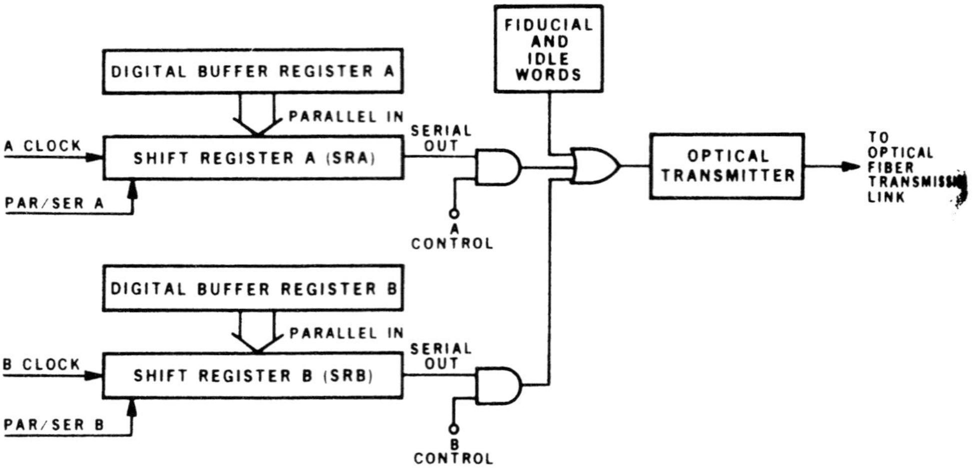

Figure 12.66 Time division multiplexer. (From Leskovar, B., *IEEE Trans. Nucl. Sci.,* 37, 271, 1990. With permission.)

Fiducial words are transmitted to alert the receiver system that the words immediately following are real data. In addition, fiducial words are transmitted periodically to pull the receiver system back into step in the event that data have become garbled and unrecoverable.[34]

REFERENCES

1. LeCroy Corporation Research Instrumentation Catalog, 1992.
2. **Bhat, P. N. et al.,** *Nucl. Instrum. & Methods,* A292, 494, 1990.
3. **Nagoshima, Y. et al.,** *Nucl. Instrum. & Methods,* 206, 147, 1983.
4. **Iqbal, M. Z. and Gimlett, J. L.,** *Nucl. Instrum. & Methods,* A248, 434, 1986.
5. **Alde, D. et al.,** *Nucl. Instrum. & Methods,* A240, 343, 1985.
6. **Abdel-Aal, R. E.,** *Nucl. Instrum. & Methods,* A309, 331, 1991.
7. **Conetti, S. and Kuchela, K.,** *IEEE Trans. Nucl. Sci.,* NS32, 1326, 1985.
8. **Freytag, D. R. et al.,** *IEEE Trans. Nucl. Sci.,* 33, 81, 1986.
9. **Hendricks, R. W. et al.,** *Nucl. Instrum. & Methods,* 201, 261, 1982.
10. **Iqbal, M. Z. et al.,** *Nucl. Instrum. & Methods,* A263, 387, 1987.
11. **Joyce, D. et al.,** *Nucl. Instrum. & Methods,* A279, 189, 1989.
12. **Allport, P. P. et al.,** *Nucl. Instrum. & Methods,* A273, 630, 1988.
13. **Crittenden, R. R. et al.,** *Nucl. Instrum. & Methods,* A270, 99, 1988.
14. **Omata, K. et al.,** *IEEE Trans. Nucl. Sci.,* 39, 143, 1992.
15. **Conetti, S. et al.,** *IEEE Trans. Nucl. Sci.,* NS32, 1318, 1985.
16. **Akimoto, T. et al.,** *IEEE Trans. Nucl. Sci.,* 37, 1594, 1990.
17. **Verweij, H.,** *IEEE Trans. Nucl. Sci.,* NS31, 211, 1984.
18. **Beconcini, F. et al.,** *Nucl. Instrum. & Methods,* A277, 222, 1989.
19. **Arai, Y. and Uehera, S.,** *Nucl. Instrum. & Methods,* A301, 497, 1991.
20. **Battaiotto, P. et al.,** *Nucl. Instrum. & Methods,* A301, 265, 1991.
21. **Sobczynski, C. W. et al.,** *IEEE Trans. Nucl. Sci.,* 36, 1989.
22. **Kurashige, H. et al.,** *IEEE Trans. Nucl. Sci.,* 35, 257, 1988.
23. **Bilei, G. M. and Castaldi, R.,** *IEEE Trans. Nucl. Sci.,* 35, 282, 1988.
24. **Ikeno, M. et al.,** *IEEE Trans. Nucl. Sci.,* 33, 779, 1986.

24a. **Walz, H. V.,** *IEEE Trans. Nucl. Sci.,* NS31, 205, 1984.

25. **Adam, W. et al.,** *IEEE Trans. Nucl. Sci.,* 39, 166, 1992.
26. **Nelson, O. R. et al.,** *IEEE Trans. Nucl. Sci.,* 37, 310, 1990.
27. **Hovander, B. and Bohm, C.,** *IEEE Trans. Nucl. Sci.,* 37, 2191, 1990.
28. **Alke, P. et al.,** The Programmable Gate Array Data Book, XILINX Inc. P/N 001004802 1986–1989.
29. **Belusevic, R. et al.,** *Nucl. Instrum. & Methods,* A295, 391, 1990.
30. **Belusevic, R. and Nixon, G.,** *Nucl. Instrum. & Methods,* A307, 110, 1991.
31. **Hoelflich, J. et al.,** *IEEE Trans. Nucl. Sci.,* 38, 348, 1991.
32. **Adarkar, H. et al.,** *Nucl. Instrum. & Methods,* A308, 574, 1991.
33. **Guazzoni, P. et al.,** *Nucl. Instrum. & Methods,* A305, 442, 1991.
34. **Leskovar, B.,** *IEEE Trans. Nucl. Sci.,* 37, 271, 1990.

Chapter 13

High-Voltage Supply

CONTENTS

13.1 Introduction ... 233
13.2 Basic Methods for Direct Current Supply ... 234
13.2.1 Full-Wave Power Supplies ... 234
13.2.2 Full-Wave Bridge Rectifier Power Supplies ... 235
13.2.3 Voltage-Multiple Power Supplies ... 235
13.2.4 Regulated Power Supplies ... 236
13.2.5 Shunt-Regulated Power Supplies ... 237
13.2.6 Series-Regulated Power Supplies with Zener Reference ... 237
13.2.7 Series-Regulated Power Supplies with Feedback ... 238
13.3 High-Voltage Supply Circuits ... 238
References ... 252

13.1 INTRODUCTION

High-voltage supplies are based on amplification of AC voltage with a transformer, rectifying, filtrating, and stabilizing. One way is to increase the AC line voltage with a transformer; another way is to produce high-frequency voltage using an oscillator or a flip-flop and then to multiply it with a transformer.

In the first method the AC voltage is fed into a power transformer and then to a high-voltage rectifier. From the rectifier the voltage is transferred to an error amplifier through a reference tube. The supply output voltage is determined by a voltage divider (at the error amplifier input, as in Figure 13.1). The voltage is kept constant by changing the reference tube conductivity which is fed back by the error amplifier output.

The voltage control system includes a reference tube which delivers a constant current to a system of reference resistors which cause a voltage drop. The difference between this voltage drop and the output voltage is amplified in the error amplifier and causes a change in the tube conductance, thus leading to voltage correction. The voltage supply accuracy is 0.85%. The ripple is ~100 μV RMS (<1 mV peak-to-peak [ptp]) and the stability is ±0.01% per day.

In the second voltage supply (Figure 13.2), an external DC is needed for the electronic circuits. In parallel 12 V is applied to the diode and the voltage divider; these are used as a variable reference circuit. The reference voltage which is set according to the required output voltage is fed into a difference amplifier which compares it with a fraction (or a sample) of the output voltage. The difference is transferred to a chopper operated by a stable multivibrator. The error voltage is now chopped at a frequency of 20 kHz. This voltage is amplified by a power amplifier and a transformer. The transformer output is rectified, filtered, and passed to the output.

This voltage supply is made so that there is protection to the preamplifier first stage from large high-voltage changes. This is done by a limited change in the reference voltage, to dv/dt < 100 V/sec. The stability of the voltage supply is <0.01%/h; the ripple is 5 mV ptp.

Methods generally used for DC production follow.

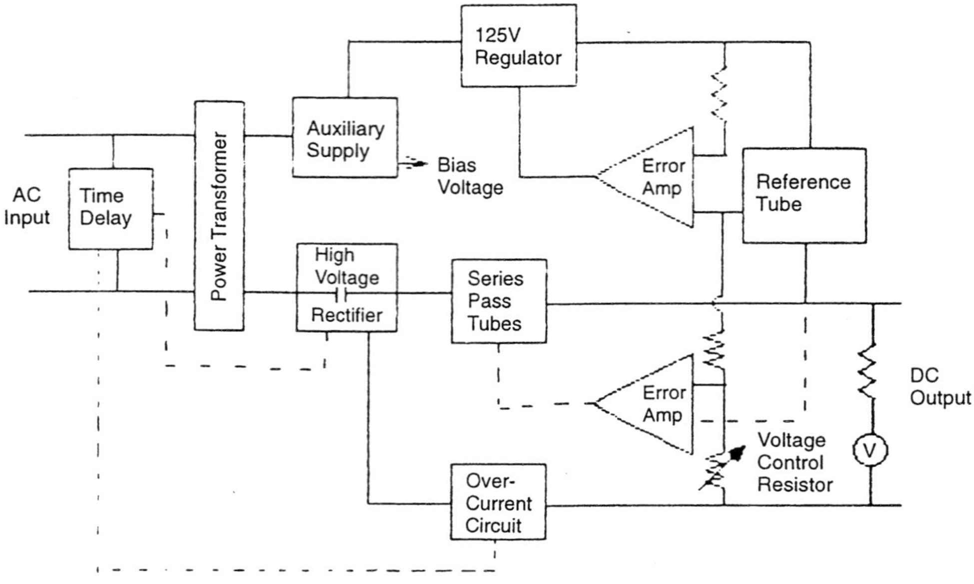

Figure 13.1 High-voltage supply.

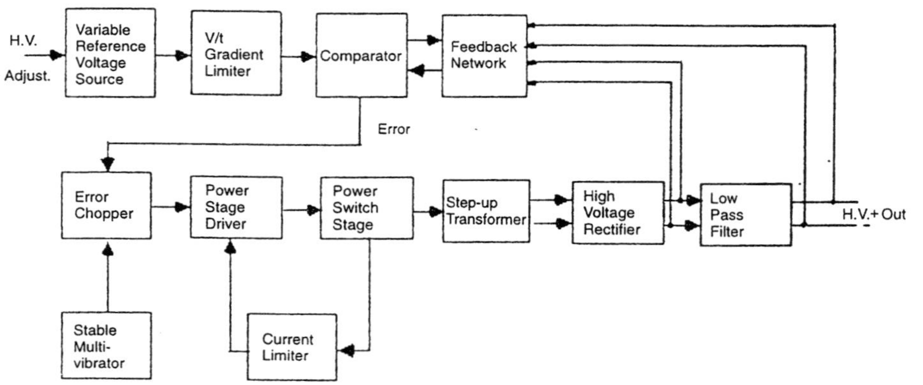

Figure 13.2 High-voltage supply.

13.2 BASIC METHODS FOR DIRECT CURRENT SUPPLY

13.2.1 FULL-WAVE POWER SUPPLIES

The V_{sec} peak of each half of the center-tapped transformer is shown in Figure 13.3. With two diodes C_1 is recharged during every half cycle of the 60-Hz waveform. This reduces the period of every charging cycle to 8.35 msec. In a full-wave rectified power supply at full load, each diode typically conducts for approximately 3.35 msec during alternate cycles. C_1 therefore discharges into the load for 5 msec per cycle. This is shown in Figure 13.4(b) using straight lines for simplicity.

The output ripple is found by substituting values into Equation 13.1 I_L and C are given in Figure 13.4(a) as 2 A and 2000 µF, respectively, and T is 5 msec.

$$V_{pp} = \frac{I_L \times t}{C_1} \tag{13.1}$$

$$\text{Ripple} = V_{pp} = \frac{2\,\text{A} \times 5 \times 10^{-3}\,\text{s}}{2000 \times 10^{-6}\,\text{F}} = 5\ \text{V} \tag{13.2}$$

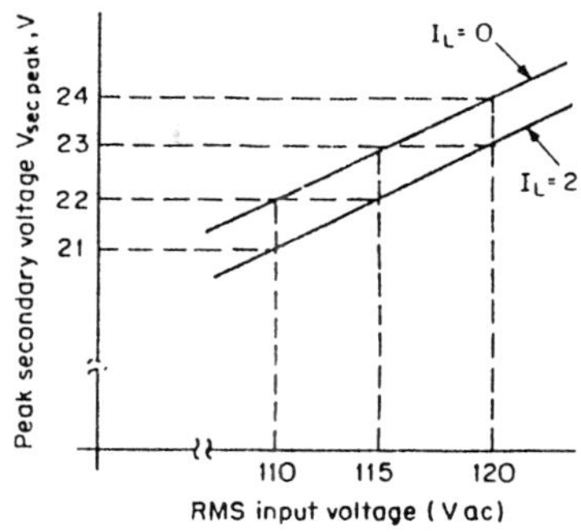

Figure 13.3 Transformer secondary voltage vs. input voltage.

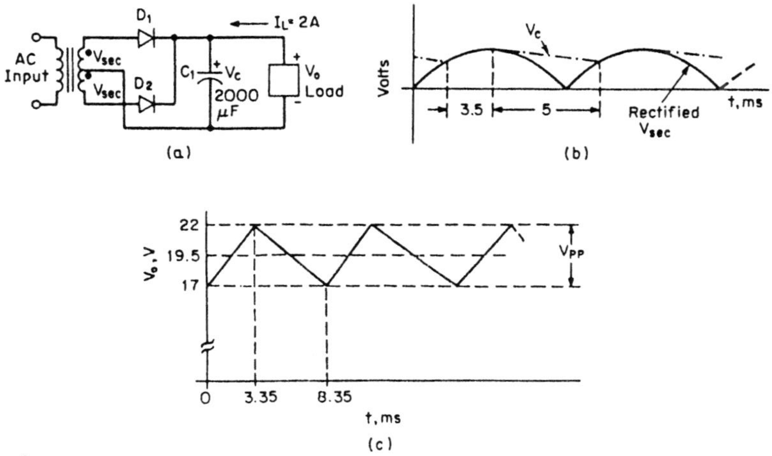

Figure 13.4 (a) Full-wave rectifier power supply; (b) V_c and rectified V_{sec} for full-wave rectifier power supply; (c) output voltage waveform for full-wave rectifier power supply.

13.2.2 FULL-WAVE BRIDGE RECTIFIER POWER SUPPLIES

Diodes D_1 and D_2 (Figure 13.5a) conduct simultaneously when V_{sec} is greater than V_c and in the correct polarity. Diodes D_3 and D_4 conduct during the next half cycle. As in the full-wave rectifier example, the discharge time of C_1 is approximately 5 msec, but this circuit offers two advantages. One advantage is that the transformer has no center tap and is therefore simpler. The other advantage is the diodes have to withstand less reverse voltage. To illustrate this, Figures 13.5(b) and 13.5(c) show the reverse voltage across the nonconducting diodes in a full-wave and bridge circuit, respectively. The conducting diode is replaced by a short circuit in both cases. In Figure 13.5(b) the reverse voltage across diode D_2 is 2 V_{sec}. In Figure 13.5(c), the reverse voltage across either D_3 or D_4 is V_{sec}. The lower reverse voltage in the bridge configuration allows the use of diodes with lower reverse-voltage capability, which are generally less expensive than their higher voltage counterparts. The cost saving of the simpler transformer and less expensive diodes, especially when the diodes are encapsulated in a single module, usually outweighs the cost of the two extra diodes needed for the bridge and has made the bridge rectifier the configuration most commonly used where stiff regulation is not a factor. The bridge integrated circuit (IC) module is also frequently used to feed a three-terminal IC regulator.

13.2.3 VOLTAGE-MULTIPLE POWER SUPPLIES

Figure 13.6(a) shows a half-wave doubler circuit, which operates in the following manner. During one half cycle, D_1 conducts, charging C_1 to V_{sec} peak. During the next half cycle, D_2 conducts, charging C_2 to V_{sec} peak plus V_{C1} which equals 2-V_{sec} peak. Figure 13.7(a) shows a full-wave doubler circuit, which operates as follows. During one half cycle, D_1 conducts, charging C_1 and providing load current through C_2. On the next half cycle D_2 conducts, charging C_2, and providing load current through C_1. The total

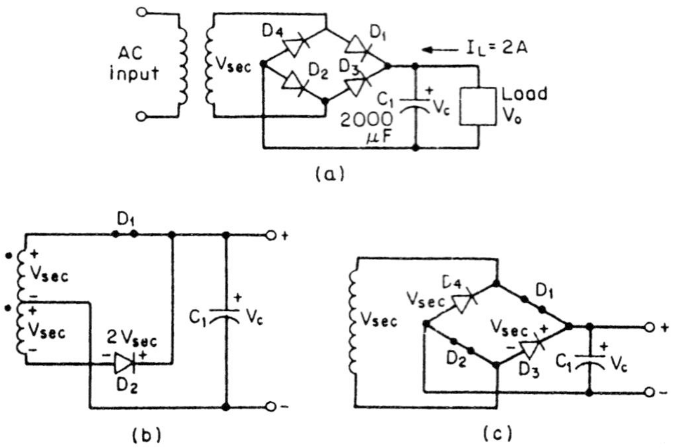

Figure 13.5 (a) Full-wave bridge rectifier power supply; (b) voltage across nonconducting diode in a full-wave rectifier power supply; (c) voltage across nonconducting diodes in a full-wave bridge rectifier power supply.

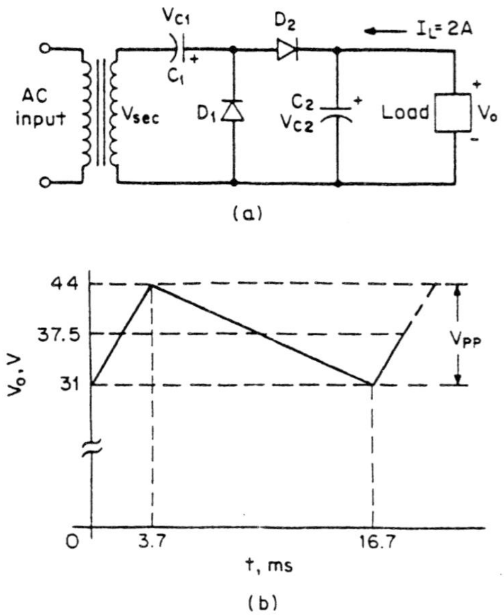

Figure 13.6 (a) Half-wave doubler power supply; (b) output voltage waveform for half-wave doubler power supply.

voltage appearing across the load is either V_{sec} plus V_{C1}, or V_{sec} plus V_{C2}. In either case, the total voltage equals 2 V_{sec}. Since C_1 and C_2 are charged on alternate half cycles, the output ripple is as in the full-wave rectifier.

Figure 13.8 illustrates the extension of the half-wave doubler circuit to provide a five-times multiplier. High-voltage diodes are used, and the entire assembly is usually encapsulated in an insulating material.

13.2.4 REGULATED POWER SUPPLIES

In many applications, the changes in power supply voltages obtained with unregulated power supplies cannot be tolerated. Ripple and regulation specifications of less than 100 mV are quite common. To achieve this, some form of voltage regulation must be used.

Sometimes it is helpful to think of a regulator as being a filter since the idea of holding an output voltage or current constant involves filtering out the variations. The two main types of regulators are series and shunt regulators, with the series type being more commonly used. The following are the most common circuits: (1) shunt-regulator, (2) series-pass regulator with zener reference, (3) series-pass regulator with feedback, and (4) three-terminal IC.

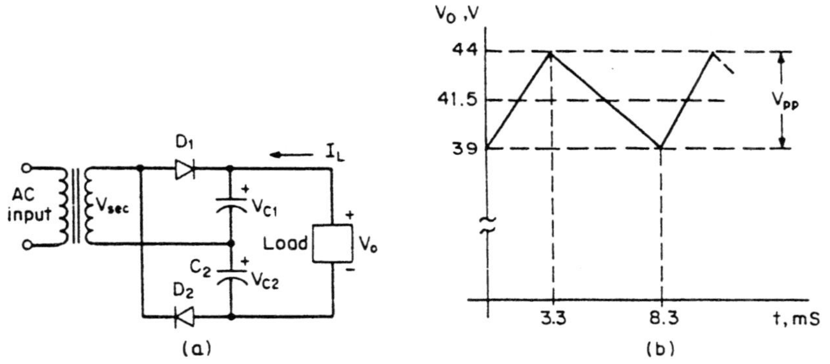

Figure 13.7 (a) Full-wave doubler power supply; (b) output voltage waveform for full-wave doubler power supply.

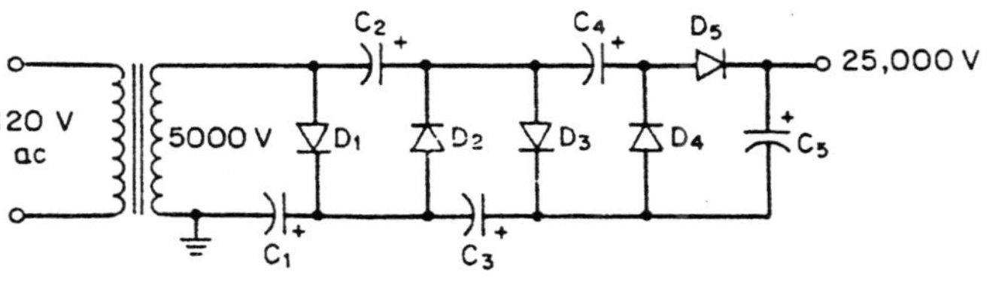

Figure 13.8 Five-times multiplier power supply.

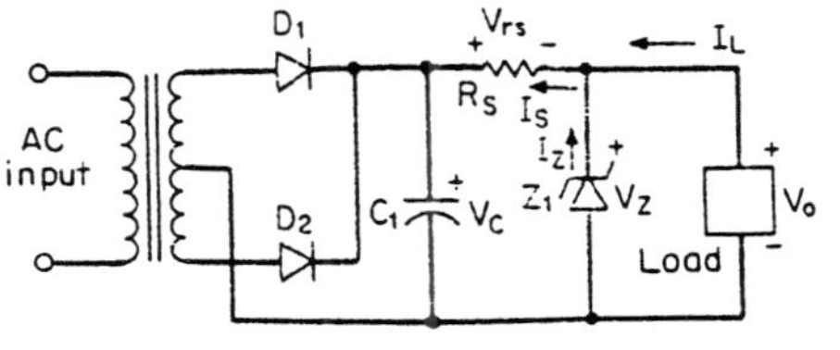

Figure 13.9 Zener shunt regulator.

13.2.5 SHUNT-REGULATED POWER SUPPLIES

The circuit in Figure 13.9 is used in many low-current applications in which the input voltage is fairly constant. When the zener diode is operated in the zener region ($I_z > I_{zt}$), it provides essentially constant voltage, even though the voltage does vary with current depending on the impedance. The minimum zener current I_{zt} is specified by the manufacturer. The equations applicable to this circuit are

$$V_o = V_z = V_c - I_s R_s \tag{13.3}$$

$$I_s = I_z + I_L \tag{13.4}$$

$$V_z = V_{z,nom} + (I_z - I_{zt}) R_z \tag{13.5}$$

where V_o = output voltage, V_z = zener voltage, V_c = capacitor voltage, $V_{z,nom}$ = nominal zener voltage, I_s = shunt current, I_z = zener current, I_{zt} = minimum zener current, I_L = load current, R_z = zener impedance, and R_s = shunt resistance.

13.2.6 SERIES-REGULATED POWER SUPPLIES WITH ZENER REFERENCE

The circuit in Figure 13.10 uses a transistor to absorb the difference between the input and output voltages and is more efficient and capable of providing more load current than the previous circuit. Equation 13.5 defines V_z, and the other equations applicable to this circuit are

$$V_o = V_c - V_{ce} = V_z - V_{be} \tag{13.6}$$

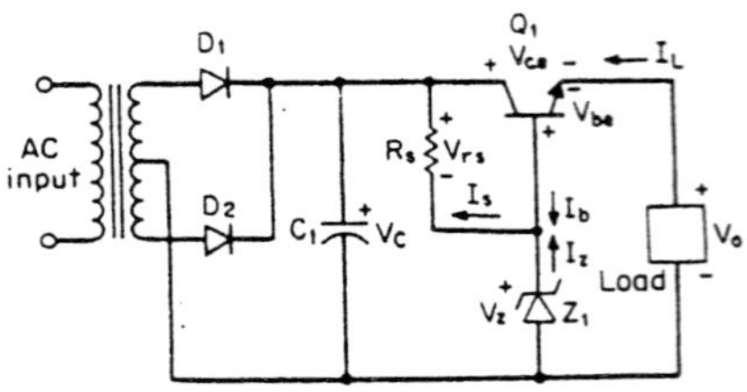

Figure 13.10 Series-regulated power supply.

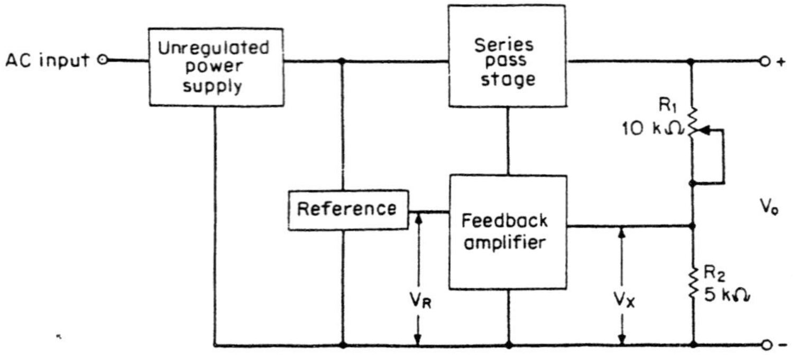

Figure 13.11 Basic diagram of a series-regulated power supply with feedback.

$$I_s = I_z + I_b = (V_c - V_z)/R_s \quad (13.7)$$

$$I_b = I_L/H_{FE} \quad (13.8)$$

At full load in this circuit, the zener diode absorbs only the base current, which is considerably less than the load current. Thus there is less power in the zener diode than in the zener of Figure 13.9. As V_c increases, V_{ce} will increase to keep V_o constant. Since I_z also increases with V_c, V_z and V_o will also increase slightly, affecting line regulation. As I_L increases, V_{be} increases due to the transconductance G_m of the transistor. Since I_b also increases, I_z decreases and the resultant decrease in V_z adds to the increase in V_{be}, affecting load regulation.

13.2.7 SERIES-REGULATED POWER SUPPLIES WITH FEEDBACK

The regulated power supplies discussed above have two drawbacks: regulation is still relatively poor, and the output voltage cannot be adjusted. To improve regulation, a high-gain feedback loop is used. A potentiometer is inserted in the feedback loop to vary the output voltage.

A generalized block diagram of such a power supply is shown in Figure 13.11. The feedback amplifier compares V_x with a reference voltage V_R and adjusts the output voltage to make them equal. If the feedback amplifier has high gain, it draws negligible input current. Thus R_1 and R_2 form a voltage divider, and:

$$V_x = V_o R_2/(R_1 + R_2) \quad (13.9)$$

13.3 HIGH-VOLTAGE SUPPLY CIRCUITS

A few examples of DC supplies designed by investigators will be cited. A low-power consumption DC converter was designed by Nirschl.[1] It is shown in Figure 13.12. The primary AC voltage is created by a push-pull oscillator. The +12 DC V obtained was by a rectifying bridge and the +120 V, by a half-wave rectifier. Both are filtered by an RC network. A zener diode feedback is used as a regulator. A 1.5-V BA-3 battery is primary voltage supply.

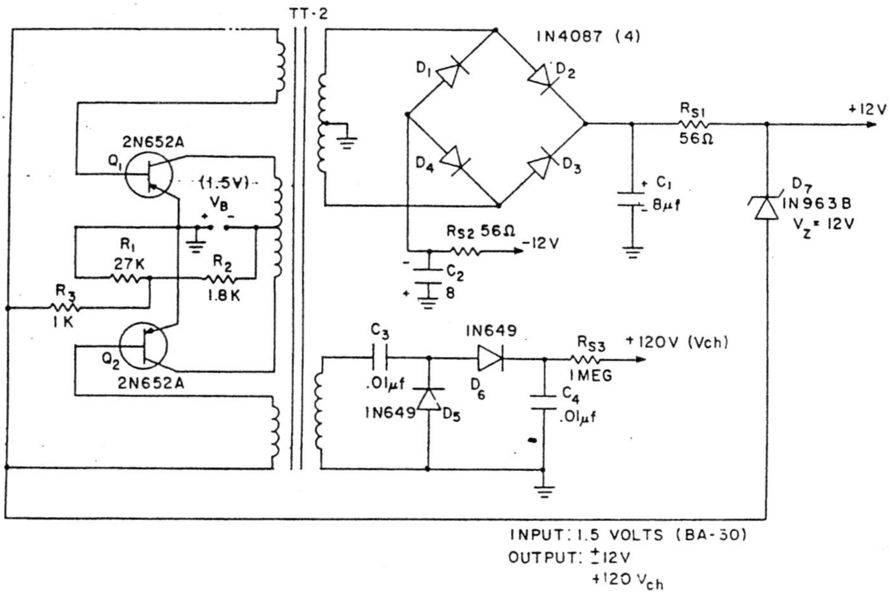

Figure 13.12 Low-power DC converter (for combination dose-rate dosimeter). (From Nirschl, J. C., *Nucl. Instrum. & Methods,* 82, 77, 1970. With permission.)

A bias supply for surface-barrier detectors (300, ±15 V) was described by Brendle and Enge.[2] It stabilized the detector bias voltages against changes of the load-resistor voltage drops. Figure 13.13 shows the circuit diagram of the regulated power supplies. The principal design ideas were the standard plus of the line-power failure detector consisting of transistors Q_4, Q_5, and Q_6, with Q_4 and Q_5 making up a bistable current regulator. The bias voltages were prevented from rising too fast. The power supply was connected to a bias regulator which limited the detector current.

Bologna et al.[3] described circuitry and performance of a programmable system (FRAM 77 HV system) to supply and control high voltages in a large scintillation counter system. Sixty-four HV channels were housed in a single crate preserving modular construction. A programmable CAMAC unit remotely controlled the HV system through a high-speed serial link. Manual control was provided by a special purpose unit. In the HV crate the analog-digital interface used a time-multiplexed technique. Each HV generator was a switching DC-DC converter with pulse-width modulation. It supplied up to 2 mA from 0.3 to 3 kV. The system block diagram is shown in Figure 13.14.

The FRAM 77 system is composed of eight HV modules with eight independent generators each, a conversion module (MCS), a timing ON/OFF module, a control module (RCU), and a CAMAC interface called master control unit (MCU).

The unit housed eight independent HV generators. Each one contained a DC-DC converter built around transistor Q_3, transformer T_1, and voltage doubler (D_3, D_4, and related capacitors) as shown in Figure 13.15. When Q_3 is driven into saturation, the supply voltage (185 V) is applied to the primary winding of T_1. The core magnetic energy is then transferred to the output, via D_4, when Q_3 opens. This is the "flyback" operating mode, but here it is somewhat modified by addition of diode D_3, giving rise to an asymmetrical voltage doubler.

The output power is modulated by varying the width of the pulse driving Q_3 from 0.45 to 4.5 μsec (a monostable, externally triggered at a constant frequency [33 kHz], has its pulse width controlled by transistor Q_4). The high-voltage resistor R_6 is used to monitor the output voltage. Amplifier IC3 yields a scaled-down buffered positive signal (VS) used by the MCS unit in readout. In addition this signal is fed to the amplifier IC4, closing the feedback loop (with control input VC).

A simple high-voltage supply used in portable radiation monitors is shown in Figure 13.16. To start with it uses a 15-V battery. A simple oscillator built around the transistor Q provides the primary AC, which is transformed into high AC voltage. A half-wave rectifier with a filtering network (C_2, C_3, R_4) provides the high DC voltage (2000 V). A zener diode is used for regulation. A ten-stage voltage doubler

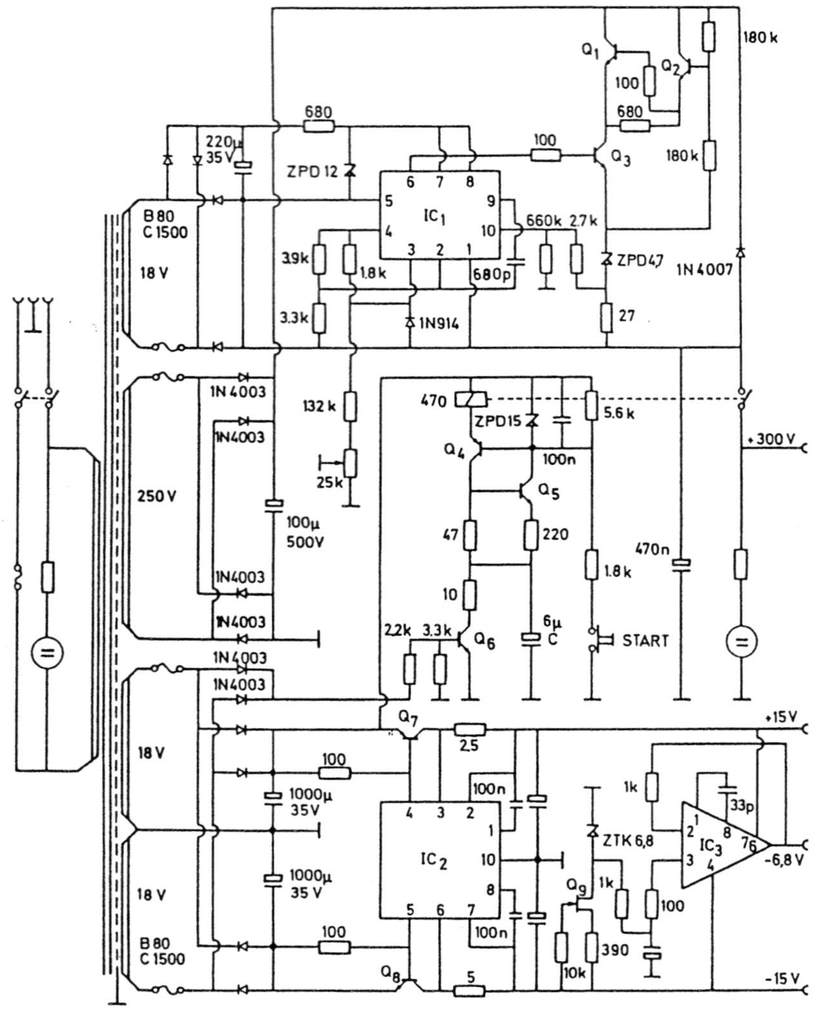

Q1, Q2, Q3 : MJE 340
Q4 : 2N 2905 A
Q5 : BC 182 B
Q6, Q8 : BD 135
Q7 : BD 136

Q9 : BC 264
IC1 : SG 723 C
IC2 : SG 4501
IC3 : SG 301A

Not explicitly denoted condensors are 10µF, 25 V tantalum

Figure 13.13 Power supply. (From Brendle, M. and Enge, R., *Nucl. Instrum. & Methods,* 129, 279, 1975. With permission.)

was designed by Baldwin and Modey,[4] as shown in Figure 13.17. It supplied 2200, 1760, 1320, and 440 V for several photomultiplier tubes. It was regulated by a self-starting saturable core oscillator. The active feedback controlled regulator kept the voltage output constant under all load conditions.

A tunnel-secondary sine wave oscillator combined with a Cockcroft-Walton voltage multiplier has been developed by Peletier,[5] which produced 40 µA at 4500 v from 12 VDC with 80% efficiency. The Cockcroft-Walton voltage multiplier method was used to obtain high DC voltages. Its voltage ripple and attenuation can affect the oscillator and was considered in the design.

An unloaded two-stage Cockcroft-Walton multiplier is shown in Figure 13.18. C_3 is charged through CR4 on each negative cycle of the sine wave generator to 1/2 $V_{S_{pp}}$; on the positive cycle, charge is transferred through C_3 to C_4 and also through C_1 to C_2. Since no charge is lost and stray capacitance is neglected, the output voltage rapidly approaches the ideal voltage:

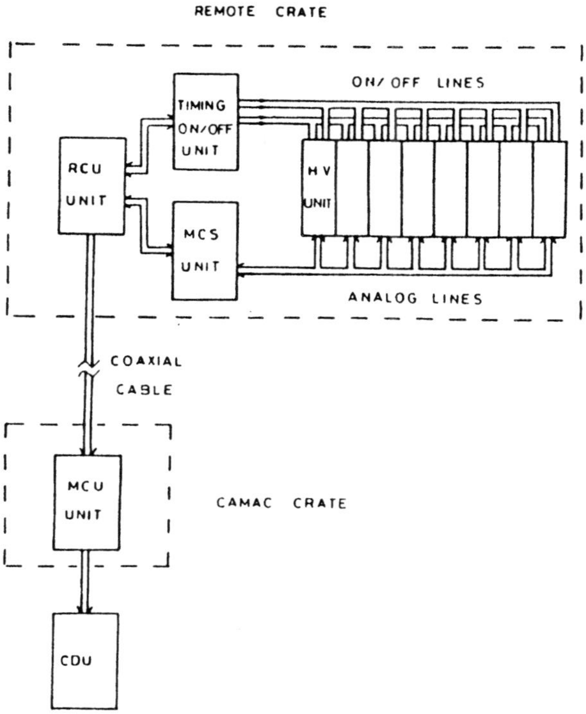

Figure 13.14 General block diagram of the system. (From Bologna, G. et al., *Nucl. Instrum. & Methods,* 192, 315, 1982. With permission.)

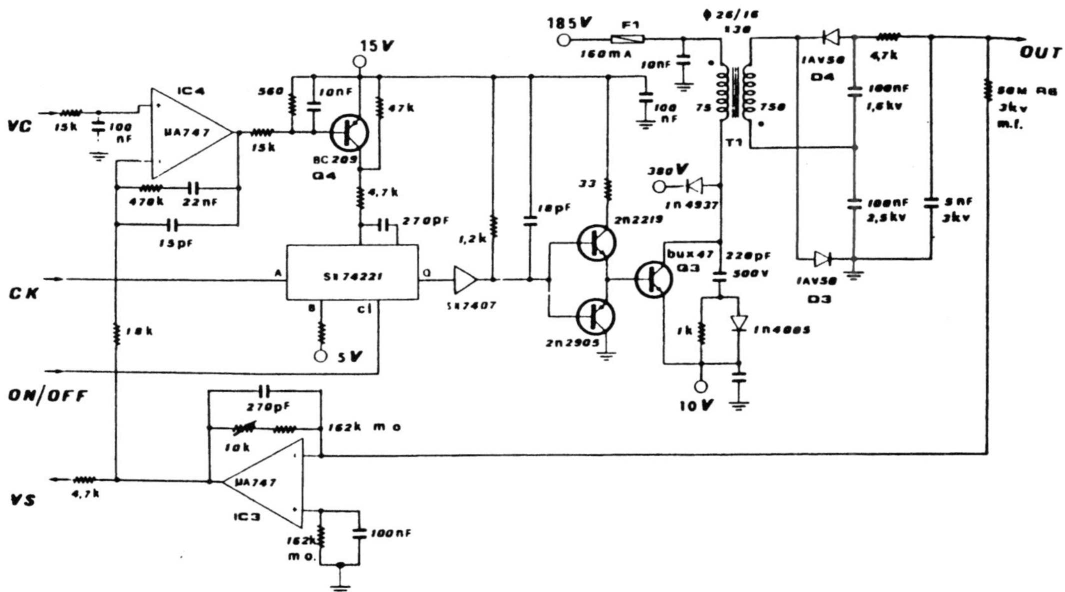

Figure 13.15 High-voltage generator circuit. (From Bologna, G. et al., *Nucl. Instrum. & Methods,* 192, 315, 1982. With permission.)

$$E_{o\ ideal} = 2V_{S_{pp}} \tag{13.10}$$

For an M-stage multiplier:

$$E_{o\ ideal} = MV_{S_{pp}} \tag{13.11}$$

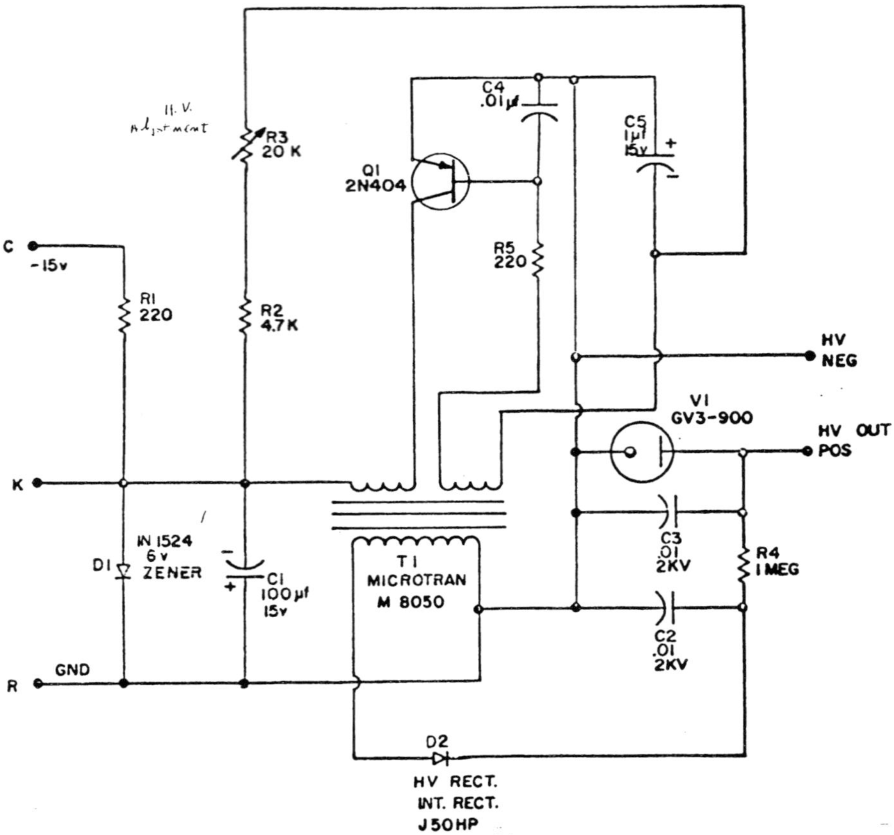

Figure 13.16 A simple high-voltage supply circuit.

The effect of stray capacitance is as follows: assume a stray capacitance per stage is C/b^2 and let each multiplier capacitor have value C. For an M-stage multiplier:

$$\frac{E_o}{E_{o\ ideal}} = \frac{b}{2M}\tanh\left(\frac{2M}{b}\right) \tag{13.12}$$

The effect of a finite load is calculated by considering the steady-state situation in which the net charge transfer per cycle through each capacitor is zero. Consider the three-stage multiplier in Figure 13.19.

The load current is supplied from C_6, C_4, and C_2 at an essentially constant rate resulting in a charge loss per cycle, ΔQ, of:

$$\Delta Q = \frac{I_L}{f} \tag{13.13}$$

This charge is restored on the positive generator cycle through the C_5-C_3-C_1-CR_1-C_2-C_4-C_6 path. C_1 is recharged on the negative generator cycle through the C_1-C_3-C_5-CR_2-C_4-C_6 path causing an additional Q loss from C_4 and C_6. The complete discharge and recharge paths are shown in Figure 13.19: the charge transfer per cycle through each capacitor is C_1, $C_2 \pm \Delta Q$; C_3, $C_4 \pm 2\ \Delta Q$; and C_5, $C_6 \pm 3\ \Delta Q$. In the case of M stages, C_{2M} and C_{2M-1} have a charge transfer per cycle of $\pm M\ \Delta Q$.[5]

Since the currents are in phase, the output ripple voltage, E_{OR} is[5]

$$E_{OR} = \Delta V_{C2M} + \Delta V_{C2M-2} + \cdots \Delta V_{C2} \tag{13.14}$$

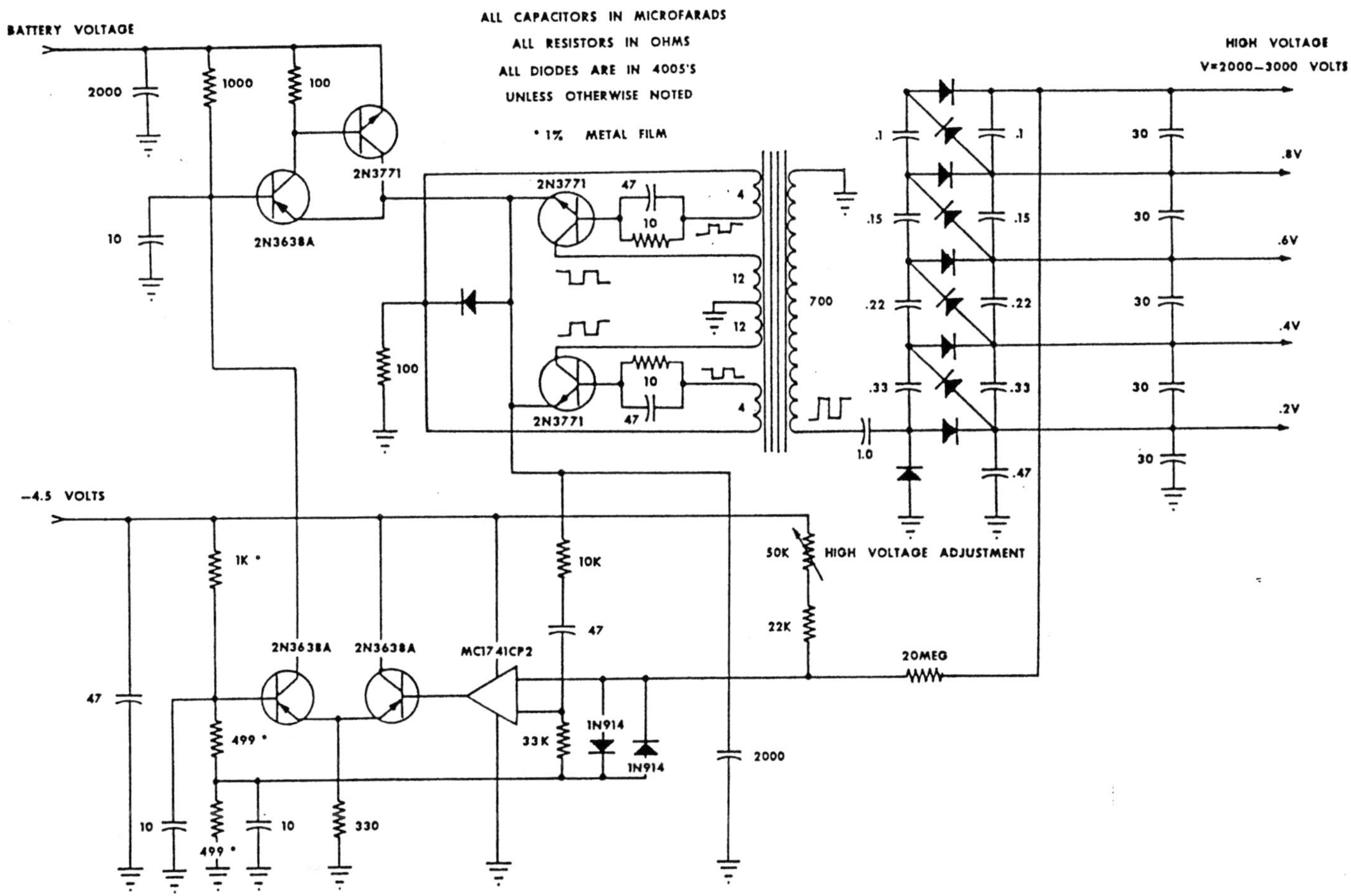

Figure 13.17 High-voltage power supply. (From Baldwin, A. L. and Modey, R., *Nucl. Instrum. & Methods,* 129, 257, 1975. With permission.)

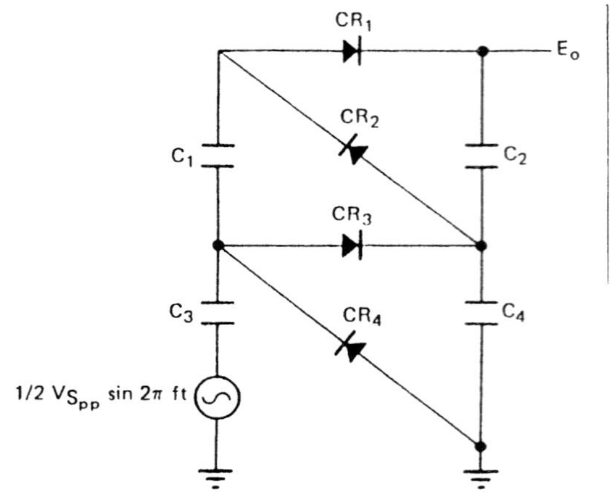

Figure 13.18 A two-stage Cockcroft-Walton multiplier. (From Peletier, D. P., *IEEE Trans. Nucl. Sci.*, NS20(1), 107, 1973. With permission.)

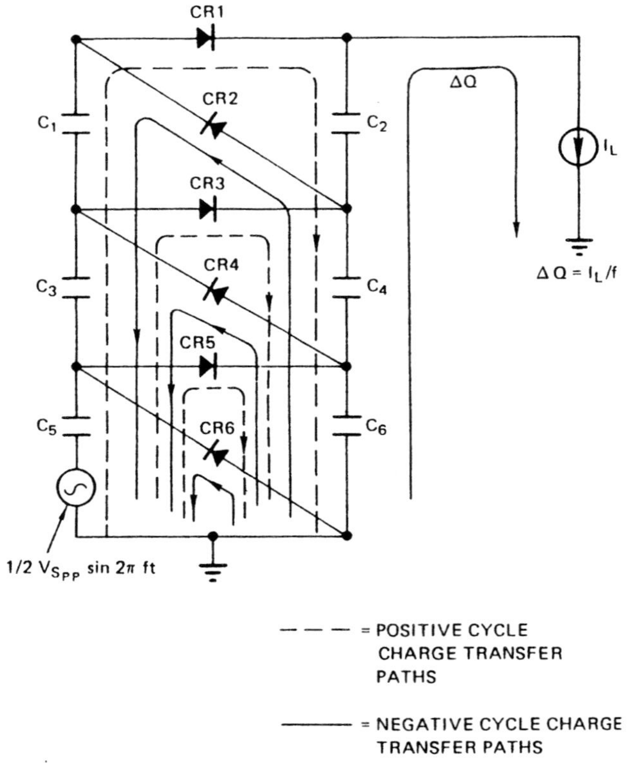

Figure 13.19 A three-stage Cockcroft-Walton multiplier showing the charge transfer paths. (From Peletier, D. P., *IEEE Trans. Nucl. Sci.*, NS20(1), 107, 1973. With permission.)

If each multiplier capacitor has value C:

$$E_{OR} = \frac{M\Delta Q}{C} + \frac{(M-1)\Delta Q}{C} + \cdots + \frac{2\Delta Q}{C} + \frac{\Delta Q}{C} \tag{13.15}$$

or in terms of load current:

$$E_{OR} = \frac{I_L}{fC}\{M + (M-1) + \ldots + 2 + 1\} = \frac{I_L}{fC}\frac{M(M+1)}{2} \tag{13.16}$$

Furthermore the peak-to-peak drive waveform for each stage is reduced by the ripple of the previous stage. Consequently, the DC output:

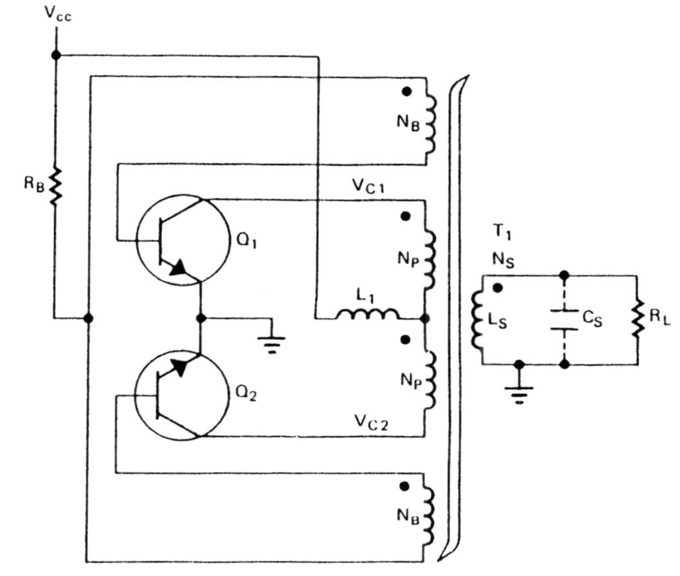

Figure 13.20 A tuned secondary linear core sine wave oscillator. (From Peletier, D. P., *IEEE Trans. Nucl. Sci.*, NS20(1), 107, 1973. With permission.)

$$E_{ODC} = V_{C_{2M}} + \ldots + V_{C_4} + V_{C_2} \tag{13.17}$$

can be rewritten in terms of the input peak-to-peak voltage, the ripple voltages, and the forward diode voltages.

It was shown by Peletier[5] to be

$$E_{ODC} = MV_{S_{pp}} - \frac{I_L}{fC}\sum_{i=1}^{M}(i)^2 - \frac{M(M+3)}{2}V_D \tag{13.18}$$

The sine wave oscillator used by Peletier is shown in Figure 13.20. The collector waveforms are half-wave rectified sine waves whose peak voltage is

$$\frac{1}{2}V_{c_{pp}} = \pi V_{cc} \tag{13.19}$$

The secondary waveform is a sine wave whose period is

$$\frac{1}{f} = 2\pi\sqrt{L_S C_S} \tag{13.20}$$

and whose peak-to-peak voltage, $V_{S_{pp}}$, is related to the primary waveform by the transformer turns ratio, a, as follows:

$$V_{S_{pp}} = a\frac{1}{2}V_{c_{pp}} = a\pi V_{cc} \tag{13.21}$$

The primary power loss mechanism in a properly designed tuned-secondary oscillator is the transformer core loss. The multiplier capacitance is chosen so as not to significantly distort the drive waveform and thereby lose the voltage-current phase advantage of the sine wave oscillator. The required number of stages is then calculated from Equation 13.21. If the number of stages is too large, a second iteration

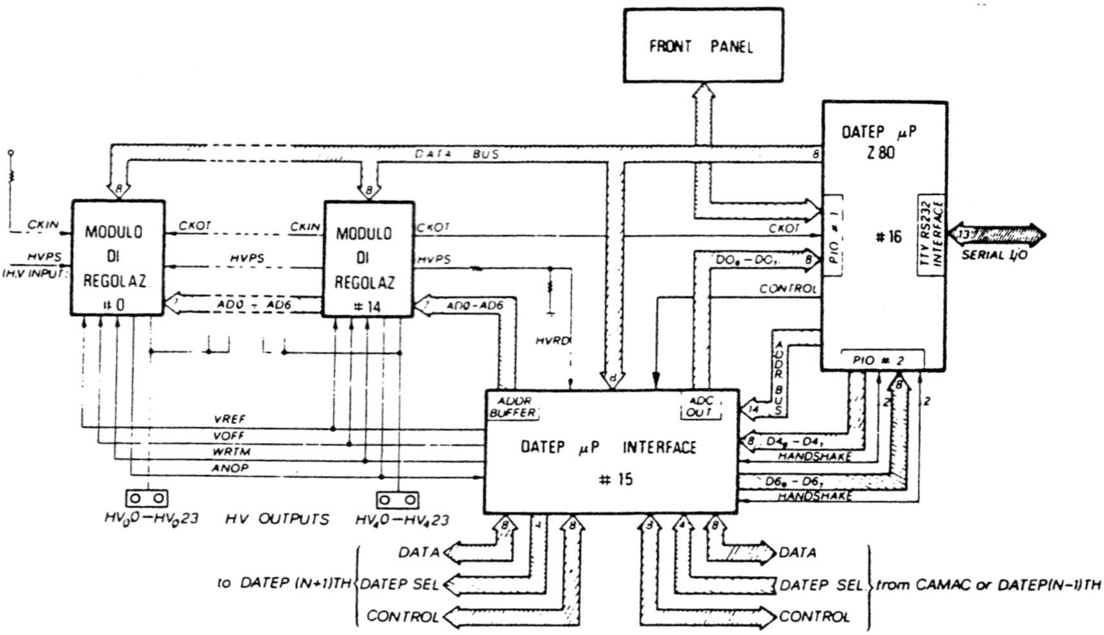

Figure 13.21 Block diagram of 120-channel high-voltage regulator. (From Centro, S. and De Giorgi, M., *Nucl. Instrum. & Methods,* 202, 439, 1982. With permission.)

of transformer design is required to increase the frequency, increase the secondary peak-to-peak voltage, or increase the allowable multiplier capacitance.

A 120-channel PM high-voltage regulator (DATEP) has been designed by Centro and De Giorgi[6] to distribute high voltage to some 1500 PMs. Each unit had its own microprocessor controller which performed continuous checking of the 120 channels and allowed operator interaction. Channel regulators were based on a specially developed thick film hybrid circuit that allowed an overall temperature coefficient better than 50 ppm/°C. Each DATEP unit supplied up to 120-HV channels with voltage that could range from about –200 to –2048 V in steps of 2 V. The maximum difference allowed among the channels was 512 V. An overall representation of the system is given in Figure 13.21. The block diagram of the regulator is given in Figure 13.22. The data lines (DD0–DD7) came directly from the microprocessor and through the bus; and the microprocessor can therefore write a byte into any channel buffer. Selection of the module and the channel was made by address lines AD0–AD6; the address lines went to the bus buffered by the interface. The actual output of a channel is given by:

$$V_n = V_{off} + V'_n \tag{13.22}$$

where the offset voltage V_{off} is supplied to every module by the interface and can be varied in steps of 8 V from about –200 to –2048 V, and V'_n is the analog value corresponding to the byte written by the microprocessor into the buffer of the nth channel; V'_n can vary from 0 to –512 V in steps of 2 V.

The measured Z_{out} impedance was about 50 Ω from DC to 100 Hz and becomes about 1 kΩ at 1 kHz. Ripple from the main power supply was attenuated 500 times at 100 Hz.

Stabilization for a high-voltage generator up to 60 kV has been developed by Georgi.[7] The voltage was stable within ±0.1 V. The load seen by the stabilizer consisted of current pulses having rise times of 2 to 10 μsec and fall times (roughly exponential) of several milliseconds. It was found convenient to distribute the stabilizing action between two stabilizers: the first deals with fast variations due to the pulsed load and sets the output voltage within a few volts; the second stabilizes against slow variations such as thermal drift and the residual 100-Hz ripple remaining after the first stabilizer. The first stabilizer was a tube-series stabilizer having up to 2.5 V ptp ripple due to 50-mA peak load current. The long time

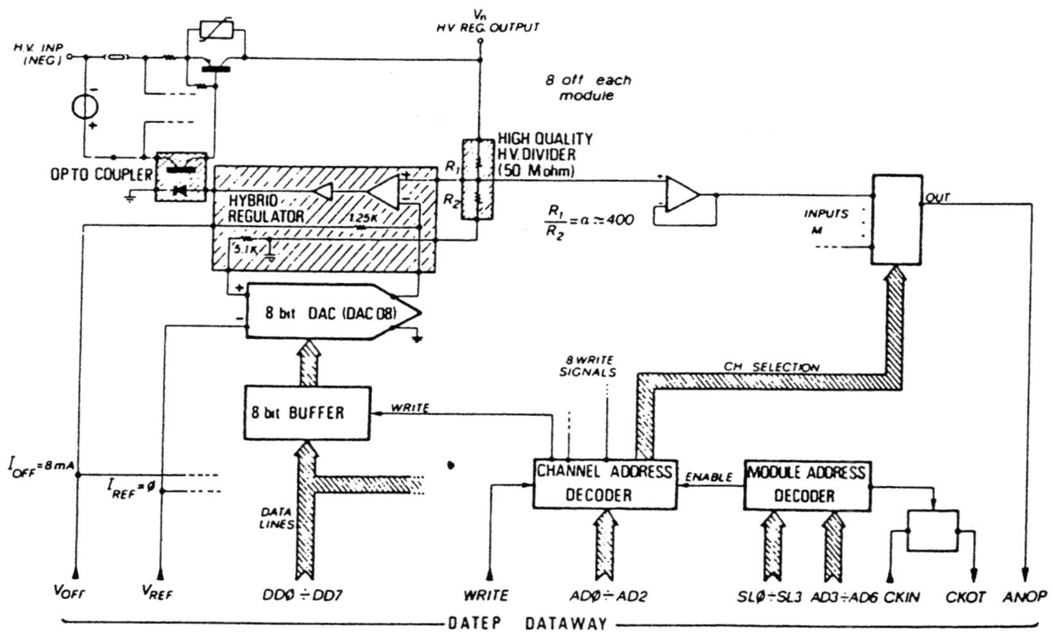

Figure 13.22 High-voltage regulator module. (From Centro, S. and De Giorgi, M., *Nucl. Instrum. & Methods,* 202, 439, 1982. With permission.)

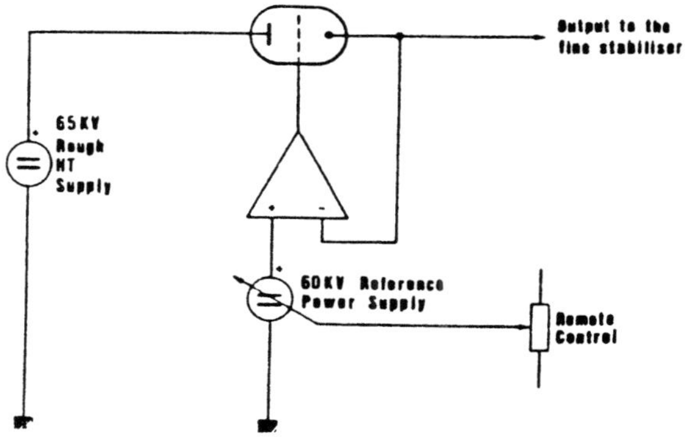

Figure 13.23 Fast stabilizer. (From Georgi, K. H., *Nucl. Instrum. & Methods,* 186, 271, 1981. With permission.)

drift and 100-Hz ripple were mainly due to the reference DC voltage generator and amount to 10 V for the drift and the ripple together. Figure 13.23 shows the circuit.

The principle of the DC regulator is shown in Figure 13.24. An accurate resistor R_o = 100 MΩ measured the stabilized voltage, and the resulting current was compared with the reference current I_{ref}. The quality of the stabilization only depends on the quality of the resistor R_o and the reference element if the amplification product V_1V_o is very high and the amplifier has nearly ideal input characteristics. With these premises:

$$U_o = I_{ref} R_o \tag{13.23}$$

The error signal at the output of the first amplifier was digitized and transmitted via fiber optics to the power regulator using V/F converters. In this way the resistor R_o performs the only direct link to the high voltage.

Assuming that the D/A converter has a real source resistance R and that R_o is much greater than R, the regulation of changes in the output can be expressed by the following formula:

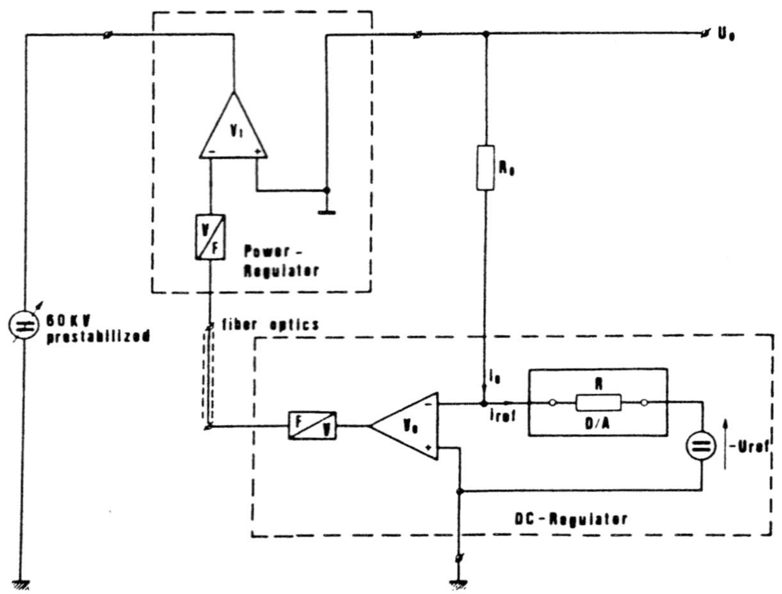

Figure 13.24 Principle of DC regulation. (From Georgi, K. H., *Nucl. Instrum. & Methods,* 186, 271, 1981. With permission.)

$$\Delta U_o = \frac{\Delta U_B}{1+(R/R_o)V_1V_o} + \frac{\Delta U_{ref}}{(R/R_o)+(1/V_1V_o)} \tag{13.24}$$

Substituting $R_o \approx 10^4\Omega$ and $V_1V_o \geq 10^7$ in the above equation results in:

$$\Delta U_o \approx 10^{-3}\,\Delta U_B + 10^4\,\Delta U_{ref} \tag{13.25}$$

The variations ΔU_B of the generator are smoothed within 0.1%, while the changes ΔU_{ref} of the reference voltage are amplified by a factor of 10^4.

The DC regulation had a finite bandwidth. The resistor R_o = 100 MΩ had the effect that a stray capacitance of about 10 pF reduced the bandwidth to 160 Hz. Moreover, the very high open loop gain requires a low 1/f roll off to stabilize the whole circuit which drives long shielded cables. Therefore a dynamic AC regulation has been added.

Circuitry which permits good chamber voltage regulation during normal operation and still provides the protection of a large series resistance to soften the voltage source in case of sparking and other discharges was described by Kerns.[8] It used the principle of fold-back current limiting. In design and application of wire chambers the series resistance between the incoming power supply voltage source and the wires is sometimes a problem.

For efficient normal operation of the chamber one would prefer that the series R be fairly small so that good voltage regulation is maintained at the chamber wires over a reasonable working range of counting rates and the ensuing average chamber current variation.[8] On the other hand, during abnormal operation, such as sparking or other damaging discharges, the chamber designer would desire a large series R to soften the voltage source.

The virtue of fold-back current limiting is that the chamber voltage is maintained constant from zero current up to some set current limit, such as 100 μA; beyond the set limit the chamber voltage falls very steeply with increasing current. Figure 13.25 shows circuitry that has been added between the cable and the chamber. During normal chamber operation, the keep-alive supply provides, via the large R, a substantially constant current (i.e., 100 μA) which at the junction of the diode and chamber wire splits and flows two ways. Depending on the count rate of the chamber, some fraction of the total constant

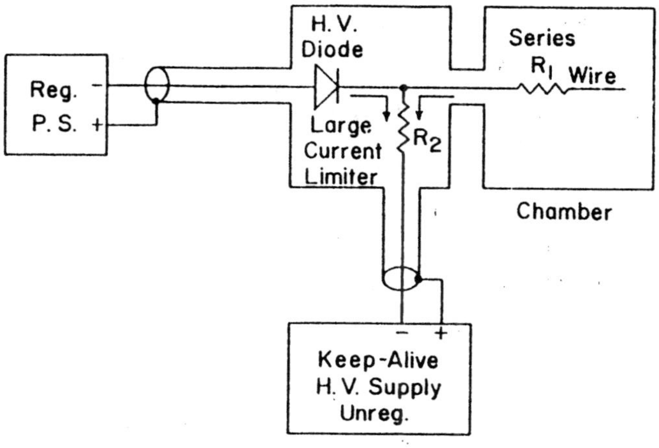

Figure 13.25 Fold-back principle. (From Kerns, C. R., *IEEE Trans. Nucl. Sci.*, NS24, 205, 1977. With permission.)

current is provided by the chamber with the remainder provided for by the diode leg. Supposing the keep-alive current to have been chosen as 100 μA, then for any chamber current from zero to 100 μA, the diode is turned ON, and the chamber voltage is constant. For any additional current δ above 100 μA, the chamber voltage falls with increasing current at the rate of $\delta \times R_2$.

The overload range can be made to fall as steeply as required, by making R_2 extremely large.

A high-voltage power supply controlled by an IBM PC XT has been constructed by Manfredotti et al.[9] to work in connection with a HP-Ge detector. The high-power supply is a half-wave cascade voltage multiplier. The highest voltage required to bias our HP-Ge is $V_0 = 5000$ V.

The maximum ripple value thought *a priori* tolerable was taken equal to 0.1% of the minimum working V_0 that is $\delta V = 1$ V. To optimize the number of stages, because the multiplication efficiency decreases as the number of stages is increased a number of six was chosen. Setting the operating frequency, they had to deal with the following problems:

1. The voltage drops due to the load ΔV_0 theoretical voltage ripple δV_0 are affected by the operating frequency, i.e., the higher the frequency, the lower the voltage drop and the ripple due to the load.
2. The rectifier efficiency due to the recovery time decreases and the circulating current increases with increasing frequency.

They have fixed an operating frequency around 5 kHz for the input AC voltage. The value of the coupling capacitors was determined, according to the ripple voltage δV_0 and the circulating current I, by the following relationship:

$$C = \frac{1}{f\delta V_0}\frac{N(N+1)}{2} = 10^{-8}\,F \tag{13.26}$$

The stray capacitance along the rectifiers, Cs, which affects the multiplication efficiency and the circulating current in the voltage multiplier, was measured to be less than 10 pF. The voltage drop ΔV_0 due to the DC load I is equal to:

$$\Delta V_0 = \frac{1}{fC}\frac{N}{3} = \left(2N^2 + 1\right) \cong 7\ V \tag{13.27}$$

where N is the number of stages.

The required input voltage to the voltage multiplier for 5000-V output is

$$V = \frac{V_0'\sqrt{C_s/C}}{\tan k\left(2N\sqrt{C_s/C}\right)} \cong 8.3\ V \tag{13.28}$$

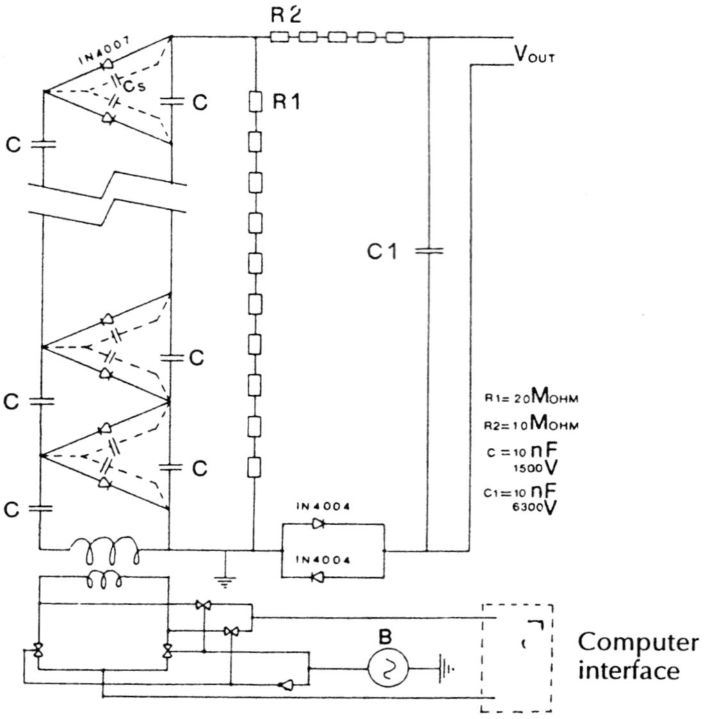

Figure 13.26 Simplified diagram of six-stage voltage multiplier. The symbol ?? means a digital switch. (From Manfredotti, C. et al., *Nucl. Instrum. & Methods,* A271, 571, 1988. With permission.)

The total voltage drop across the rectifier is about 100 V, and this is due to the voltage multiplier drive system. The drive system is a transformer in connection with an op-amp that works like an oscillator (B in Figure 13.26).

A portable HV supply was described by Kunz et al.[10] The HV is based on a tuned-circuit oscillator (Meissner oscillator) in push-pull configuration with a frequency of sinusoidal oscillation between 40 and 44 kHz, determined by the primary inductance LR and the parallel capacitor CR (Figure 13.27.). The collector voltage is coupled by the first secondary winding of the transformer (turns ratio 10:1) to the basis of the transistor to achieve positive feedback. The second secondary winding (turns ratio 1:40) is associated with a voltage-doubling circuit and delivers the required DC output voltage, ranging from 450 to 900 V. The feedback control voltage (2.5 to 5 V) regulates linearly the high-voltage output.

A very low-power consumption of less than 80 mW of the HVS is achieved by using HV diode UHVP208 with a leakage current lower than 1 μA and fast reverse recovery time of 50 nsec in the doubling rectifier part. A ripple voltage lower than 3 mV with no line distortion is achieved using a simple pi-filter circuit (C-R combination) at the output. The high voltage has, at the level of 650 V, a long-term drift of ±0.03% in 24 h and a temperature coefficient of 0.02%/°C in the range of –10 to +70°C. To avoid spark-over, the high voltage increases slowly with a rise time of 5 sec.

A thyristor device constructed for the generation of pulses having a voltage on the order of 1.5 kV and a variable frequency between 50 Hz and 1 kHz was described by Gippetto.[11] By increasing the number of stages and especially by using different components (thyratron or ignitron), it was possible to obtain high voltages.

The functioning principle of this device was based on the charging of capacitors in parallel and the discharging of the same in series. (The same method is used in the Marx device.) The circuit is shown in Figure 13.28.

As can be seen in Figure 13.28, no resistance but inductor resonance was used for the charging of the capacitors at double the power supply voltage; and their discharge was possible, without appreciable energy dispersion, using diodes. During discharge the capacitors were arranged in series and can be assimilated to a single capacitor of capacity 1/N (where N is the number of identical capacitors) and to a single inductance. During discharge the circuit behaves like the pulse-shaping line illustrated in Figure 13.29.

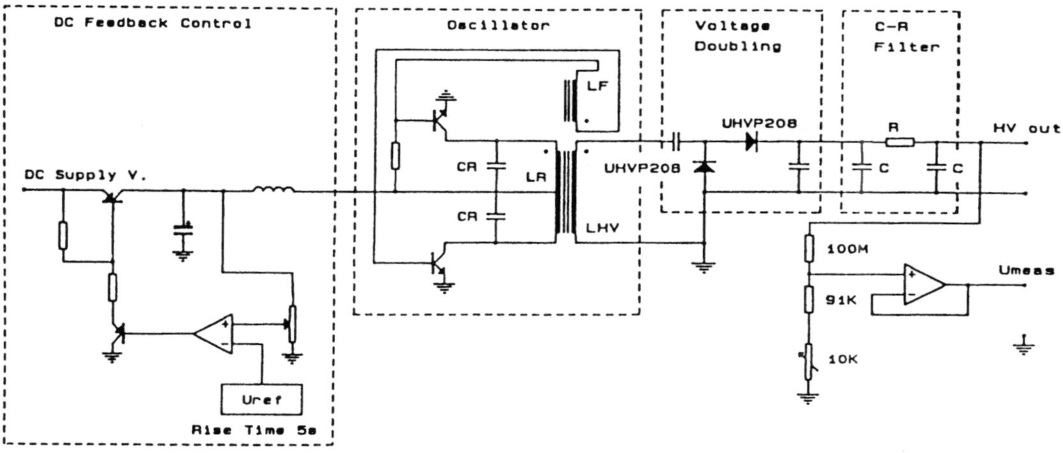

Figure 13.27 Diagram for high-voltage power supply (HVS) developed for the HANDI system to provide a stable HV ranging between 450 and 800 V to the anode of the TEPC. The HVS requires 5-V DC input supply. The feedback control circuit determines and regulates the output HV. (From Kunz, A., *Nucl. Instrum. & Methods,* A299, 696, 1990. With permission.)

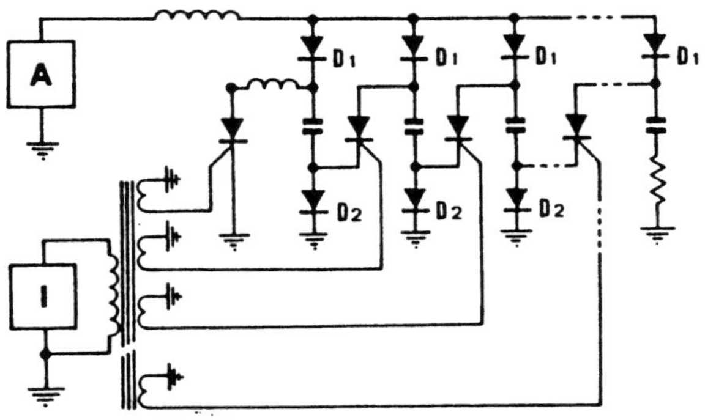

Figure 13.28 High-voltage pulse generator circuit diagram. (From Gippetto, C., *Nucl. Instrum. & Methods,* 187, 447, 1981. With permission.)

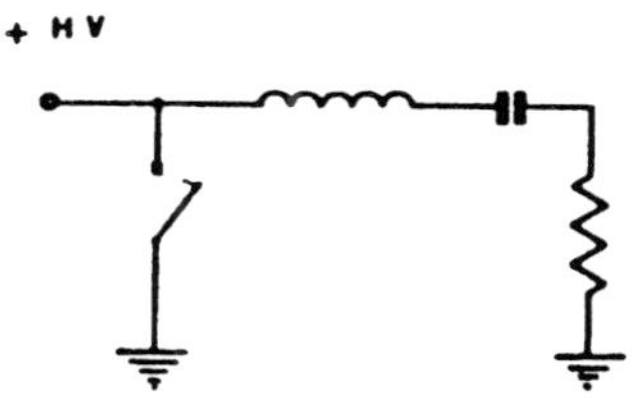

Figure 13.29 Circuit equivalent. (From Gippetto, C., *Nucl. Instrum. & Methods,* 187, 447, 1981. With permission.)

It should also be noted that during discharge of the capacitor energy is prevented from grounding by diodes D_2 while diodes D_1 should apparently not prevent the passage of current. This in fact does not occur, as the turn-on time of the thyristors is much shorter than the reverse recovery.

Edmunds et al.[12] described a system used to monitor the output of 640 high-voltage pulsers in a flash chamber neutrino detector. The device was designed to detect the most common failure modes of spark gap-switched high-voltage pulsers: failed capacitors and misfiring spark gaps.

The high-voltage pulsers (shown schematically in Figure 13.30) on receiving a trigger, switch the charge stored in their capacitors (99 nF) across the chambers. The pulse monitor system then (1) checks whether the front edges arrive in proper synchronism with a trigger, (2) determines whether there is spurious pulsing, and (3) measures the total charge in each pulse.

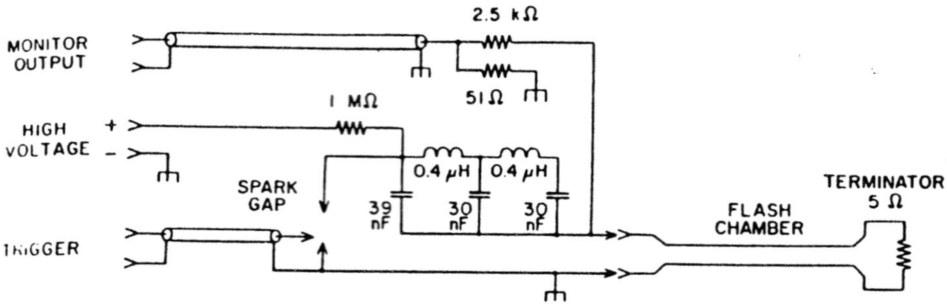

Figure 13.30 Schematic drawing of one of the 608 high-voltage pulsers. (From Edmunds, D. L. et al., *Nucl. Instrum. & Methods,* 192, 325, 1982. With permission.)

REFERENCES

1. **Nirschl, J. C.,** *Nucl. Instrum. & Methods,* 82, 77, 1970.
2. **Brendle, M. and Enge, R.,** *Nucl. Instrum. & Methods,* 129, 279, 1975.
3. **Bologna, G. et al.,** *Nucl. Instrum. & Methods,* 192, 315, 1982.
4. **Baldwin, A. L. and Modey, R.,** *Nucl. Instrum. & Methods,* 129, 257, 1975.
5. **Peletier, D. P.,** *IEEE Trans. Nucl. Sci.,* NS20(1), 107, 1973.
6. **Centro, S. and De Giorgi, M.,** *Nucl. Instrum. & Methods,* 202, 439, 1982.
7. **Georgi, K. H.,** *Nucl. Instrum. & Methods,* 186, 271, 1981.
8. **Kerns, C. R.,** *IEEE Trans. Nucl. Sci.,* NS24, 205, 1977.
9. **Manfredotti, C. et al.,** *Nucl. Instrum. & Methods,* A271, 571, 1988.
10. **Kunz, A. et al.,** *Nucl. Instrum. & Methods,* A299, 696, 1990.
11. **Gippetto, C.,** *Nucl. Instrum. & Methods,* 187, 447, 1981.
12. **Edmunds, D. L. et al.,** *Nucl. Instrum. & Methods,* 192, 325, 1982.

Chapter 14

Transmission Lines and Delay Lines

CONTENTS

14.1 Transmission Lines 253
14.2 Delay Lines 270
References 275

14.1 TRANSMISSION LINES

Homogeneous transmission lines consist of an assemblage of at least two conductors, arranged parallel to an axis x′x. The geometric and physical parameters of the line (the nature of the conductors and of the dielectric) are assumed to be constant everywhere along the line (homogeneity of the line). The current through the forward conductors is equal, but in the opposite sense, to the current through the return conductors. The total resistance of the conductors is proportional to the length of the line. A resistance per unit length R is defined. In the same way, a self-inductance per unit length L is defined, which is the result of the true self-inductance of the conductors and the mutual inductance between the two conductors. It is thus possible to define a transverse conductance per unit length G. The ratio of the charge on a unit length conductor element to the voltage between the two conductors at the element considered is the capacitance per unit length C.

Consider a line element of length Δx (Figure 14.1). This element can be compared to a quadruple (AA′, BB′) with elements R Δx, L Δx, G Δx, and C Δx. The voltage rise from A to B is

$$\Delta v = v_B - v_A = -R\Delta x i - L\Delta x \frac{\partial i}{\partial t} \tag{14.1}$$

The current Δi flowing into B from B′ is

$$\begin{aligned} \Delta i &= -G\Delta x(v + \Delta v) - C\Delta x \frac{\partial}{\partial t}(v + \Delta v) \\ \Delta i &= -G\Delta x v - C\Delta x \frac{\partial v}{\partial t} + G\Delta x^2\left(Ri + L\frac{\partial i}{\partial t}\right) + G\Delta x^2\left(R\frac{\partial i}{\partial t} + L\frac{\partial^2 i}{\partial t^2}\right) \end{aligned} \tag{14.2}$$

The first fundamental equation is thus

$$\frac{\partial v}{\partial x} = -Ri - L\frac{\partial i}{\partial t} \tag{14.3}$$

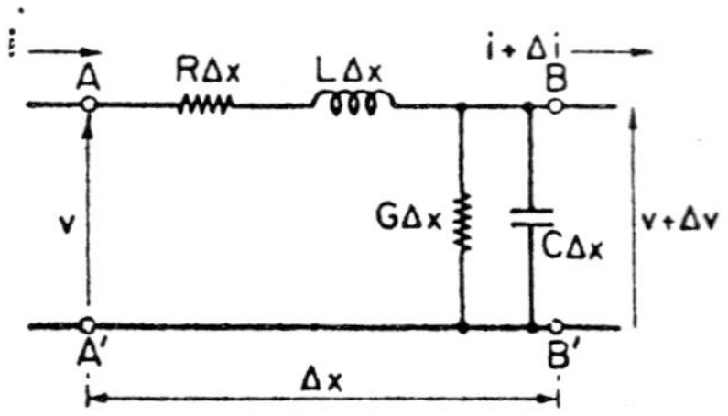

Figure 14.1 An infinitesimal line element.

and the second fundamental equation is

$$\frac{\partial i}{\partial x} = -Gv - C\frac{\partial v}{\partial t} \tag{14.4}$$

Differentiating Equation 14.3 with respect to x and replacing $\partial i/\partial x$ in this by its value in Equation 14.4 leads to

$$\frac{\partial^2 v}{\partial x^2} = RGv + (RC + LG)\frac{\partial v}{\partial t} + LC\frac{\partial^2 v}{\partial t^2} \tag{14.5}$$

This is the telegraphers' equation.

If the parameters G and R can be neglected, the fundamental Equations 3 and 4 simplify to

$$\frac{\partial v}{\partial x} = -L\frac{\partial i}{\partial t} \tag{14.6}$$

$$\frac{\partial i}{\partial x} = -C\frac{\partial v}{\partial t} \tag{14.7}$$

These relations lead to the equation for the propagation of plane waves, (setting R = 0, G = 0, in Equation 14.6):

$$\frac{\partial^2 v}{\partial x^2} = LC\frac{\partial^2 v}{\partial t^2} \tag{14.8}$$

Setting

$$u = \frac{1}{(LC)^{1/2}}\frac{1}{\delta} \tag{14.9}$$

in which u is the propagation velocity and δ is the propagation constant, or the delay per unit length:

$$\frac{\partial^2 v}{\partial x^2} = -\frac{1}{u^2}\cdot\frac{\partial^2 v}{\partial t^2} = 0 \tag{14.10}$$

The solution of Equation 14.10 is of the form:

$$v(x,t) = g(x - ut) + h(x + ut) \tag{14.11}$$

in which g and h are arbitrary functions. The current in the line can now be found from relations in Equations 14.6, 14.7, and 14.11 which yield:

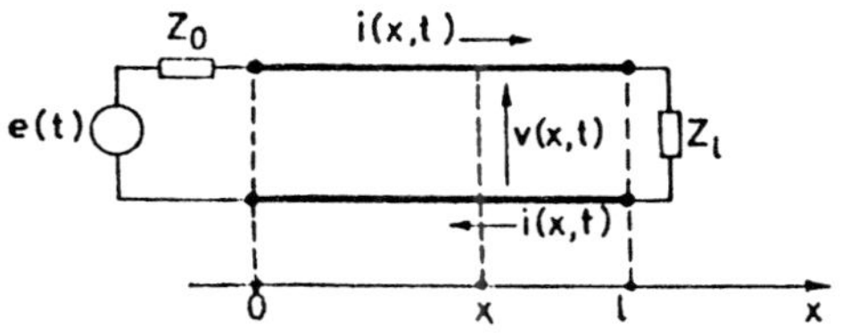

Figure 14.2 The transmission line to be studied.

$$i(x,t) = \frac{1}{R_c}\left[g(x-ut) - h(x+ut)\right] \tag{14.12}$$

where

$$R_c = (L/C)^{1/2} \tag{14.13}$$

is the characteristic resistance of the line.

Consider a transmission of length 1, as shown in Figure 14.2. At the instant t = 0, an electromotive force e(t), arising from a generator with internal impedance Z_o, is applied to the left end, or input, of the line. The right end, or output, of the line is terminated in an impedance Z_ℓ. Using the Laplace transforms of the current and voltage functions:

$$V(x,p) \leftrightarrow v(x,t); \;\; I(x,p) \leftrightarrow i(x,t)$$

Equations 14.3 and 14.5 in the transform domain become:

$$\frac{\partial V(x,p)}{\partial x} + (R+Lp)I(x,p) = 0 \tag{14.14}$$

$$\frac{\partial I(x,p)}{\partial x} + (G+cp)V(x,p) = 0 \tag{14.15}$$

Differentiating Equation 14.14 with respect to x and using Equation 14.15 to eliminate ∂I(x, p)/∂x from the result:

$$\frac{\partial^2 V(x,p)}{\partial x^2} - (R+Lp)(G+Cp)V(x,p) = 0 \tag{14.16}$$

An analogous relation can be found for the current:

$$\frac{\partial^2 I(x,p)}{\partial x^2} - (R+Lp)(G+Cp)I(x,p) = 0 \tag{14.17}$$

substitute

$$\gamma(p) = \left[(R+Lp)(G+Cp)\right]^{1/2} \tag{14.18}$$

This is called the propagation function:

$$\frac{\partial^2 V(x,p)}{\partial x^2} - \gamma^2 V(x,p) = 0 \tag{14.19}$$

$$\frac{\partial^2 I(x,p)}{\partial x^2} - \gamma^2 I(x,p) = 0 \tag{14.20}$$

The general solutions of Equations 14.19 and 14.20 are thus:

$$V(x,p) = V_1 e^{-\gamma x} + V_2 e^{\gamma x} \tag{14.21}$$

$$I(x,p) = \frac{V_1 e^{-\gamma x} - V_2 e^{\gamma x}}{Z_c} \tag{14.22}$$

where

$$Z_c(p) = \left(\frac{R + Lp}{G + Cp}\right)^{1/2}$$

The functions V_1 and V_2 are determined using the conditions at the ends of the line. Let $E(p) \leftrightarrow e(t)$ be the transform of the generator voltage. At the input to the line ($x = 0$):

$$E(p) = Z_o I(o,p) + V(o,p) = \frac{Z_o}{Z_c}(V_1 - V_2) + V_1 + V_2 \tag{14.23}$$

using Equations 14.21 and 14.22 or:

$$V_1\left(\frac{Z_o}{Z_c} + 1\right) - V_2\left(\frac{Z_o}{Z_c} - 1\right) = E(p) \tag{14.24}$$

or again

$$V_1 - V_2 \frac{Z_o - Z_c}{Z_o + Z_c} = E(p)\frac{Z_c}{Z_o + Z_c} \tag{14.25}$$

Let

$$\Gamma_o = \frac{Z_o - Z_c}{Z_o + Z_c} \tag{14.26}$$

This is called the voltage reflection coefficient at the input to the line. $|\Gamma_o|$ is always between 0 and 1:

$$V_1 - \Gamma_o V_2 = E(p)\frac{Z_c}{Z_o + Z_c} \tag{14.27}$$

In the same way, at the output of the line ($x = 1$):

$$Z_1 = \frac{V(\ell,p)}{I(\ell,p)} = Z_c \frac{V_1 e^{-\gamma\ell} + V_2 e^{\gamma\ell}}{V_1 e^{-\gamma\ell} - V_2 e^{\gamma\ell}} \tag{14.28}$$

or

$$\frac{Z_\ell}{Z_c}\left(V_1e^{-\gamma\ell} - V_2e^{\gamma\ell}\right) = V_1e^{\gamma\ell} + V_2e^{\gamma\ell} \tag{14.29}$$

$$V_1e^{-\gamma\ell}\left(\frac{Z_\ell}{Z_c} - 1\right) - V_2e^{\gamma\ell}\left(\frac{Z_\ell}{Z_c} + 1\right) = 0 \tag{14.30}$$

or finally

$$V_1e^{-\gamma\ell}\frac{Z_\ell - Z_c}{Z_\ell + Z_c} - V_2e^{\gamma\ell} = 0 \tag{14.31}$$

Let

$$\Gamma_\ell = \frac{Z_\ell - Z_c}{Z_\ell + Z_c}$$

This is the voltage reflection coefficient at the output of the line. $|\Gamma_\ell|$ is always between 0 and 1:

$$\Gamma_\ell V_1e^{-\gamma\ell} - V_2e^{\gamma\ell} = 0 \tag{14.32}$$

For the determination of V_1 and V_2 we have the two equations:

$$V_1 - \Gamma_o V_2 = E(p)\frac{Z_c}{Z_o + Z_c} \tag{14.33}$$

$$\Gamma_\ell V_1e^{-\gamma\ell} - V_2e^{\gamma} = 0 \tag{14.34}$$

Solving these yields:

$$V_1 = E(p)\frac{Z_c}{Z_o + Z_c}\frac{e^{\gamma\ell}}{e^{\gamma\ell} - \Gamma_o\Gamma_\ell e^{-\gamma\ell}} = E(p)\frac{Z_c}{Z_o + Z_c}\frac{1}{1 - \Gamma_o\Gamma_\ell e^{-2\gamma\ell}} \tag{14.35}$$

$$V_2 = \Gamma_\ell V_1e^{-2\gamma\ell} = E(p)\frac{Z_c}{Z_o + Z_c}\frac{\Gamma_\ell e^{-2\gamma\ell}}{1 - \Gamma_o\Gamma_\ell e^{-2\gamma\ell}} \tag{14.36}$$

which when substituted into Equations 14.21 and 14.22 result in:

$$V(x,p) = E(p)\frac{Z_c}{Z_o + Z_c}\frac{e^{-\gamma x} + \Gamma_\ell e^{-\gamma(2\ell - x)}}{1 - \Gamma_o\Gamma_\ell e^{-2\gamma\ell}} \tag{14.37}$$

$$I(x,p) = E(p)\frac{1}{Z_o + Z_c}\frac{e^{-\gamma x} - \Gamma_\ell e^{-\gamma(2\ell - x)}}{1 - \Gamma_o\Gamma_\ell e^{-2\gamma\ell}} \tag{14.38}$$

By series expansion:

$$\frac{1}{1-\Gamma_o\Gamma_\ell e^{-2\gamma\ell}} = \sum_{n=0}^{\infty} \Gamma_o^n \Gamma_\ell^n e^{-2n\gamma\ell} \tag{14.39}$$

For a matched line where $Z_\ell = Z_c$ the reflection coefficient Γ_ℓ is zero.

Formulas for the values of the per unit length parameters follow. The units of International System (MKS System) are used.

1. The constants of vacuum, or free space

   Dielectric constant:

$$\varepsilon_o = \frac{10^{-9}}{36\pi} \text{(F/meter)} \tag{14.40}$$

   Absolute permeability:

$$\mu_o = 4\pi \times 10^{-7} \text{ (F/m)} \tag{14.41}$$

   Speed of light:

$$c = \frac{1}{(\varepsilon_o\mu_o)^{1/2}} = 3\times 10^8 \text{ m/sec} \tag{14.42}$$

   Wave impedance:

$$\eta_o = \left(\frac{\mu_o}{\varepsilon_o}\right)^{1/2} = 120\mu = 377\Omega \tag{14.43}$$

2. Constants of the material used
   Relative permeability: μ_r
   Absolute permeability: $\mu = \mu_o\mu_r$
   Relative dielectric constant ε_r
   Absolute dielectric constant: $\varepsilon = \varepsilon_o\varepsilon_r$
   Speed of propagation: $\mu = c/(\varepsilon_r\mu_r)^{1/2}$
   Resistivity (ohm-meter): ρ
   Tangent of the loss angle: $\tan \phi$
3. The coaxial line

The dielectric is assumed nonmagnetic, $\mu_r = 1$, and the two conductors are assumed to have the same resistivity (Figure 14.3):

$$L = \frac{\mu}{2\pi} \text{Log} \frac{d_2}{d_1} = 2\times 10^{-7} \text{Log} \frac{d_2}{d_1} = 4.6\times 10^{-7} \log \frac{d_2}{d_1} \text{ (H/m)} \tag{14.44}$$

(Here Log is the natural logarithm, base e; and log is the common logarithm, base 10):

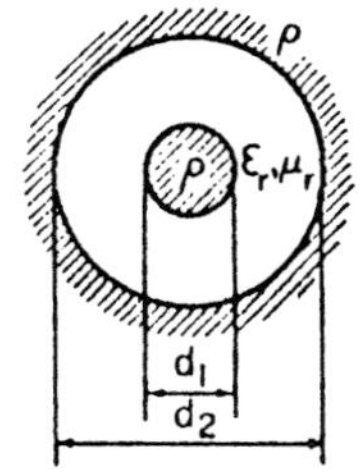

Figure 14.3 A coaxial line: parameters.

$$R = \left(\frac{\mu f \rho}{\pi}\right)^{1/2} \frac{d_1 + d_2}{d_1 d_2} \; 0.632 \times 10^{-3} (\rho f)^{1/2} \frac{d_1 + d_2}{d_1 d_2} \; (\Omega/\text{m}) \tag{14.45}$$

In the case of copper conductors:

$$R = 8.3 \times 10^{-8} f^{1/2} \frac{d_1 + d_2}{d_1 d_2} \; (\Omega/\text{m}) \tag{14.46}$$

$$C = \frac{2\pi\varepsilon}{\text{Log}\,\frac{d_2}{d_1}} = \frac{10^{-9}}{18} \times \frac{\varepsilon_r}{\text{Log}\,\frac{d_2}{d_1}} = 0.24 \times 10^{-10} \frac{\varepsilon_r}{\log \frac{d_2}{d_1}} \; (\text{F/m}) \tag{14.47}$$

$$G = \text{cotan}\,\varphi = 2\pi f C \tan\varphi \; \left(\Omega^{-1}/\text{m}\right) \tag{14.48}$$

$$R_c = \left(\frac{L}{C}\right)^{1/2} = \frac{1}{2\pi}\left(\frac{\mu}{\varepsilon}\right)^{1/2} \text{Log}\,\frac{d_2}{d_1} = \frac{1}{2\pi}\eta_o\left(\frac{\mu_r}{\varepsilon_r}\right)^{1/2} \text{Log}\,\frac{d_2}{d_1} \tag{14.49}$$

$$R_c = \frac{60}{\varepsilon_r^{1/2}} \text{Log}\,\frac{d_2}{d_1} = \frac{138}{\varepsilon_r^{1/2}} \log \frac{d_2}{d_1} \; (\Omega) \tag{14.50}$$

Examples of other lines are shown in Figure 14.4.

The ratio between the voltage and current amplitudes of the same traveling wave is the characteristic resistance $R_c(Z_c)$ (v_i, i_i and v_r, i_r are two waves traveling through in opposite directions):

$$\frac{v_i}{i_i} = R_c; \; \frac{v_r}{i_r} = -R_c \tag{14.51}$$

Each voltage wave sees an impedance $R_c(Z_c)$ in its propagation.
The reflected voltage is given by:

$$v_r = \frac{R_\ell - R_c}{R_\ell + R_c} v_i \tag{14.52}$$

and the reflected current:

$$i_r = -\frac{R_\ell - R_c}{R_\ell + R_c} i_i \tag{14.53}$$

	Characteristic resistance (Ω)	Capacitance per unit length (pF/meter)	Self-inductance per unit length (μH/meter)
Asymmetric two-wire line 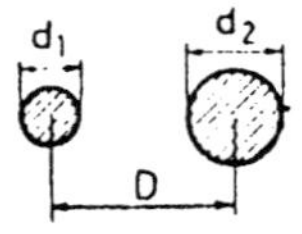	$R_c = 120\,\text{Log}\,\frac{2D}{\sqrt{d_1 d_2}}$	$C = \frac{27.5}{\text{Log}\,\frac{2D}{\sqrt{d_1 d_2}}}$	$L = 0.4\,\text{Log}\,\frac{2D}{\sqrt{d_1 d_2}}$
Symmetric two-wire line 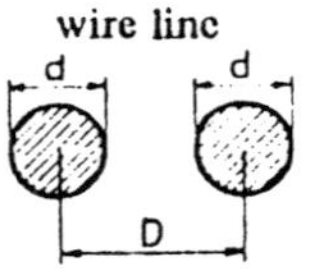	$R_c = 120\,\text{Arc cosh}\,\frac{D}{d} = 120\,\text{Log}\,\frac{2D}{d}$	$C = \frac{10^3}{36\,\text{Arc cosh}\,\frac{D}{d}} = \frac{27.5}{\text{Log}\,\frac{2D}{d}}$	$L = 0.4\,\text{Arc cosh}\,\frac{D}{d} = 0.4\,\text{Log}\,\frac{2D}{d}$
Shielded two-wire line 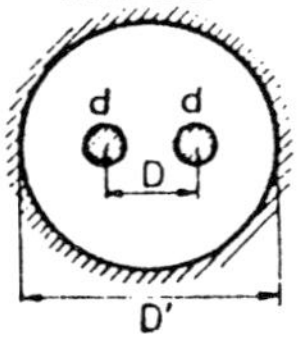	$R_c = 120\,\text{Log}\left(\frac{2D}{d}\,\frac{D'^2-D^2}{D'^2+D^2}\right)$	$C = \frac{27.5}{\text{Log}\left(\frac{2D}{d}\,\frac{D'^2-D^2}{D'^2+D^2}\right)}$	$L = 0.4\,\text{Log}\left(\frac{2D}{d}\,\frac{D'^2-D^2}{D'^2+D^2}\right)$
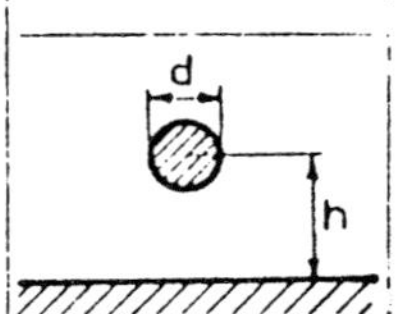	$R_c = 60\,\text{Log}\,\frac{4h}{d}$	$C = \frac{55}{\text{Log}\,\frac{4h}{d}}$	$L = 0.2\,\text{Log}\,\frac{4h}{d}$
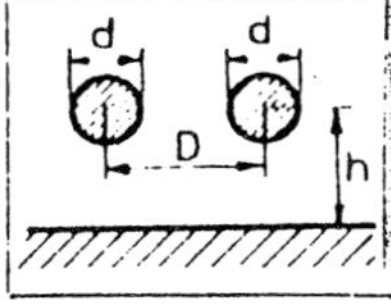	$R_c = 120\,\text{Log}\,\frac{2D}{d\sqrt{1+\left(\frac{D}{2h}\right)^2}}$	$C = \frac{27.5}{\text{Log}\,\frac{2D}{d\sqrt{1+\left(\frac{D}{2h}\right)^2}}}$	$L = 0.4\,\text{Log}\,\frac{2D}{d\sqrt{1+\left(\frac{D}{2h}\right)^2}}$

Figure 14.4 Parameters for various line configurations.

or

$$v_r = -\Gamma_\ell v_i;\ i_r = -\Gamma_\ell i_i \tag{14.54}$$

Consider a transmission line Figure 14.5. If the incident wave encounters an impedance R_c at the output end (matched load), it is not reflected, $v_r = \Gamma_l v_i$. If $R_\ell = R_c$, then $\Gamma_\ell = 0$, so that:

$$\begin{aligned} v_r &= 0;\ i_r = 0 \\ V_B &= v_i + v_o;\ I_B = i_i + i_o \end{aligned} \tag{14.55}$$

The impedance R_c which so terminates the line can be either lumped, or distributed (a transmission line with the same R_c). It thus results as a special case that the impedance seen looking into a semi-infinite line at any point is R_c.

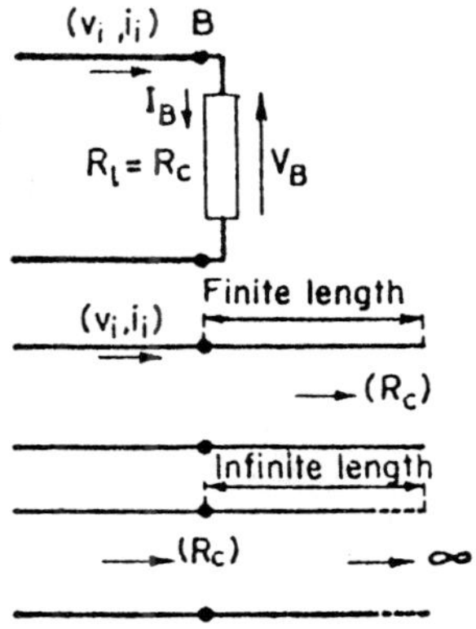

Figure 14.5 Transmission line (matched load).

In the open-circuited line:

$$R_\ell = \infty; \ \Gamma_\ell = +1$$

so that

$$v_i = v_r; \ i_i = -i_r$$

There is thus total reflection, without change of sign for the voltage, and with change of sign for the current, so that:

$$V_B = 2v_i + v_o; \ I_B = 0$$

In the short-circuited line:

$$R_\ell = 0; \ \Gamma_\ell = -1$$

so that

$$v_i = -v_r; \ i_i = i_r$$

Thus there is again total reflection, but with change of sign for the voltage, and without change of sign for the current, so that:

$$V_B = 0; \ I_B = 2i_i + i_o$$

Figure 14.6 shows the responses of lines, each of which is lossless, of length ℓ, and characteristic impedance R_c, fed by a source with internal impedance R_c supplying a unit step of voltage, but each of which is terminated in a different impedance.

The relations in the case of the transition from one transmission line to another can be obtained by writing the Kirchhoff laws at the junction. Consider two lines with characteristic impedances R_{c1} and R_{c2}, shown in Figure 14.7. Suppose an incident step (v_i, i_i) propagates from left to right. When it arrives at the discontinuity of characteristic impedance, it gives rise to a reflected step (v_r, i_r) and a transmitted step (v_t, i_t).

Three relations result from the fundamental property of transmission lines:

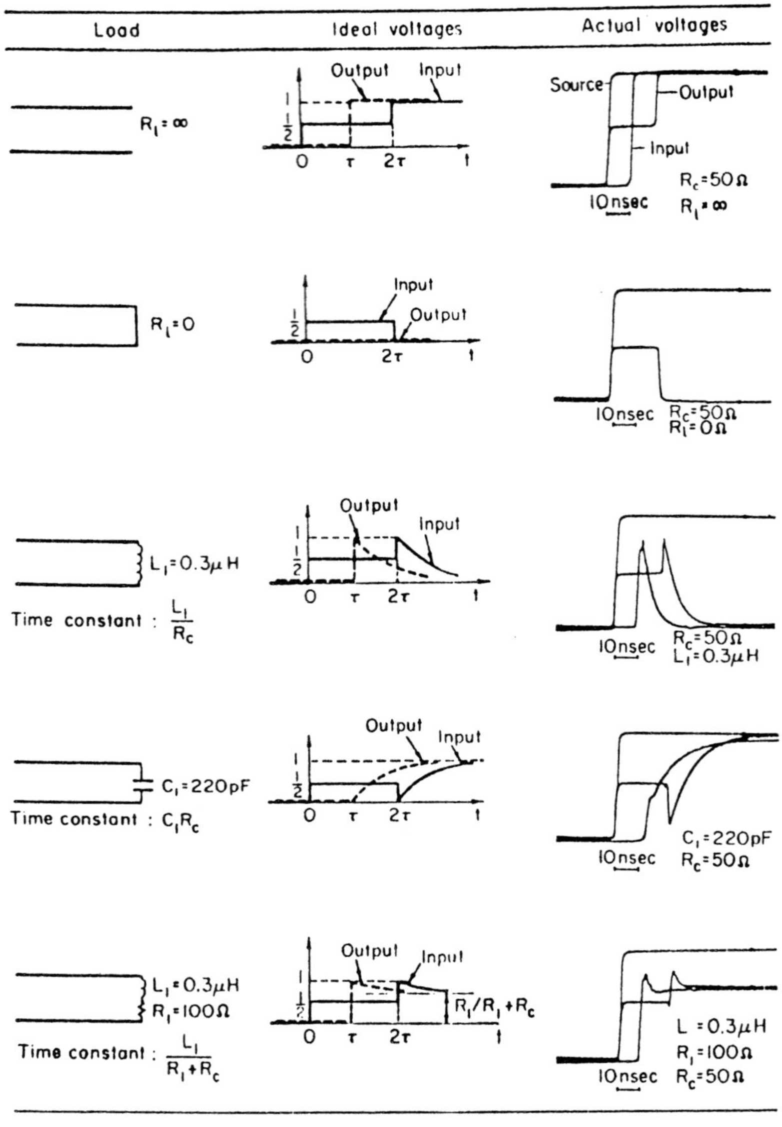

Figure 14.6 Voltage step response of various lines. Length 2 m, lossless, characteristic resistance 50 Ω, matched source.

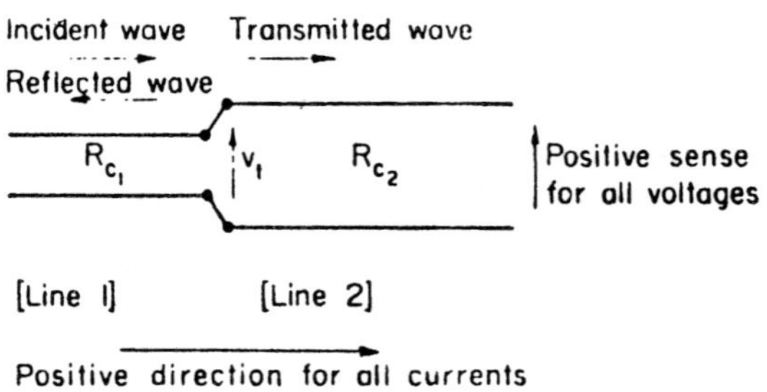

Figure 14.7 Line with discontinuous structure.

$$v_i/i_i = R_{c1}$$
$$v_t/i_t = R_{c2} \qquad (14.56)$$
$$v_r/i_r = -R_{c1}$$

The waves v_i and i_i have the same sign, as do v_t and i_t, but v_r and i_r have the opposite sign.

At the junction of the two lines:

$$i_t = i_i + i_r$$
$$v_t = v_i + v_r \qquad (14.57)$$

The relation in Equation 14.57 states the fact that at the junction the voltage across line 1, $v_i + v_r$, is the same as that across line 2, v_t. Relations in Equation 14.57 express the fact that energy is conserved at the junction of the two lines:

$$\left|v_i i_i\right| = \left|v_r i_r\right| + \left|v_t i_t\right| \qquad (14.58)$$

or

$$v_i i_i = v_t i_t + \left|v_r i_r\right| \qquad (14.59)$$

which yields

$$\frac{v_i^2}{R_{c1}} = v_t\left(i_i + i_r\right) + \frac{v_r^2}{R_{c1}}$$
$$= v_t\left(\frac{v_i}{R_{c1}} - \frac{v_r}{R_{c1}}\right) + \frac{v_r^2}{R_{c1}} \qquad (14.60)$$

from which

$$v_i^2 - v_r^2 = v_t\left(v_i - v_r\right);\; v_i + v_r = v_t \qquad (14.61)$$

The reflection and transmission coefficients are as follows: a reflection coefficient Γ defined by:

$$\Gamma = \frac{R_{c2} - R_{c1}}{R_{c2} + R_{c1}} \qquad (14.62)$$

Writing v_r and v_t as functions of v_i, R_{c1} and R_{c2}:

$$v_r = v_t - v_i = R_{c2} i_t - v_i = R_{c2}\left(i_i + i_r\right) - v_i$$
$$v_r = R_{c2}\left(\frac{v_i}{R_{e1}} - \frac{v_r}{R_{e1}}\right) - v_i \qquad (14.63)$$

These yield:

$$v_r\left(1+\frac{R_{c2}}{R_{c1}}\right)=v_i\left(\frac{R_{c2}}{R_{c1}}-1\right)$$

$$v_r = v_i\cdot\frac{R_{c2}-R_{c1}}{R_{c2}+R_{c1}} \tag{14.64}$$

$$v_r=\Gamma v_i$$

Thus Γ is the voltage reflection coefficient from line 1 to line 2.

In addition:

$$v_t = v_i + v_r = v_i(1+\Gamma)$$

$$v_t=(1+\Gamma)v_i \tag{14.65}$$

Thus $(1 + \Gamma)$ is the voltage transmission coefficient from line 1 to line 2. The voltage reflection coefficient from line 2 to line 1 is

$$-\Gamma=\frac{R_{c1}-R_{c2}}{R_{c1}+R_{c2}} \tag{14.66}$$

The voltage transmission coefficient from line 2 to line 1 is $(1 - \Gamma)$:

$$i_r=-\Gamma i_i \tag{14.67}$$

$$i_t=(1-\Gamma)i_i \tag{14.68}$$

Using the Laplace transformation the transmission line expressions are shown below. In case of a line terminated in a lumped impedance equal to the characteristic impedance Z_c, applying the above formulas with $Z_\ell = Z_c$ yields $\Gamma_\ell = 0$, so that:

$$V(x,p)=E(p)\frac{Z_c}{Z_o+Z_c}\,e^{-\gamma x} \tag{14.69}$$

$$I(x,p)=E(p)\frac{1}{Z_o+Z_c}\,e^{-\gamma z} \tag{14.70}$$

At each instant t, and at each point x of a line terminated in its characteristic impedance, the ratio of the voltage across the line to the current through the line is equal to the characteristic impedance. This result is of theoretical interest only, except in a few special cases, such as lossless lines and lines at the beginning of the transient regime, since it is not possible to realize a lumped-parameter circuit with impedance:

$$Z_c=\left(\frac{R+Lp}{G+Cp}\right)^{1/2} \tag{14.71}$$

In the semi-infinite line ($\ell = \infty$) Equations 14.69 and 14.70 hold. Thus the result is that a line of infinite length behaves like a line of finite length terminated in its characteristic impedance.

A line for which R = 0 and G = 0 is said to be lossless. The impedance given by Equation 14.71 reduces in this case to a pure resistance, called the characteristic resistance:

$$R_c = (L/C)^{1/2} \tag{14.72}$$

At each instant and at each point of a lossless line, either terminated in its characteristic resistance R_c or of infinite length, the ratio of the voltage across the line to the current through the line is R_c.

The general expression for the propagation constant:

$$\gamma = \left[(R + Lp)(G + Cp)\right]^{1/2} \tag{14.73}$$

becomes for the lossless line:

$$\gamma = p(LC)^{1/2} = p\delta \tag{14.74}$$

where we define $\delta = (LC)^{1/2}$; γ is called the propagation constant, and has the dimensions of an inverse velocity. The equations for the lossless line are

$$\begin{aligned} V(x,p) &= E(p)\frac{R_c}{Z_o + R_c}\frac{e^{-p\delta x} + \Gamma_\ell e^{-p\delta(2\ell - x)}}{1 - \Gamma_o\Gamma_\ell e^{-2p\delta}} \\ &= E(p)\frac{R_c}{Z_o + R_c}\left[e^{-p\delta z} + \Gamma_\ell e^{-p\delta(2-x)}\right]\sum_{n=0}^{\infty}\Gamma_o^n\Gamma_\ell^n e^{-2np\delta\ell} \end{aligned} \tag{14.75}$$

$$\begin{aligned} I(x,p) &= \frac{E(p)}{Z_o + R_c}\frac{e^{-p\delta x} - \Gamma_\ell e^{-p\delta(2\ell - x)}}{1 - \Gamma_o\Gamma_\ell e^{-2p\delta\ell}} \\ &= E(p)\frac{1}{Z_o + R_c}\left[e^{-p\delta x} - \Gamma_\ell e^{-p\delta(2\ell-x)}\right]\sum_{n=0}^{\infty}\Gamma_o^n\Gamma_\ell^n e^{-2np\delta\ell} \end{aligned} \tag{14.76}$$

Consider a lossless line with a matched load ($Z_l = R_c$). Then $\Gamma_l = 0$, and the general equations become:

$$V(x,p) = E(p)\frac{R_c}{Z_o + R_c}\, e^{-p\delta x} \tag{14.77}$$

$$I(x,p) = E(p)\frac{1}{Z_o + R_c}\, e^{-p\delta x} = \frac{V(x,p)}{R_c} \tag{14.78}$$

Hence $i(x,t) = v(x,t)/R_c$, and it suffices to calculate V(x,p).

The term $e^{-p\delta x}$ in V(x,p) corresponds to a delay δx. The transform of v(0,t), the voltage at the input to the line is

$$E(p)\frac{R_c}{Z_o + R_c} \tag{14.79}$$

hence

$$v(0,t) = e(t)\frac{R_c}{Z_o + R_c} \tag{14.80}$$

Knowing the voltage at the input to the line, the voltage at any abscissa x along the line follows from:

$$v(x,t) = v(0,t-\delta x)\,y(t-\delta x) \tag{14.81}$$

Here $y(t - \delta x)$ is the unit step function beginning at time δx.

The voltage and current at time t and abscissa x along the line are the same as the voltage and current at the input $x = 0$ at time $t - \delta x$. The velocity of propagation is

$$u = \frac{1}{\delta} = \frac{1}{(LC)^{1/2}} \tag{14.82}$$

If the line is surrounded by a dielectric medium with relative dielectric constant ε_r, disturbances propagate along the line with the same velocity as electromagnetic waves in the dielectric medium:

$$u = \frac{c}{\varepsilon^{1/2}} \tag{14.83}$$

from which

$$\delta = \frac{1}{u} = \frac{\varepsilon^{1/2}}{c} \tag{14.84}$$

Since

$$R_c = (L/C)^{1/2},\ \delta = (LC)^{1/2}$$

$$R_c^2 = L/C,\ \delta^2 = LC$$

then

$$L = R_c\delta \tag{14.85}$$

$$C = \frac{\delta}{R_c} \tag{14.86}$$

For a coaxial cable of the diameters d_1 and d_2:

$$C = \frac{2\pi\varepsilon_o\varepsilon_r}{\text{Log}\,\dfrac{d_2}{d_1}};\ R_c = \frac{1}{2\pi}\left(\text{Log}\frac{d_2}{d_1}\right)(\mu_o/\varepsilon_r\varepsilon_o)^{1/2} \tag{14.87}$$

from which

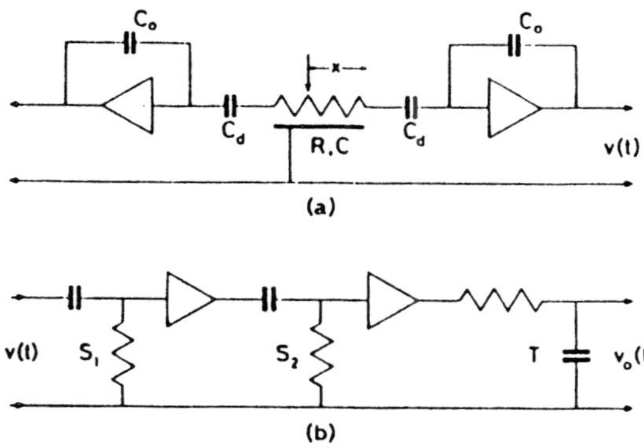

Figure 14.8 (a) Transmission line, total resistance R, and total capacitance C symmetrically terminated in capacitance C_d used as input elements to shunt feedback amplifiers, feedback value C_0. (b) The filter configuration used in this work with S_1 = S_2 = T. (From Mathieson, E. et al., *Nucl. Instrum. & Methods,* 126, 199, 1975. With permission.)

$$R_c C = \frac{\varepsilon^{1/2}}{3\times10^8} \tag{14.88}$$

$$L = \frac{\sqrt{\varepsilon_r}}{3\times10^8} R_c \tag{14.89}$$

For cables with Teflon insulation (ε_r = 2.25), this becomes:

$$C = \frac{5\times10^{-9}}{R_c} = \frac{5000}{R_c} 10^{-12}\ (\mathrm{F}), \quad L = 5\times10^{-9} R_c\ (\mathrm{H})$$

Mathieson et al.[1] described the characteristics of a charge-signal location system, comprising a uniform RC line in near open-circuit termination and conventional waveform-shaping filters, for operation in the rise-time mode. Figure 14.8 shows the essential features of the rise-time mode system under consideration. The line, total resistance R and total capacitance C, is terminated at each end with capacitance C_d ($\ll$C). The examination has been restricted to the case of the three-element filter:

$$S_1 = S_2 = T = T_a/\left(RC/\pi^2\right) \tag{14.90}$$

where T_a is an element time constant in seconds. Charge q_o is injected at a position $x(\leq 1)$ along the line and t(x) is the zero-cross time, expressed in units of RC/π^2, of the filter output waveform $v_o(t)$. The difference zero-cross time is

$$T(x) \equiv t(x) - t(1-x) \tag{14.91}$$

and the positional sensitivity is $S(x) \equiv dT(x)/dx$. Two figures of merit describing system performance are δx, the RMS nonlinearity in T(x), and $(\Delta x)_m$, the RMS positional uncertainty due to system noise. The subscript m indicates that a mean value has been taken over a central position of the line.

The RMS voltage fluctuation Δv_o in the output waveform $v_o(t)$ due to noise in the system results in uncertainty in the crossover time of amount:

$$\Delta t(x) = \Delta v_o / g(x) \tag{14.92}$$

where g(x) is the (negative) slope at crossover. Ignoring correlation between the two noise outputs, the mean square fluctuation in difference crossover time, T(x), is

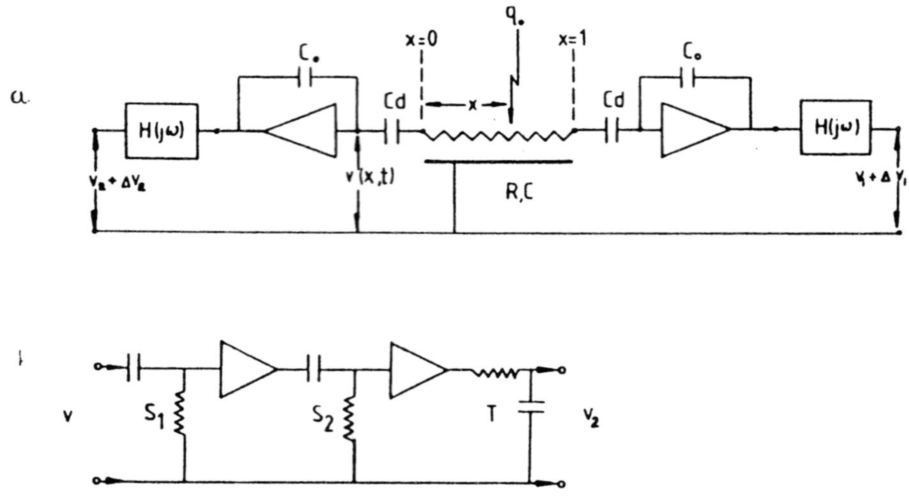

Figure 14.9 Transmission line, total resistance R, and total capacitance to ground C symmetrically terminated in capacitances C_d which act as input elements to shunt feedback preamplifiers of feedback capacitance C_0. The shaping filters have transfer function H(jω). (b) Three-element doubly differentiating, singly integrating shaping filter with normalized time constants S_1, S_2, τ. (From Fraser, G. W. et al., *Nucl. Instrum. & Methods,* 180, 255, 1981. With permission.)

$$\Delta^2 T(x) = \Delta^2 v_o\left[1/g^2(x) + 1/g^2(1-x)\right] \tag{14.93}$$

The RMS uncertainty in position determination due to electronic noise is then:

$$\Delta x = \Delta T(x)/S(x) = \Delta v_o\left[1/g^2(x) + 1/g^2(1-x)\right]^{1/2}/S(x) \tag{14.94}$$

It is convenient to refer system noise to the line termination; thus let q_n/C_d be that voltage step which, applied to the line termination, yields an output pulse of height equal to the RMS noise. That is

$$\Delta v_o = 0.231 A_v q_n / C_d \tag{14.95}$$

C_d is the preamplifier input capacitance (the line-terminating capacitance) used. A_v is the voltage gain of the system. The factor 0.231 is simply the amplitude response to unit step function of a passive CR-RC-CR filter. The expression, therefore, defines an equivalent noise charge q_n.

The total noise q_n may be calculated as the resultant of line noise q_l, and amplifier noise q_a:

$$q_n^2 = q_l^2 + q_n^2 \tag{14.96}$$

A dimensionless line-noise parameter μ may be formally defined by the expression:

$$q_l = C^{1/2}(kT_e)^{1/2}(T, C_d/C) \tag{14.97}$$

The parameter μ, which is on the order of C_d/C, is a function of T and C_d/C only.[1]

A quantitative description of the nonlinearity and resolution performance of a capacitively terminated RC line position encoder using difference zero-cross timing was given by Fraser et al.[2] All configurations between open- and short-circuit termination were examined for a doubly differentiating, singly integrating filter with equal time constants.

The model of a capacitively terminated RC line is illustrated in Figure 9. Charge q_o is injected as a temporal delta function at normalized coordinate x ($0 \le x \le 1$). The resultant preamplifier output voltage is V(x,t) which may be calculated using heat conduction theory. Convolution of V(x,t) with the impulse response of the shaping filter yields the system output voltage $V_2(x,t)$ in a closed form. The filter function of the filter CR-CR-RC is, in its unmatched form:

$$H(j\omega)^2 = \frac{\omega^4 (S_1 S_2)^2}{(1+\omega^2 T^2)(1+\omega^2 S_1^2)(1+\omega^2 S_2^2)} \qquad (14.98)$$

The zero-cross time of $V_2(x,t)$ is t(x) and its gradient at zero cross is $(q_o/C_o)g(x)$. The difference zero-cross time is T(x) = t(x) – t(1 – x). Determination of t(x) and t(1 – x) is perturbed by the noise signals ΔV_1 and ΔV_2, thus limiting the positional resolution. The positional sensitivity is S(x) = dT(x)/dx. The RMS positional uncertainty Δx, normalized by the signal-to-noise ratio, is then calculable from a knowledge of $V_2(x,t)$ and a knowledge of the correlation between line-end noise signals:

$$\Delta x = D(x,T,C_d/C)q_n/q_o \qquad (14.99)$$

D is the dimensionless resolution parameter, which for difference zero-cross timing has the form:

$$D(x,T,C_d/C) = \frac{a}{S(x)}\left[\frac{1}{g^2(x)} + \frac{1}{g^2(1-x)} - \frac{2r_{12}(x)}{g(x)g(1-x)}\right]^{1/2} \qquad (14.100)$$

where a is the peak amplitude of the shaping filter response to a unit step (0.231 for the matched CR-CR-RC filter). The noise correlation coefficient r_{12} is given by:

$$r_{12} = (\Delta V_1 \Delta V_2)\,\Delta V_o^2 \qquad (14.101)$$

where the time-averaged mean of the product of noise signals is understood in the numerator, ΔV_o^2 is the mean square value of either noise signal, and r_{12} expresses the correlation between the system output noise signals at the instant of zero cross.

q_n is the total equivalent noise charge. It is calculated as the sum of uncorrelated contributions from the line thermal noise and from noise generated in the preamplifiers.

Borkowski and Kopp[9] have calculated that for the capacitively terminated line of Figure 14.9:

$$S(x) = \pi^2(1+2C_d/C) \qquad (14.102)$$

so that in the asymptotic case of open-circuit termination, $S(x) = \pi^2$. Mathieson has determined the short-circuit asymptote:

$$S_\infty = \pi^2/3 \qquad (14.103)$$

Many instruments provide the slow linear output signals through a very low-source impedance, typically <1 Ω. The low impedance allows connection of almost any load without loss of signal amplitude. For example, a 100-Ω load may be driven to the full 10-V span. A low-impedance outputs are simple to use, because they permit paralleled multiple loads without loss of span. A 93-Ω coaxial cable, such as RG-62A/U, is normally used to connect slow linear signals between modules.

A potential problem with the low-impedance output in oscillation is caused by reflections from unterminated cables more than 1.5 m in length. For this reason, long 93-Ω coaxial cables should be terminated at the receiving end by a 93-Ω load. This is usually accomplished by adding a Tee connector and a 100-Ω terminator on the input to the module at the receiving end. The 100-Ω terminator in parallel with a 1000-Ω or larger input impedance in the module provides adequate termination of the 93-Ω cable.

An alternative solution to the oscillation problem is to use the 93-Ω output. This slow linear output is provided, in addition to the low-impedance output, on many instruments. The receiving end can be left unterminated. The 93-Ω output provides termination of the cable at the signal source. The 93-Ω output can be used for full-span signal transfer only if the impedance of the load at the receiving end is very large compared to 93 Ω. If the 93-Ω source must drive a 93- or 100-Ω load, half the span will be lost. The chief virtue of the 93-Ω output is stability against oscillation for variable cable conditions.

14.2 DELAY LINES

Standard cable delay units are often used to delay fast nuclear instrument module (NIM) pulses. These units are a series of switches to insert or remove binary weighted lengths of coaxial cable. The practice of sending fast NIM pulses through six or seven mechanical switches sometimes results in unreliable operation because the switches do not always close with negligible impedance. The resulting reflections (at these switches) can cause a degradation in the timing and the width of subsequent pulses.

A programmable delay line has been developed by Brown and Neumann[3] that uses microstrip transmission line and temperature compensated ECL gates. The instrument delayed standard NIM pulses and had settings of 0 to 64.5 nsec. The output rise time was = 1 nsec, independent of the delay setting. The delay stability at maximum delay was ≈130 ppm/°C, which is better than for RG-55 B/U coaxial cable. The minimum pulse width was less than 3 nsec, and the difference in pulse width between the output and the input was less than 150 psec.

The advantages of the system over standard coaxial cable delay boxes are (1) in the standard delay box, one must choose switches that have a low inductance (slide switches are often used) to pass the fast NIM pulses with minimum reflections; (2) changes in switch contact resistance from milliohms to more than 100 Ω have negligible effect on the delay, thus further ensuring reliable switch operation; and (3) the system can be adapted for remote operation either by manual switches or under CAMAC control.

For a given delay, the microstrip line is about the same length as coaxial cable but occupies less volume. It also has the advantage of being fabricated on the same printed circuit board (PCB) as the gates used to switch the delay elements. Thus, assembly is facilitated and the reflections at each point are well controlled.

The method used to generate the various delays is shown in Figure 14.10. After the input protection and level shift circuit, the pulses go through four different delay sections. Each delay section has four possible gates for the pulse to pass through. The gates are emitter-coupled logic (ECL) and the output emitter followers are OR connected. Only one gate is open at one time. A DC voltage of about –1.7 V (or more negative) on the control input will enable the selected gate while the other three gates of each section are disabled (with a voltage of –0.7 V). The minimum delay is obtained by setting all the thumbwheel switches to the 0 position. The delay for this 0000 delay setting is about 6 nsec and is due mainly to the delay in the gates and in the cables to the front panel.

The delays are in the form of microstrip transmission lines which meander back and forth on a PCB. About 19.3 cm/nsec of delay are needed for all the delay lines. The PCB material was Teflon with 1-oz copper on both sides. Teflon was the chosen dielectric because losses are lower with Teflon than with epoxy fiberglass. To minimize the reflections caused by the capacitive loading of each gate input, the value of the characteristic impedance of the transmission line, Z_o, should be as low as possible.

The circuit details of the first 16-nsec delay increment are shown in Figure 14.11. Resistors R_{A0} and R_{A1} (plus the associated 100-Ω resistors) reduce the difference in pulse width caused by different gate thresholds. U3A is used to reshape the pulse. R_{A1} was added to U3A (rather than U2B) because the rise time at U3A is slower and thus more susceptible to errors in gate threshold. The purpose of U3B was to provide a complement of the AO control input to U3A.[3]

Flat helical delay lines were developed by Okuno et al.[4] for avalanche position determination along each anode wire, for unambiguous readout of multiparticle events in multiwire drift or proportional chambers. In most applications, the positions of the avalanches are projected on two orthogonal delay lines for two-dimensional readout. If more than one particle is present, the readout is ambiguous and requires additional coordinate readout.

The structure of the delay line and its position on a MWPC are shown in Figure 14.12. The delay line consisted of a wire helix wound around a thin plastic strip. A narrow straight ground conductor was cemented to one exterior surface. Characteristic properties of the line depend largely on the geometry. The length of the line was about 30 cm. Three kinds of cross sections of the core strip were tested:

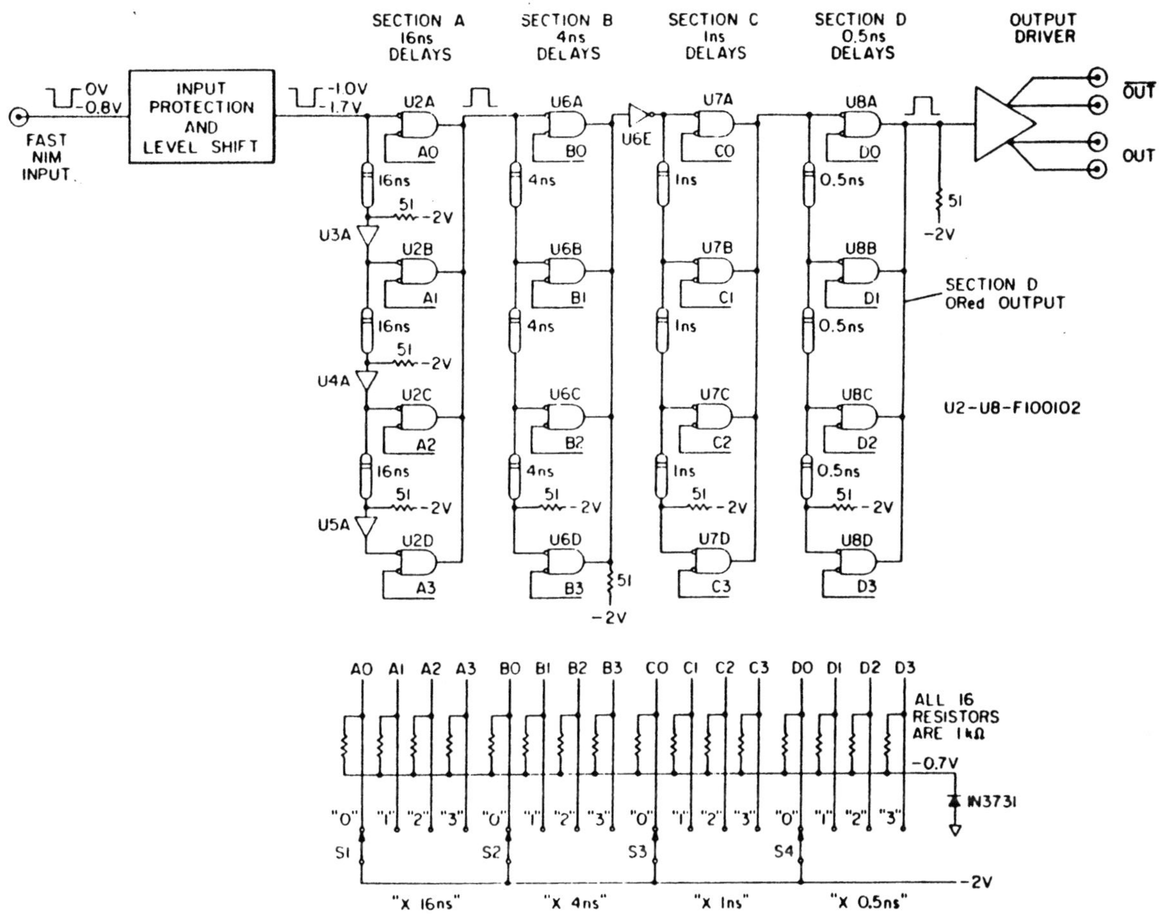

Figure 14.10 Simplified schematic of programmable delay line. (From Brown, D. and Neumann, U., *Nucl. Instrum. & Methods,* 158, 281, 1979. With permission.)

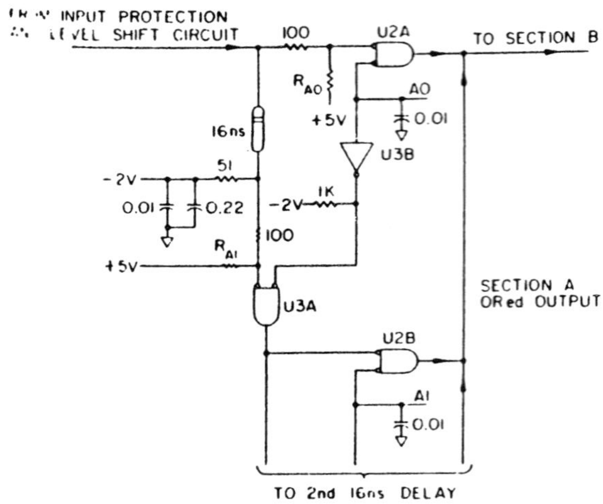

Figure 14.11 Circuit details of the first 16-nsec delay increment. (From Brown, D. and Neumann, U., *Nucl. Instrum. & Methods,* 158, 281, 1979. With permission.)

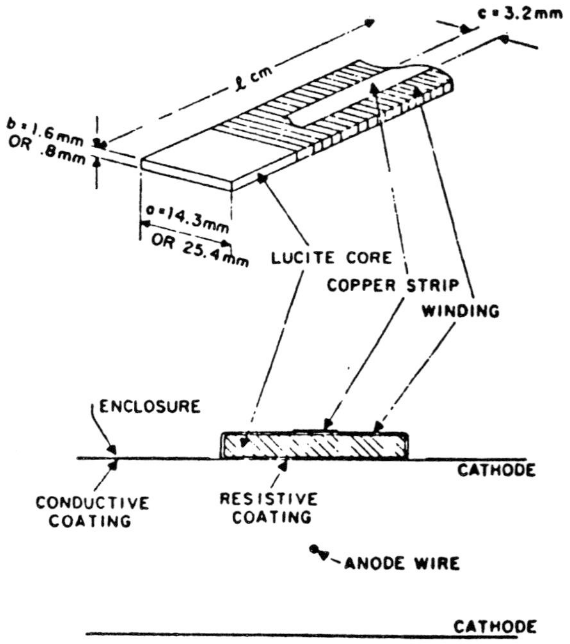

Figure 14.12 Structure of the delay line and its position on a MWPC. (From Okuno, H. et al., *IEEE Trans. Nucl. Sci.,* NS24, 213, 1977. With permission.)

14.3 × 1.6 mm², 14.3 × 0.8 mm², and 25.4 × 1.6 mm². The winding ranged from 11 to 38 turns/cm, with wire diameters of 75, 100, and 125 μm. Characteristic impedances and specific propagation delays were in the region between 270 and 1000 Ω; and between 1.3 and 8 nsec/cm, respectively.

Attenuation of the signal along the total length of the line, 30 cm, was less than 10%, due to the ratio of total resistance and characteristic impedance. The total delay-to-rise time ratio was around 6.

An analog processor to be used with silicon detectors belonging to active targets in high-energy physics has been implemented by Cisotti et al.[5] A trapezoidal signal can be obtained at the output of an operational integrator when a bipolar current pulse of the shape shown in Figure 14.13 is injected into its virtual ground.

A way of transforming a step into a rectangular pulse by using a delay line matched at both ends is based on the transversal filter of Figure 14.14. If this operation of the delay line is iterated by a further transversal filter which employs a capacitor as a feedback bipole of the operational amplifier, the final result would be the desired trapezoidal pulse, according to Figure 14.15. At high frequencies, the virtual ground of the first operational amplifier loses it properties and no longer ensures an adequate low-

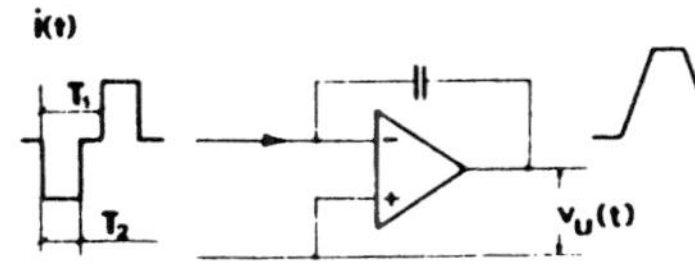

Figure 14.13 Shaping a trapezoidal signal from a bipolar current pulse. (From Cisotti, M. et al., *Nucl. Instrum. & Methods,* 159, 235, 1979. With permission.)

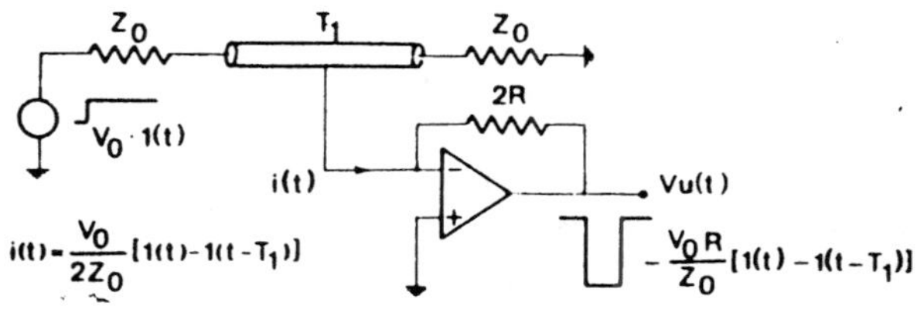

Figure 14.14 Signal clipping with a delay-line transversal filter. (From Cisotti, M. et al., *Nucl. Instrum. & Methods,* 159, 235, 1979. With permission.)

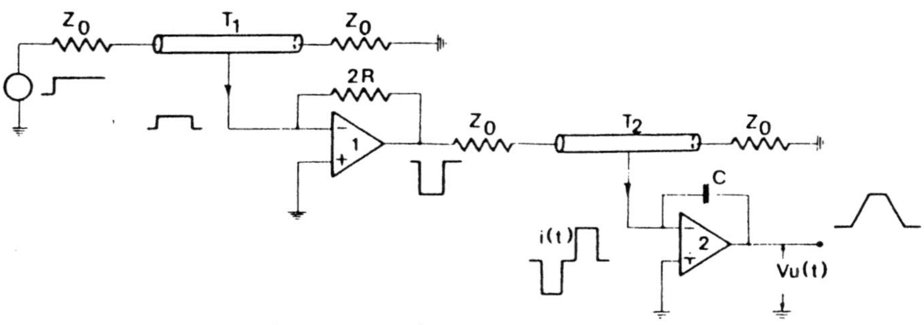

Figure 14.15 Complete circuit to generate a trapezoidal signal starting from a step. (From Cisotti, M. et al., *Nucl. Instrum. & Methods,* 159, 235, 1979. With permission.)

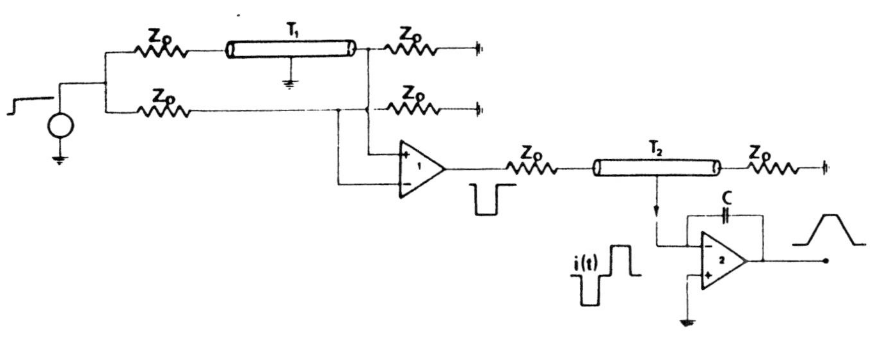

Figure 14.16 Actual circuit employed in the processor as a trapezoidal shaper. (From Cisotti, M. et al., *Nucl. Instrum. & Methods,* 159, 235, 1979. With permission.)

impedance connection for the outer conductor of the delay line T_1. This behavior results in a distortion of the edges of the shaped pulse. Such an effect does not occur to the delay line T_2, for which the low impedance connection of the outer conductor is guaranteed, when the gain of the operational amplifier drops, by the capacitor C connected across it.

For this reason, the diagram shown in Figure 14.16 has been preferred to that of Figure 14.15. The second shaper which includes the integrator has been maintained, while the first has been replaced with a difference arrangement which is free from high-frequency distortions on the signal edges.

The use of distributed and lumped-constant delay lines as substitutes for coaxial cable to delay analog and digital signals was discussed by Sandoval et al.[6]

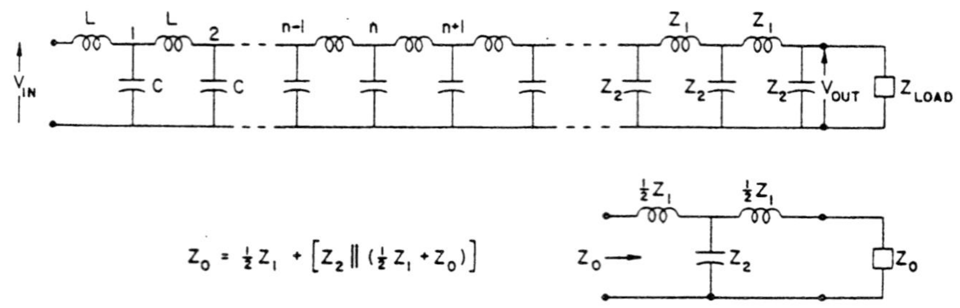

Figure 14.17 A schematic diagram of a lumped-constant delay line and its image section model. The image impedance concept is outlined in the lower part of the figure. (From Sandoval, J. P. et al., *Nucl. Instrum. & Methods,* 216, 171, 1983. With permission.)

For cables commonly in use, the high-frequency attenuation, expressed in decibels, increases approximately as $\omega^{1/2}$. Because of these effects, the pulse response of a coaxial cable is strongly waveform dependent. It can be shown[7] that the 0 to 50% rise time, T_0, response to a step-function input is given by:

$$T_0 = 5.4 \times 10^{-6}\,\alpha^2 L^2 \text{ (nsec)} \tag{14.104}$$

where α = attenuation at 1 GHz in units of decibels per 30.5 m and L = cable length in meters. For 61 m of RG-58 cable, α = 20 dB/30.5 m so that T_0 = 7.3 nsec. The 10 to 90% rise time is roughly 29 times longer.

A lumped-constant delay line (LC delay) consists of a series of low-pass reactive filters composed of discrete series inductors and shunt capacitors. A typical LC delay network is shown in Figure 14.17. An analysis of this network can be performed using either classical image filter theory[8] or by considering it as a discrete approximation to a continuous system (e.g., a transmission line). Each section is a K-filter (i.e., the product of its series impedances, Z_1, and its shunt impedances, Z_2, is independent of frequency) with an image impedance given by:

$$Z_0 = \left(\frac{Z_1^2}{4} + Z_1 Z_2\right)^{1/2} = \left(\frac{L}{C}\right)^{1/2}\left(1 - \frac{\omega^2}{\omega_c^2}\right)^{1/2} \tag{14.105}$$

where ω_c is the cutoff frequency given by $\omega_c = 2(LC)^{-1/2}$. (The impedance is the input impedance for a network with an infinite number of sections. It is the network analog of the characteristic impedance of a coaxial cable.) For input signals, frequency components less than ω_c propagate along these networks according to the functions:

$$\exp\left[i\omega t - i2n \sin^{-1}(\omega/\omega_c)\right] \tag{14.106}$$

where n labels the section. The frequency components greater than ω_c are attenuated according to the functions:

$$\exp\left[i\omega t - 2n \cosh^{-1}(\omega/\omega_c)\right] \tag{14.107}$$

For pulse waveforms with frequency components well below the cutoff frequency, the propagating modes may be approximated by:

$$\exp\left[i(t - 2n/\omega_c)\omega\right] \tag{14.108}$$

which corresponds to a delay given by:

$$T_d = n(LC)^{1/2} \tag{14.109}$$

These pulses propagate with good fidelity and little attenuation. In this approximation, the image impedance is nearly constant and a finite series of sections may be terminated with a resistor of value $(L/C)^{1/2}$.

REFERENCES

1. **Mathieson, E. et al.,** *Nucl. Instrum. & Methods,* 126, 199, 1975.
2. **Fraser, G. W. et al.,** *Nucl. Instrum. & Methods,* 180, 255, 1981.
3. **Brown, D. and Neumann, U.,** *Nucl. Instrum. & Methods,* 158, 281, 1979.
4. **Okuno, H. et al.,** *IEEE Trans. Nucl. Sci.,* NS24, 213, 1977.
5. **Cisotti, M. et al.,** *Nucl. Instrum. & Methods,* 159, 235, 1979.
6. **Sandoval, J. P. et al.,** *Nucl. Instrum. & Methods,* 216, 171, 1983.
7. **Kerns, Q. et al.,** LRL Counting Note CC2-1B, Lawrence Radiation Laboratory, UC, Berkeley, CA, 1966.
8. **Karakash, J. J.,** *Transmission Lines and Filter Networks,* Macmillan, New York, 1950.
9. **Borkowski, C. J. and Kopp, M. K.,** *IEEE Trans. Nucl. Sci.,* NS-17, 340, 1970.

Index

A

AC/C, see Associated anticoincidence/coincidence logic
Accelerator, tandem, 137
ACM, see Analog control module
Acquisition
  module (AM), 66
  system, overall, 226
ADC, see Analog-to-digital converter
ADCC, see Analog-to-digital converter control
Address
  counter, 185
  generator, 219
  increment register (AIR), 191
ADSM, see Analog data storage module
AGEE, see Alpha gamma equivalent energy
AIR, see Address increment register
Alpha gamma equivalent energy (AGEE), 146
ALU, see Arithmetic logic unit
AM, see Acquisition module
Amplifier
  CRID hybrid, 227
  positive feedback, 27
  sensing, 17
Amplitude
  -and-rise-time-compensated (ARC) timing, 83, 99, 101
  variations, walk due to, 98
Analog control module (ACM), 226
Analog data storage module (ADSM), 226
Analog-to-digital converter (ADC), 44, 46, 51, 72
  charge-integrating, 125
  control (ADCC), 65
  costs, 73
  fast successive approximation, 164
  flash, 183
  parallel encoding, 55
  READY signals, 181
  system, 144
Analog store and multiplexer unit (ASMU), 75
AND
  gate, 84, 135, 149
  output, 69
Anticoincidence, 149, 153
Antiwalk discriminator, 10, 116
Arbitration timing control card (ATC), 202
ARC, see Amplitude-and-rise-time-compensated timing
Arithmetic logic unit (ALU), 71
ASMU, see Analog store and multiplexer unit
Associated anticoincidence/coincidence (AC/C) logic, 120, 121
ATC, see Arbitration timing control card
Automatic handshaking, 200

B

Base current, 1
Baseline restorer (BLR), 10, 23, 96
BCD, see Binary coded decimal
BEB, see Board event buffer
BG, see Bus grant
Bias
  circuit, 1
  forward, 38
Binary coded decimal (BCD), 29, 47
  counting, 30
  output, 33
Binary counter
  parallel synchronous, 31
  series synchronous, 31
Bipolar pulses, 3
BLR, see Baseline restorer
Board event buffer (BEB), 213, 214
BR, see Bus request
Bridge rectifier, full-wave, 235
Buffer, 1
  amplifier, 118
  memory, 185
  -tape technique, 51
Bus
  grant (BG), 205
  request (BR), 205

C

Calibration, 146
Calorimeter data unit (CDU), 166
CAMAC, 175

dataway, 53
to FASTBUS interface (CFI), 204
interface, 184
module, 181, 187
smart, 188, 196
Capacitor
charge, 119
potential, 174
Cathode ray tube (CRT), 65
CC, see Crate controller
CDU, see Calorimeter data unit
CE, see Chip-enable
CEB, see Crate event buffer
Central processing unit (CPU), 199
CF, see Constant-fraction
CFA, see Constant fraction amplifier
CFD, see Constant-fraction discriminator
CFI, see CAMAC to FASTBUS interface
CFT, see Constant-fraction trigger
CH, see Controller head
Channel counts, standard deviation s of, 56
Charge-to-digital converter (QDC), 112, 125
Charge-to-time converter (QTC), 184
Cherenkov ring imaging detector (CRID), 226
Chip-enable (CE), 41
Circuit
auxiliary, 149–174
linear gate, 159–164
pulse stretcher, 173–174
sample and hold circuits, 164–173
timing circuits, 149–159
bias, 1
charge-to-time converter, 184
clock, 62
coincidence, 153
data handling, 182
discriminator, 9
external, 104
integrated, 154
monolithic, 169
monostable, 19
peak-detect and hold, 167
pick-off, 130
pulse shape discriminator, 127
timing, 81
tunnel diode test, 16
zero-crossing detector, 101
CLD, see Complementary level discriminator
Clock
circuit, 62
pulse, 28
CMOS, see Complementary metal oxide semiconductor
CMPX, see Control single multiplexer
Coaxial cable, 266
Coaxial line, 258
Cockcroft-Walton voltage multiplier, 240
Coincidence
circuits, 150, 151, 153, 154
counting, shift-register, 155
data, 47
logic, 154
of pulses, 154
Collector voltage, 1
Comparator, 87, 125
constant-fraction, 101
input-offset voltage, 103
timing, 135
zero-crossing, 113
Compensation resistance, 21
Complementary level discriminator (CLD), 7, 8
Complementary metal oxide semiconductor (CMOS), 7, 102, 171, 172, 195
CFD, monolithic, 103
logic, 72
memory cells, 42
transmission gate, 173
Computer-controlled phase (CPH), 69
Constant fraction (CF)
amplifier (CFA), 88, 89
circuitry, 88
discriminator (CFD), 22, 85, 87, 98, 102, 105, 112, 122
shaping, 88
timing, 79
trigger (CFT), 10, 88, 111
Controller head (CH), 209
Control single multiplexer (CMPX), 65
Control and status register (CSR), 206
Counting, 29
high capacity, 34
electronic, 27
Count rate, 6
CP, see Crate processor
CPH, see Computer-controlled phase
CPU, see Central processing unit
Crate
controller (CC), 188, 217, 220, 221, 224
event buffer (CEB), 214
flow, 186
processor (CP), 214
CRID, see Cherenkov ring imaging detector
CRT, see Cathode ray tube
CSR, see Control and status register
Current
reflected, 259
regulator, 239
-switching pairs, 154
Cutoff frequency, 274

D

DAC, see Digital-to-analog converter
DAS, see Data acquisition system
Data
collection, VME-based, 196, 197
handling, 182
input bus (DIB), 180
multiplexer (DMPX), 65
processing system, 217
transfer

control (DTC), 65
methods of, 199
Data acquisition, 51, 53, 175–232
CAMAC, 175–201
basic concepts, 175–177
systems, 177–201
fast, 188, 196
FASTBUS, 201–214
applications, 203–214
introduction to, 201–203
IBM PC AT interface for, 214
multiparameter, 66
other data acquisition systems, 214–232
rate, 55
system (DAS), 178, 179, 184, 206
DELPHI, 213
GAMS-4000, 186
for recording multiparameter digital data, 191
TTL outputs in, 193
versatile, 196
Dataway
cycle generator, 190
lines, 177
operation, 176
wiring, 176
DCCFD, see Dual-channel constant-fraction discriminator
DDSM, see Digital data storage module
Decay constant
Decimal readout, 32
Delay
lines, 10, 270–275
distributed, 273
flat helical, 270
lumped-constant, 273, 274
programmable, 270, 271
structure of, 272
time, propagation, 60
trigger pair, 116
Demultiplexer, 70
Detector, large photon, 184
DIB, see Data input bus
Dielectric constant, 258
Digital-to-analog converter (DAC), 46, 185
Digital data storage module (DDSM), 226
Digital pulse shape discriminating, 125, 126
Digital rate meter, 33
Diode
bridge switch, 165
PIN silicon, 5
resistance, 174
snap-off, 94, 95
switching, 156
Direct memory access (DMA), 46, 51, 217
channel, 218
operations, 219
Discrimination
low-level, 3
threshold, 5
Discriminator(s), 1–26
antiwalk, 10
channel, dual-clocked, 58
circuit, 9, 18
components, 2
constant-fraction, 14, 86, 96, 102
constant-fraction timing, 79–106
differential, 18–26
block schematic of, 20
channels of, 24
circuit diagram of, 20
pulse height, 19
fast, 4
general considerations, 1–2
high dynamic range, 11
hysteresis, 10
IC, 5
input amplifier, 16
integral, 2–10, 91
leading-edge, 13, 17, 83, 86, 88, 101
level, stability of, 3
monolithic constant-fraction, 103
pulse-height, 1
pulse shape, 107–130
other methods, 113–130
zero crossing, 107–113
Schmitt-type, 3, 19
slow, 4
subnanosecond time, 17
timing, 10–18
zero-crossing, 12, 14, 15, 16, 81
Display mode, 184
Divider chain, 134, 140
elements used in, 144
voltage, 143
DMA, see Direct memory access
DMPX, see Data multiplexer
DPM, see Dual-port memory
Dt_{coll}, 12
Dt_{PMT}, 12
Dt_{walk}, 12
Dual-channel constant-fraction discriminator (DCCFD), 92, 93
Dual-parameter time spectrometry, 66
Dual-port memory (DPM), 217

E

ECL, see Emitter-coupled logic
Electronics, front-end, 203
Emitter-coupled logic (ECL), 9, 59, 86, 105, 113
Energy spectrometry, 66
ERC, see Extended range counting
Ethernet controller, 68
Event rate, 6
Extended range counting (ERC), 6, 7

F

FA, see Fast amplifier
FAPH, see Fast autonomous acquisition phase
Fast amplifier (FA), 127

Fast autonomous acquisition phase (FAPH), 69
Fast bipolar amplifier (FBA), 122
FASTBUS, 201
  backplane, 202
  digital readout module, 210
  enable (FBEN), 205
  interface, 212
  intersegment processor (FIP), 214
  master, 211
  module, 207
  readout module, 204
  slave interface (FSI), 214
  time-to-digital converter, 208
Fast coincidence, 152
Fast data digitizers and processors (FDDP), 206
Fast multichannel scaler, 69
Fast pulse shape discriminator, 111
FBA, see Fast bipolar amplifier
FBEN, see FASTBUS enable
FBG, see Feedback gain
FDDP, see Fast data digitizers and processors
FEB, see Front-end buffer
Feedback
  decimal scaler, 30
  gain (FBG), 7
  negative, 6
  positive, 1, 2
FEMC, see Forward electromagnetic calorimeter
FEP, see Front-end processor
FET, see Field effect transistor
FF, see Flip-flop
Fiber-optic transmission links, 231
Field effect transistor (FET), 46, 166
  input stage, 170
  high input-impedance, 195
FIFO, see First-in-first-out
Figure of merit, 108, 110
FIP, see FASTBUS intersegment processor
First-in-first-out (FIFO), 34, 35, 56, 66, 185, 216, 233
Flash ADC, 183
Flip-flop (FF), 16, 27, 28
  circuit
    basic, 38
    bistable, 27
  D-type, 57
  J-K, 29
  LAM, 181
  R-S, 30
  T, 28
Forward electromagnetic calorimeter (FEMC), 206
Front-end
  buffer (FEB), 206, 213
  processor (FEP), 205, 210, 226
  shift-register, 205
FSI, see FASTBUS slave interface

G

GaAs, see Gallium arsenide
Gain
  adjustments, 134
  stability, 145, 162
Gallium arsenide (GaAs)
  integrated circuit, 42
  memory, 42, 43
Gate
  array, programmable, 214
  generators, 123
  pulse, 124
GEB, see Global event buffer
Generator, stabilization for high-voltage, 246
GES, see Global event supervisor
Global event
  buffer (GEB), 214
  supervisor (GES), 214
GONG, see Go-or-no-go module
Go-or-no-go (GONG) module, 210

H

HDRD, see High dynamic range discriminator
Height to time converter (HTC) circuit, 127
High dynamic range discriminator (HDRD), 10
High voltage digital-to-analog converters (HV-DAC), 146
High-voltage pulsers, 251
High-voltage supply, 233–252
  circuits, 238–252
  portable, 250
  basic method for direct current supply, 234–
    full-wave bridge rectifier power supplies, 235
    full-wave power supplies, 234
    regulated power supplies, 236
    series-regulated power supplies with feedback, 238
    series-regulated power supplies with zener reference, 237–238
    shunt-regulated power supplies, 237
    voltage-multiple power supplies, 235–236
HTC, see Height to time converter circuit
HV-DAC, see High voltage digital-to-analog converters

I

IC, see Integrated circuit
Impedance, characteristic, in semi-infinite line, 264
Input
  current error, 21
  pulse, 2, 10
  signal, 15, 22, 98
  synchronizer, 62
Integral discriminator, 91
Integrated circuit (IC), 4, 154, 235

J

J-K counting system, 30

L

LAM, see Look-at-me
LCA, see Logic cell array
LE, see Leading edge
Leading-edge (LE), 17, 83
  extrapolation, 12
  comparator, time walk of, 97
  discriminator, 101, 102
  method, 79
  timing, 14, 84, 136
  trigger technique, 10
  walk, 85
Least significant bit (LSB), 60
LED, see Light-emitting diode
LES, see Local event supervisor
LG, see Linear gate
Light-emitting diode (LED), 75, 139, 140, 146
Line
  capacitively terminated, 269
  -noise parameter, 268
Linear gate (LG), 159, 160
  active, 162
  passive, 163, 164
Linear postamplifier (LPA), 127
List processor, 190
LL, see Low-level channel
LLD, see Lower level discriminator
LLLE discriminator, see Lower level leading-edge discriminator
Local event supervisor (LES), 214
Local trigger supervisor (LTS), 206
Logic
  cell array (LCA), 216
  pulse, 13
Look-at-me (LAM)
  interrupts, 182
  request, 177
  signal, 193
Loop gain, 2
Lossless line, 265
Low-gain stage, 9
Lower level discriminator (LLD), 22
Lower level leading-edge (LLLE) discriminator, 84
Low-level (LL) channel, 91
LPA, see Linear postamplifier
LSB, see Least significant bit
LTS, see Local trigger supervisor

M

Magnetic disk cartridge, 47
Magnetic optical (MO) disk, 196
Master
  computer, 200
  control unit (MCU), 239
  reset (MR), 190
MBNIM, see Multibit nuclear instruments module
MC, see Mode control
MCA, see Multichannel analyzer
MCS, see Multichannel scaler
MCU, see Master control unit
MEB, see Multievent buffer
Megachannel analyzer, 51, 52
Memory, 40, 45
  cells, CMOS, 42
  controller, 54
  model of, 41
  partner (MP), 209
  read requests, 191
  read-write, 42
  register, 65
Metal oxide semiconductor (MOS), 133
Microprocessor, 46
Miller integrator, 23
Mirror transistors, 9
Mixer unit for data acquisition (MUDA), 226, 228
MO, see Magnetic optical disk
Mode control (MC), 65
Monitoring, automated, 226
MOS, see Metal oxide semiconductor
MP, see Memory partner
MPA, see Multiparameter acquisition
MR, see Master reset
MUDA, see Mixed unit for data acquisition
Multibit nuclear instruments module (MBNIM), 157, 158
Multichannel analysis, multiparameter, 52
Multichannel analyzer (MCA), 37–77, 122
  CAMAC
    -aided, 49
    -based, 51
    -composed, 49, 53
  dual-parameter, 199
  fast, 54
  hardwired, 53
  high-duty cycle, 59
  memory, 37–43
    bistable multivibrator, 37–40
    circuits, 40–43
  microprocessor-controlled, 47, 48
  modular, 54
  multiparameters, 65–69
  multiplexer, 72–76
  multiscaler, 69–72
  printing-type, 48
  single-parameter, 44–64
    CAMAC-aided, 49–64
    early models, 44–48
  versatile, dual-parameter, 66
Multichannel discriminator, 57
Multichannel scaler (MCS), 53, 71
Multievent buffer (MEB), 214
Multiparameter acquisition (MPA), 66
Multiparameter system, 67

Multiple window coincidence module (MWCM), 92
Multiplexer, nuclear spectroscopy, 74
Multiplier
  three-stage, 242
  voltage, 249
Multiscaling, 69
Multivibrator, 1
  bistable, 37
  monostable, 3, 4, 5
MWCM, see Multiple window coincidence module

N

Neutron-gamma pulse shape discriminator circuit, 115
NIM, see Nuclear instruments module
Noise, electronic, 268
Nonlinear mixers, 153
Nontiming output, 25
NOR gates, 9, 39
Nuclear instruments module (NIM), 91, 112, 128, 175, 270
Nuclear spectroscopy multiplexer, 74

O

On-line method, 51
Open-circuited line, 261
Operation system (OS), 199
OS, see Operating system
Output circuit, 62
Output
  nontiming, 25
  peak-sensing, 25
  pulse, 173
  signal, timing edge of, 106
  zero-crossing, 26
Overlap
  converter, 150
  determination, 149

P

Particle identification signal (PID), 112
PDP-11 computer, 51
Peak
  -detect and hold circuit, 167
  -detect mode, 169
  detector, 127
  hold capacitor, 171
Permeability, absolute, 258
PHA, see Pulse-height analysis
Photomultiplier (PM), 131, 134, 141, 142
  cable clipping pulses from, 151
  fast timing with, 135
  stabilization of, 144
  tube (PMT), 91, 92, 133
    current, 140
    voltage divider, 131–146
PID, see Particle identification signal
Pilot lamp, 32
Plane waves, propagation of, 254
PM, see Photomultiplier
PMCCM, see Programmable multichannel correlation module
PMI, see Precision Monolithic Incorporated
PMT, see Photomultiplier tube
Positron annihilation, 66
Power dissipation, 9
Power supplies
  full-wave, 234
  high-voltage, control by IBM PC XT, 249
  regulated, 236
  series-regulated, 237, 238
  shunt-regulated, 237
  voltage-multiple, 235
PPR, see Pulse-pair rejector
Preamplifier, 129, 142
Precision Monolithic Incorporated (PMI), 164, 170
Programmable multichannel correlation module (PMCCM), 60, 61, 63, 64
Programmable read-only memory (PROM), 62, 64, 209
Programmable system, circuitry of, 239
PROM, see Programmable read-only memory
Propagation
  constant, 265
  velocity, 254, 266
PS, see Pulse shape
PSA, see Pulse-shaping amplifier
PSD, see Pulse shape discriminator
Pulse
  bipolar, 3
  -counting system, 6
  gamma full-energy, 145
  -height analysis (PHA), 45, 53
    multichannel, 65
    system, megachannel, 47
  height analyzer, 4096-channel, 44
  -height trigger, 80, 81
  -pair
    rejector (PPR), 122
    resolution, 122
  points, timing of, 124
  resolving, 149
  shape (PS), 111
    analyzer, 117
    digitizer, 183
  -shaping amplifier (PSA), 5
  stretcher, 173
  walk, 2
Pulse shape discriminator (PSD), 107, 112, 113, 125
  circuit, 127
  decay-time, 119
  digital, 125
  fast, 111
  gate, 130
  low-power, 117
  zero-crossing, 119, 122

Q

QDC, see Charge-to-digital converter
QTC, see Charge-to-time converter

R

Radiation monitors, high-voltage supply in portable, 239
RAM, see Random access memory
Random access memory (RAM), 40, 42
  dynamic, 42
  static, 42
Rate meter, 32, 33, 34, see also Scalers, rate meters and
RC, see Readout card
READ command, 176
Read-only memory (ROM), 41, 208
Readout
  card (RC), 217, 219, 222, 225
  controller, 223
Read-write memory, 42
Real-time clock (RTC), 180, 181
Reduced instruction set (RISC) microprocessor, 206
Reflection
  coefficient, 263, 264
  total, 261
Regulator, 120-channel high-voltage, 246
Resolution time, 149, 153
Ripple voltages, 245
RISC, see Reduced instruction set microprocessor
Rise-time
  discriminator (RTD), 120
  protection (RTP), 228
ROM, see Read-only memory
RTC, see Real-time clock
RTD, see Rise-time discriminator
RTP, see Rise-time protection

S

Sample-and-hold
  amplifiers, 165, 166
  circuits, 164, 165
  clock, 173
Sampling digitizer, 74
SB, see Surface barrier
SCA, see Single-channel analyzer
Scalers, rate meters and, 27–35
  counting systems, 29–30
  data readout, 32
  rate meters, 32–35
  synchronous counters, 31–32
  two states circuits, 27–29
Schmitt trigger, 2, 10
Scintillation
  -counter time resolution, 131
  crystals, 125
SEB, see Spy event buffer
Set-reset flip-flop (SRFF), 37, 38, 40
Shape analyzer, 122
Shift register (SR), 31, 35, 155, 231
  coincidence, 157
  front-end, 205
  programmable, 62, 63
  scaler, 32
Short-circuited line, 261
Signal formation, 100
Single-channel analyzer (SCA), 19, 21, 22, 91
Slew rate, 106, 165
Sliding scale, 44
Slow rise-time (SRT) reject circuit, 85
Smart crate controller, 189
SMI, see Standard multipurpose interface
Snap-off diode, 94, 95
Software handshaking, 200
Spectrometer circuitry, 122
Spy event buffer (SEB), 214
SR, see Shift register
SRAM, see Static random access memory
SRFF, see Set-reset flip-flop
SRT, see Slow rise-time reject circuit
Stabilization, 146
Standard multipurpose interface (SMI), 180
Static random access memory (SRAM), 42
Strobe pulse, 57
Surface barrier (SB), 79
Switch, high-voltage, 144
Switched capacitor array, 172
Switching time, 162
Synchronous counter, decimal, 31

T

TAC, see Time-to-amplitude converter
TC, see Transfer count register
TCF, see True-constant-fraction timing
TCM, see Timing and control module
TD, see Tunnel diode
TDC, see Time-to-digital converter
Thevenin equivalent circuit, 21
Threshold network, 9
Time
  -to-amplitude converter (TAC), 107, 112, 208
  -to-digital converter (TDC), 180
  -digitizing system, 211
  discriminator, circuit diagram of, 18
  division multiplexer, 231, 232
  -to-pulse height converter (TPHC), 122, 123
  resolution, 12, 131
  spectroscopy, 150
  walk, of leading-edge comparator, 97
Timing
  circuit, 81, 149
  comparator, 135
  constant-fraction, 82, 137
  and control module (TCM), 226
  discriminator, dual-channel constant-fraction, 92
  error, 84, 97
  fast, 135
  leading-edge, 84, 136

properties, 131
system, 82
threshold generator, 118
zero-cross, 268
TPHC, see Time-to-pulse height converter
Trailing edge, 15, 23
Transfer count (TC) register, 191
Transfer function, 82
PKD-01, 170
total, 6
Transistor-to-transistor logic (TTL), 35, 46, 69, 75, 133, 171, 183
Transmission
coefficients, 263
lines, 253–270
True-constant-fraction (TCF) timing, 99
TTL, see Transistor-to-transistor logic
Tunnel diode (TD), 5, 16, 27, 79, 151, 153
circuit, 28
discriminator, 5
monostable multivibrator, 4
test circuit, 16
triggering of, 108

U

UART, see Universal asynchronous receiver transmitter
ULD, see Upper level discriminator
ULLE, see Upper level leading-edge discriminator
Universal asynchronous receiver transmitter (UART), 48
Upper level discriminator (ULD), 22
Upper level leading-edge (ULLE) discriminator, 84

V

Variable phase gate, 44
Very large system integration (VLSI), 195
Video random access memory (VRAM), 199
VLSI, see Very large system integration
VME system, 198
Voltage
chains, 131, 132
comparator, 15
control system, 233
distribution circuit, 140
divider, 134, 141, 143
multiplier, Cockcroft-Walton, 240
reflection coefficient, 256, 257, 264
ripple, 245
sources, 3
stop channel TAC, 208
transition, 16
transmission coefficient, 264
VRAM, see Video random access memory

W

Wave impedance, 258
Waveform, 18, 116
peak-to-peak drive, 244
pulse shape analyzer, 123
sampler module, 188
-shaping filters, 267
timing, 187
triple polar, 128
WE, see Write-enable
Work memory, 68
WRITE command, 176
Write-enable (WE), 41

Z

ZCD, see Zero-crossing discriminator
Zener
diode, 239
dynodes, 133
reference, 236, 237
-zap trimming, 169
Zero crossing, 23, 80, 107, 128, 151
AND gate, 84
detector, 82, 112, 113
discriminator (ZCD), 12, 151
edge, 16
output, 26
pick off, 88
point, 16
pulse, 81
Zero-crossover
method, 79
time discriminator, 135, 136
Zero-scale error, 171